A CONCISE DICTIONARY OF MILITARY BIOGRAPHY

A CONCISE DICTIONARY OF

MILITARY BIOGRAPHY

Two hundred of the most significant names in land warfare, 10th–20th century

MARTIN WINDROW
FRANCIS K. MASON

WINDROW & GREENE LTD.
LONDON

This revised edition
published 1990 by
Windrow & Greene Ltd.
5 Gerrard Street
London W1V 7LJ

First published in 1975 by
Osprey Publishing Ltd.,
Reading, Berkshire
ISBN 0 85045 199 X

British Library Cataloguing in Publication Data
Windrow, Martin, *1944–*
A concise dictionary of military biography:
two hundred of the most significant names in
land warfare, 10th–20th centuries.
1. Military leaders, history—Biographies—collections
I. Title II. Mason, Francis K. (Francis Kenneth), *1928–*
355'.0092'2

ISBN 1–872004–20–2

Printed and bound in Great Britain by
Butler & Tanner Ltd., Frome and London

FOREWORD

Since war has proved to be man's second most popular pastime throughout recorded history, any selection of a few hundred names from the rolls of history's warriors must automatically be vulnerable to a charge of arbitrary choice. To this charge the authors at once plead guilty. While we are confident that many of the names included in this book are beyond all challenge, it is quite inevitable that some readers will feel that others do not deserve their place, and that more suitable candidates have been excluded. There is no objective answer to this kind of argument; since significance is ultimately a matter of personal judgement, a writer presented with this type of brief can only set his teeth and stand by his own instincts.

Significance, clearly, is far from being the same thing as human or even professional excellence. Many of the two hundred warriors, leaders and soldiers described in these pages are men who gave such clear evidence of genius as to set themselves apart from their contemporaries. A few 'great captains' – Genghiz Khan, Gustavus Adolphus, Frederick the Great, Napoleon – earned reputations which will last as long as history is recorded. Their conquests astonished the world, and changed its maps; in many cases their professional innovations marked decisive steps forward in the ways in which men waged war. While not in the same class of timeless brilliance, we presume that nobody will cavil at the inclusion of such great leaders of troops as Wellington and Lee, Cromwell and Rommel, Wolfe and Eisenhower. But we have deliberately spread our net wide, and have also included several categories of significance which have little to do with excellence on a grand scale.

There are some men in this book whose claim to immortality rests firmly upon their incompetence, or indeed upon their villainy – Budenny and Keitel, Custer, even Benedict Arnold (a soldier is no less significant for being a traitor). There are some whose individual exploits were not particularly impressive *sub specie aeternitatis*, but whose careers represent with great clarity the military climate of their times – some of the medieval figures belong to this category, and some of the British generals of Victoria's reign. There are some who never held major field command, but whose administrative or theoretical mastery earns them a place – Clausewitz and Liddel-Hart, Steuben and Carnot. There are some whose careers were relatively obscure, but whose local innovations may be judged significant in the light of later developments – Jan Žižka, Robert Rogers, David Stirling, and T. E. Lawrence. There are also some whom historians will never be able to slot neatly into any grand scheme of things, but whose names are household words – Jeanne d'Arc and *El Cid* present obvious examples. And because historical importance is a relative matter, we have been at some pains to include, however briefly, some of the great men of other cultures – Pontiac, Hideyoshi, and Shaka among them.

One of the most difficult choices concerned those great names of history who were primarily rulers – some of the Oriental kings and sultans, in particular – and whose personal contribution to the victories of their armies is somewhat obscure. We have chosen with care, but these are deep and far-off waters, and anyone who disagrees with our selection may do so with our good will.

It would be impossible, and of no great value to the reader, for us to list here all the hundreds of sources which we have consulted while compiling this book. We simply wish to record our gratitude to all those friends and colleagues who have helped us with their patient advice.

M.C.W.

December 1989 F.K.M.

The entries in this book are arranged alphabetically for the convenience of the reader; but a chronological listing, by date of birth, is given below.

941–1014	Brian Boru
966?–1025	Boleslav I
971–1030	Mahmud of Ghazni
1015–66	Harald III, called Hardraade
c. 1022–66	Harold II, Godwinson
1027–87	William I, the Conqueror
c. 1030–72	Alp-Arslan
c. 1043–99	*El Cid*
c. 1057–1111	Bohemund I
c. 1123–90	Frederick I, called Barbarossa
1138–93	Saladin
1157–99	Richard I, called *Cœur de Lion*
c. 1160–1218	Simon de Montfort-l'Amauri
1167?–1227	Genghiz Khan
fl. 1190–1242	Subatai Ba'adur
1220–63	Alexander Nevski
1239–1307	Edward I, King of England
c. 1270–1305	Sir William Wallace
1274–1329	Robert I, surnamed Bruce
1312–77	Edward III, King of England
c. 1323–80	Bertrand Du Guesclin
1330–76	Edward Plantagenet, called the Black Prince
1336–1405	Timur-i-Leng (Tamerlane)
c. 1354–*c.* 1416	Owain Glyn Dŵr
c. 1366–1421	Jean Le Meingre Boucicaut
c. 1376–1424	Jan Žižka
1387–1422	Henry V, King of England
c. 1387–1456	János Hunyadi
1404–53	Constantine XI
1405–68	*Skanderbeg*
1412–31	Jeanne d'Arc
1432–81	Mohammed II, called Fatih
1453–1515	Gonzalo de Córdoba
1464–1534	Philippe Villiers de l'Isle Adam
c. 1470–1520	Selim I, called Yavuz
1485–1547	Hernán Cortés
1494–1568	Jean Parisot de La Valette
1494–1566	Suleiman I, called Kanuni
1507–82	Fernando Alvarez of Alva
1536–98	Hideyoshi Toyotomi
1542–1605	Akbar
1545–92	Alessandro Farnese of Parma
1553–1610	Henry IV, King of France
1559–1632	Johann Tserclaes Tilly
1567–1625	Maurice, Prince of Orange and Count of Nassau
1583–1634	Albrecht von Wallenstein
1594–1632	Gustavus II, Adolphus
1599–1658	Oliver Cromwell
1603–51	Lennart Torstensson
1611–75	Henri de Turenne
1612–71	Thomas Fairfax
1612–50	James Graham of Montrose
1618–1707	Aurangzeb
1619–82	Prince Rupert of the Rhine
1621–86	Prince de Condé
1629–96	John III, Sobieski
1633–1707	Sébastien de Vauban
1650–1722	John, Duke of Marlborough
1653–1734	Claude de Villars
1654–1712	Louis Joseph de Vendôme
1663–1736	Prince Eugène
1682–1718	Charles XII, King of Sweden
1696–1750	Herman Maurice, Comte de Saxe
1712–59	Louis-Joseph Montcalm
1713–86	Frederick II, called the Great
1715–89	Jean Baptiste Gribeauval
c. 1720–69	Pontiac
1721–65	William, Duke of Cumberland
1721–73	Friedrich Wilhelm von Seydlitz
1722–92	John Burgoyne
1724?–58	George Augustus Howe
1725–74	Robert Clive
1725–1807	Jean Baptiste Rochambeau
1726–83	Sir Eyre Coote
1727–59	James Wolfe
1730–1800	Alexander Vasilevich Suvarov
1730–94	Friedrich Wilhelm von Steuben
1731–95	Robert Rogers

1732–99	George Washington
1734–1801	Sir Ralph Abercromby
1738–1805	Charles, Lord Cornwallis
1741–1801	Benedict Arnold
1742–1819	Gebhard von Blücher
1744–1808	Gerard Lake
1745–1813	Mikhail Ilarionovich Kutuzov
1745–96	Anthony Wayne
1746–1817	Tadeusz Kosciuszko
1747–79	Kazimierz Pulawski
1748–1843	Thomas Graham
1753–1815	Louis-Alexandre Berthier
1753–1823	Lazare Nicolas Carnot
1755–1813	Gerhard von Scharnhorst
1757–1834	Marie Joseph de Lafayette
1758–1817	André Masséna
1758–1815	Sir Thomas Picton
1760–1831	August Wilhelm von Gneisenau
1761–1809	Sir John Moore
1764–1812	Robert Craufurd
1767–1815	Joachim Murat
1768–1854	William Carr Beresford
1768–1854	Sir Henry Paget
1769–1821	Napoleon I, surnamed Bonaparte
1769–1815	Michel Ney
1769–1851	Nicolas-Jean de Dieu Soult
1769–1852	Arthur, Duke of Wellington
1770–1823	Louis Nicolas Davout
1770–1826	Louis-Gabriel Suchet
1772–1842	Rowland Hill
1778–1850	José de San Martín
1780–1831	Karl von Clausewitz
1783–1830	Simón Bolívar
1784–1849	Thomas Bugeaud de la Piconnerie
1784–1850	Zachary Taylor
1786–1866	Winfield Scott
c. 1787–1828	Shaka
1788–1855	Fitzroy, Lord Raglan
1792–1863	Colin Campbell
1795–1857	Sir Henry Havelock
1797–1868	James, Earl of Cardigan
1798–1854	Jacques Leroy de Saint-Arnaud
1800–91	Helmuth von Moltke
1803–63	Sir James Outram
1807–70	Robert Edward Lee
1808–83	Abd-el-Kader
1808–93	Marie Edmé de MacMahon
1809–95	François Canrobert
1811–88	François-Achille Bazaine
1820–91	William Tecumseh Sherman
1822–85	Ulysses Simpson Grant
1824–63	Thomas Jonathan Jackson
1829–90	George Crook
c. 1829–1909	Geronimo
1831–88	Philip Henry Sheridan
1832–1914	Frederick Sleigh Roberts
1833–85	Charles George Gordon
1833–64	James Ewell Brown Stuart
1833–1913	Garnet Wolseley
1838–1919	Sir Henry Evelyn Wood
1839–76	George Armstrong Custer
1840–1904	Chief Joseph
1850–1916	Horatio Herbert Kitchener
1851–1929	Ferdinand Foch
1852–1931	Joseph Joffre
1854–1934	Louis Hubert Lyautey
1856–1924	Robert Georges Nivelle
1856–1951	Henri Philippe Pétain
1857–1941	Robert Baden-Powell
1860–1948	John Joseph Pershing
1861–1936	Edmund Henry Hynman Allenby
1861–1928	Douglas Haig
1865–1937	Erich von Ludendorff
1867–1951	Carl Gustav Emil von Mannerheim
1870–1964	Paul Emil von Lettow-Vorbeck
1871–1956	Pietro Badoglio
1872–1958	Maurice Gustav Gamelin
1875–1953	Karl Rudolf Gerd von Runstedt
1880–1964	Douglas MacArthur
1880–1959	George Catlett Marshall
1881–1963	Abd-el-Krim
1881–1938	Mustafa Kemal Atatürk
1882–1955	Rodolfo Graziani
1882–1946	Wilhelm Keitel
1883–1963	Viscount Alanbrooke
1883–1973	Semyon Mikhailovich Budenny
1883–1946	Joseph W. Stilwell
1883–1950	Archibald Percival Wavell
1884–1981	Sir Claude Auchinleck
1884–1948	Tōjō Hidecki
1885–1960	Albrecht Kesselring
1885–1945	George Smith Patton
1885–1946	Yamashita Tomoyuki
1886–1976	Chu Teh
1887–1973	Fritz Erich von Manstein

1887–1976 Viscount Montgomery
1888–1954 Heinz Guderian
1888–1935 Thomas Edward Lawrence
1889–1952 Jean de Lattre de Tassigny
1890–1970 Charles de Gaulle
1890–1969 Dwight David Eisenhower
1891–1969 Earl Alexander
1891–1945 Walther Model
1891–1944 Erwin Rommel
1891–1970 Viscount Slim
1892–1966 Josef Dietrich
1892–1980 Tito
1893–1981 Omar Nelson Bradley
1895–1966 Tadeusz Bór Komorowski
1895–1970 Sir Basil Henry Liddell-Hart
b. 1895 Matthew Bunker Ridgway
1895–1970 Semion Konstantinovich Timoshenko
1896–1968 Konstantin Rokossovsky
1896–1974 Georgi Konstantinovich Zhukov
1897–1973 Ivan Stepanovich Konev
1899–1984 Raoul Salan
1902–47 Philippe Leclerc de Hautecloque
1903–44 Orde Charles Wingate
b. 1909 Vo Ngyuen Giap
1915–1981 Moshe Dayan
b. 1915 David Stirling
b. 1916 Marcel Bigeard

Abd-el-Kader (*1808–83*)

The Algerian Muslim leader and military commander Abd-el-Kader was the most persistent thorn in the flesh of the French colonial forces during the early years of their occupation of North Africa.

He was born near Mascara in north-western Algeria on 6 September 1808, the third son of Mahi-ed-Din, head of the Kadria Muslim sect, and grew up in a strict environment. It seems that he soon displayed a fine gift of horsemanship. A devout Muslim, he received a sound education and at the age of nineteen made the pilgrimage to Mecca.

In 1832 Abd-el-Kader was proclaimed amir of Mascara, but later the same year his father declared a holy war against the French in Algeria. Abd-el-Kader carefully built up his power, taking full advantage of French military mistakes, so that when they were obliged to negotiate with him in 1834 he made it clear to the Muslims that he was regarded by the French as the Muslim leader. In time the war broke out again, and although he beat the French at La Macta on 28 June 1835 he was unable to prevent General Bertrand Clauzel from destroying Mascara in December that year. He was defeated by General Thomas Bugeaud at Sikkah on 6 July 1836. Always resourceful in adversity, he negotiated once again with the French, and by the Treaty of Tafna managed to preserve his prestige just when it seemed likely that most of the tribes were about to desert him. Notwithstanding, he continued to experience difficulty in exercising his authority owing to the persistent rivalries between the various Muslim brotherhoods.

In October 1837 Constantine, which had withstood a year's siege, was stormed by a French assault led by Marshal Damremont. Abd-el-Kader, declaring that the Treaty of Tafna had been violated, took the field within two months with an army of 60,000 – almost entirely composed of cavalry and irregular horse – and engaged the French in a prolonged campaign of bitter fighting, laying waste the fertile plain of the Mitidja and destroying numerous isolated French units. In 1840 France decided to take decisive steps to revenge her loss of prestige at the hands of Abd-el-Kader. Bugeaud, now Governor, was given command of an army of 108,000, and in December 1840 moved against the Algerian commander. In 1842 Tlemcen was captured; and on 10 May 1843, at the Battle of the Smala, Abd-el-Kader's army was humbled by a serious defeat when a French flying column of less than 2,000 men under Duc Henri d'Aumale surprised and dispersed the 40,000 Berbers of the amir's mobile village and headquarters.

During 1843 and 1844 Bugeaud systematically drove Abd-el-Kader west into Morocco, while French naval squadrons discouraged Moorish support by bombardments of Tangier and Mogadir. On 14 August 1844 was fought the decisive Battle of Isly. Abd-el-Kader had encamped his army of 45,000 on the banks of the Isly river; Bugeaud launched his 'boar's head' square formation across the river, and followed this with two simultaneous enveloping attacks by flying columns which overran Abd-el-Kader's camp. Abd-el-Kader's forces broke and retreated, and their success at Sidi Brahim in September 1845 only served to delay slightly the final French victory. The Muslim leader eventually surrendered on 23 December 1847, his sole condition being a safe conduct for himself and his family. This was formally agreed by the French, who promptly arrested him and sent him into imprisonment in France, where he languished for five years, first at Toulon, then at Pau, and eventually at Amboise.

He was released on 16 October 1852, and after a stay of nearly three years at Bursa in Turkey, he settled in Damascus in 1855. There in 1860 he was instrumental in saving 12,000 Christians from a mob of fanatic Muslims – an act for which he was awarded the *grand cordon* of the *Légion d'Honneur*. During a final visit to Paris in 1863–5 he tried in vain to influence Napoleon III's Algerian policy. He died in Damascus during the night of 25–26 May 1883.

Abd-el-Krim (Mohammed Abd-al-Karim al-Khattabi) (*1881–1963*)

Abd-el-Krim was the Berber-Muslim leader of the Riff revolt against Spanish and French implantation of Morocco in 1921–5.

Born at Adjir, the son of a caid in the Berber Riff tribe of the Uriaghel, Abd-el-Krim was given a Spanish education and acquired a certain amount of culture. As a result his father was able to obtain for him appointments as 'chief Muslim judge' at Melilla and editor of a Riff newspaper. During the first forty years of his life he remained on good terms with the Spanish authorities and, indeed, assisted the administration to consolidate its control of the treaty zone. However, he finally took offence at a slight offered by an arrogant Spaniard, was imprisoned, and in 1920 fled to his birthplace, Adjir, near Alhucemas, where he embarked on a nationalist revolutionary career against the Spanish administration by inciting and organising armed resistance by the Beni Uriaghel tribe.

The Spanish had for some time been experiencing trouble from guerrilla tribes throughout Morocco, but managed to pacify the western area with forces under the command of General Dámaso Berenguer. In the east, however, Abd-el-Krim assembled a formidable army of tribesmen which confronted a Spanish army of 20,000 under General Fernandes Silvestre as it strove to enter the Riff mountains. The Spanish general had apparently made little effort to reconnoitre ahead – if he had he would have learnt that Abd-el-Krim's forces had destroyed the frontier post for which the Spanish army was making, and the garrison force had fled from another post. As Silvestre reached the settlement at Anual on 21 July 1921 Abd-el-Krim's skirmishers opened a devastating fire on the Spanish flanks; Silvestre's troops panicked and were slaughtered. About 12,000 Spaniards were killed, including Silvestre himself, and about 4,000 others were captured; Abd-el-Krim pursued the remnants almost into the outskirts of Melilla. Previous Spanish atrocities were now answered in kind, greatly demoralising surviving Spanish garrisons in the area.

This catastrophe brought about a political upheaval in Spain and the downfall of the government. On 13 September 1923 'strong man' General Primo de Rivera became – with the passive approval of the King – virtual dictator of Spain, while Abd-el-Krim declared a 'Republic of the Riff' and prepared to expel the French and Spanish from Morocco; to support his intentions he raised an army of 20,000, composed of tribesmen and foreign mercenaries, and well equipped with modern rifles, machine guns, artillery captured from the Spanish in 1921, and even with two aircraft flown by mercenary pilots.

Realising that such preparations would bring about swift reaction from Spain and France, Abd-el-Krim determined to seize the initiative and, stealthily crossing the mountainous border, overwhelmed almost all the French frontier posts between Taza and Fèz. The French commander Lyautey (q.v.), with slender forces at his disposal, could do little more than check the rebel army when it had arrived almost in sight of Fèz itself.

Putting behind them their previous rivalry in Morocco, France and Spain decided to mount a joint counter-offensive, and sent powerful expeditions. A successful landing was made on the coast of Alhucemas by a Spanish force, and an army of 160,000 French under Marshal Pétain moved up from the south. The Spanish force of 50,000 men under General José Sanjurjo captured Adjir on 2 October 1925, and Pétain moved six converging columns into Riff-held territory between Tafrant and Taza – thereby initiating a huge pincer movement. As the latter advanced over the valley of the Wargla – on which Abd-el-Krim depended for most of his supplies – the Riff commander again took the initiative, and in fact almost captured Fèz.

Ultimately, however, as a result of defection by a number of tribesmen who, though unbeaten in their inaccessible strongholds, were tired of fighting, Abd-el-Krim was forced to surrender to the French; he was sent as an exile to Réunion island, where he remained for twenty years. He had displayed outstanding qualities of both a military and an administrative nature during his struggle against the Spanish, and had the technical disparity between Morocco and Europe been no greater than in previous ages, he might well have succeeded in expelling the foreigners. But times had changed, and military methods had progressed with unprecedented rapidity during the First World War. As a result the methods of waging war traditional to the Spanish and French prior to 1914 had disappeared by the time Abd-el-Krim took up arms against them. There was thus no possibility of restoring Moroccan independence until a new generation had grown up, versed in modern schools and methods of war.

After just over twenty years spent in exile, the sixty-six-year-old Moroccan was given residential rights in France in 1947, but was persuaded to take refuge instead in Egypt, where he presided over a national sympathisers' organisation known as the Maghrib Office. He soon realised, however, that he was no longer sympathetic to the violent intentions of the nationalist movement in North Africa; but when, after Moroccan independence had been restored, Sultan Mohammed V invited him to return to his native land, he refused to do so as long as foreign troops remained on North African soil. He died in Cairo on 6 February 1963.

Abercromby, Sir Ralph (*1734–1801*), *Major-General*

Although his career was crowned by several notable victories, Abercromby is remembered more as the restorer of high professional standards in the British Army than as a master of tactics. A man of unflinching honesty and high principle, he deserves much of the credit for transforming the neglected, corrupt and inefficient Army of the late eighteenth century into the excellent weapon which Moore and Wellington (qq.v.) were able to bring to a peak of effectiveness during the late stages of the Napoleonic Wars. He was also one of those British generals who have endeared themselves to the Army by their genuine concern for the welfare of the common soldiers.

Abercromby was born on 7 October 1734 at Menstry near Tullibody in Clackmannanshire, the son of a prominent Whig landowner. He was educated at Rugby and later read law at the universities of Edinburgh and Leipzig, but found the legal profession so unappealing that in the end his father 'bought him a pair of colours' – a cornetcy in the 3rd Dragoon Guards, which he took up in 1756. In 1758 he travelled with his regiment to Germany to serve in the English force under the command of Prince Frederick of Brunswick, the victor of Minden. Abercromby was subsequently appointed aide de camp to General Sir William Pitt; he became a lieutenant in 1760 and a captain in 1762, and saw much active service. The efficiency produced by Prussian discipline impressed him during these campaigns. After the Treaty of Hubertusburg he returned home and was stationed in Ireland for several years – experience which was to stand him in good stead later. He rose to the rank of major in 1770 and to lieutenant-colonel three years later, but shortly thereafter he abandoned his military career to fight a parliamentary election. This was against his inclination, but family duty and pressure from local Whig interests persuaded him. It was a bitter contest against one Erskine, a man who commanded strong Jacobite sympathies, and the extraordinary standards of electoral behaviour tolerated in those days led to a duel between the two candidates – happily, the result was bloodless. Abercromby was elected, largely due to the influence of his powerful relative Sir Lawrence Dundas; but once in Parliament he refused to vote according to the private wishes of this magnate, and lost much useful influence. He strongly opposed the American War, and thus lost any chance of professional advancement at that time, although he longed to return to active service. Eventually he retired from Parliament, disgusted by political life, and devoted himself to his family. One of his brothers fell in action in America, and another greatly distinguished himself.

Abercromby became a major-general in 1787, and on the outbreak of war with France he had no difficulty in obtaining a command. He was ordered to take a brigade to Flanders in 1793, and distinguished himself during the Duke of York's expedition. He was conspicuous at Furnes, Valenciennes, and Roubaix. The generals who replaced the Duke of York after the latter had sailed home were more than usually incompetent, and as commander of the rearguard during the eventual dreary retreat Abercromby did much to save the expedition from a worse disintegration than in fact occurred. He had every opportunity during this campaign to take to heart the appallingly low level to which the Army had sunk in thirty-five years. Discipline was poor, financial and political corruption was rife among the officers, a long series of defeats had destroyed confidence and pride in the ranks, and the physical standards were so low that an observer compared the Army to Falstaff's band of ragged greybeards and witless boys. For the rest of his career Abercromby worked tirelessly and effectively to remedy these abuses.

Rather to his surprise he was greeted on his return to England as the greatest general of the day; considering that Lord Cornwallis was the nearest Britain could boast to a tolerable general at that time, the compliment was no wild exaggeration. He was made a Knight of the Bath in 1795, and in November of that year he was sent to the West Indies with 15,000 men to capture the French sugar islands. Reaching Jamaica in January 1796, he conducted a most able campaign. St Lucia, with its strong fortress of Morne Fortunée, fell to him, and he left the island in the charge of his brilliant protégé Moore. He then took Demerara, and relieved St Vincent, reorganising the defences of the latter and of Grenada. During this campaign he took intelligent steps to protect the health of his soldiers and to make service in the punishing Caribbean climate as bearable as possible. He also encouraged pride, self-reliance, and discipline by a number of reforms and systems of rewards for good service. He went home for the summer of 1796, but returned later in the year and took Trinidad, where he installed Picton (q.v.) as Governor. He failed at Porto Rico, but this was due to the insufficent force under his command. He returned home in 1797, having resigned his command on account of ill-health, and was received with great acclaim. He was sent to Ireland in that same year, to command the troops there at a time of great unrest. His native tolerance and common sense enabled him to see far beyond the ostensible reasons for the instability – the activities of demagogues and French agents – and having seen the truth he was too honest to acquiesce in the rascally activities of Dublin Castle. He knew that the outside agitators were only gaining a following because of the injustices committed against the mass of the population, and he recognised that the excesses of the troops were responsible for much of the ill feeling. These troops were very largely poorly disciplined English and Scottish militia, not regulars; and on 26 February 1798 Abercromby issued a famous general order in which he declared that the militia were more dangerous to their friends than to their enemies. A bitter quarrel ensued with the Lord Lieutenant, Lord Camden, as a result of which Abercromby resigned his command – not the first, nor by any means the last Englishman to discover that an attachment to the truth is of little service to a man condemned to play a part in Irish affairs.

Abercromby was well received on his return to England, and was immediately given command of the troops in Scotland, where he became very popular. In 1799 he was ordered to sail for Holland with a division of 10,000 men; and on 27 August he landed successfully in the Helder. On 10 September he successfully fought off a Franco-Dutch army under General Daendels, but three days later was superseded in the command of the expedition by the Duke of York. Reinforcements included a Russian contingent. The Duke's attack on Bergen failed on 19 September, and was renewed on 2 October. Abercromby's column advanced through the sand dunes at Egmont-op-Zee, seeing heavy fighting, and was completely successful – but the other columns were not, and the attack failed once again. On 20 October Britain agreed to the Convention of Alkmar; her naval objectives had been secured, and she was now willing to hand back the prisoners taken in return for an undisturbed re-embarkation. Abercromby was hailed as a hero once more and offered a peerage, but he was disgusted with the conduct of the campaign and refused to have his name associated with it.

After a few quiet months in his Scottish command he was appointed to succeed Sir Charles Stuart as commander of British troops in the Mediterranean theatre. He reached Minorca in June 1800, but was unable to obey his orders to land in Italy because of the French victory at Marengo. He awaited orders on Minorca, meanwhile using the time to improve the morale and efficiency of his troops. He was ordered to proceed to Gibraltar, pick up more troops, and land at Cadiz. The expedition, 20,000 strong, arrived off Cadiz on 3 October 1800, but no major landing was made. Abercromby would not commit his troops without the assurance that Lord Keith's naval squadron would stay and support him for a fortnight. Keith refused, as he held that the anchorage was unsafe. Both were probably right, and the expedition was abandoned as impractical – a conclusion which should have been reached about many pointless landing operations by the British Army at that time.

On 24 October Abercromby was ordered to Egypt to operate against the French army which Napoleon had installed there, in concert with the Turks. He reached Malta on 19 November and departed again on 13 December, anchoring in the Bay of Marmorice on 27 December. During a pause of many weeks he established that the Turks were useless as allies and that he should proceed independently; he also used the time to exercise his command over and over again in the intricacies of rapid disembarkation. These drills served their purpose; on 8 March he landed 14,000 infantry, 1,000 cavalry, and 600 artillery at Aboukir Bay in a single day, an example of an efficiency unknown at that time. Opposition was vigorous, and he advanced cautiously towards Alexandria, fighting skirmishes on 13 and 18 March. On 21 March General Menou came out of Alexandria with a strong force and attacked the British army with great determination The day was saved by the tenacity of Moore's division on the British right, particularly the 28th Foot – the Glosters – who won their famous back-badge in this action. Menou's force was eventually beaten off with heavy casualties, including three dead generals. Among the 1,464 British casualties was Abercromby. He habitually exposed himself to danger in front of his troops, and a musket-ball struck him in the thigh. Under the unsophisticated medical conditions of 1801, and in that climate, the wound was very serious. He was carried on board the flagship *Foudroyant*, seemed to rally for a while, but died on 28 March. He was buried in Malta.

Abercromby was a man greatly loved and greatly missed by all ranks. His contribution to the regeneration of the British Army at a difficult time was very considerable, and not the least of the legacies he left behind him was the group of subordinate commanders whom he consciously trained and groomed for senior command – of whom the most famous was, of course, Sir John Moore. Abercromby was an able soldier, if no tactical genius, and the kind of commander who inspires scores of affectionate anecdotes in the ranks; many were probably connected with his notorious short-sightedness. His character is best summed up by his words on recovering consciousness after receiving the wound which was to kill him. 'What is it you have placed under my head?' he asked an aide, to be told that it was only a soldier's blanket. 'Only a soldier's blanket? Make haste and return it to him at once!'

Akbar (*1542–1605*), *Emperor of India*

This greatest of the Mogul Emperors brought the Empire to its first great flowering by building on the foundations laid by his mighty grandfather Babur, and by his father's enemy Sher Khan. Although his historical significance extends far beyond military matters, and although he often conquered through generals rather than by taking the field himself, still his military organisation and his successful campaigns earn him a place in this book.

The Mogul Empire – the word is a corruption of 'Mongol' – was established in April 1526 by Babur, King of Kabul and reputedly a descendant of Genghiz Khan (q.v.). Encouraged by anarchy in the Moslem states of northern India, he launched several campaigns, culminating in that month in a great victory at Panipat over Ibrahim, Sultan of Delhi. Babur occupied Delhi a week later, and the Empire was established. Babur's son Humayun, however, was defeated and deposed in 1539 by Sher Khan, an old Afghan-Turk who had risen to the governorship of Bihar. This wise old man ruled until his death in an accidental gunpowder explosion in 1545, and during this time laid firm foundations for the military organisation of the later emperors. His standing army was comprised of absolutely trustworthy garrisons of paid troops in key hill forts, and a large army of manœuvre. Humayun travelled a long road of exile, during the course of which, in October or November 1542 at Umarkot in Sind, his son Jalal-ud-Din-Mohammed (the future Akbar) was born. Humayun rose to be the ruler of Afghanistan, watching and waiting while the Mogul dominions slipped into anarchy after the death of Sher Shah (as he had proclaimed himself on taking the throne). The minister and general Hemu was ruling as a dictator from Delhi when, in 1555, Humayun decided his time was ripe. He marched through the Punjab and seized Delhi – but died in an accident the following year. The fourteen-year-old Akbar was in the Punjab with his capable adviser Bairam at the time; they raised an army and marched on Delhi, which Hemu had retaken with an army of Afghan-Turks raised in the Ganges Valley. On 5 November 1556, thirty years after his grandfather's triumph there, Akbar met Hemu at Panipat. Hemu had the larger army, and his victory appeared certain when he fell, wounded by an arrow; in the moment of hesitation Akbar and Bairam pressed home a counter-attack, and won the day. For four years Bairam ruled as Regent, but in 1560 Akbar took the reins of power into his own hands, and began a forty-year career of conquest.

The core of his field army was a force of artillery and some 12,000 infantry armed with matchlock muskets. Most of the remainder of the army was composed of highly mobile cavalry, which after his conquest of Rajputana was largely recruited from among the Rajput lancers. This army was led in several major campaigns by a most able Hindu general, Raja Todar Malla, but the inspiration behind the Mogul conquests came from Akbar. In 1561 and 1562 he conquered Melwa, and by the end of the latter year his Empire covered the Punjab and Multan, the basin of the Ganges and Jumna from Panipat to Allahabad, Gwalior, Ajmere, Malwa, and the province of Kabul in Afghanistan. The conquest of Rajputana occupied him from 1562 to 1567, culminating in his capture of Chitor in 1567. He confirmed the Rajput princes in power, and they were soon among his most loyal supporters. A distraction in 1566 was a raid into the Punjab from Afghanistan by Akbar's half-brother Mohammed Hakim, who withdrew again when threatened. In the early 1570s Akbar extended his dominion over Gujerat, Bihar, and Bengal. In 1576 his son Murad led an expedition into the Deccan, but this was fiercely resisted by the Moslem sultans of the northern Deccan. Intermittent warfare took place on this front for many years. In 1581 Mohammed Hakim again raided into the Punjab; Akbar repulsed him, and then marched into Afghanistan and conquered it.

In the period 1586 to 1595 Kashmir, Sind, Baluchistan, Kandahar, and parts of Orissa were added to Akbar's Empire. With his power firmly established in the north, he turned south for a major effort in the Deccan in 1596. His campaigns between 1596 and 1600 gained him dominion over Khandesh, Berar, and Ahmadnagar. In 1601 his son Salim rose in revolt in the Ganges Valley, and Akbar turned north to deal with the threat. Having defeated and captured Salim, the old Emperor pardoned him in 1603. It is believed that when Akbar finally died at Agra on 16 October 1605, it was from poison administered at his son's order. He left an Empire stretching from the Hindu Kush to the Godavari river, and from Bengal to Gujerat, divided into fifteen *subahs* or provinces. The framework of his administration can still be traced in the Indian and Pakistani provincial systems; it was run by imperial officers, both civil and military, appointed by the central government and frequently transferred. It was their task to recruit the cavalry for the field army, and this system provided an army loyal, effective, and quite large enough for the needs of seventeenth-century India. Akbar's conquests were underwritten by a complete reform of the tax system, of which the most striking feature was the replacement of the flat one-third levy of all products previously enforced, by a cash payment assessed according to local yield and prices.

Alanbrooke, Alan Francis Brooke, 1st Viscount, of Brookeborough (*1883–1963*), *Field-Marshal*

Born at Bagnères de Bigorre in France on 23 July 1883 of Irish stock, Alan Brooke was educated abroad, but attended the Royal Military Academy, Woolwich, between 1900 and 1902, when he was commissioned in the Royal Field Artillery. For the next four years he served in Ireland, and then sailed for India, where he transferred to the Royal Horse Artillery in 1909. On the outbreak of the First World War he accompanied the Secunderabad Cavalry Brigade to France, landing at Marseilles in command of an ammunition column. His active service in France was exemplary but not outstanding, although he was mentioned in dispatches six times and twice awarded the D.S.O.; he was appointed adjutant in the 2nd Indian Royal Horse Artillery Brigade in February 1915, and brigade major in the 18th Divisional Artillery in November the same year.

Between the two world wars Brooke held a succession of staff appointments, which included Commandant of the School of Artillery from 1929 to 1932, and Inspector of Artillery in 1935–6. In 1934–5 he commanded the 8th Infantry Brigade; in 1936–7 he became director of military training at the War Office, and afterwards was appointed to command Anti-Aircraft and Southern Commands.

On the outbreak of the Second World War, Brooke assumed the command of the British 2nd Army Corps, and to him was due much of the credit for the successful extrication of the British Expeditionary Force from Dunkirk – although in view of the highly fluid nature of the withdrawal, the tactical responsibility devolved upon divisional commanders such as General Harold Alexander (q.v.). Winston Churchill, who became Britain's Prime Minister on 10 May 1940, was quick to recognise Brooke's qualities; and in his determination to render continuing assistance to France after Dunkirk he sanctioned the reconstitution of the British Expeditionary Force, comprised of the remaining British forces in France, with Brooke in command. But the situation was soon seen to be irredeemable, and no coherent defence against the invading Germans could be organised before the armistice was signed on 21 June. The following month Brooke was appointed Commander-in-Chief of British home forces, and as such was responsible for the overall preparation by the Army for the seemingly inevitable German invasion.

When Field-Marshal Sir John Dill took up his appointment as Churchill's personal representative in Washington, Alan Brooke was appointed Chief of the Imperial General Staff (December 1941) – the British Army's senior post, and one which Brooke was to occupy until 1946. At once he not only displayed an extraordinary grasp of Churchill's broad strategy and identified himself wholly with it; but at the same time demonstrated superb tact when representing British strategic planning to the Americans, who in 1942 were not yet unanimous on the importance of eliminating Germany and Italy from the war before turning Anglo-American strength against the Japanese in the Far East. It is probably true to say that the absence of any serious conflicts in the Chiefs of Staff Committee was largely attributable to the personality of Brooke; at the same time he was able constantly to represent the military views of this Committee firmly and ably both to Churchill and to Roosevelt (via Dill).

Brooke was also recognised as a brilliant field commander, and there is no doubt that he would dearly have liked to accept one of the very high commands overseas during the War. When it was decided to replace Auchinleck and Ritchie after Rommel's advance into Egypt in July–August 1942, Churchill wanted to give the Middle East command to Brooke. Typically, Brooke declined because he felt that he had not occupied the post of C.I.G.S. long enough to ensure smooth continuity if he left; he also believed that his relationship with Churchill was one that another man might find it difficult to achieve. It was Brooke himself who recommended Alexander for the Middle East Command, and General Gott for the Eighth Army (when Gott was killed Montgomery took command).

Brooke's relations with Stalin were interesting in that he earned an initial distrust when, having at first supported an early proposal to invade northern Europe in 1942, as demanded by the Russians, he (in company with all other Western military planners) realised that no such landings could be made until 1943 at the earliest. It was characteristic of Brooke that he himself explained and justified the necessary delays at the various staff meetings with the Soviet leaders. Thereafter he enjoyed the trust of the Russian military command.

The enormous increase in demands made upon British military reserves during 1943 threatened to cause the abandonment of one or more of the promising offensives at that time in train. The withdrawal of forces (in particular armoured units) from the Mediterranean in preparation for the forthcoming Normandy landings threatened so to denude Alexander's forces as to suggest to the Americans that the invasion of Italy and Sicily would not be possible with the forces available. It was on the specific direction of Brooke that the

British First Armoured Division, two Polish divisions, the New Zealand Division, and the Fourth British-Indian Division were re-created and sent to the Mediterranean. The provision of these 100,000 men was critical in the continuation of a substantial campaign in the theatre, and was instrumental in holding vital enemy forces which would otherwise have been redeployed in Russia or in the West.

It was his absolute concentration on and faith in the planning of the invasion of northern Europe during 1943 that suggested to Churchill that Brooke would be the ideal supreme commander; indeed, it is likely that Brooke himself expected to be offered the command. Nevertheless, Churchill deferred to American opinion that in the view of the two-to-one numerical superiority of American forces participating in the invasion, the supreme command should be assumed by an American; Eisenhower's (q.v.) nomination was approved unanimously. Yet it should never be overlooked that the greater part of the strategic planning for the invasion of Normandy (indeed of France) had been completed by General Sir Frederick Morgan under Brooke's direction before Eisenhower and Montgomery arrived in Britain to take up their commands.

Brooke was made a baron in 1945 at the end of the war, and created Viscount Alanbrooke of Brookeborough the following year. An idea of his universal recognition as one of the greatest of Allied military strategic planners can be gained from the unique succession of honours showered upon him from all over the world. No military commander ever received a comparable list of foreign honours: Grand Cross of the Order of Polonia Restituta (Poland); First Class Order of Suvarov (U.S.S.R.); Grand Cordon of the Order of Leopold (Belgium); Grand Officer of the *Légion d'Honneur* (France); First Class of the Military Order of the White Lion of Czechoslovakia; First Class Order of the Redeemer (Greece); Grand Cross of the Lion of the Netherlands; Knight Grand Cross of the Order of Dannebrog (Denmark); Knight Grand Cross of the Order of the North Star of Sweden; First Class Order of Trinity (Ethiopia); Grand Cross of the Order of Christ (Portugal); and *Croix de Guerre* (France, Belgium, and Czechoslovakia).

Alanbrooke died at Hartley Wintney, Hampshire, on 17 June 1963. Despite his yearning as a professional soldier for operational or field command, he had brought to bear an unsurpassed understanding of the qualities demanded in high staff appointments – and the tact required to harness the military strength and aims of a polyglot alliance. If General Alexander achieved the glamour of successful campaigning on three continents, Alanbrooke deserves no smaller recognition for his staff direction that made those successful campaigns possible.

Alexander, Harold Rupert Leofric George, 1st Earl Alexander of Tunis (*1891–1969*), *Field-Marshal*

Unquestionably the greatest British field commander of the Second World War, Alexander played the leading role in almost every campaign in which British troops were involved, displaying a unique tactical and strategic aptitude in defeat as well as in victory.

Born in London on 10 December 1891 of Irish stock, the third son of the fourth Earl of Caledon, Harold Alexander was educated at Harrow and the Royal Military Academy, Sandhurst. He joined the Irish Guards, serving with gallantry and distinction during the First World War, and was in command of the Baltic German troops fighting the Communists in northern Russia in 1919. During the 1930s he took part in the campaigns on the northern frontier of India, commanding the Nowshera Brigade, and being promoted major-general in 1937. In 1938 he assumed command of the British First Division, and on the outbreak of the Second World War accompanied the British Expeditionary Force to France. After the onslaught in the West by the German armies of Bock and Runstedt (q.v.) in May 1940, and during the retreat to the coast, Alexander was charged with fighting the difficult rearguard actions to cover the B.E.F.'s escape from Dunkirk.

Following Hitler's controversial intervention to halt Guderian's devastating armoured thrust from the south, Alexander seized the opportunity to entrench three divisions and await renewed German attacks. With the remains of the French First Army to the south and some

Belgian remnants to the north, this zonal defence bought a critical delay of a week, during which the amazing evacuation of forces from the beaches of Dunkirk was achieved. In the eight days between 28 May and 4 June more than 338,000 British, French, and Belgian troops were brought off safely from under the very gun-muzzles of the German armies. On the final night General Alexander personally toured the beaches on foot to satisfy himself that no Allied soldier remained who could be evacuated; only then did he himself step aboard a boat to return to England. In the midst of so much military ineptitude, Alexander's brilliant conduct of the delaying rearguard stood alone as the salvation of the British army in 1940.

Back in England during the dark months in which Britain awaited the seemingly inevitable German invasion, Alexander was appointed General Officer Commanding Southern Command; his troops would have met that invasion, had it come. After the Battle of Britain the threat of invasion receded, and by mid-1941 it had disappeared.

The entry of Japan into the war at the end of 1941 and the subsequent Japanese invasion of Burma threatened to lead to headlong British retreat from that country. Alexander was immediately sent out to take command of the British and Commonwealth forces, but on arrival found that the situation was beyond recovery: the ill-equipped troops, with almost non-existent lines of communication, were quite unable to stem the powerful, well-organised Japanese forces. Nevertheless, with the able General Slim (q.v.) in command of the First Burma Corps providing a tough centre, Alexander was able to direct a slow but organised retreat northwards past Mandalay into Assam, maintaining in being the essential components of the army that would eventually take part in the reconquest of Burma two years later.

These two brilliant campaigns of survival so impressed the Allied High Command that when, in August 1942, Rommel (q.v.) stood poised at El Alamein to strike at Cairo, Alexander was rushed to Egypt to take over supreme command of the Middle East from General Auchinleck (q.v.). At the same time command of the British Eighth Army passed to General Montgomery (q.v.). Although the victory in battle won at El Alamein was that of Montgomery, the overall Middle Eastern strategy was that of Alexander, and it was to him that Churchill gave the order to clear North Africa of all Axis forces.

After the Anglo-American landings in Algeria under General Eisenhower (q.v.), Alexander was directly responsible for the co-ordination of the simultaneous convergence from east and west on the Axis forces withdrawing into Tunisia. Following a setback suffered by the American forces attacking from the west (for which Eisenhower must take some responsibility, since he had become embroiled in the Franco-American political turmoil), all Allied ground forces in Tunisia were grouped into the 18th Army Group under Alexander's command. Almost simultaneously General Jürgen von Arnim assumed command of all German forces in the theatre. Despite quickly-stiffening resistance in the country, Alexander's armies linked up from east and west, and inexorably forced the Axis troops into the north-eastern tip of Tunisia; by use of superior air support he finally compelled the enemy to surrender on 13 May 1943. A telegram from Alexander to Churchill was typical: 'Sir: It is my duty to report that the Tunisian campaign is over. All enemy resistance has ceased. We are masters of the North African shores.'

The invasion of Sicily followed, again under the overall command on the ground of Alexander. The assault featured both airborne and seaborne landings, and was the first on a major scale against a heavily-defended coast in the European theatre, and the first involving the joint subordinate command by British and American generals of forces opposed by a unified enemy command. It was carried out by elements of Alexander's 15th Army Group – General Patton's (q.v.) American Seventh Army and Montgomery's British Eighth Army – and the occupation of the island was completed in five weeks.

Alexander's next and most frustrating campaign was the invasion of Italy, for, at the moment when Italy herself surrendered to the Allies and thereby provided an opportunity to roll up the German forces in the country, not only were many of Alexander's forces redeployed for the forthcoming invasion of Normandy, but many of the most experienced commanders were moved as well. While the Eighth Army crossed the Straits of Messina to land on the Italian mainland, General Mark Clark's American Fifth Army landed south of Naples in an attempt to cut off the German forces in the south. However, the element of surprise was lacking at Salerno, and the German forces reacted swiftly, pinning the invaders to their beach-head while reinforcements were rushed southwards. The Eighth Army managed to overcome resistance in the toe and heel of Italy, but its advance was slower than had been hoped, so that by the time Fifth and Eighth Armies joined forces the Germans had established the first of their strong lines of defence from the Volturno river on the west coast to Termoli on the east. Against this line the Allies came on 8 October 1943.

Eisenhower returned to England in December and Alexander was appointed supreme Allied commander in the Mediterranean theatre. More of his forces were withdrawn to England: the 15th Army Group now consisted of Indian, French, American, Italian, New Zealand, and British divisions. Soon Polish, Greek, and Czech forces were to join it, as well as Brazilian units. Determined to avoid an enervating stalemate which would give his opponent Kesselring (q.v.) an opportunity to assemble superior forces, Alexander decided to try to break the Volturno–Termoli line before winter and, using the American Fifth and British Eighth Armies in the west, broke through on the Volturno – only to come up against a prepared line fifty miles further on. The Eighth Army, now under General Leese, swung north beyond Termoli, but found that Kesselring had created another strong coast-to-coast line, known as the Gustav Line. This was the position of the opposing armies when winter broke in mid-January 1944.

Once again Alexander determined to allow Kesselring no respite, and he launched another seaborne landing behind Kesselring's lines at Anzio on 22 January 1944 with 50,000 British and American troops under U.S. General John P. Lucas. Within two days almost all the troops were ashore facing little organised resistance, but the American commander failed to pursue his advantage and press inland, preferring to await the landing of his heavy support equipment. The result was that 50,000 valuable Allied troops became locked in a temporarily useless bridgehead. On the Gustav Line the main Allied armies tried in vain to assault the strongpoint at Monte Cassino, and the whole front lapsed into exhausted stalemate.

Although this failure to maintain the offensive must be accounted Alexander's responsibility, it must be conceded that his forces had by this time become critically depleted through the withdrawals already mentioned – to which should be added other wasteful obligations in the Aegean and operations against the Dodecanese. Many of the amphibious craft which could have landed the Anzio equipment more rapidly had been brought back to England, and others were being prepared for the landing in the south of France six months ahead.

Eventually, after several months' build-up of forces, the bastion at Monte Cassino fell to a gallant assault by Polish troops, and this key battle led to a general breakthrough of the Gustav Line on a twenty-mile front in mid-May. The link-up with the Anzio forces followed quickly as the Germans fell back, and Rome, the first of the Axis capital cities to fall, was entered by the Allies on 4 June.

The remainder of the Italian campaign consisted of a lengthy pursuit of the retreating German armies northwards to the Gothic Line – extending from Bologna to Ancona. When Clark's Fifth Army failed to capture Bologna a new winter stalemate ensued. The final phase was played out in April 1945 when, with Alexander in supreme command and Clark as army group commander, the Eighth Army attacked behind flame-throwing tanks and broke through south-east of Bologna; four days later the Fifth Army broke into the Po Valley. Despite committal of all reserves the Germans were unable to stem the Allied armour, and on 29 April Alexander received the unconditional surrender of all German forces in the Italian theatre – the first unconditional surrender signed by Germany.

Although never quite 'forgotten' to the same extent as the British Army in Burma, Alexander's forces in Italy tended to fade into the background when compared with the massive operations being mounted against Germany in the West and East. But it was the very demands, both in men and material, made by the great operations in France that slowed the pace of Alexander's campaign. It must be said that throughout 1944 he was to some extent handicapped by his necessary dependence upon relatively inexperienced American corps and divisional commanders, and these men would have been the first to pay tribute to the British commander's magnificent tact in his avoidance of open disarray at command level. In short, he enjoyed with friend and foe alike a unique reputation for integrity. He himself attributed his successes both in defeat and victory to an accurate assessment of his enemy's most formidable qualities. He retains one distinction quite unequalled by any other senior commander of any combatant nation of the Second World War: that even thirty years later, in an age of cynicism and unconcealed squabbling between former generals, not a single man can be found to say a harsh word about Alexander's character and conduct. He must have been the most universally liked British officer since 'Daddy' Hill (q.v.).

Alexander Nevski (*1220–63*), *Grand Prince of Vladimir and Kiev*

At the very time when the Mongols were extending their rule over Russia, and Subatai's (q.v.) horsemen were poised to invade western Europe, Alexander of Novgorod was winning renown as the defender of the western frontiers – a striking comment on the complete fragmentation of Europe which provided the Mongol hordes with such a tempting prize.

Alexander was the son of Yaroslav Vsevolodovich, at one time the ruler of Novgorod. In 1240, the year in which Subatai burnt Kiev, Alexander was leading an army to stem a Swedish invasion of northern Novgorod. He defeated the Swedes, who were led by the King's brother-in-law Birger Magnusson, on the Neva river; so decisive was his victory that Swedish plans for expansion in this direction were abandoned, and Alexander was honoured with the title Nevski – 'of the Neva'. He is believed to have quarrelled with the princes of Novgorod shortly after the victory, and spent a year at Pereyaslavl; but he returned in 1241 on receiving an urgent summons. The Teutonic Order, the military order of German knights, was at that time extending its influence eastwards through pagan Prussia and Lithuania, and pressing on Novgorod's western frontiers. In a brilliant campaign Alexander recovered all that had been captured by the Order, and carried the war into the territories of the Teutonic Knights in Estonia. His climactic victory was won on the ice of frozen Lake Peipus in April 1242, and the Grand Master of the Order subsequently renounced all conquests on Russian territory. A vigorous simultaneous campaign against the Lithuanians ensured that the western frontiers would have peace for some time – at a period when the Mongol expansion was flooding over Russia from the East.

In 1243 Alexander's father had become Grand Prince of Vladimir and Kiev. He died in 1246, and it was a pointed reminder of the new realities that Alexander and his brother Andrei had to travel to Karakorum to be confirmed in their inheritance by the Mongols. Alexander became Prince of Kiev and Andrei Grand Prince of Vladimir; but in 1252 Andrei was crushed by the Mongols for disobedience, and Alexander became Grand Prince. He co-operated with the Khan of the Golden Horde, Batu, who spread his suzerainty over all Russia from his capital at Sarai on the Volga, and in return received courtesy and consideration from the Mongols. He assisted in the administration of a great census which caused considerable unrest, even going so far as to arrest his own son when he became involved in disturbances over the census in Novgorod. In 1262 there was a general uprising against the Mongols in Suzdal, and it was as a result of Alexander's direct intercession that Berke Khan, Batu's successor, decided to withhold the massive and ruthless punitive measures which usually followed immediately upon any show of defiance. While in a military sense Alexander's later life was a sad anti-climax after his early brilliance, this successful mission of mercy must be recognised as an achievement as fine as any victory on the battlefield. It was while returning from the Golden Horde after succeeding in this plea that Alexander died in November 1263, at Gorodets on the Volga. In 1380 he was canonised by the Russian Orthodox Church.

Allenby, Edmund Henry Hynman, 1st Viscount (*1861–1936*), *Field-Marshal*

Born on 23 April 1861, Allenby was educated at Haileybury and the Royal Military College, Sandhurst, and was commissioned in the Inniskilling Dragoons in 1882. It was with the Inniskillings that he served on the Bechuanaland expedition of 1884–5 and in Zululand in 1888; he was appointed adjutant during 1889–93. During the South African War of 1899–1902 he won a sound reputation as a column commander; he was mentioned in dispatches and appointed C.B., and ended the war as brevet lieutenant-colonel. Between 1902 and 1905 he commanded the 5th Royal Irish Lancers, and from 1905 to 1910 the Fourth Cavalry Brigade. In that year, as a fifty-year-old major-general, he became Inspector-General of Cavalry.

As the senior surviving and most experienced cavalry staff officer, it was inevitable that in 1914 Allenby should assume senior command of the British horse – committed as Britain was to depend upon an arm that was already demonstrably outmoded in modern warfare – and shortly after the beginning of the First World War he was given command of the cavalry division that was sent to France in the first few weeks. Later in 1914 he assumed command of the much-enlarged Cavalry Corps and distinguished himself at the Battle of Ypres at the head of the Fifth Corps in 1915. Thereafter he was given command of the Third Army, but was not engaged in major operations until the Battle of Arras in April 1917. This great offensive, like so many others on the Western Front, opened with a crashing bombardment, clouds of gas, and much promise; but an advance of some 10,000 yards in six days north of the Scarpe was not exploited, and the front line stabilised with nothing to show for it but 84,000 British casualties and 75,000 German.

Meanwhile the war against Turkey in the Middle East had been marked in 1917 by apparent ineptitude in the British command in Palestine. Early in the year British forces had succeeded in clearing the Sinai Peninsula of Turkish troops, and operations had begun against Gaza in March. Using a mixed force of infantry and cavalry, General Charles Dobell had attacked on 26 March, but following a breakdown in his communications within his command, the offensive had failed and he had been forced to retire. Unfortunately his commanders' report of the engagement gave the impression of a British victory, and Dobell launched a further attack the following month on strong Turkish positions at Gaza – this time being hurled back with nearly 4,000 casualties. Dobell was promptly relieved of his command, and his place was taken by General Allenby on 17 April 1917. Now widely respected for his gift of leadership and his tactical ability, and with instructions 'to capture Jerusalem before Christmas', Allenby set about an immediate reorganisation of the British front, moving his headquarters from a Cairo hotel into the front line and demanding immediate reinforcements. Within two months he had assembled a force of 88,000 men, organised in seven divisions and an effective cavalry element – the Desert Mounted Corps of camel- and horse-mounted troops. By the end of September he had increased his force by a further two divisions – both of horsed cavalry.

By the end of October he was ready to take the offensive, and, leaving three divisions to demonstrate in front of the Gaza positions, he delivered a powerful surprise attack on Beersheba on 31 October. In classic style Allenby threw his infantry against the Turks in direct frontal assault, while his Desert Mounted Corps swept round to the east and into the city from the rear. Wearing down the Turkish defences with repeated infantry charges, Allenby waited until dusk and then discharged his Australian cavalry brigade at the exhausted defenders. The horsemen swept over the Turkish wire into the city itself, capturing the vital water supplies. Having abandoned the city, the Turkish Seventh Army then lay with an exposed flank, at which Allenby hurled his cavalry, driving a wedge between the Seventh and Eighth Turkish Armies and opening up a path to the coast. Realising the nature of the trap, the Turkish Eighth Army hastily abandoned Gaza and fled northwards. Pursuing closely, despite a chronic shortage of water, Allenby's cavalry advanced northwards along the coast, and early in December swung east, to capture Jerusalem on 9 December.

Owing to demands then made upon his reserves for men to be sent to France for the 1918 offensives, Allenby could sustain no further campaigns for nearly nine months, having to content himself with the systematic training of his raw reinforcements. He had, in effect, to create an entirely new army. But his final battle, when it came, was brilliant and must stand among the finest feats of arms by the British Army – albeit anachronistically, for it was the very last major victory made possible by the classic use of massed cavalry.

Facing him, the Turko-German fortified line extended from the Mediterranean shores north of Jaffa to the valley of the Jordan, skilfully sited and defended by nearly 40,000 men and 350 guns. Allenby's army numbered 57,000 men (of whom 12,000 were horse-mounted) and 540 guns. His plan was to smash a gap in the enemy lines on the western flank while using elaborate deceptive measures in the east (deploying huge dummy camps and horse lines) to distract the Turks' attention. On 19 September 1918 he opened a devastating barrage in preparation for the infantry attack. Within the first day this great attack (reminiscent of an attack on the Western Front and quite unprecedented in the Middle Eastern theatre) tore a five-mile-wide gap in the Turkish lines, and through this the Desert Mounted Corps poured. Providing all the elements of a latter-day film epic, Allenby's breakthrough was spectacular, with cavalry pursuit on the ground and R.A.F. bombers attacking from the air. Within the next two days the entire British line completed a huge eastward wheel and the cavalry reached the Sea of Galilee; they pushed on to Damascus, entering Aleppo on 25 October. In thirty-eight days Allenby's army advanced 360 miles, fighting almost continuously, capturing or killing 80,000 Turks, Germans, and Austrians, and suffering no more than 853 killed and 4,482 wounded itself. Three Turkish armies were utterly destroyed, and at once Turkey sued for peace; an armistice was signed at Mudros on 30 October. In recognition of his superb campaign Allenby was promoted field-marshal; later, when created viscount, he chose as part of his title the name of Megiddo – the point at which he achieved his breakthrough on 19 September 1918.

The remainder of Allenby's career in public service was spent as High Commissioner in Egypt which, after the First World War, underwent a long period of civil unrest. He was frequently at odds with the British Foreign Office, whose overseas policies reflected Lloyd George's utter incomprehension of the nature of the war's effects abroad; for all Allenby's tough reputation, it should be said that his policies were far more conciliatory than those of the Foreign Office. Nevertheless he did exact stern penalties following the murder of the sirdar, Sir Lees Stack. He retired in 1925, and at the time of his death in 1936 was Rector of Edinburgh University. He was buried in Westminster Abbey.

Viscount Allenby was a big man, intimidating and irascible by nature, but scholarly and well-read. He must surely rank as one of the greatest of cavalrymen – on a level with Seydlitz – and the last great British leader of horsed cavalry.

Alp-Arslan (*c. 1030–72*)

Alp-Arslan, second Sultan of the great Seljuk line, helped to inflict one of the most damaging defeats the Byzantine Empire ever suffered. During the ninth, tenth, and early eleventh centuries the Empire had enjoyed a remarkable revival of vigour and stability after the exhausting Persian Wars. This revival was brought to an abrupt end by Alp-Arslan (and by Byzantine treachery) at Manzikert in 1071, when he destroyed the imperial army. Within weeks Anatolia, the heartland and chief recruiting source of the Empire, was almost entirely overrun. So devastating was the effect of the Turkoman occupation that even when such able later emperors as Alexius Comnenus (see under Bohemund I) recovered much of the lost territory, they found a population so decimated that they were unable ever again to raise significant forces in the area. The long-term effects of Manzikert on the structure and capability of the imperial armies were as important as the short-term casualties.

The Seljuks, who took their name from their first great leader, were a collection of Turkoman peoples who migrated from the east into western Asia in the first half of the eleventh century, embracing Islam on the way. They were to found a great empire comprising much of modern Iran, Syria, and Turkey; their occupation of Anatolia and their massacres of the native population were to lay the foundations of the modern state of Turkey. In 1048–9, during a period when the Byzantine forces were suffering from neglect, the Seljuks invaded Armenia and defeated the Armenians at Kars. They were later repulsed at Stragna and Manzikert, but the respite was to be brief.

Alp-Arslan ('the lion hero' – his given name was Mohammed Ibn Da'ud) was born in about 1030, son of Da'ud Chagri Beg of Khurasan, and great-grandson of Seljuk. He succeeded to his father's territories in about 1060, and three years later to those of his uncle Togrul Beg, Sultan of western Iran and Mesopotamia. He thus became the ruler of a unified Seljuk empire. In 1064 his armies overran much of Armenia, and in subsequent years raided deep into Anatolia; some of his bands even penetrated as far as Georgia. In 1067 the Emperor Constantine X was succeeded on the Byzantine imperial throne by Romanus Diogenes, an able soldier-emperor picked by the aristocracy in view of the threatening situation in the east. He spent a year reorganising the Byzantine armies, and then took the field against Alp-Arslan in the winter of 1068–9. He defeated the Seljuks at Sebastia, forcing their advanced elements to fall back into Armenia and Mesopotamia, and then carried out a rapid campaign of reprisal in Syria. Romanus continued to manœuvre his forces with skill and imagination; he besieged the Turkish fortress of Akhlat on Lake Van, and seriously defeated Alp-Arslan at Heraclea. The Seljuks were forced to withdraw to Aleppo, leaving only a few small enclaves in Armenia.

In 1070 the Seljuks moved westwards once again. Alp-Arslan led one army in person, and his brother-in-law Arisiaghi led a second. Alp-Arslan captured Manzikert but was repulsed from Edessa; Arisiaghi beat a Byzantine army under Manuel Comnenus near Sebastia. In the spring of 1071 Romanus himself led an army of between 40,000 and 50,000 men eastwards. Covered by an advanced detachment of some 10,000 under Basilacius, he recaptured Manzikert and advanced on Akhlat. Alp-Arslan marched to meet him with some 50,000 men. Unknown to the Byzantine Emperor, Basilacius was party to a widespread conspiracy against him. His covering force was easily brushed aside by the Seljuks, and Basilacius withdrew south-westwards without warning the Emperor. Romanus had to fall back on Manzikert once again, and during the retreat many of his light mercenary cavalry deserted him, reducing his total strength to less than 35,000 men.

The doomed emperor fought his final battle at Manzikert on 19 August 1071. He gave command of his second line to another traitor, Andronicus Ducas. He pushed the Seljuks hard all day, and Alp-Arslan's forces fell back before him and allowed their camp to be captured rather than let themselves be pinned down. As dusk came on, the Byzantine army was in an exposed position several miles from their own camp, and Romanus began to withdraw. Alp-Arslan's horse-archers harassed the retreat, and the mercenary elements in the imperial army began to lose their order. The Emperor ordered a halt, and faced about with his first line – but the treacherous Andronicus continued to fall back with the second line and the flank guards. There is no suggestion that there was any collusion between the plotters and the Seljuks, but Alp-Arslan was swift to take advantage. Before Romanus could organise his command for all-round defence they were surrounded and overwhelmed, and he himself was captured. In due course the conspiracy flamed into open rebellion in the capital, and Romanus eventually perished miserably at the hands of the traitor Andronicus; while rival factions struggled for power in Byzantium, Alp-Arslan's horsemen overran the whole of the Empire's Asian provinces. Vast numbers of civilians were put to the sword, and the greater part of Anatolia was reduced to a wilderness. Alp-Arslan was, by all accounts, a leader uninterested in either administration or culture. He died the following year, in November, at the hands of a prisoner taken in a central Asian campaign against the Qarakhanids.

Alvarez de Toledo, Fernando, Duke of Alva (*1507–82*)

The Duke of Alva was one of the most formidable generals of his century; he gave the Holy

Roman Emperor Charles V most distinguished service, and rose high in his favour. Under his son Philip II, however, he had less success, and he finally died far from the centre of events. In the meantime he had ensured for himself an infamous immortality in the annals of the Netherlands by his ruthless suppression of nationalist aspirations.

He was born at Piedrahita on 29 October 1507, the offspring of a rich family with a long tradition of service to the Castilian crown. He joined the imperial forces in 1524, and so distinguished himself at Fuenterabbia in the fighting against the French that when the town fell he was made Governor. In various subsequent campaigns he developed his skill and experience, rising to command of the imperial forces in Germany in the war against the Schmalkaldic League in 1546–7. At Mühlberg on 24 April 1547 he led the army of Charles V to victory over the Elector of Saxony and the other dissident Protestant nobles of the League; this was one of the first major battles in which the use of cavalry armed with pistols was significant. Alva's strengths were many: he was a strong disciplinarian, and insisted on good training and order among his troops. His grasp of logistics was impressive. The Spaniards long enjoyed a supremacy in the tactical development of firearms, and Alva was in the forefront of this movement. Above all, perhaps, he had a monumental confidence, and was never deflected from his aim. He was made Commander-in-Chief of the imperial forces in Italy in 1552, and after the succession of Philip II of Spain to the Holy Roman Empire's possessions in that area, he became Viceroy of Naples in 1556. In the last stages of the Franco-Spanish war in Italy he outmanœuvred the Duke of Guise, and was instrumental in forcing Pope Paul IV to come to terms with Spain in 1557. After the Peace of Câteau-Cambrésis in 1559 Alva became one of Philip's two chief ministers, and pressed for a vigorous foreign policy. In 1566 political and religious unrest swept through the Spanish Netherlands, and in August 1567 the Duke of Alva arrived with a Spanish army to hunt out the heretics, restore respect for the King's authority, and punish the rebels.

His methods included the setting aside of all established legal principles and the trial and condemnation of thousands of religious and political opponents of the régime by a new tribunal, soon known as the 'Council of Blood'. So efficient was Alva's reign of terror that when William the Silent and his brother Louis of Nassau invaded the Netherlands from Germany (1568), not a town rose to support them. Louis, with 15,000 men, was crushed by Alva at Jemmingen on 21 July; although the forces were roughly equal Alva overwhelmed Louis by superior fire-power and experience, inflicting some 6,000 fatalities for the loss of only about 100 Spaniards. The rebellion was stamped out in the north-east for several years. In October, William himself led 25,000 men into the southern provinces; Alva got behind him with 16,000 men and destroyed his rearguard at Jodoigne, forcing him to retreat southwards into France and then back into Germany. For the next three years the Netherlands suffered under Alva's harsh rule; two aristocratic and moderate leaders of the popular movement, Egmont and Hoorn, went to the block. Nevertheless the determined civil disobedience of the common people prevented Alva from placing his regency on a sound financial basis. It was impossible to collect the taxes he needed to pay his troops, who became mutinous and unreliable.

In 1572 the raids of the 'Sea Beggars' on the coastal areas culminated in the capture of Bril on 1 April. This was the signal for a new round of rebellions to break out all over the Netherands, and within a short period the Spanish had been swept out of the major part of Holland and Zealand. William led another invasion from Germany, and Louis a second force from France. Louis was besieged by Alva in Mons during the summer, and William failed in his efforts to relieve him. Slowly Alva won back control of the south and east, advancing from town to town and capturing them one by one. His skilled military movements were accompanied by appalling atrocities committed by the Spanish soldiers and mercenaries upon the civil population – notably at Mechelen, Zutphen, and Naarden. In July 1573 Haarlem fell after a heroic seven-month resistance, and Alva had the garrison butchered to a man. On the coasts and waterways, however, he found himself unable to subdue the 'Sea Beggars', and a Spanish squadron was defeated on the Zuider Zee in October 1573. Despite all his efforts he could not recapture the greater part of Holland and Zealand, where the population were prepared to cut the dykes rather than submit. He was also hampered by lack of funds, and by the machinations of his enemies in the Spanish court. In November 1573 he resigned his post.

Alva went through a period of disfavour, and at one stage was under house arrest. Nevertheless, largely due to the intercession of Cardinal Granvelle, he was given command of the Spanish forces in the invasion of Portugal in 1580. King Sebastian I of Portugal had been killed in battle against the Moroccans two years before, and civil war had broken out between no less than seven claimants to the throne. One was Philip II of Spain – and that he was successful was due entirely to Alva's most able campaign of 1580. On 25 August he smashed the army of Don Antonio de Crato at Alcantára, and within weeks of the invasion had occupied Lisbon. Strangely enough he never regained his king's favour, despite this final service. He died in Lisbon on 11 December 1582.

Arnold, Benedict (*1741–1801*), *Major-General*

An American militia officer of the War of Independence who intrigued for military favour, Arnold was disgraced in his own army, deserted to the British, and eventually died ostracised and despised by all.

Born of a distinguished family in Norwich, Connecticut, on 14 January 1741, Arnold served with the militia at the age of sixteen in the French and Indian War. For fifteen years after this he engaged in successful trading with Canada and the West Indies, becoming the prosperous owner of several ships and a thriving business. Towards the end of 1774 he became captain in a militia company, and on the outbreak of hostilities at Lexington the following year he marched this force to Cambridge, with the idea of seizing Fort Ticonderoga. Promoted colonel, he was authorised to raise a regiment to pursue his plan. However, Ethan Allen had already conceived the same idea and, realising that his own progress could not compare with Allen's, Arnold abandoned his own claim and joined Allen as a volunteer in the successful attack.

Soon after this Washington (q.v.) gave Arnold command of a force of 700 men with which he was to capture Quebec. After an extraordinary march through the Maine forest, Arnold realised that his force was inadequate to take the city, and waited to be joined by another column under General Richard Montgomery. The combined assault, carried out on 31 December 1775, failed; Montgomery was killed and Arnold badly wounded. Nevertheless, he continued to stand before Quebec under conditions of great hardship, and was rewarded for his steadfastness with promotion by Congress to brigadier-general. Finally the appearance of strong British forces obliged Arnold to retreat but, building a small flotilla of vessels on Lake Champlain, he inflicted considerable losses on the superior British fleet on 11 October 1776. Arnold returned home a hero – only to find that he had been passed over in the promotion of senior officers. For what he considered an unjustified snub, engineered by jealous fellow officers, he never forgave Congress. Even when he attacked and drove back the British force under General William Tyron at Danbury he was only belatedly promoted major-general, and without restoration of seniority. Still aggrieved, Arnold determined to resign, and was only forestalled by a fresh invasion from Canada under General Burgoyne (q.v.). Instead he marched up the Mohawk Valley in August 1777 against Colonel Barry St Leger, and raised the siege of Fort Stanwix. In the Battle of Freeman's Farm on 19 September Arnold commanded the American left wing which hurled back Burgoyne's attack, but he was denied reinforcements with which to counterattack. In the Second Battle of Saratoga (at Bemis Heights) on 7 October, Arnold was deprived of his command, but nevertheless galloped into action, took charge of the leading battalions, and thrust the British back against the river, inflicting heavy casualties. He received wounds that left him a cripple, but at least he was then promoted, with backdated seniority.

Later, while living in semi-retirement with a nominal command at Philadelphia, Arnold violated a number of minor regulations to secure personal wealth, and after foolishly arousing official suspicions he found himself facing charges which were referred to Congress. Socially ostracised, Arnold now pursued a plan (either of his own volition or as a result of pressure from his young wife) to seek favour with and desert to the British. In May 1779 he made secret overtures to the enemy, who suggested that he should remain where he was until he could betray an important position or vital stronghold. Haggling over recompense, Arnold allowed the negotiations to lapse while his own court martial was conducted. Found guilty of only minor charges by a lenient court, Arnold nevertheless displayed unwarranted indignation, and resumed negotiations with the British. His plans to deliver up West Point, of which he expected to be appointed commander, failed; and Arnold deserted to the British by escaping in a vessel that had brought a British officer to negotiate under a flag of truce. To achieve his own ends, Arnold abandoned this officer, John André, to be executed as a spy. Found with Arnold's plans of the fort on his person when captured, André behaved with the utmost dignity and correctness during his captivity, trial and execution, and deeply impressed his American captors.

The odious nature of Arnold's defection brought him no popularity among the British, even after he had gained permission to sail to England. Derisory pensions were settled upon him, his wife and family, and to escape the neglect and scorn he encountered in England he returned to his trading activities in Canada and the West Indies. But no matter where he travelled his notoriety preceded him to the detriment of his business, and he returned to relative obscurity in London in 1792. Wholly ostracised and in failing health, he survived until 14 June 1801.

Atatürk, *see* Mustafa

Auchinleck, Sir Claude John Eyre
(1884–1981), Field Marshal

British commander in the Second World War; Auchinleck's series of campaigns in the Middle East was ended by the near-catastrophic advance of Rommel's (q.v.) *Afrika Korps* in 1942.

Born in Northern Ireland of an Ulster-Scottish family on 21 June 1884, Claude Auchinleck was educated at Wellington College and at the Royal Military College, Sandhurst, being commissioned in the Indian Army (unattached) in 1903. The following year he joined the 62nd Punjabis. His entire service during the First World War was spent in the Middle East. Arriving in Egypt in 1914, he moved on to Aden the following year and to Mesopotamia in 1916, where he took part in operations against the Turks before transferring to Kurdistan. His war services brought him brevet promotion to lieutenant-colonel, and the award of the D.S.O., the O.B.E., and the French *Croix de Guerre*.

After attending the Imperial Defence College in 1927, he was given command of the 1st Battalion, 1st Punjab Regiment in 1929, and the following year became senior instructor at the Quetta Staff College. Appointed commander of the Peshawar Brigade in 1933, he took part in operations against the Upper Mohmands in that year and again in 1935. From 1936 to 1938 he held the post of Deputy Chief of the Indian General Staff.

Returning to Britain at the beginning of the Second World War, Auchinleck was placed in command of the second Allied expeditionary force, sent to Narvik in northern Norway – an enterprise that augured well but was short-lived, owing to logistic failures and to the impossibility of maintaining prolonged and adequate air support of the ground forces. For all his experience in mountain warfare the situation deteriorated rapidly with the build-up of German forces, and the decision was taken to evacuate the British force – an operation skilfully achieved and one that enhanced Auchinleck's reputation among the British general staff. On his return to England he was appointed General Officer Commanding Southern Command for a short period before returning to India as Commander-in-Chief, Indian Army.

In July 1941 Auchinleck was ordered to Egypt to succeed General Wavell (q.v.) as Commander-in-Chief, Middle East. Here, after a year's operations against the Italians (and latterly against the Germans), General

Wavell's forces were critically weak, having been called upon to pursue repeated campaigns throughout the theatre – campaigns which had nevertheless achieved astonishing success. Axis pressure was increasing against Egypt, however, and the port of Tobruk was besieged when Auckinleck arrived to take up his command.

On 18 November 1941 he launched a new offensive against the still-predominantly-Italian forces in the Western Desert, won an important victory at Sidi Rezegh, relieved Tobruk, and in five weeks recaptured Benghazi. Total Axis casualties were about 60,000 compared with 18,000 British, and General Rommel (q.v.) had lost almost 400 tanks and 850 aircraft. However, the dispatch of reinforcements to South East Asia which followed Japanese entry into the War, severely weakened the British desert army, while a temporary loss of British naval supremacy in the Mediterranean allowed a rapid reinforcement of the German *Afrika Korps*. As a result the British advance through Cyrenaica was poorly sustained, air superiority was not fully exploited, and supply lines – already critically extended – were badly administered; it was the final blow when the British advanced columns ran against and were held by well-prepared defence lines at El Agheila. An immediate German counter-attack caught the British forces off balance, and rout followed. All the newly-won territory was lost, together with large quantities of stores and equipment.

Auchinleck and General Ritchie (whom Auchinleck had placed in command of the Eighth Army in place of General Alan Cunningham) hastily prepared a line of resistance at Gazala, and this temporarily held against the advancing *Afrika Korps*. After some months of quiescence, Rommel attacked once more with overwhelming strength, employing superior tanks and the highly-mobile and deadly 88mm. general-purpose gun, which annihilated the critically vulnerable tanks with which the British forces were equipped. Having previously lost so much manpower and equipment, the Eighth Army was again hurled back; Tobruk fell, and the whole of northern Egypt – and the vital Suez Canal – was threatened.

The Battle of the Gazala–Bir Hakeim line was almost a disaster, and only the heroic defence of the Free French on the Allied left flank at Bir Hakeim prevented the entire Eighth Army from being enveloped and destroyed. As it was, the delay so dearly bought not only allowed much of the Allied strength to be withdrawn, but critcally exhausted Rommel's superb *Afrika Korps*. Between 14 and 30 June the Allies fell back to the Egyptian border, at which point Auchinleck assumed personal command of the situation and rallied his troops for a bold delaying action at Mersa Matruh; he then retired to a strongly fortified line anchored on the Alam-el-Halfa ridge between El Alamein and the Qattara Depression, only sixty miles west of Alexandria. Here the Eighth Army stood and fought off probing attacks by the critically-extended Italo-German forces.

It was at this point that Churchill decided to replace the generals in command of the Allied forces in Egypt: Auchinleck was succeeded by Alexander (q.v.), and Ritchie by Montgomery (q.v.).

Ineptitude there had been in North Africa, though Churchill was quick to exonerate Auchinleck. Most British commanders were unable to counter the brilliance of the German commander Erwin Rommel, whose aptitude for desert warfare was second to none – until eventually he was matched against such 'safe' generals as Alexander and Montgomery. Remembering that the British desert army had never enjoyed a sustained build-up of men and really modern equipment, and had been decimated by demands from Greece, Crete, Iraq, Palestine, Syria, and the Far East, it was to Auchinleck's lasting credit that he had achieved some successes during the winter of 1941–2 and that he managed to halt Rommel's advance at Alam-el-Halfa.

His military prestige thus relatively undamaged, Auchinleck was again appointed Commander-in-Chief in India – a post he held until 1946. Though it is no part of this book's purpose to dwell on his political and diplomatic activities, one cannot but remark that this period of his career was probably that in which he contributed his greatest services to the British cause. India was in a ferment of nationalist and factional dissatisfaction, by no means fully understood by certain British wartime politicians. It was Auchinleck's knowledge of the Indian religious structure and the political aims of such men as the Hindu nationalist leader Mahatma Mohandas Gandhi that won the confidence of the Indian Army, and made possible the relatively smooth reorganisation of the forces that were to confront and defeat the Japanese in north-east India and Burma in 1944–5.

After the War the 'Auk', as he was popularly and universally known, was promoted field-marshal. Although appointed to command the combined force operating in the Punjab in 1947, he retired shortly after the declaration of Indian independence on 15 August that year. He died in Morocco on 23 March 1981.

Aurangzeb (Mohi-ud-Din Mohammed), Alamgir I (*1618–1707*)

Last of the great Mogul Emperors of India, a despotic ruler and an intriguer of the utmost duplicity, Aurangzeb was the conqueror of Assam, Jodhpur, Bijapur, and Golconda.

Born during the night of 2–3 November 1618 at Donat, Gujerat, Aurangzeb was the third son of Shah Jahan by Mumtaz Mahal. He was expensively educated and grew up a bigoted Sunni, determined to extirpate Hinduism and idolatry. At an early age he demonstrated outstanding military gifts and before the age of twenty was entrusted by his father with a number of expeditions. When his father became seriously ill his eldest brother, Dara, seized the throne, whereupon Aurangzeb embarked upon ruthless manœuvres to displace him; he seized Shah Jahan and imprisoned him at Agra (where he died in 1666), slaughtered his brothers, and so became sole ruler.

In 1645 he had been made Governor of Gujerat, and two years later became Commander-in-Chief in Balkh and Badakhstan. He took the field against the Uzbeks, but the countryside over which he operated had been utterly impoverished by twenty years of war between the Moguls and the Persians and was quite unable to support his armies, so that this campaign – like so many others later – made intolerable demands on the imperial treasury without achieving any decisive conclusion.

After destroying opposition to his own ambitions by the seizure of Shah Jahan, Aurangzeb crowned himself Emperor on 31 July 1658, taking the title 'Conqueror of the World', and embarked on a reign that was to last for forty-nine years of despotic rule. Regarded by orthodox Mohammedans as the 'crowned saint of Islam', Aurangzeb was undoubtedly a deeply religious man. By embarking on a ruthless consolidation of power in northern India, however, he embittered a wide range of hitherto acquiescent factions and, while territorial gains continued to mount, the internal unification so methodically achieved by his father rapidly disappeared from the central power.

He seized Assam in 1662 (but lost it five years later), quelled the pirates of Chittagong in 1666, pacified the Pathan tribes of Kabul during the 1670s, and conquered the kingdom of Rajputana. But his adventures in the Deccan led to a widespread revolt of the Marathas, so that much of the second half of his reign was spent in war with them. His persecution of the Sikhs turned that peace-loving and religious group into a community of fanatical warriors absolutely dedicated to the destruction of Moslem rule, and under the Hindu general Durgadas the Rajputs destroyed the majority of the Mogul garrisons between 1675 and 1679. In 1681 Aurangzeb's son, Akbar, joined the Rajputs, but was defeated by his father and forced to flee to Persia.

By his conquest of the kingdoms of Bijapur (1686) and Golconda (1687) Aurangzeb extended the Mogul Empire to a size it had not attained before, stretching from the Coromandel Coast to Chittagong and the Hindu Kush. But his long series of military adventures was undertaken with little heed to cost, and his successes resulted, not from the excellence of his generalship or the prowess of his warriors, but from the very size of his armies, whose logistic demands were immense. With huge columns of camp followers to support, his conquests were a crippling burden on the central treasury, and his summary increases in taxation resulted in his authority being constantly in dispute. Although he overran the Maratha territory in 1689 with his Grand Army of the Deccan (and captured and executed the Sivaji's son, Sambhuji), the Marathas refused to acknowledge defeat and simply took to the hills to conduct prolonged guerrilla warfare – thereby rendering their country unconquerable.

When his final illness came upon him the Marathas, whom he had never succeeded in crushing, rose in open revolt – ostensibly in opposition to Aurangzeb's re-introduction of the poll-tax on non-Muslims – and were joined in northern India by the Jats, Rajputs, and Sikhs. He died in Ahmednagar on 3 March 1707, leaving the throne critically vulnerable to internecine quarrels. Indeed, so bitter was the rivalry between his sons Bahadur Shah and Muazim in their claims on the throne that their alliances with supporting factions still further exhausted the resources of Aurangzeb's Mogul administration. British and French influence in the sub-continent was growing, but the once-proud and mighty Mogul Empire was largely destroyed from within.

Baden-Powell, Robert Stephenson Smyth, 1st Baron Baden-Powell (*1857–1941*), *Lieutenant-General*

Although Baden-Powell has become a household name through his sponsorship of the Scout movement, his purely military record also deserves attention; he was one of the more flexible and imaginative senior officers holding command in the British Army at a time of rapid, not to say traumatic, transition.

Born in London, Robert Baden-Powell was the seventh son of the Savilian Professor of Geometry at Oxford University. He entered the Army in 1874, and joined the 13th Hussars in 1876 in India, where he began developing his natural aptitude for fieldcraft into an ability to use the terrain over which he operated. He quickly became known for his scouting techniques – a military art which spread among the British forces in India, especially in the difficult terrain of the north-west provinces. For the first time scouting became a branch of regular military training, and this resulted in Baden-Powell's becoming recognised as its leading exponent. Having been awarded the India General Service Medal with five bars, he was appointed to special service in Africa during 1895–6, taking part in the Expedition of Ashanti to suppress slavery and human sacrifices, and to punish the recalcitrant King Prempeh for his refusal to abide by the treaty of 1874. The force had a very difficult march through swamps and dense jungle before the King was compelled to render public submission to the Governor, and many deaths occurred in the British force as a result of the appalling climate. Baden-Powell was awarded the Ashanti Star for his participation in this campaign. He was also awarded the British South Africa Company's Medal for field operations in Rhodesia and Matabeleland in 1896; the following year he returned to India to take command of the 5th Dragoon Guards, Princess Charlotte of Wales's.

On the eve of the South African War, in response to appeals by the Natal garrison for reinforcement, the 5th Dragoons (with the 9th Lancers and 19th Hussars) left India for the Cape, arriving in September 1899, and forming part of General George White's main force of under 10,000 men based on Ladysmith. Colonel Baden-Powell was sent 250 miles to the north-west to take command of the small garrison town of Mafeking in the Bechuanaland – a decision much questioned at the time, in view of the size of the opposing Boer forces and the remoteness of the township. On 11 October, when war with the Boers broke out, Baden-Powell's garrison contained no regular soldiers at all, but was under the command of a small group of good officers which included Lord Edward Cecil (son of the Prime Minister), Colonel Gould Adams, and Colonel Hoare. His forces comprised 340 members of the Protectorate Regiment, 170 police, and 200 volunteers – mostly adventurers and irresponsible sportsmen. A local militia was also formed, so that the total defence force amounted to no more than about 900 men. On 13 October a Boer contingent appeared before the town, and within a week this had grown to between 5,000 and 6,000 men, led by the formidable Cronje. The subsequent siege was a far-from-passive operation for the defenders; every detail of the defensive system showed the ingenuity of the controlling mind of Baden-Powell. An intricate network of wire entanglements, trenches, forts, and shelters was built, and despite sporadic bombardment by large siege guns, casualties in the garrison were remarkably light. Numerous sorties were made by the encircled men, invariably utilising the tactics of stealth and surprise advocated in Baden-Powell's training.

The Siege of Mafeking lasted until mid-May 1900, when two relief columns, the larger under Colonel Plumer and the smaller under Colonel Mahon, reached the town, which had withstood investment for 217 days. The garrison had suffered about 200 killed, but had inflicted five times these casualties upon the enemy; infinitely more important, they had tied down a fifth of the entire mobilised Boer forces and a quarter of their artillery during the first six months of the war. The garrison's survival prevented the invasion of Rhodesia and provided a rallying point for whites and natives from Kimberley to Bulawayo. The lifting of the siege was a signal for celebrations throughout the British Empire – the first glimmer of good news in a War that had seemed to bring nothing but disaster upon disaster. Baden-Powell became a national hero, and was promoted major-general and deputed to raise and train the South African Constabulary. A year after the South African War ended he returned home to the appointment of Inspector-General of Cavalry in the British Army.

About this time Baden-Powell discovered that his book *Aids to Scouting* (which had been published just before the outbreak of the South African War) was being used to train boys, and in 1907 he decided to institute a trial scouting camp at Brownsea Island, Dorset; shortly afterwards he published an outline scheme for organised boy scout troops. At once such troops sprang up all over the country, their underlying philosophy being comradeship and self-sufficiency. Baden-Powell was promoted lieutenant-general in 1907, but he left the army in 1910 to give more time to promoting the Scout movement, and the same year was born the Girl Guide movement. Both spread rapidly throughout the Empire and in 1920, at the first international scout jamboree, Baden-Powell was acclaimed Chief Scout; his wife and sister did much to promote the Girl Guides.

Baden-Powell was created a baron in 1929 and was appointed to the Order of Merit in 1937. He died at Nyeri in Kenya on 8 January 1941.

CCD

Badoglio, Pietro, Duke of Addis Ababa (*1871–1956*), *Marshal*

Born at Grazzano Monferrato, Piedmont, on 28 September 1871, Badoglio entered the Italian Army as an artillery officer; he saw action during the Italo-Ethiopian War of 1896–7, and as a colonel in the Italo-Turkish War of 1911–12. During the First World War he rose to the rank of general, and participated in the interminable planning that was such a feature of the repeated Battles of the Isonzo (twelve in all) against the Austrians. In particular he planned and put into execution the assault on Monte Sabotino which preceded the fall of Gorizia in August 1916. Later he led the negotiations for an armistice at the Villa Giusti, and eventually signed on behalf of Italy.

Badoglio rose to become Chief of the Italian General Staff from 1919 to 1921, and was ambassador to Brazil in 1924–5. Promoted field-marshal in June 1925, he again became Chief of the General Staff until 1928, when he was made Governor-General of Libya.

In December 1934 Italian and Ethiopian troops clashed at Ualual in a disputed zone on the borders of Italian Somaliland, but after a year-long investigation by the League of Nations' arbitral board, responsibility could not be apportioned. Italy nevertheless demanded reparations. On 3 October 1935, without any formal declaration of war, Italian colonial forces invaded Ethiopia and within three days captured Aduwa. In command of the Italian expeditionary force, sent from Italy through the Suez Canal, was Pietro Badoglio. A united Italy backed this force, and Badoglio employed all the products of a totalitarian Western economy, ranging from a powerful air force to poison gas, against a gallant but hopelessly outmatched tribal army whose deployment and tactics were frankly medieval in concept. On 5 May 1936 Addis Ababa fell to the Italian invaders; the Emperor Haile Selassie I fled the country, and all Ethiopian resistance collapsed. Pietro Badoglio was appointed Viceroy of the country – an office he occupied only for a few months, at his own request – and was created Duke of Addis Ababa.

During the Second World War he again occupied the post of Chief of Staff from June

(Italy's entry into the war) until December 1940, when he resigned in protest at Mussolini's 'ill-advised and unnecessary' invasion of Greece. Though apparently a cautious supporter of Italy's seizure of Albania, he did not consider Italian forces able to maintain sufficient strength to overcome likely British resistance in Greece as well as launching an offensive (under Graziani, q.v.) in the Western Desert.

Though he was obviously careful to hide his opposition to Mussolini's continued support of the German war strategy, his attitude was not unknown to the Allies. The cautious King of Italy decided to ask Badoglio to head an Italian Republic in 1943; the arrest of Mussolini was planned (25 July), and negotiations were conducted between Badoglio and the Allies for the unconditional surrender of Italy (8 September). In this extremely difficult period, during which powerful German forces occupied almost the entire country, Badoglio acted with great courage, dissolving the Fascist Party and, on 13 October, declaring war on Germany. He resigned as Prime Minister in June 1944 and went into retirement. He died at Grazzano Monferrato – his birthplace – on 31 October 1956.

Not an outstanding field commander by any standards, Badoglio achieved nothing more than notoriety by his campaign against Ethiopia in 1935–6, only perhaps expiated by his opportunist volte-face in 1943. As periodic Chief of General Staff between the wars he must also bear his share of responsibility for the poor showing of Italian troops when faced by those of the Allies.

Bazaine, François-Achille (*1811–88*), *Marshal of France*

Bazaine is one of the tragic figures of military history – an officer who showed distinguished courage and ability throughout a long and eventful career, only to come to utter ruin at the end. His disgrace is a sad commentary on the state of France in the aftermath of the Franco-Prussian War – not a period when the highest standards prevailed in French public life.

Bazaine was born on 13 February 1811 at Versailles, the son of an engineer who had recently emigrated to Russia. His mother died young, and he was brought up by a guardian. He failed to pass the entrance examination of the École Polytechnique and, unattracted by trade, enlisted at the age of nineteen as a private in the *37e Ligne*. By the following year (1832) he had risen to sergeant; and in 1833 he was commissioned *sous-lieutenant*. In that year he transferred from the Line to the Foreign Legion in order to see service in Algeria. He distinguished himself in the disaster of General Trézel's column at Macta in 1835, being wounded and mentioned in dispatches. He was promoted lieutenant and admitted to the Legion of Honour. When the Foreign Legion was transferred to Spain to fight in the Carlist War Bazaine went with his unit, and with the local rank of captain led the *voltigeurs* of the 5th Battalion, distinguishing himself in the defence of Pons. In 1837, as chief-of-staff to the commanding officer, Colonel Conrad, Bazaine was again prominent at the bloody Battle of Barbastro. After returning to France, Captain Bazaine served in 1841 in the newly-raised *8e Chasseurs à pied*, seeing action in Algeria in the flying columns of General Bugeaud (q.v.). In 1844 he was promoted *chef de bataillon* in the *58e Ligne*, in 1847 lieutenant-colonel, in 1850 colonel in the *55e Ligne*, and in 1851 colonel and commanding officer of the 1st Foreign Legion. (Despite these regimental appointments he in fact spent much of the 1840s as head of the *Bureau arabe* at Tlemcen, a useful experience for an ambitious officer.)

Bazaine led his *légionnaires* to the Crimea, where he fought with distinction in the assault on Sebastopol. After the capture of the fortress he was named as its commandant and promoted *général de division*. He led his division in 1859 in the Franco-Austrian War in Italy, fighting with conspicuous gallantry at the head of his men and being wounded at Melegnano, and taking a key position at Solferino. He was posted to Mexico in 1862 to command a division of the French expeditionary force which was attempting to set Maximilian on the throne, and to keep him there. Bazaine, with his political experience and his fluent Spanish, was an excellent choice for this appointment, and immediately attracted favourable notice by his contribution to the capture of Puebla. He succeeded Forey as commander of the whole expedition in August 1863. It was an impossible campaign, fraught with political, logistic, and operational difficulties, but he handled his command with real skill. The results he achieved were genuinely impressive; and when his political masters admitted the failure of the adventure and ordered a withdrawal to Vera Cruz in 1867, Bazaine executed that most demanding of all operations with consummate expertise. He joined the ranks of the marshalate in September 1864.

At the outbreak of the Franco-Prussian War Bazaine commanded one of eight French corps spread along the frontier. The French army was badly under strength, and extremely weak as regards transport, artillery, munitions, logistics, and particularly intelligence. Napoleon III held the chief command in person, and no realistic strategy existed. In August 1870 the corps were split between two new armies, of which Bazaine was given the five-corps Army of Lorraine, and MacMahon (q.v.) the Army of Alsace. In the next few weeks Bazaine – like so many other French generals – was to demonstrate that a brilliant career as a 'frontier general' did not necessarily fit a man for the supreme command of an army in a modern, large-scale, fast-moving European war. Whatever the detailed criticisms that have been levelled at his leadership, it is permissible to summarise simply by saying that he was a fine commander promoted above his level and entrusted with a mission beyond his powers. His handling of his army was not helped by the fact that no effective army staff and command structure existed. His corps were spread too widely for mutual support, and on 6 August his II Corps was battered slowly into defeat by the Prussian First and Second Armies while its pleas for reinforcement went unanswered. For a week Bazaine retreated, and the fast advanced units of the Prussians penetrated deeply between his command and that of MacMahon. Moving back over the Moselle, he made an attempt to reach Verdun and effect a junction there; a fight at Borny on 15 August convinced him that Verdun was effectively out of his reach. He concentrated between the Moselle and the Orne, and after heavy fighting on 16 August at Vionville, Mars-la-Tour, and Rezonville, he apparently gave up hope of breaking westwards. He took up a long westwards-facing ridge position, where Moltke (q.v.) attacked him on 18 August. Again one of his corps performed miracles of stubborn defensive fighting, but its urgent requests for support went unanswered. At one stage of this Battle of St Privat it is clear that a general counter-attack had a good chance of success, but Bazaine remained passive. His army was forced back into the entrenched camp at Metz; and there it stayed, 173,000 strong, invested by the Prussian First and Second Armies, while the Third Army and the Army of the Meuse charged on towards Paris. On 27 October, after fifty-four days of little action but growing hunger, Bazaine surrendered Metz and his army.

This was a time of political, social, and military ferment in France, and it was inevitable under the circumstances that Bazaine should be reviled as a traitor by some parties. When he returned to France he demanded a commission of inquiry, and this began proceedings in April 1872 and concluded by issuing a severe censure. The terms of this persuaded Bazaine – and he was right – that he was being blamed not only for the Metz débâcle but also for matters over which he had had no control; he rejected the report of the commission, and demanded that he be court-martialled so that he could clear his name once and for all. It was a demand in which he was encouraged by various political circles, and a cloud of intrigue and counter-intrigue rose around the issue. On 10 December 1873 the court sentenced the simple, dignified marshal to degradation and death. He accepted the sentence, and declined to appeal. His old comrade MacMahon, now President of the Republic, commuted the death sentence to twenty years' confinement – a poor enough gesture.

Bazaine was imprisoned on the island of Sainte-Marguerite in the Mediterranean, and was allowed to have his wife and children with him. With the help of his wife – a spirited girl far younger than the sixty-three-year-old marshal – and of his faithful soldier-servant, Bazaine made a daring escape down the cliffs on the night of 9 August 1874. He fled to

Italy, and later to Spain; and on 28 September 1888 he died in penury in Madrid.

Beresford, William Carr, 1st Viscount Beresford (*1768–1854*), *General, Marshal of the Portuguese Army*

A tall and massively-built Anglo-Irishman of great courage and strength, Beresford was also a masterly administrator. However, when faced with the problems of an independent command far from the eye of his Commander-in-Chief, Wellington (q.v.), he became prey to a certain lack of flexibility and of inspiration, which was to give the British infantryman his bloodiest battle-honour for decades.

Born on 2 October 1768, the natural son of the 1st Marquis of Waterford, Beresford joined the Army in 1785 and distinguished himself as a junior officer at Toulon in 1793. He became commanding officer of the 88th Foot – the Connaught Rangers, a famous fighting regiment recruited largely in Ireland – in 1795. Beresford served under Sir David Baird in Egypt in 1801 and at the Cape of Good Hope in 1806. In that year Baird allowed him to sail on an unauthorised raid against one of Spain's colonial possessions, Buenos Aires in South America; Spain was at that time allied with France. As brigadier commanding 1,200 infantry, Beresford crossed the South Atlantic with Sir Hope Popham's naval squadron, and Buenos Aires fell to the surprise assault. Inevitably, however, the lack of support and of intelligent higher direction of the raid enabled the local forces to recover, and the British force capitulated on 12 August 1806. After six months' captivity Beresford escaped. In 1807 he was sent to Madeira, which he seized and commanded in the name of Britain's ally, Portugal. Recalled to the British Army in Portugal, he was for a time Commandant of Lisbon, before marching into Spain with the unlucky army of Sir John Moore (q.v.). Beresford was prominent at the Battle of Corunna on 16 January 1809.

In February 1809, in the hiatus between the evacuation of Moore's army and the reinforcement of the troops still in Portugal by Wellesley, the Portuguese authorities offered Wellesley the task of reorganising the ruinous Portuguese army, with the rank of Marshal. Wellesley, his eyes on greater prizes perhaps, declined but recommended his fellow Anglo-Irishman and friend Beresford, whom he knew to be qualified for the task. It was an excellent choice, and Beresford struck the Portuguese military establishment like a whirlwind – a whirlwind speaking fluent Portuguese, incidentally. He found an Army with excellent human material, but ill-trained, ill-organised, ill-supplied, ill-paid, ill-treated, and appallingly ill-led. The command system was rotten with corruption and privilege. Recognising the fine potential of the Portuguese peasant as a steady, courageous infantryman, Beresford set about liberating that potential by the most vigorous means. The supply, procurement, training, organisation, and distribution systems were taken over by British experts, and British officers and N.C.O.s were drafted into Portuguese units in large numbers. Beresford dismissed mercilessly all rogues and fools, regardless of their connections, and promoted men of promise, regardless of their lack of them. Outraged *fidalgos* who pityingly pointed out to him the subtle patterns of local influence which had for so long governed appointments and promotions in the army suffered immediate disillusionment. He could be terrifying in his anger; his irregular, rough-hewn features were distinguished by a milky and ruined left eye, damaged in a shooting accident, whose basilisk glare shrivelled many a quaking victim during his reformation of the Portuguese chain of command. The best testimonial to his efficiency was paid by the Portuguese light and line battalions interspersed with British units along Bussaco Ridge on 27 September 1810; under Masséna's (q.v.) assaults they acquitted themselves spectacularly, and mixed Anglo-Portuguese divisions were a feature of the Peninsular War from that time forward.

In the spring of 1811 the illness of 'Daddy' Hill (q.v.) and the rigid seniority system brought Beresford the independent command of an Anglo-Portuguese corps operating in Estremadura with the Spanish. Beresford invested the French-held fortress city of Badajoz, and in due course Marshal Soult (q.v.) came up from Andalusia to relieve the garrison. Beresford went forward to meet him, and the two armies met at Albuera on 16 May 1811. Soult sent heavy detachments of horse and foot in a wide swing around the Allied right flank, and through some misunderstanding that sector had not been strengthened as Beresford had intended. Out of a hail-storm which cut down visibility and damped powder came a regiment of Polish lancers and another of French hussars, and inside five minutes an entire British infantry brigade had been virtually annihilated. British infantry reinforcements rushed to the right to block the French hook, and both sides suffered terrible losses in a bloody and unimaginative killing-match which raged for about an hour. The intelligent counter-attack led by the British Fusilier Brigade which eventually broke the French pressure on the right was initiated not by Beresford, who was said to be considering retreat, but by a divisional commander at the urgent suggestion of a young staff captain. The day was saved, but not before British casualties reached 4,000 out of some 10,000 engaged.

Beresford showed great personal courage during the battle, particularly in hand-to-hand combat with a Polish lancer; but his control of his units and his grasp of the progress of the action were clearly faulty. He was shattered by the casualties suffered at Albuera, and his tribute to the dead, lying in their ranks as they had fallen, reveals a man of some sensitivity beneath the roaring exterior. Indeed Wellington, who could ill afford the loss of so many of his veteran infantry, considered that the man who had thrown them away for him was too downcast by the incident, and attempted in his brusque way to cheer him. All the same, he lost no time in bringing Hill back to his old command and sending Beresford to look after his Portuguese once more.

Beresford was at the storming of Badajoz, and was wounded at Salamanca in 1812. In 1814 he was created Baron Beresford of Albuera and Cappoquin, and he remained in the Portuguese service until the British officers were finally dismissed in 1819. In 1823 he was created Viscount Beresford, and in Wellington's cabinet of 1828 he was given the appointment of Master General of Ordnance. In 1830 he retired from public life to his estate at Bedgebury, in Kent. He died on 8 January 1854.

Berthier, Louis-Alexandre, Prince de Neuchâtel et Valangin, Prince de Wagram (*1753–1815*), *Marshal of France*

Napoleon's closest associate among his marshals was a small, rather portly man who bit his nails and was quite incapable of commanding large bodies of troops in the field. His considerable contribution to the success of his Emperor lay purely in the field of staff work, in which department he was a phenomenon of his age.

Berthier was born on 20 November 1753 at Versailles, and joined the army at the age of thirteen. It is significant that he initially served in a technical branch, as an *ingénieur géographe*, but over the following fifteen years he saw service in both foot and horse, and was by no means a stranger to the battlefield. (In later life he was twice wounded, at Marengo and at Brienne.) He sailed to America during the War of American Independence, and served on the staff of Lieutenant-General Count Jean de Rochambeau, the commander of the expeditionary force which fought alongside the American colonists. From this point onwards his career was to advance exclusively through his activities in the administrative sphere, with a single disastrous exception. He was a lieutenant-colonel by the time of the Revolution, and in the confused years which followed he served under Rochambeau once again, subsequently becoming chief of staff to Lafayette, and later to Luckner. This was a period of great uncertainty and peril for soldiers of rank who had served the Old Regime, and Berthier was probably grateful to have survived the Terror with no worse misfortune than suspension from rank and duties between September 1792 and March 1795. He was reinstated in that month, with the rank of general of brigade, and took up the duties of chief of staff of the Army of the Alps and of Italy. Only three months later he was promoted general of division. From his arrival in Italy as General Bonaparte until his first abdication as Emperor in 1814, Napoleon employed Berthier as his chief of staff. In this capacity the latter earned a considerable fortune, a sovereign princedom and numerous other honours, and a wife of royal blood. The two men were professionally inseparable, and the common soldiers, with rough humour, dubbed Berthier 'The Emperor's Wife'. In a sense there was a good deal of truth in this crude metaphor; sound housekeeping was as essential as strategic vision and tactical brilliance in the campaigns which carried Napoleon to the heights.

Most schoolboys know that Napoleon's armies lived off the country over which they campaigned, and it is sometimes superficially assumed for this reason that they must have been as administratively neglected as a medieval horde. Nothing could be further from the truth. Napoleon had, among his other skills, a very sound appreciation of the material needs of his vast commands, and logistical planning was brought to a peak of efficiency unprecedented in history – thanks largely to Berthier's brilliance in translating the stated requirements of his chief into actual stocks, depots, and distribution systems on the ground. While the staff departments of the *Grande Armée* were primitive and unwieldy by today's standards, and tended towards ill-defined responsibilities and duplicated effort, it is important to remember that they represented an enormous advance over the immediate past, and over most of their foreign contemporaries. Like most great visionary conquerors and statesmen, Napoleon was chronically bad at delegating authority; it says much for Berthier and his small and harassed team of generals and colonels that, by and large, the Emperor always had under his hand an army logistically as well as militarily capable of marching rings around his enemies. The Austerlitz campaign of 1805 was a prime example. Some 200,000 troops were marched from northern France to Austerlitz, via Ulm and Vienna, in just five weeks, and arrived in a condition to win one of the great victories of history. While they were billeted on the populations through which they marched, who were required to feed them, the troops carried several days' rations for emergencies, and were closely followed by supply trains carrying a further provision. Intermediate depots followed the army to provision these supply trains, and were in their turn supplied from the main base depots hundreds of miles away. The size, speed, and scope of this logistical operation was probably unprecedented in recorded history.

The one occasion on which Berthier was given an important field command was in the spring of 1809. The Austrians invaded the German states while Napoleon, newly returned from the pursuit of Sir John Moore (q.v.) in Spain, was in Paris. When the Emperor arrived at Stuttgart in mid-April he found that Berthier had contrived to get the 176,000-strong *Grande Armée* split up over a front of seventy-five miles, with its appallingly weak centre menaced by the enemy; the staff genius quite openly declared that he was unable to think what to do next, and appealed to his Emperor to extricate French fortunes!

Though efficient to the end, Berthier was increasingly crushed by the experience of the 1812 and 1814 campaigns, and was one of the group of marshals who accepted the first abdication with relief. He was honoured by the Bourbon monarchy upon its restoration, and when Napoleon landed in France in 1815 it was Berthier who escorted Louis XVIII to safety. He then retired to Bamberg in Bavaria; his failure to rally to his old chief was a severe shock to Napoleon, and the unwise appointment of Soult (q.v.) as chief of staff in his stead had damaging consequences during the Waterloo campaign.

Berthier died after falling from a high window under suspicious circumstances on 1 June 1815 at Bamberg. Various theories have been advanced, including murder by mysterious masked assailants, suicide out of remorse at seeing Allied troops marching towards the France he had abandoned, and a straightforward accident brought on by ill-health. The truth will never be known, but the murder theory is considered unlikely. Napoleon was apparently prostrated by news of his death.

Bigeard, Marcel (*b. 1916*), *Général de Corps d'Armée*

Marcel Bigeard, one of France's most heavily decorated serving soldiers, is closely, perhaps uniquely, associated with the evolution of a certain type of warfare. He has been called 'the world's best paratrooper', and his passionate involvement in the perfection of France's airborne forces over a period of twenty years lends credence to the claim. He has pursued a most distinguished career at all levels of command during a period of great strain for the French Army; it is a measure of his character that he has reconciled a philosophy of unrelentingly aggressive leadership with complete loyalty to his political masters.

He was born of humble parents at Toul, Meurthe-et-Moselle, on 14 February 1916. He had an indifferent formal education, being forced to leave school at twelve years of age to find work. He did several jobs, including working in a bank, and had no early ambition for a military career. He was called up for military service in September 1936, serving with the 23rd Regiment of Fortress Infantry in the Maginot Line, and as a reserve sergeant was recalled to his unit on the outbreak of the Second World War in September 1939. He rose to the rank of *adjudant* before being captured by German forces on 28 June 1940. He was imprisoned in a camp at Limburg, and made two escape attempts on dates which he chose for luck – 14 July 1941, Bastille Day; and 11 November 1942, the anniversary of the 1918 Armistice. The second attempt was successful; he travelled via Nice to Senegal in French West Africa, and was commissioned *sous-lieutenant* in the Free French Forces. He remained in Senegal until the summer of 1944, when he was parachuted into France to reorganise the Maquis in the Ariège sector; the resistance groups in this area were scattered and weak, and the presence of Spanish guerrillas in this region was an added complication. With the temporary rank of major and the *nom-de-guerre* 'Commandant Aube', he distinguished himself during the remaining months of occupation. He was heavily decorated for his resistance work, earning five citations in all and receiving, among other medals, the British Military Cross. It was from his radio call sign of this period – and ever since – that he took his nickname of 'Bruno'.

In the immediate post-war period he volunteered for service in Indo-China, and as a captain in the 23rd Colonial Infantry sailed for the Far East in October 1945. He now began the first of three tours of duty in Indo-China which were characterised by great energy, a deep study of local conditions, and an extremely aggressive tactical doctrine. The French command in Indo-China suffered from many shortcomings; but Bigeard's many successes in the field were based on the unswerving conviction that European troops could only defeat the Viet Minh by making themselves as physically tough and as resourceful as the enemy. He specialised in operations deep in the hilly jungle, carrying the war to the enemy by his own methods, and quickly attracted attention by successful pacification of insecure areas. Between May and October 1947 he commanded in the Phu Yen–Lai Chau region; by the time of his return to France at the end of the year he had achieved spectacular results throughout Son La province. He returned for a second tour in Tonkin in November 1949, and the following year raised the 3rd Thai Battalion and led it to gain an *élite* reputation. During his third tour, beginning in July 1952, he commanded the 6th Colonial Parachute Battalion with great distinction, and was named a Commander of the *Légion d'Honneur* the following October. His unit suffered heavy losses, and he himself was wounded twice; on one occasion he was isolated in Viet Minh-controlled jungle, but fought his way back to rejoin his command. After brief home leave he led his battalion in the drop, on 20 November 1953, which seized control of the Dien Bien Phu area. For reasons which have caused raging controversy ever since, the French High Command established a relatively weak entrenched camp in this valley far behind Communist lines, supposedly as a base for wide-ranging harassing operations by airborne and Foreign Legion battalions based inside the perimeter. General Giap's (q.v.) regular divisions soon invested the camp and confined French movement to a small area. Supported by unexpectedly heavy artillery and anti-aircraft batteries, Giap then proceeded to tighten the stranglehold. Several of the units in the garrison proved unreliable; and during the agonising battles of spring 1954, when the Viet Minh took strongpoint after strongpoint in bitter fighting, the handful of airborne and Legion battalions bore the brunt of the ordeal. Bigeard played a most active part in the defence, far beyond his official status as a battalion commander, and together with such officers as Langlais, Lalande, Botella, and Seguins-Pazzis, inspired the doomed garrison to hold out longer than was thought possible. When the camp fell he went into Communist captivity, and survived both the death-march and the prison camp itself, to be repatriated in September 1954. He had been promoted lieutenant-colonel by radio in April 1954.

In 1958 Bigeard was appointed director of the Centre for Instruction in Subversive and Guerrilla Warfare. Despite his advocacy of a spartan military training to produce soldiers capable of enduring great hardship, exhaustion, and heavy casualties without losing their aggressive morale, he is by no means simply the hard-eyed, crew-cut killer sometimes depicted by a hostile Press. His practical experience of counter-insurgency fighting, perhaps unparalleled in the Western world, has given him considerable sophisticated insight into the fundamentals of this type of warfare, and he has formulated effective training doctrines accordingly. In the later months of 1958, promoted colonel and given command of the 3rd Colonial Parachute Regiment, he took part in operations against the F.L.N. in Algeria. Again his successes and his luck became legendary; on one occasion three terrorists tried to assassinate him as he was walking on a beach, off duty and unarmed, and although one shot broke his arm he put them to flight. In 1959 he commanded the Saida operational sector in South Oran Province, again with great distinction. He held this command until the generals' *coup* of May 1960, in which he played no part, although several colleagues belonging to the much-exaggerated group dubbed by the Press 'the paratroop Mafia' were deeply implicated. The confidence placed in him at this traumatic moment in French military history was underlined by his appointment in July 1960 to a new command, that of the 6th Overseas Mixed Regiment, an all-arms combat group based at Bangui in the Central African Republic.

In December 1963 Bigeard took up a technical advisory post with the Central African Republic, and in August 1964 he joined the staff of the *élite* 11th Intervention Division at Pau. He commanded first the 25th Parachute Brigade, and later, from January 1966, the 20th Brigade. In that year he was promoted to general rank. From 1967 to June 1970 he served on the staff of the French forces at Dakar in Senegal, and in 1970–1 with the Defence Ministry in Paris. He was named in August 1971 as commander-in-chief of the French forces in the Indian Ocean area, based on Diego Suarez in Malagasy. In June 1973 Bigeard was transferred to the post of deputy to the Military Governor of Paris, commanding the 1st Military Region. In March 1974 he was promoted to the rank of *Général de Corps d'Armée* – four-star general – and took over command of the 4th Military Region with headquarters at Bordeaux.

General Bigeard served as State Secretary to the Defence Ministry (deputy defence minister) under the Chirac administration, from January 1975 to August 1976; and sat as deputy for the department of Meurthe et Moselle, 1978 to 1988. A Grand Officer of the *Légion d'Honneur* since 1971, and the winner of twenty-five citations in all, he is one of the best-known and most admired soldiers in France.

Black Prince, *see* Edward Plantagenet

Blücher, Gebhard Leberecht von, Prince of Wahlstadt (*1742–1819*), *Field-Marshal*

Born at Rostock in 1742, Blücher entered the Swedish army and gained a commission at the age of sixteen, but shortly afterwards was taken prisoner by the Prussian Colonel von Belling. By all accounts Belling much influenced his officers by his instinctive love of fighting; he soon recognised in young Blücher considerable military promise, and took him into his own regiment. But Blücher displayed rather too much high-spiritedness, being something of a rake and fond of high living. This was not tolerated in the proud Army of Frederick the Great (q.v.), and the young officer appears to have overstepped the mark. It seems that he was suspected of excesses committed against a captive Polish priest when stationed in Poland at a time when Frederick was anxious to appease that country. Blücher was passed over for promotion and, feeling the futility of further service under the cloud of Frederick's displeasure, impetuously resigned his commission. As time passed in boredom, Blücher became contrite and sought re-entry into the Army – but in vain. Only when Frederick died in 1786 was Blücher successful in gaining readmittance to the Prussian army. He was recommissioned as a major in his old regiment, and by 1793 had reached the rank of colonel under the Duke of Brunswick. The following year, at fifty-two, he was promoted major-general, and in 1801 he was made a lieutenant-general.

When in 1802 the bishopric of Münster passed into the military control of Prussia, Blücher was appointed to the command of the occupying forces and, together with the great liberal reformer Baron Stein, performed the unpleasant task of administration with such honesty and fairness that the ecclesiastical authorities petitioned the King of Prussia to appoint Blücher governor.

Following the 'Act of Mediatization' of 1803, which deprived numerous small German states of their independence, Prussia and Austria were left apparently all-powerful in the tottering Holy Roman Empire, their power only balanced by several strong, but secondary states – Bavaria, Baden, Württemberg, Saxony, Hanover, and Hesse-Darmstadt. When, later that year, the French under Mortier occupied Hanover, Blücher hastened back to Berlin only to find his Government displaying an attitude of complete indifference to the situation. These changes were but a prelude to a decline in the power of Prussia and Austria in Germany. Austria in particular suffered defeat and loss of territory throughout 1805, and on 6 August 1806 Francis II of Austria renounced the imperial dignity and became instead simply the Emperor of Austria. In the meantime, as deprecated by Blücher, proud Prussia had remained aloof from the struggle against France and, by the treaty of Schönbrunn (December 1805), received Hanover. Now faced with the French-created Confederation of the Rhine, and apprehensive lest she become the next victim, Prussia declared war on France. In a campaign of only three weeks, the military prestige bequeathed by Frederick the Great was shattered in the battles of Jena and Auerstadt; at the latter Blücher had his horse shot under him while leading his squadron, but immediately afterwards commanded forces covering the withdrawal of Hohenlohe. When Hohenlohe surrendered at Prenzlau, Blücher managed to fight his way through to Lübeck, but he in turn was forced to surrender to the French. He was not long a prisoner, soon being exchanged for the French General Victor.

The Treaty of Tilsit (July 1807) further humbled Prussia: all territories west of the Elbe were absorbed into a new kingdom of Westphalia, Beyreuth was ceded to France, most of Prussia's Polish lands were formed into the new grand duchy of Warsaw, some territory went to Russia, and Danzig became a free city. At a blow Prussia was reduced in size by about a half. When in 1809 a Prussian major, Schill, rose against the French but was defeated and killed at Stralsund, Blücher publicly disavowed the action as foolhardy, but privately gave protection to many of Schill's survivors; this action brought a severe reprimand from the King. Considering that his patriotism had been affronted, Blücher sought to resign from the Army a second time. The King hurriedly reassured Blücher of his continuing confidence; and somewhat mollified, the sixty-seven-year-old veteran accepted the post of general of cavalry and was placed in command at Colberg. This command was prematurely terminated when the French consul at Stettin discovered that his force was 7,000 over the number permitted. Following a demand for Blücher's removal by Napoleon's ambassador, the King sheepishly retired him from the appointment.

Spending the winter of 1811–12 at Stargard, Blücher was wholly unemployed during the French invasion of Russia, but as news arrived of the growing misfortunes being suffered by the *Grande Armée*, there was no restraining the old warhorse. At first King Frederick William sought to preserve a nervous peace with France, but as a result of nationwide demands for retribution – voices led in unison by Blücher, Scharnhorst (q.v.) and others – the King wrote to Blücher: 'I have determined to place you in command of those troops that are the first to take the field. I order you accordingly to mobilise here as speedily as possible.'

The Army of Silesia, under Blücher, straightway set about eliminating the French outposts; Blücher crossed the Elbe and established his headquarters at Dresden. In the first battle of the campaign, at Lützen (2 May 1813), the French won a hollow victory, suffering 22,000 casualties compared with half that number lost by the Allies. During the battle the seventy-one-year-old Blücher led a cavalry charge on the corps of Ney (q.v.) and Marmont, had his horse shot under him, received a bullet wound in his side, and spent eleven hours in the saddle.

Blücher was adored by his army which, in spite of a pitiful lack of training, possessed a fiery spirit and would have followed its leader into the jaws of hell. Critics frequently charged Blücher with a fondness for drinking and gambling, but his personal bravery was acknowledged by all; moreover Blücher possessed in General August von Gneisenau a brilliant chief-of-staff. 'Gneisenau, being my chief of staff and very reliable, reports to me on the manœuvres that are to be executed and the marches that are to be performed. Once convinced that he is right I drive my troops through hell towards the goal and never stop until the object has been achieved – yes, though officers of the old school may sulk and bellyache to the point of mutiny.' Affable though the old man may have been, he could be utterly ruthless, as the Saxons found to their cost after the mutiny of 1815.

1813 was the year in which Blücher achieved his military greatness. Following the defeat at Lützen, Blücher succeeded Count Ludwig Wittgenstein and set about an extensive reorganisation of the army, checking further advances by the French army meanwhile. During a short armistice between 4 June and 16 August 1813 Napoleon reorganised his cavalry, but on 12 August Austria declared war on France, so that when hostilities reopened there were three Allied armies – the Bohemian Army of a quarter of a million men under Schwarzenberg, the Silesian army with 195,000 men under Blücher, and the Northern army with 110,000 men (including a small British contingent) under Bernadotte.

Disaster followed disaster for the French. The Allied strategy was to avoid direct battle with Napoleon's main army, but to attack the forces of his lieutenants. Bernadotte gained a success against Oudinot in the battle of Grossbeeren (23 August), and Blücher defeated Macdonald at Katzbach (26 August) In the battle of Kulm (29–30 August) a French corps under Vandamme was virtually annihilated by armies of Austria, Russia, and Prussia, totalling about 100,000 men. When Napoleon recovered his scattered forces early in September and attempted to entice Blücher into battle in Silesia, the cunning Prussian declined combat in unfavourable circumstances and retired rapidly.

Following the defeat of Marshal Ney by Bernadotte at Dennewitz on 6 September the main French army, exhausted and dispirited, was reduced to 200,000 men and had been almost hemmed in between Leipzig and Dresden. Blücher, abandoning his lines of communication, crossed the Elbe north of Leipzig to threaten Napoleon's rear, intending to link up with Schwarzenberg coming up from the south. Sensing indecision by Bernadotte, Napoleon's main army attacked Blücher in the north on 15 October and, during the four days following, simultaneously fought the Russians and Austrians in the south – Bernadotte moving in from the east on 18 October. In this battle of Leipzig, Napoleon was driven back into the city and should have been encircled and decisively beaten; Blücher and Schwarzenberg had, however, failed to cut the French escape route to the west and Napoleon managed to extricate the main body of his forces – though he had lost 60,000 men, 150 guns, and great quantities of equipment.

After Napoleon's foolish rejection of peace offers in 1813, the Allies decided to invade France with three armies, of which Blücher's force of 75,000 men was to advance in Lorraine. The objective of all these armies was to be Paris. Blücher's advance was pressed with mixed success, but despite humiliating defeats at Craonne and Rheims (7 and 13 March 1814), the Prussian and Austrian armies linked up before Paris on 28 March. Three days later the French capital surrendered to the Allies, and a week later Napoleon abdicated.

After the return of Napoleon to France in March 1815 and the resurgence of the French Army under his command, the Allies again concentrated armies on the French border. By

1 June five such armies, totalling more than half-a-million men, were assembling from Belgium in the north to Italy in the south. Believing that the Allies could not exercise their combined power for at least two months, Napoleon determined to defeat the opposing armies in detail, and on 11 June moved out of Paris to face the nearest menace – Wellington's Anglo-Dutch army of 95,000 men and Blücher's 124,000 Prussians, assembled in Belgium. Although taken by surprise, Blücher reacted promptly, and by the evening of 15 June had positioned part of his forces at Sombreffe, while Wellington was concentrated fifteen miles to the west. The key point linking the two armies was the small crossroads village of Quatre Bras. On the following day Napoleon fell on Blücher at Ligny, rocking him back; believing that Ney had captured Quatre Bras, he ordered him to strike Blücher's right flank and thereby complete the victory. But following a gallant defence by a British brigade and procrastination by the French marshals, the centre of the Allied line was temporarily held. Blücher, initially defeated and himself wounded, retreated north to Wavre, but the French commander, Grouchy, failed to maintain contact with the retreating Prussians and gave Blücher time to reorganise his forces.

Wellington, in the time bought at Quatre Bras, had concentrated his main army of 68,000 men at Waterloo, ten miles to the north, and it was here that Napoleon with 72,000 men attacked at noon on 18 June. In four hours the British were forced back all along the line, and had Ney's cavalry attack been supported by the Imperial Guard infantry, the British centre would have been pierced. In the event, worried by the appearance of Blücher's leading elements advancing from Wavre on his right flank, Napoleon held back his infantry. As the strength of Blücher's main force of 67,000 men was felt and the Imperial Guard fell back, Wellington counter-attacked and Napoleon's army collapsed into a rabble of refugees. Meanwhile Blücher had left a rearguard at Wavre to cover his move to assist Wellington, and at last Grouchy attacked this and gained a limited victory; but the defeat at Waterloo brought final collapse to the French Empire. Three days later Napoleon abdicated for the last time.

'Marshal Forwards', as he was known, by his timely arrival at Wellington's side at Waterloo set the seal on Prussian military resurgence. The path back from the gloomy humiliation suffered after the death of Frederick the Great had lasted only five years, and the Prussian Army had only been effectively capable of meeting the might of France during the last two of those years. Blücher had more to offer than simply dash and courage in battle, for he assembled about him a select band of military genius – men like Gneisenau and Clausewitz – who together gave new life to the Prussian military legend. This legend was to survive to Prussia's advantage in the war of 1870–1; indeed, many historians manage to trace the 'blood and iron' strategy through to the great wars fought by Germany in the next century.

Bohemund I, de Hauteville (*c. 1057–1111*), *Count of Taranto, Prince of Antioch*

An outstanding figure of the First Crusade, Bohemund de Hauteville stands as a symbol of the awesome confidence and hardiness of the handful of Norman barons who marched thousands of miles into the unknown east to carve out for themselves crusader kingdoms among the infidel.

The patriarch of the Hautevilles was Tancred, an obscure Norman knight with holdings near Coutances. He was visited with the extremely mixed blessing of twelve sons, and since his fief could not provide for half that number, they scattered to seek their fortunes with their swords. Four went to Italy, and by 1060 one of them, Robert, had risen to become Duke of Apulia. Robert was recognised as the leader of the Normans of Southern Italy, and crowned a career not notable for its mildness or morality by becoming the secular vassal and influential protector of the Pope. He was named 'Guiscard', on account of his guile in war. By his first marriage to one Alberada he had a son christened Mark, who was apparently nicknamed Bohemund (after a giant of legend) because of his great size and strength. Robert Guiscard had the marriage with Alberada annulled subsequently and married one Sicelgaeta, daughter of a powerful Lombard house. This remarkable woman presented him with several more sons, and seems to have had a fondness for accompanying her lord into battle in full mail.

Bohemund served with his father in his campaign against the Byzantine Empire in 1081, at Corfu and Durazzo. During Guiscard's absence in Italy in 1082–4 Bohemund commanded the Norman army, advancing into Thessaly with some success but eventually being repulsed by the armies of the Byzantine Emperor, Alexius Comnenus. On Guiscard's death in 1085 Sicelgaeta's son Roger succeeded to Apulia, and the landless Bohemund plunged into a family war against him. Pope Urban II mediated between them in 1098, but by that time Bohemund had almost as much land in Apulia and Calabria as he would have had if he had succeeded to the duchy in the first instance. He remained the true leader of the Normans of Italy, with the title Count of Taranto.

The Byzantine Empire had lost much of Anatolia in 1071, and despite the religious schism which divided Rome and Constantinople, the loss of the Holy Land to the Turks and other Islamic peoples gave weight to the Emperor's requests for aid from Western European knights. Eventually, at Clermont in 1095, Pope Urban declared holy war against the infidel, and promised remission of sins to all who died on the Crusade. The response was remarkable but ill-organised; between 25,000 and 30,000 crusaders had crossed the Bosphorus by early summer 1097, but there was no appointed leader, and chaos reigned.

Apart from many civilian pilgrims who looked to the knights for protection, there were hundreds of knights and men-at-arms motivated by a simple greed for loot, and by the Norman's constant desire for battle. Some came with purer motives, for religious faith was crude but strong in Norman Europe; some were landless younger sons seeking permanent fiefs; and some were from the noblest families in Christendom, pledged to stay with the Crusade at least until Jerusalem was secured. Among the leaders, who managed the military direction of the Crusade by a brawling council of touchy near-equals, were the brother of the King of France, Hugh of Vermandois; Robert, Duke of Normandy, son of William (q.v.) the Conqueror; Robert of Flanders; Raymond of Toulouse; the pious and able Godfrey de Bouillon, Duke of Lower Lorraine, who was to become Advocate of Jerusalem; and Bohemund, with his mettlesome nephew Tancred, leading a large contingent from Italy. In council Bohemund, who was of much less noble birth than many of his fellows, gradually came to exert considerable influence through his acknowledged military skill; his main opponent was Raymond of Toulouse.

Bohemund paid homage to his old enemy Alexius Comnenus at Constantinople before the crusaders marched east and south; but there was to be constant argument over the exact implications of such vows among the crusaders, especially when the Emperor's promised help dried up suddenly and left the Europeans to their own devices. Bohemund distinguished himself at the Battle of Dorylaeum, when the crusaders nearly met disaster at the hands of a Seljuk Turkish army (1 July 1097); he displayed a more flexible approach than his colleagues to the tactical problems presented by the unfamiliar methods of warfare in the Middle East, and ensured that the Christian advantages of heavy mail, heavy horses, and heavy charges with the lance were not thrown away through lack of discipline or foresight. He was also most prominent during the bitter Siege of Antioch (21 October 1097–3 June 1098), which eventually fell to the crusaders after Bohemund had bribed a garrison officer named Firuz to open a gate. Almost immediately after the fall of the city – which was marked by the sort of savage excesses which the crusaders apparently felt quite able to reconcile with the declared purpose of the expedition – they were in their turn besieged in the starving city by a belated army of relief some 75,000 strong under the Amir of Mosul. Only 1,000 of the 15,000 crusaders were still mounted; nevertheless on 28 June Bohemund led them out in a sortie against the Moslems, and by superior planning and desperate fighting the crusaders triumphed at the Battle of the Orontes. Bohemund now claimed Antioch for himself and ruled it as a principality, against the bitter opposition of both the Byzantine Emperor and his vassal Raymond of Toulouse, whose coastal territory in the County of Tripoli blocked Antioch's southern expansion. The rest of the crusaders pushed on to Jerusalem, and took the city, under Godfrey de Bouillon's leadership, on 18 July 1099; Godfrey also smashed a relieving army of 50,000 men from Egypt at Ascalon on 12 August, with a mere 10,000 crusaders.

Tancred held Antioch for his uncle in 1100–3 while Bohemund was a captive of the Danishmend Turk Amir of Sivas. On his ransom being paid he resumed his campaigns against the Moslems of north-east Syria, but was handicapped by lack of manpower. After a serious defeat at Rakka on the Euphrates in May 1104 he returned to Europe to recruit help, and was welcomed as a hero of Christendom. He married Constance, daughter of King Philip I of France – a remarkable achievement for a man of his birth – and raised a new army. The warmth of his reception seems to have made him overconfident, however; instead of returning to Antioch to secure his base, he took on his old enemy the Emperor openly by laying siege to Durazzo – for the second time in his life. The campaign was long and hard, but Bohemund could not overcome the numbers arrayed against him. In 1108 he was forced to submit to Alexius in a humiliating treaty signed at the Devoll river; he became a vassal of the Empire, holding Antioch for Alexius, and was forced to admit a Patriarch of the Greek Church to the city.

In fact Bohemund never returned to Syria; he went home instead to Italy, where he died three years later. He was buried at Canosa di Puglia in Apulia. So great was his fame as a crusader hero, despite his final humiliation, that his nickname was taken by six lineal descendants before the end of the thirteenth century.

Boleslav I, called the Brave *(966?–1025), King of Poland*

This contemporary of Brian Boru (q.v.) was the first King of Poland, and only the second Christian leader of a people who had been pagan until the 960s. His reign was marked by political vision and military prowess, and saw the beginning of Poland's rise to prominence as a European power.

Poland had been a pagan Slav state when contact was first made with the Germans in the 960s. Mieszko, the first Christian leader of the Poles, expanded his territory westwards as far as the Oder, and consequently became involved in ten years of warfare with the Germans. Eventually he voluntarily accepted German suzerainty, in order to accelerate his people's progress towards civilisation. Boleslav, Mieszko's son, came to power on his father's death in 992. After taking the prudent precaution of exiling his stepmother and brothers, he began a career of diplomacy and conquest coupled with a steady improvement of the internal administration of his realm. His goal was to rule as the recognised king of a strong, civilised, well-governed and independent Poland, and he achieved it. He placed some strain on the relationship with the Holy Roman Emperor, Otto III, by emphasising Poland's strength, wealth, and dignity. He brought about the raising of the capital, Gniezno, to the status of an archbishopric, and improved the organisation and independence of the Polish church. By able military campaigns he added land in eastern Pomerania, Silesia, Moravia, and Cracow to what had been his father's territories. He also ceased paying certain established taxes to Otto.

In 1002 Otto died, and Boleslav immediately provoked a confrontation with his successor, the German King Henry II. Boleslav seized the German provinces of Lusatia and Misnia, and then turned his attention to Bohemia. Here the Polish ruler intervened on the side of the exiled Bohemian prince, Boleslav the Red, during internal troubles in 1003. Having defeated the opposition and restored his protégé briefly to power, Boleslav of Poland had Boleslav of Bohemia blinded and imprisoned – a miserable fate, as he lived until 1037 – and took the throne himself. The Emperor tried to regularise the situation by offering Bohemia to Boleslav as a fief of the Empire, but the Polish ruler refused to accept this arrangement, and began open warfare against Henry. Three distinct campaigns were fought; in 1005 Boleslav lost all his western conquests to the Germans, but in 1010–13 he regained Lusatia and Misnia. Eventually the Peace of Bautzen, concluded in January 1018, confirmed Boleslav's rule over all his conquests and his suzerainty over Bohemia.

Turning eastwards, Boleslav then defeated Yaroslav, Grand Prince of Kiev, in the Battle of the Bug river in 1020. He occupied the territory along the Upper Bug which had been the cause of the dispute, and briefly held Kiev itself; he also placed his son-in-law Svyatopolk, Yaroslav's brother and rival, on the throne, although Yaroslav soon regained it. Boleslav's ambition to rule as king over a sovereign state was gratified in 1025, when he was crowned in Gniezno. He died the same year, leaving an entirely new state stretching from the Elbe to the Bug and the Baltic to the Danube.

Bolívar, Simón (*1783–1830*)

Bolívar, the liberator of the northern Spanish-American colonies, was perhaps the most remarkable man to emerge from either North or South America. He was certainly a political philosopher of genuine vision, a leader of great character and tenacity, and a most able soldier.

Bolívar was born of aristocratic creole parents in Caracas, Venezuela, on 24 July 1783. His grounding in libertarian philosophy came early – one of his tutors was a disciple of Rousseau. Bolívar spent his youth in Spain, and brought a bride back to Caracas when he was nineteen, but she died of yellow fever within a year, and in 1804 the young man returned to Europe and immersed himself in the study of political philosophy. He travelled in France and Italy, and was impressed by Napoleon's demonstration of what one determined man could achieve. He returned to Venezuela via the United States, and became active in circles dedicated to the overthrow of Spanish rule. The French invasion of Spain in 1808, followed by the creation of a puppet monarchy and the national rising, sparked off the independence movement in South America. Bolívar participated in the events which led to the replacement of the viceroy in Venezuela by a *junta* and the declaration of national independence in July 1811. He was given an army command at Puerto Cabello, a vital port; a treacherous subordinate betrayed the fortress to the Spanish forces sent to restore royal authority, and from this point on the revolutionary leader Miranda lost heart. The movement was temporarily crushed, but Bolívar escaped to Cartagena and continued to agitate for independence. In May 1813 he took the field again with a small army of rebels, defeating General Monteverde at Lastaguanes and entering Caracas in early August. He beat the royal forces at Araure in December 1813, at La Victoria in February 1814, at San Mateo in March 1814, and at Carabobo in May 1814. However there was no true unity among the liberated population, and bitter internecine strife led to atrocities being committed by various factions. The savage irregular cavalry of General Boves defeated the rebels at La Puerta, and took Caracas back into the royalist fold with great cruelty. The second republic of Venezuela was crushed, and Bolívar was forced to flee once again. He spent four indecisive years mounting two main operations and some minor raids with very mixed success. The end of the Napoleonic Wars released Spanish troops for the colonies; but even so Bolívar's tireless efforts led to his general recognition as the patriots' leader in the northern provinces, and by 1819 he had a secure base at Angostura (Ciudad Bolívar). He stiffened his local forces with a mercenary corps of British and Irish veterans of the Napoleonic Wars, and in June 1819 began his epic march on Bogotá.

The route took Bolívar's well-armed but ragged army across a low plain crossed by seven major rivers in the middle of the rainy season. For days on end they marched in water up to the waist, crossing the rivers in cowhide canoes. When they left the plain Bolívar led them up into the icy Andes Mountains, where many perished from exposure. They crossed the range by the pass of Pisba, which the Spanish had considered impassable, and came down into the valley of Sagamose with surprise on their side. On 7 August 1819 Bolívar defeated General Barreiro at Boyaca, opening the road to Bogotá. Barreiro had about 2,000 royalist infantry and 400 cavalry; Bolívar outmanœuvred him, and placed himself between Barreiro and the city. While his right held Barreiro's left he dispatched the British force in a frontal assault, and this repulsed the royalist cavalry and won the day. Barreiro lost a hundred dead and 1,600 prisoners, and three days later Bolívar entered Bogotá – a key day in the history of South America. He then proclaimed the Republic of Great Colombia, including the Spanish-occupied Venezuela and Ecuador; this brave boast was to be realised within a few short years. After some indecisive campaigning the Spanish General Morillo agreed to a six-month truce, and Bolívar used this to improve his army. When it ended in April 1821 he had 6,000 men, whom he led south-eastwards to defeat General de la Torre at Carabobo on 25 June. This was another battle in which the British force distinguished itself; the royalists were driven from their blocking position at the southern end of the pass in complete rout, and Bolívar entered Caracas. Cartagena fell in October after a long siege, and other Spanish garrisons began to collapse. In early 1822 Bolívar and his trusted lieutenant Antonio de Sucre led two separate armies into Ecuador; at Pichincha on 24 May Sucre defeated the Spanish Governor-General of Quito, and Bolívar joined up with him the following month.

Only Peru now remained in Spanish hands; after his controversial meeting with the liberator of the southern colonies, San Martín (q.v.), at Guayaquil in July 1822 there was nothing to prevent Bolívar from moving south. He arrived at Lima in September 1823, and set about organising the local forces and the administration of the new republic. The Spanish army was established in the mountains to the east, and remained in its strong positions throughout the winter. In the spring Bolívar and Sucre led 9,000 men into the high country. Only some 2,000 were engaged against roughly equal enemy forces at Junin on 6 August 1824 – a victory for the republicans which was won, it is said, entirely with sword and lance, for neither side fired a shot throughout the engagement. The bulk of the Spanish army retreated into the mountains south-east of Lima; whither Bolívar returned to Lima to deal with political problems, leaving the rest of the campaign to Sucre. The 7,000-strong republican army finally beat the 10,000 royalist troops of General de La Serna at Ayacucho on 9 December 1824, ending for ever any hope that Spain could regain her South American colonies. The campaign in Upper Peru, renamed Bolivia in Bolívar's honour after a last victory in April 1825, was simply a matter of 'mopping up'.

It was Bolívar's genius and unflagging tenacity which had created the league of new republics; it was his tragedy that he lived long enough to see his brightest hopes disappointed. Although he was an arrogant man with a strong belief in authoritarian forms of government betrayed by the terms of his draft constitutions, Bolívar had a genuine vision of international co-operation. He worked ceaselessly to set up institutions which would stand the stress that was bound to come during the early years of these young countries, and was a passionate believer in negotiation and arbitration as a means of avoiding conflict between nations. His last years were darkened by constant squabbling and outbreaks of warfare among the new states; in the end he even decided that his own presence on the South American continent was an obstacle to compromise, and prepared to go into voluntary exile in Europe. His health had suffered greatly during his jungle and mountain campaigns, and the disappointment of his hopes undermined it further. The final blow was the news of the assassination of Sucre, whom he had been grooming to take his place. He died of tuberculosis on 17 December 1830, in a private house near Santa Marta.

Bonaparte, *see* Napoleon I

Bór, General, *see* Komorowski

Boucicaut, Jean Le Meingre (*c. 1366–1421*), *Marshal of France*

Boucicaut, son of another marshal of the same name, was a famous French soldier of the 'old school' – the splendid but inflexible chivalry who died in their thousands under the English arrow-storm in the battles of the Hundred Years' War. He was a renowned champion of the tourney, a crusader, and the founder of a chivalric order pledged to the defence of the families of knights absent from their lands – the order of *Dame blanche a l'écu vert*.

Boucicaut served with distinction in several of the Valois campaigns during his youth, notably in the suppression of the Flemish revolt led by Philip van Artevelde which culminated in the French victory of Roosebeke in 1382. In 1391 he was made a marshal by King Charles VI. Five years later he was one of the leaders of the so-called 'Crusade of Nikopolis'; a mainly-French army advanced into the Danube valley to check Ottoman Turkish expansion in the Balkans in response to an appeal by Pope Boniface IX. The Turks won a decisive victory at Nikopolis in Bulgaria, with the aid of the Serbians under Stephen Lazarevitch; Boucicaut was captured in this battle, but was later ransomed. Since 1391 the Turks had been exerting extreme pressure on Constantinople, under the Sultan Bayazid I. In 1398–9 Boucicaut distinguished himself in the defence of the city, arriving with a force of 1,400 soldiers and a strong naval flotilla. He defeated the Turkish fleet off Gallipoli, and arrived in the Golden Horn in time to prevent the imminent capture of Galata by the Turks. For a year he led the French expeditionary force with great success, repulsing many Turkish attacks, before returning to France to seek more volunteers. In point of fact he did not return to the East, but was sent instead to Italy to strengthen France's hold on Genoa. He was generally successful, but while he was involved in a campaign against Venice in 1409 the Genoese threw off French domination.

Boucicaut led one of the two main sections of the French army in the campaign of Agincourt in 1415 under Constable d'Albret, harrying the starving English army of King Henry V (q.v.) up the Somme. In the climactic battle of 25 October he was captured by the English, and was one of the few French noblemen to be granted quarter after Henry ordered the slaughter of the prisoners. He never returned to France, dying in English captivity in 1421.

Bradley, Omar Nelson (*1893–1981*), *General of the Army*

Omar Bradley commanded the United States 1st Army, the formation which landed in Normandy on D-Day (6 June 1944). That summer he was given command of the largest single formation in American military history – the 12th Army Group, of 1,300,000 men – and led it across France and into Germany with great success. Unostentatious, homely in appearance, soundly grounded in the disciplines of his profession, Omar Bradley will probably be remembered by those who served with him above all else for his monumental common sense. He was a total professional.

He was born on 12 February 1893 at Clark, Missouri. He graduated from West Point in the class of 1915 – together with Marshall and Eisenhower (qq.v.) – and was commissioned into the infantry. He passed up through the ranks between the wars, graduating from the advanced course at the Infantry School in 1925; from the Command and General Staff College in 1929; and from the War College in 1934. The opening of the Second World War found him the Commandant of the Infantry School at Fort Benning, Georgia. He was then given command of the 82nd Infantry Division, and later of the 28th. When America sent an expeditionary force to French Morocco under Eisenhower, that officer and General Marshall picked their old classmate Bradley for a staff post. At first he was 'Ike's' trusted 'eyes and ears' in the fighting line, but when General Patton (q.v.) was given the task of stabilising the situation and going on to the offensive after the American defeat at Kasserine, he picked Bradley for his deputy commander of the U.S. 2nd Corps. Bradley soon succeeded to the command of this formation – which represented the whole of the American ground combat forces in the theatre – and led it with notable success in the final stages of the Tunisian campaign in the spring of 1943. On 7 May troops under his command captured Bizerta and some 40,000 Axis prisoners.

Bradley continued to command the 2nd Corps as part of Patton's 7th Army in the subsequent invasion of Sicily; his units landed at Gela and Scogliti and made a rapid northwards advance across the island before turning eastwards and encountering heavier opposition. They entered Messina five weeks after the initial landings. In the autumn of 1943 Bradley was selected on the basis of his fine combat record to command the American troops who were to land in Europe to open the Second Front. He was transferred to Britain, where he was involved both in the training and preparation of his 1st Army, and in the high-level planning of the Normandy landings. On 6 June 1944 his divisions stormed ashore on

Utah and Omaha Beaches. After hard fighting, particularly at Omaha, his troops established the beach-head, and then advanced to capture St Lô and Cherbourg. Bradley must take great credit for the decisive American breakthrough in Normandy; he pushed his army at great speed through the gap between Mortain and Avranches. On 1 August he was promoted to command the 12th Army Group, consisting of the 1st, 3rd, 9th, and 15th Armies. This huge command raced across France (spear-headed by the tanks of Patton's 3rd Army), liberated Paris, and drove the Germans back hundreds of miles without allowing them to establish any major defensive lines. In the winter of 1944–5 Bradley's imperturbability paid dividends when the Germans launched their major counter-offensive in the Ardennes sector. He made rapid and practical changes to the chain of command, against some opposition, and succeeded in a policy of outflanking the German penetration.

Breaching the Siegfried Line on Germany's western frontier, Bradley's troops established a bridgehead over the Rhine at Remagen and advanced so quickly that some 335,000 German troops were penned in the Rühr pocket and eventually forced to surrender. Promoted full general in March 1945, Bradley presided over the American drive across Germany which linked up with Soviet units on the Elbe in the following month. Thirteen days later Germany surrendered.

Between 1945 and 1947 Bradley headed the Veterans' Administration, a fitting post for a general who had always shown respect for his men. In 1948–9 he was U.S. Army Chief of Staff. Between 1949 and 1953 he was the first chairman of the Joint Chiefs of Staff, and in 1950 he was promoted General of the Army. After his retirement he accepted a number of industrial appointments, and published his memoirs. He died on 8 April 1981.

Brian Boru (*941–1014*), *High King of Ireland*

During the first half of the ninth century Viking raids on the Irish coast were followed by the establishment of permanent outposts. In the middle years of the century the Norsemen penetrated deep into the interior, seizing and fortifying several strong positions including Dublin and Annagassan. By his death in 845 the chieftain Thorgest was the ruler of half Ireland. In the 860s and 870s the Irish High King Ael Findliath gradually drove the Vikings from their enclaves in the north, but they strengthened their permanent settlements, including Dublin, Waterford, and Limerick, in the eastern and central regions.

Brian was born in 941, a younger son of the petty king Cennedig, who ruled a small region known as Dal Chais. He was surnamed 'Boru' after the ford of Béal Boruma on the Shannon near Killaloe, where he was born. Cennedig's tiny kingdom grew steadily stronger under the stimulus of constant warfare against the Norsemen who were encroaching on his territory from their settlement at Limerick, seized in 922. At the same time the neighbouring kingdom of Munster, less successful in its wars against the Vikings of the Suir river settlements, declined in power. In 964 Brian's half-brother Mathgamain overthrew the Eoganachta rulers of Munster, seized the strong position of Cashel, and declared himself King of Munster. He was an energetic war-lord, and in 968 he defeated the Limerick Vikings and captured the town. In 976, however, he was murdered – not by the Norsemen, but by the displaced Eoganachta rulers of Munster. One of them, Maelmuadh, took his place on the throne.

Brian's career of conquest started with a war of revenge against Maelmuadh, whom he defeated and killed at Belach Lechtha, in the north of the present county of Cork, in 978. He now became King of Munster, and began a steady process of extending his power. In numerous actions he defeated Viking bands, and built a fleet of ships to enable him to harry them from the Shannon. In gaining power over the neighbouring Irish kings he showed both military boldness and diplomatic skill, and under his rule Munster became stronger and more unified than ever before. In 983 Brian invaded Ossory, thus starting a simmering feud with the High King, Maelsechlainn II, which was to last almost twenty years. In 997 Brian won control of the whole of southern Ireland from Maelsechlainn; and two years later, at the battle of Glen Mama, he defeated the Irish King of Leinster, who was allied with the Norsemen of Dublin. In 1002 he finally defeated Maelsechlainn and forced him to name him, Brian, as his successor, with immediate effect. Within a few years all the petty kings had acknowledged Brian Boru as High King.

Brian was a military leader of great skill and hardihood, and a king of remarkable ability; Ireland did not produce another man of his firmness and strength of character for many years (as witness the chaos of internal warfare which lasted for some 150 years after his death, and which prevented any unified resistance to the Norman conquest of the island in the 1160s).

It was in 1013 that the men of Leinster and their Norse allies in Dublin leagued against Brian once again; this time they summoned help from other Viking communities, notably the earldom of Orkney and the Isle of Man. The final battle was fought outside Dublin, at Clontarf, on 23 April 1014. Brian at seventy-three was now too old to lead his warriors in the field, and his son Murchad took his place; Brian waited in his tent on the edge of the battlefield, and it may be assumed that the disposition of the army was made according to his instructions. The battle of Clontarf raged for twelve hours, and at the end, after terrible losses on both sides, the Norsemen broke and ran. A small group of Viking fugitives came upon Brian's tent by chance during their flight; they killed his small bodyguard, and cut down the aged High King. He was buried in the chief Christian church in Ireland, St Patrick's at Armagh. The victory of Clontarf is generally credited to Brian Boru; although Magnus of Norway attempted an invasion early in the next century, Clontarf is seen as the end of paganism and Viking settlement in Ireland.

Brooke, Alan, *see* Alanbrooke

Bruce, *see* Robert I

Brudenell, James Thomas, *see* Cardigan

Budenny, Semyon Mikhailovich (*1883–1973*), *Marshal of the Soviet Union*

Budenny was a Russian ex-peasant who rose to command Soviet armies on the South-West Front in 1941, when they suffered their most devastating defeats.

Born near Rostov on 25 April 1883, Semyon Budenny was the son of a peasant; conscripted into the Tsarist army in 1903, he rose to the rank of cavalry sergeant-major prior to the Revolution in 1917. Gaining meteoric promotion thereafter, he had by 1920 risen to command the First Soviet Cavalry Army of about 16,000 men, and fought with some distinction in the Russo-Polish campaign of that year. In June 1920 he threatened to outflank and encircle the Polish Third Army in the Ukraine, but the Poles successfully disengaged and followed up with an equally successful counter-offensive. Budenny's handling of his command in the latter stages of the campaign has been criticised, but his lack of success can scarcely be laid to his charge alone, as there was much high-level disagreement, and Budenny himself was supported by Stalin and Voroshilov.

During the early 1920s Budenny was closely associated with the anti-Trotsky 'Tsaritzyn' group which enjoyed Stalin's support. In 1924 he was appointed a member of the Revolutionary Military Council, and in 1935 he became one of the five Marshals of the Soviet Union who were first created that year. He had maintained impeccable credentials as a 'Stalin man', and he survived the purges of the late 1930s which cost the Soviet Army three marshals, thirteen army commanders, and over 400 corps, divisional, and brigade commanders.

The most flamboyant of the Soviet generals, he affected mahogany-butted revolvers and handlebar mustachios, and constantly betrayed his origins in his longing for a war of traditional manœuvres and cavalry charges. He owed his promotion far more to political expediency than to any real aptitude for command. To pit this hard-drinking, hard-riding, but simple old man against the honed brilliance of such Panzer leaders as Guderian (q.v.) was little short of cruelty.

After Hitler's invasion of June 1941 the Soviet State Committee for Defence appointed Budenny Commander-in-Chief of armies in the Ukraine and Bessarabia; this sector was given priority in supplies and reinforcements by Stalin, who determined to hold Kiev at all costs. Budenny thus had overwhelming numerical superiority over the German forces opposing him, but he was hopelessly outgeneralled by Runstedt (q.v.) and Kleist; moreover much of the Russian equipment was outdated by as much as five years in comparison with that of the invading German forces. By 10 July 1941 the Germans had cut the north-to-south lines of communication between Budenny's two major concentrations at Uman and Kiev, but the Soviet commander did not grasp the importance of the threat and failed to react, deploying sluggishly and incoherently. By 3 August the Germans had surrounded the Uman concentration, and Budenny withdrew across the Dneiper, destroying all industrial capacity in the area, as ordered by Stalin. By the time the German armies had surrounded and decimated the forces at Kiev as well (entering the city on 19 September), Budenny's incompetence had helped to lose the Soviet Union no less than one and a half million Russian soldiers dead, disabled, and captured. At that time this figure represented more than half the active strength of the Red Army. Awarded the title of Hero of the Soviet Union, Budenny was relieved of command and given the job of supervising recruit training; he took no further part in the conduct of operations.

During his last years Budenny compiled a book, *The Road Travelled*, in which he described in some detail the formation of the Red Army – in particular the Soviet First Cavalry Army – and the fighting during the Russo-Polish War of 1919–20. His incompetence during the Second World War forgotten, he was widely revered for his pioneering command in the early days of the Red Army, and was honoured on every anniversary of its formation. He died on 26 October 1973 at the age of ninety.

Bugeaud de la Piconnerie, Thomas Robert, Duc d'Isly (*1784–1849*), *Marshal of France*

Bugeaud, the most effective French commander during the early years of the conquest of Algeria, was above all a 'soldier's soldier'. Though of aristocratic birth, he had through force of circumstance served in the ranks of Napoleon's (q.v.) infantry, and rose to command by hard front-line experience. He was autocratic, arbitrary, given to outbursts against intellectuals and politicians, and ruthless in action; he was also an extremely approachable and practical commander of fighting troops, loyal to his subordinates and genuinely concerned for the welfare of his men – provided they met his exacting standards of discipline.

Born at Limoges on 15 October 1784, young Bugeaud had a hard childhood. His father was a nobleman of Périgord who was imprisoned by the Revolutionary authorities in 1793. His mother, of Irish extraction, died in 1796, and the boy was brought up by his sisters among the Dordogne peasants. He enlisted in the *vélites* of the Imperial Guard as a lad of fifteen, and fought at Austerlitz as a corporal. He was commissioned *sous-lieutenant* in 1806, and saw much hard fighting in the Peninsula. He distinguished himself in July 1808 after the disaster of Baylen, when General Dupont surrendered some 17,600 French troops to the Spaniards. Bugeaud's regiment, the *116e Ligne*, was the only organised unit to break out of the encirclement and rejoin the French army, and Bugeaud won great credit in command of the rearguard. He had a distinguished career under Suchet in eastern Spain during the latter years of the campaign, and rose to colonel. He declared for the Bourbons in 1814, but during the Hundred Days was obliged by his troops to come out for Bonaparte; as a consequence he was put on half pay after the second Restoration. An advantageous marriage gave him the means to buy back his family estates and he immersed himself in farming, becoming greatly interested in agricultural matters. The revolution of July 1830 allowed him to resume his career under Louis Philippe. He was elected deputy for Excideul, and for a brief interlude in 1833 was the jailer of the Duchesse de Berry, fiery daughter-in-law of the exiled Charles X. In 1834 he helped to put down rioting by the Paris mob.

France's adventure in Algeria had begun with the seizure of Algiers and a narrow coastal strip in 1830, but had then become bogged down. There was no concensus of opinion among French politicians as to the direction to take – outright annexation, the creation of a client state under native rulers, withdrawal – all were canvassed by their partisans. The area had no unified native power-structure with which the French could negotiate a long-term agreement, and the small enclave held by French troops was under constant harassing attack by completely independent native tribes. In 1836 Bugeaud was sent to Algeria to treat with the influential Emir Abd-el-Kader, and in May 1837 he concluded the Treaty of Tafna. This ill-judged agreement was supposed to further France's interests by 'drawing the teeth' of the amir, who was recognised as ruler of certain territories in return for an undertaking not to upset the *status quo* in other areas. It was practically unworkable, the more so since Abd-el-Kader signed it for the most cynical motives. Using his French sponsorship to enhance his power over neighbouring tribes, he built up a larger and more cohesive army against the time when war would inevitably break out once again. This occurred in November 1839, and the amir achieved some notable early successes. He drove the French back to the very walls of Algiers, slaughtering colonists and garrisons of small outposts alike. France now abandoned all thought of limited sovereignty in Algeria, and Bugeaud was sent out again, this time with strong reinforcements at his disposal. He took up the post of Governor-General in February 1841, and set about the conquest with impressive energy and common sense.

Arguing that the highly-mobile Arab striking forces would retain the initiative as long as French strategy was tied to large static positions, he instituted a programme of counter-raids by light columns. The soldiers' equipment was reduced, their clothing was rationalised, and unwieldy baggage-trains were abolished. Each man carried his own rations for a few days, and the bare necessities of heavy equipment were carried on mules. Bugeaud also obtained mule-borne mountain guns to give the columns close support, and recruited native and French units of a new and practical kind to act as the scouts and flank-guards of his columns – *spahis*, *Turcos*, and *chasseurs d'Afrique*. The typical operation was the *razzia*, a simple adaptation of the Arab raids themselves. If the fighting men of a village could be caught unprepared they were killed; whether or not they were present, the village was razed to the ground and the crops and food-stores were destroyed. Bugeaud never shrank from the implications of his policy, or denied the misery thus caused to native communities, but pointed out that if pacification had to be achieved, then the quicker it was done, the less would be the suffering in the long run. He made genuine attempts to minimise the deaths among Arab women and children, but was under no illusions about the effectivness of his orders under battle conditions. It is clear that his experience of the Peninsular fighting had hardened him to warfare among civilians. Nevertheless, once an area was pacified Bugeaud was notably responsive to appeals from the natives against the rapacities and stupidities of the French civilian administrators, who were not initially of a high calibre, and whom the Governor-General despised.

'Père Bugeaud' won fame and popularity during this campaign; his troops worshipped him for giving them a chance to hit back after years of pin-pricks, and the government was impressed by his success. His army, finally numbering some 108,000 men, was adequately funded and aggressively led. By late 1843 Abd-el-Kader was forced to take refuge in Morocco, whose Sultan supported him. When Moroccan troops totalling about 40,000 moved up to the border on the Oued Isly in August 1844 Bugeaud, a Marshal of France since the previous year, led 8,000 French in a devastating night attack which opened a battle lasting several hours. The Moroccans were smashed, and in September the Sultan was forced to sign the Treaty of Tangier, promising to intern Abd-el-Kader if he took refuge in Morocco again and allowing the French the right of 'hot pursuit'. This victory brought Bugeaud his dukedom; Abd-el-Kader was to remain at large until 1847, however.

Bugeaud travelled back and forth between France and Algeria several times in 1845–7, pressing his view that victory was pointless without colonisation. He favoured a system similar to that of the old Roman *colonia*, whereby old soldiers were given land and encouraged to settle down in villages of tough, self-sufficient veterans. He had a deep feeling for the land, and was frustrated at the lack of official interest in his scheme. In September 1847 he resigned over the issue. The following year, on the outbreak of revolution in Paris, he commanded the Army for Louis-Philippe but was unable to save the monarchy. Under the Second Republic he was a prolific writer of pamphlets, and accepted command of the Army of the Alps, before dying of cholera on 10 June 1849. His many writings include a most effective and often-quoted description of what it felt like when Napoleonic infantry made a frontal attack on British infantry in a good defensive position – in this he drew on his Peninsula experience.

Burgoyne, John (*1722–92*), *Lieutenant-General*

One of the most attractive characters to emerge from a dismal period of British military history, 'Gentleman Johnny' Burgoyne was perhaps the only senior military commander who pursued a successful simultaneous career as a dramatist. He was by no means the worst of Britain's commanders in the War of American Independence and his famous capitulation at Saratoga was neither premature nor avoidable.

He was born in 1722, the son of Captain John Burgoyne, described as 'a gentleman of fashion'. Educated at Westminster School, the young Burgoyne became a close friend of the son of the eleventh Earl of Derby – a friendship which was to play an important part in his life. In 1740 he purchased a cornet's commission in the 13th Dragoons; he rose to lieutenant in 1741, and in 1743 he eloped with the Earl of Derby's daughter. This match pleased the bride's brother but infuriated her father, and she was cut off without an allowance. The lovers used her last savings to purchase Burgoyne a captaincy, and for three years they lived in London. In 1746 mounting debts obliged Burgoyne to sell his captaincy again, and the couple sailed for France, where they lived quietly on the proceeds. In time the Earl of Derby was reconciled with his spirited son-in-law, and from Burgoyne's return to England in 1756 the resources of the family, both financial and otherwise, were behind him. He was immediately secured a captaincy in the 11th Dragoons, which he exchanged in May 1758 to become a captain in the Coldstream Guards and, in accordance with the dual rank system then operating in the Guards, a lieutenant-colonel in the Army. During the Seven Years' War he saw action in two raids on the French coast, at Cherbourg in 1758 and at St Malo in 1759.

In August 1759 Burgoyne raised a regiment of light horse, the 16th (Queen's) Light Dragoons. After seeing action at Belle Isle in 1761, the Regiment was shipped to Portugal in 1762 as part of an expedition to assist the Portuguese again the Spanish. Burgoyne also sailed for Portugal, as a brigadier-general; he and his dragoons distinguished themselves at Villa Venha in 1762. In 1763 he became colonel-commandant of the Regiment, and in 1768 the Derby interest enabled him to enter Parliament as member for Preston. His contribution to the deliberations of the Commons was generally intelligent, and notably so in the case of his campaign to include in the relevant legislation some element of Government control over the Honourable East India Company's administration of its territories. In 1769 Burgoyne was made Governor of Fort William in Scotland, and his promotion to major-general followed in 1772. During this period he became an established leader of fashionable London society, a reckless gambler, a popular clubman, and an amateur actor. In 1774 his first play, *The Maid of the Oaks*, was performed, and he kept up this literary hobby until his old age.

Burgoyne was sent to Boston to reinforce General Gage in May 1775; he had been reluctant to leave his invalid wife, and was not impressed by the inactivity which he found in his new posting. Unable to stir up any constructive initiative, he returned to England in disgust, but was almost immediately sent to Canada as second-in-command to Sir Guy Carleton in June 1776. His wife's death coincided with his arrival, and his mood was not lightened by what he observed of Carleton's ineffective efforts to invade New England. He took the field with the army, and after the capture of Crown Point and the reconnoitring of Ticonderoga he submitted his own plan of action. This envisaged an army of 12,000 advancing into New England by way of Crown Point and Ticonderoga, to link up with other columns moving north from New York and east from the interior. This would have cut the central and southern colonies off from New England. The plan was approved, but Burgoyne found that he was only to be given a force of 6,400 troops and some 650 Indians with which to carry it out.

This wholly inadequate force left Three Rivers in May 1777, with the ultimate aim of linking up with other columns at Albany, New York. The quality of the troops was generally good; the expedition included some first-class German mercenary contingents. On 6 July Ticonderoga fell after a six-day encirclement, but the advance from that point was slow. The terrain was rugged, and the column was much harassed by Indians and irregulars along the route. The American General Schuyler fell back before Burgoyne, gathering strength as he went; and the other British column under St Leger, advancing eastwards from Oswego, was defeated and turned back at Fort Stanwix in late August. Short of supplies, Burgoyne pushed on; he still had no news of the progress of the New York column under General Howe; in fact this force had not even started, due to the appalling inefficiency of the Secretary of State for the Colonies. Burgoyne's army was beaten at Bennington on 16 August, and at Freeman's Farm on 19 September, but there was no alternative to trying to reach Albany before winter. His urgent appeals for support from New York brought General Clinton out in a brief diversionary raid, but no attempt was made to link up with Burgoyne. Eventually, in early October, Burgoyne was blocked and repulsed by 20,000 American troops under Horatio Gates at Bemis Heights. His 5,000 or so remaining men, worn out and starving, were surrounded, and when all food and ammunition had been expended Burgoyne surrendered to Gates at Saratoga on 17 October. The failure of his sound plan had been due to inadequate resources, lack of co-ordination of the overall campaign, and underestimation of the difficulties of maintaining a timetable while advancing through thick country.

Washington (q.v.) allowed Burgoyne to return to England, where he was attacked in the most extreme terms from all but one quarter: the Whig opposition championed his cause, and he was pushed into supporting them by the unjust criticism of the King and Government. Apart from serving as Commander-in-Chief in Ireland in 1782–3, he retired increasingly into private life. Between 1782 and 1788 he fathered four bastards on the singer Susan Caulfield, one of whom was to become a Field-Marshal. He continued to write plays, the most successful of which was *The Heiress*, performed in 1786. Burgoyne lived a full life to the end, and died very suddenly at his Mayfair house on 4 August 1792. He was buried in Westminster Abbey. A kindly and amusing fictional portrait of him is included in George Bernard Shaw's *The Devil's Disciple*.

Campbell, Colin, Baron Clyde of Clydesdale *(1792–1863), Field-Marshal*

Britain's commander-in-chief in India during the later stages of the Indian Mutiny, Colin Campbell occupied a place of great respect and affection in the hearts of the Army and the population at large. One of the more attractive characters among the Victorian generals, he took an intelligent interest in the welfare of troops under his command; his victories were characterised by careful preparation and economy of lives.

He was born in Glasgow on 20 October 1792, the son of a carpenter named Macliver and one Agnes Campbell, a relative of the Campbells of Islay. His humble background did not prevent him from rising to high command, but his limited financial means were a burden to him almost throughout his life. His patron as a young man was his mother's brother, Colonel John Campbell; it was this uncle who presented the young man to the Duke of York in 1807 as a candidate for an Army commission. The Duke assented, but put Colin's name down as Campbell rather than Macliver; and it was in the name of his mother's clan that he served for the rest of his life. He was gazetted an ensign in the 9th Foot in May 1808, and sailed for Portugal with the 2nd Battalion. Campbell saw a great deal of action in the Peninsular War, and his distinguished gallantry on several occasions was the making of his career. He fought at Rolica and Vimiero, and was on the retreat to

Corunna. In 1809 he served in the Walcheren expedition, contracting the fever which was to trouble him for the rest of his life. In 1810 he was posted to Gibraltar, and distinguished himself under the eyes of Sir Thomas Graham (q.v.) at the Battle of Barossa. Lieutenant Campbell was then attached to Ballasteros's Spanish army for a period, returning to his regiment in time for the Battle of Tarifa. In 1813 he fought at Vittoria, and later at the Siege of San Sebastian and at the crossing of the Bidassoa. In both these latter engagements he particularly distinguished himself, and was twice wounded. He returned to England, and on his recovery was rewarded with a captaincy in the 7th Battalion, 60th Rifles, then stationed in Nova Scotia.

Although he had risen from penniless ensign to 'noticed' captain in just five years, it was to take him another thirty years to reach the rank of colonel. He served patiently and with credit in a number of home and foreign postings, but after the end of the Napoleonic Wars promotions were hard to come by for a young man without wealth. His next chance to attract the attention of his superiors did not come until the Opium War of 1841–3. He had finally obtained a lieutenant-colonelcy in 1835 in his old regiment, the 9th; he moved to the 98th in 1837. It was this unit which he led out to China in 1841, and there he served with notable success under Sir Hugh Gough. Mentioned in dispatches, Campbell was made an A.D.C. to the Queen and promoted colonel. He was created a Companion of the Bath, promoted brigadier-general in January 1844, and commanded a brigade at Chusan until July 1846. He was then posted with his regiment to Calcutta.

Campbell commanded a division with conspicuous success during the Second Sikh War, particularly distinguishing himself at Ramnuggur and Gujerat; in recognition of his services he was knighted in 1849. Sir Colin now expressed a desire to return to England; he was feeling his age, and his health was not reliable. He was persuaded to stay on, however, and commanded the extremely exposed division on the frontier until 1852, with headquarters at Peshawar. He achieved an impressive series of successes against various tribal enemies in the period 1849–52; but in the latter year, incensed at the refusal of the administration to release to him reserves which he considered essential to his mission, he resigned his command. He returned to Britain after an absence of twelve years, and spent a year on half-pay leave.

In February 1854 he was given command of the Highland Brigade for the Crimea, and while at Varna in June of that year was promoted major-general. He distinguished himself yet again at the head of his Highlanders in the Battle of the Alma; in December he was given the 1st Division, and commanded at Balaclava throughout the harrowing winter siege of 1854–5. He commanded the reserve at the storming of the Redan in September 1855, and took leave to return to England in November. He was promoted lieutenant-general in June 1856, and after a brief return to the Crimea, was brought home and appointed Inspector-General of Infantry. In July 1857 news reached England of the outbreak of the sepoy mutiny in India, and of the death of the Commander-in-Chief, General Anson. Sir Colin Campbell accepted the post, and started for India immediately, arriving in August. He led reinforcements to Cawnpore, and began careful preparations for the relief of the besieged Lucknow – preparations so careful that, in view of the desperate position of the garrison, he was accused of dilatoriness. On 9 November he set out to their relief with 4,700 men and thirty-two guns; and by 30 November he had fought his way into the city, rescued 400 women and children and 1,000 sick and wounded, and brought them safely back to Cawnpore. By May of 1858 a series of wide-ranging operations under his overall control had restored peace to northern India. In that month he was promoted general, and in July he was raised to the peerage as Lord Clyde. He held the command in India until, weakened by poor health, he returned to England in June 1860. His last years were brightened by many honours and in November 1862 the one-time soldier of fortune (his own phrase) and poverty-stricken ensign of foot was accorded the ultimate promotion to field-marshal. Much loved and admired, Sir Colin Campbell finally died at Chatham on 14 August 1863; he is buried in Westminster Abbey.

Campbell's career was creditable in more ways than one. Though such views were rare in officers of his day, he always stressed the importance of the physical and mental welfare of the soldiers, and took pains to alleviate their hardships whenever he could. Tactically he was a modernist in that he favoured thorough field training, and open formation under fire; and he was a believer in the intelligent co-operation of all arms, including engineers and artillery, to bring about a victory which less imaginative men might have sought by the massed use of the bayonet. In the everyday matters of regimental life he demanded a high level of responsibility from his subordinate officers, and he was never reluctant to promote on ability rather than seniority.

Canrobert, François Certain (*1809–95*), *Marshal of France*

An able, if limited soldier whose distinguished service in North Africa led to his appointment as French Commander-in-Chief in the Crimea, Canrobert's fear of the highest responsibility resulted in subordinate command during the Franco-Prussian War.

Born at St Céré, Lot, on 27 June 1809, Canrobert studied at St Cyr, and was commissioned in the army in 1828. Sent to Algeria seven years later, he won distinction during the capture of Constantine before being recalled to Paris in 1839 to organise a battalion of the Foreign Legion; he returned to Africa in 1841. In 1847 Colonel Canrobert won further renown amongst the Zouaves at the Siege of Zaatcha; he was again recalled to Paris and, promoted brigadier-general, was appointed aide-de-camp to the Prince President Louis Napoleon. He participated in the *coup d'état* of 1851.

Promoted general of division in 1853, he sailed to the Crimea the following year. This war, while confirming his merits as a fighting soldier, also exposed his limitations. Upon the death of Marshal de Saint-Arnaud (q.v.) from cholera on 29 September 1854 Canrobert was appointed Commander-in-Chief, but at once displayed irresolution and shirked the responsibility of firm decisions. These failings he was himself well aware of, and following disagreement with the British (and on the pretext of failing health), he appealed to the Emperor to relieve him of supreme command, while begging to be allowed to retain command of a division. Recalled to Paris in 1855, he was promoted marshal of France.

In 1859 he took part in the campaign in Lombardy as a commander of French forces supporting the Piedmontese efforts to expel Austria from Northern Italy, serving with particular distinction at the Battles of Magenta on 4 June, and Solferino on 24 June. His admitted reluctance to assume the responsibility of supreme command continued to haunt his career and, on the outbreak of the Franco-Prussian War in 1870, resulted in his serving under the junior marshal, Bazaine. Commanding the French force at the battle of St Privat, he succeeded in withstanding the heavy Prussian attacks, but was eventually forced to retire when he ran out of ammunition and his reinforcements failed to arrive. He was taken prisoner at Metz.

Resuming his military career after the ending of hostilities, he became a member of the Conseil Supérieur de la Guerre, and afterwards was elected Senator for Lot (1876) and Charente (1879 and 1885), representing Bonapartist ideas under the Third Republic. He died in Paris on 28 January 1895.

Cardigan, James Thomas Brudenell, 7th Earl of (*1797–1868*)

Lord Cardigan, one of the most unattractive personalities ever to hold a senior command in the British Army, has gone down in history as the leader of the 'Charge of the Light Brigade'. A man of undoubted physical courage but very few other admirable qualities, he drew upon himself by his unpopularity a measure of blame for this epic disaster which he perhaps did not altogether deserve.

Only son of the sixth Earl of Cardigan, James Brudenell was born at Hambleden in Buckinghamshire on 16 October 1797. He was at Christ Church, Oxford, for two years. On coming of age in 1818 he was elected to Parliament as member for Marlborough, sponsored by his father's cousin, the first Marquis of Ailesbury. At the age of twenty-seven he purchased a cornetcy in the 8th

Hussars, and made up for lost time by spending vast sums to purchase his grades, becoming lieutenant in January 1825, captain in June 1826, major in August 1830, lieutenant-colonel in December 1830, and a lieutenant-colonel in the 15th Hussars in 1832. Following a difference of opinion with the Marquis of Ailesbury, Brudenell resigned his seat for Marlborough in 1829, obtaining instead the seat for Fowey.

In the Army he was notoriously unpopular with his fellow officers, having bought his way into command. A crisis occurred when he unjustly placed an officer in custody, and he was on the point of being forced to resign his commission when his father – an old friend of King William IV – managed to arrange his transfer to the 11th Hussars. He assumed this command in India during 1836, but the Regiment was soon ordered home and Brudenell found that his father had in the meantime died, leaving him the earldom, with an annual income of £40,000. Although liberal with his money, he was no more popular than before; but the 11th Hussars came to be regarded as one of the smartest cavalry regiments in the service – due very largely to the money lavished upon it by Cardigan – and was accordingly selected by Queen Victoria to be termed Prince Albert's Own Hussars.

Back in India again in May 1840, Cardigan became known for his high-handed habit of arresting officers for petty incidents. A certain Captain Harvey Tuckett had written an account of one such incident, and this was published. Cardigan challenged the junior officer to a duel and Tuckett was shot. This resulted in a considerable scandal, with much public feeling running against Cardigan.

1854 found Lord Cardigan a major-general, continuing to lead the life of a wealthy nobleman until the outbreak of the Crimean War, when he took command of a cavalry brigade in Major-General Lord Lucan's division. Cardigan and Lucan were brothers-in-law, but there was little family affection between them – old family quarrels and jealousies governed their relationship. This deep antagonism was particularly apparent at Varna and on the day before the battle of the Alma. Whilst encamped outside Balaclava Lord Lucan lived rough in the field camp, but Cardigan ate and slept aboard his own luxury yacht moored in harbour.

In the famous assault on Balaclava, after the Russians had been driven back by the 93rd Highlanders and charged in the flank by the heavy cavalry, an order was sent by the hand of Captain Nolan, aide-de-camp to Major-General Airey, for the Light Brigade to charge along the southern line of heights and drive the enemy from the captured Turkish batteries. Such an order would have been fairly simple to put into execution; it seems probable that Lord Lucan was quite clear along which line the Light Brigade was to charge, and Captain Nolan knew exactly in what direction to lead the troopers. But Cardigan could see nothing from his own position, and believed that his orders were to charge straight along the valley in front of him. Lord Lucan did nothing to correct this error, and Nolan was killed at the very moment when he realised that the Brigade was charging in the wrong direction.

Cardigan led his men straight down the valley, flanked on one side by hills lined with Russian batteries and on the other by the captured Turkish guns, right up to the Russian batteries in his front. Galloping many yards ahead of his troopers, Cardigan was first among the enemy guns at the end of the valley, and received a slight wound in the leg. Unfortunately for his reputation, he was not the last to quit the mêlée, but left behind him bewildered officers and men searching for their general and awaiting orders. Cardigan had played the part of a hero, but not of a general. Great excitement awaited him back in camp. Lord Raglan was profoundly displeased. Blame for the débâcle was distributed widely, some upon Raglan himself, some upon Cardigan, Nolan, and Airey. In fact, no single commander was wholly responsible. After all, Cardigan had faithfully obeyed what he had understood to be his orders.

He returned to England a hero in 1855, a part he was at pains to cherish. His conduct was boastful, especially so at a Mansion House banquet given in his honour. He was appointed Inspector-General for Cavalry for the usual five-year term, made a K.C.B., appointed Commander of the Legion of Honour, and Knight of the 2nd Class of the Order of the

Medjidie. Promotion to lieutenant-general followed in 1861. He had been colonel of the 5th Dragoon Guards in 1859, an appointment he exchanged for the colonelcy of the 11th Hussars in 1860.

In 1863 Cardigan applied for a libel action against Lieutenant-Colonel the Hon. J. G. Somerset, who had allegedly defamed Cardigan for his actions in the charge at Balaclava, but was nonsuited. After the trial he retired to live quietly at Deene Park, Northamptonshire, where he died on 28 March 1868 after a fall from his horse. He left no children, and his title devolved upon his second cousin, the second Marquis of Ailesbury.

Carnot, Lazare Nicolas Marguerite (*1753–1823*), *Général de Division*

Lazare Carnot's direct involvement in military affairs was brief but significant. Although he was a professional soldier, his most active years were spent in the murderously unstable world of French Revolutionary politics, and it was through political power that he was able to make a contribution to the birth of the army which Napoleon led to glory.

He was born at Nolay in Burgundy on 13 May 1753, the son of a lawyer. He passed into the Mézières school with distinction, and by the age of thirty was a captain of engineers. He published several works on military engineering, and was acquainted with Robespierre. Elected to the Legislative Assembly in 1791, he was prominent on the military committee. He was one of the first deputies sent to the armies, setting off to purge the Army of the Rhine in August 1792. He was elected to the Convention the following month, and made a study of the Spanish frontier defences. He voted for the death of the King in January 1793, and three months later barely escaped in the disorders which swept the Army of the North after the defeat of Dumouriez and false accusations of treason against him. By the time Carnot was elected member of the Committee of Public Safety with responsibility for military affairs in August 1793, France was on the verge of ruin. The British and the Hanoverians were investing Dunkirk, the hostile Netherlanders linked them with the powerful Austro-Prussian invasion of the east, and monarchist revolt blazed in the south. The Reign of Terror was hampering all efforts to set the administrative and military life of France on a sound footing; the professional officers were fleeing for their lives or retiring to sit out the Terror in their homes, and the bad habit of executing defeated generals was discouraging initiative.

A week after Carnot's appointment to the Committee of Public Safety the *levée en masse* had been decreed, conscripting the entire male population of France. The raw recruits, many of them keenly patriotic but few with any training, were hastily organised into fourteen new armies. Carnot played a great part in amalgamating these armed mobs with the remnants of the old professional army, and producing a new sort of military animal which could fight hard as long as it was not required to manœuvre. It was this aspect of the Revolutionary armies which dictated the use of the massed column as the standard French infantry attack formation – highly inefficient from the point of view of bringing fire-power to bear, but with the psychological advantage of impetus and huge, closely-packed numbers, well suited to inexperienced troops whose morale was unequal to facing trained troops in conventional battlefield formations. At Hondschoote on 6 September General Houchard beat the British by sheer weight of numbers, but failed to manœuvre the Austrians into retreating from eastern France after his victory at Menin, and was guillotined for his failure. Jourdan took over the Army of the North in late September, and Carnot came in person to join him. He ordered the relief of Maubeuge, besieged by 30,000 Austrians under Saxe-Coburg, and the 50,000-strong French army achieved this by the victory of Wattignies on 15–16 October. Since operations on this scale were outside the experience of both Jourdan and Carnot it is hard to establish who deserves the credit; both were probably responsible in some measure. Wattignies marked a turning-point in the performances of the Revolutionary armies.

Carnot's prestige soared, and he was enabled to promote his nominees among the generals – including men such as Souham, Pichegru, and Kléber. Internal political feuding continued to damage the national military effort, however. In May 1794 the Army of the North defeated an Allied army at Turcoing, and there was a drawn battle at Tournai. A French victory at Hooglede in June started an Allied retreat northwards, and Jourdan was placed in command of the combined Army of the Sambre et Meuse, some 80,000 strong. Carnot had political objections to a large unified military command, and these were brought to the fore by Jourdan's victory at Fleurus on 26 June 1794. A succession of French victories on the Rhine, in the Nertherlands, in the Alps, and along the Mediterranean coast followed in late 1794 and early 1795, but in July 1794 the coup of 9 Thermidor placed Carnot in danger. He split with Robespierre, and was active in protecting the interests of the Army and of individual officers. He was able to report to the Convention in late 1794: 'In seventeen months' campaign, twenty-seven victories, 80,000 enemy killed, 116 places taken, ninety flags . . .' In March 1795 he ceased to hold political office and was promoted major of engineers at long last, but his enemies in political life still attacked him until the voice of an unknown deputy, raised in a stormy debate on 28 May 1795, paid him the tribute which was to save him: 'Carnot has organised victory.'

In November 1795 Carnot was elected to the Directory, but he was a long time rebuilding a power-base strong enough to enable him to push through his plans. The two main operations which he sponsored – those of Jourdan in Germany and Hoche in Ireland – were failures, redeemed from a national point of view by Napoleon's Italian victories, for which Carnot could take no credit. The *coup* of 18 Fructidor (in September 1797) forced him to flee to Switzerland. He returned from exile under the Consulate, and worked at the Ministry of War with Berthier (q.v.) – for six months in 1800 he was Minister. He sat in the *tribunat* until 1807, and voted against the Life Consulate and the *Légion d'Honneur*. Carnot then returned to his first love – the study of fortification – and published several influential works, but despite several approaches, Napoleon did not manage to persuade him to return to active service until the crisis of 1814. He was then appointed general of division and Governor of Antwerp; he secured the fortress at a critical time, and held it until the abdication. Naturally enough he was cold-shouldered during the first Restoration of the Bourbons as a regicide. During the Hundred Days Napoleon made Carnot his Minister of the Interior, and apparently set store by the advice of the wise old revolutionary who had survived so many dangers; he is reported as saying after Waterloo, 'Monsieur Carnot, I have known you too late.' On the second abdication of his Emperor Lazare Carnot headed the provisional Government until ousted by Fouché; the second Restoration brought proscription and exile. He died at Magdeburg on 2 August 1823.

Charles XII (*1682–1718*), *King of Sweden*

Strong-willed and energetic, Charles devoted his short adult life to warfare and political intrigue undertaken with the aim of preserving Sweden's Baltic territories and thus her great-power status. His failure was historically inevitable, and his obstinacy may have cost Sweden a greater price than she would have had to pay under a more flexible monarch; indeed, there was a persistent rumour that his death was the work not of an enemy but of a Swedish soldier. But however limited his wisdom, he was a very remarkable field commander.

He was born on 17 June 1682, the second child and eldest surviving son of Charles XI of Sweden and Eleonora of Denmark. The death of Eleonora brought father and son particularly close together, and the King spent much time and thought on preparing his son for the responsibilities of the throne. Charles XI died in April 1697, and immediately the fifteen-year-old King began to drive himself hard. He suppressed his natural good humour and became deliberately solitary and inscrutable. He spent much time studying his father's written advice and consulting his old ministers and generals; he also embarked on a deliberate programme of physical toughening which caused anxiety to those responsible for his safety. Within three years of his succession Peter the Great of Russia, Augustus II of Poland and Saxony, and Frederick IV of Denmark leagued themselves together to attack Sweden and break her dominance in the Baltic area. The Great Northern War, which was to last until 1721, was sparked off by a Danish invasion of Sweden's ally Schleswig in April 1700; in June and August Augustus attacked Swedish Livonia, and Peter the Great invaded Ingria.

His late father's generals were largely responsible for the Swedish victories of 1700–2, but Charles took an increasing part in the military direction of the war, and his courage and energetic encouragement of the troops were good for morale. It was at his instigation, and against the advice of his generals, that a Swedish force made a bold seaborne attack on Zealand and advanced on Copenhagen, forcing Denmark to sue for peace before the end of August 1700. In the autumn he landed in Livonia and marched to relieve the besieged Narva. His 8,000 men scattered nearly 40,000 Russians in a two-hour battle on 20 November; he then marched back to Livonia and defeated a Polish-Saxon army outside Riga on 17 June 1701. From 1702 onwards Charles showed himself the superior of all his generals except Karl Rehnskiöld. He reasoned that Sweden could profit most quickly by invading Poland, where Augustus did not enjoy total support. In July 1701 he crossed the Dvina, and on the 9 July defeated a Saxon-Russian army at Dunamunde; he then occupied Courland and moved into Lithuania. On 14 May 1702 his army occupied Warsaw, and he marched westwards to force Augustus to fight a pitched battle. On 2 July he defeated a much larger Polish-Saxon army at Kliszow, and continued his march through hostile country to sieze Cracow. During the next year he made use of the anti-Saxon elements to destroy Augustus's control of Poland; he defeated yet another much stronger Saxon force at Pultusk in April 1703, and the following year placed a client king, Stanislas Leszcynski, on the throne of Poland. For another two years civil war raged throughout the country, and in the meantime Charles returned to Lithuania to defeat a Russian army which had occupied Courland. In 1706 he finally broke Augustus by invading Saxony itself and forcing the Elector to sign the treaty of Altranstadt in September; by this Stanislas was recognised as King of Poland.

During 1702–3 Peter the Great had taken advantage of Swedish involvement in Poland to improve the effectivness of his army and to seize Swedish Baltic territories piecemeal. His successes led to the foundation of St Petersburg and the massacre of the Swedish inhabitants of Narva. After Charles's success in Saxony Peter sued for peace, but Charles was determined on a dramatic revenge. In January 1708 he added his name to the list of unfortunate generals who have tried to defeat Russia by a land invasion. With an army of 45,000 he crossed the Vistula from Poland, took Grodno, sat out the spring thaw near Minsk, and on 4 July defeated a Russian army defending a river line at Holowczyn. The Russians refused battle during the late summer, withdrawing before Charles's army and using 'scorched earth' tactics as they went. In difficulties over provisions, Charles decided that rather than retreat he would turn south into the Ukraine and explore the possibility of co-operating with Ivan Mazeppa's disaffected Cossacks. He foolishly ordered General Lowenhaupt to march down to join him from Livonia with a huge supply train but only small reinforcements. In October 1708 Peter the Great burned Mazeppa's capital; only a small force of Cossacks joined the Swedes the following month. On 9–10 October, at Lyesna, Lowenhaupt's 11,000 men were beaten by much stronger Russian forces and the supply train was destroyed; Lowenhaupt and his 6,000 exhausted survivors joined Charles later that month. The Swedes were forced to fight continually throughout the coldest winter in memory, harassed by the Russians and thinned by cold, disease, and starvation. By the spring Charles had only some 20,000 men left, and little ammunition. Peter moved against him with 80,000 men, and at Poltava on 28 June the Swedish army was smashed in attacks on the Russian position. Charles, wounded a few days previously, escaped to Turkish Moldavia with a few officers and cavalrymen and some of Mazeppa's Cossacks; the surviving Swedes surrendered two days later. They had lost more than 9,200 dead and wounded and 18,700 prisoners. During 1709–10 the allies proceeded to divide the spoils of Sweden's defeat, and only the energy of General Magnus Stenbock prevented

Russia, Saxony, and Denmark from completely dismembering Swedish territory.

Charles remained in Turkish Moldavia as an exile until 1714. He was frantically involved in negotiations to bring Turkey into the War against Russia, and had some short-term success, but in the end made himself such a troublesome guest that the Turks arrested him. In November 1714 he arrived back at Stralsund, having crossed Europe with a single servant. His country was in an unhappy situation, but negotiations at this stage could have accomplished much. Charles refused to negotiate; he pursued a series of international intrigues in an attempt to gain support for the retention of Sweden's German territories, coupled with stubborn military campaigns to prevent their being seized by force. In 1716 a Danish–Norwegian invasion of Sweden was abandoned after Charles led 20,000 men into Skane in southern Sweden. In 1717–18 he led a Swedish army into Danish Norway, as much to gain territory which he could trade in future negotiations as for any immediate military reason. On 11 December 1718, while with his troops in the trenches before Frederickshald, he was shot in the head and killed instantly. The war ended with the Treaty of Nystad (August 1721), by which Sweden lost Livonia, Estonia, Ingria, part of Karelia, and many of the Baltic islands, although she retained most of Finland. She ceased to be the dominant power in the Baltic, and Russia's rise as a major power can be traced from this date.

Churchill, John, *see* Marlborough

Chu Teh (*1886–1976*), *General*

Born of a peasant family in the south-west Chinese province of Szechwan in 1886, Chu Teh attended the Yünnan Military Academy, graduating in 1911. In the same year, as a young officer, he participated in the military overthrow of the Manchu (Ch'ing) dynasty, and four years later served under Ts'ai O in the revolt against Yuan Shih-k'ai during 1915–16. His head apparently turned by military authority, he appears to have degenerated into a semi-warlord figure for about three years, but his revolutionary aspirations reappeared in 1921 when, as a military commander in Yünnan, he was forced to flee the province when a rival political faction gained power. In 1922 he came to Europe, where he joined the emergent Chinese Communist Party; he studied military history at Berlin and attended the University of Göttingen. Expelled from Germany in 1926 for his political activities, he returned to China, where civil warfare was growing between the forces of the Nationalist Government at Hankow (under General Chiang Kai-shek) and radical factions fomented by Communists and Kuomintang leftists.

Following the purge of Chinese Communists from the Hankow Government, Communist elements of the Army mutinied at Nanchang in the hope of sparking a nationwide revolution; among the leaders of the rebel faction was Chu Teh. The mutinous troops were quickly dispersed by Nationalist forces, and Chu Teh escaped into the mountains of Kiangsi with a small band of followers. On this date (1 August 1927) started the twenty-two-year-long Chinese Civil War – the date itself being regarded by the Communists as that of the birth of the Chinese Red Army. During the next six weeks another Communist uprising, under Mao Tse-Tung in Hunan, also failed, and he also fled to western Kiangsi, where he was later joined by Chu Teh. Here they together established the Kiangsi Soviet.

In July 1930 the Communist Party Chairman, Li Li-San, believing that the time was now ripe for open revolution by the Chinese urban proletariat, ordered the Kiangsi and Fukien guerrilla forces to seize the cities in central China. Despite protests by Mao Tse-Tung, who believed that Communist success lay rather in a peasant revolt than in an urban confrontation, the city of Ch'ang-sha was taken on 28 July and briefly held by Communist forces led by General P'eng Teh-Huai; but it was soon given up. Chu Teh also took the field at this time, but was savagely beaten by the Nationalists.

For four years thereafter Chiang Kai-shek's Nationalist armies maintained steady pressure against the Communists; and in 1934, their influence and survival now threatened, the Communists started the famous 'Long March'. Far from being a disordered retreat, the 6,000-mile journey made by Chu Teh's First Front

Army was accomplished over a period of thirteen months. Starting from Kiangsi in October 1934, 200,000 men set out to march across southern and western China and up into north Shensi, overcoming sporadic opposition *en route* and distributing 40,000 men along the way as underground cadres. Although they suffered about 100,000 casualties, they were joined by 50,000 recruits before reaching their destination. The Long March is generally considered to have been the longest and fastest sustained march ever undertaken by an army on foot under combat conditions. The redisposition of the Chinese Communists, although undertaken as a strategic necessity, achieved two important results: it concentrated the main Communist strength in an area relatively inaccessible to Chiang Kai-shek's main Nationalist strength, and it brought into being a substantial number of powerful Communist cells distributed over the whole breadth of China.

Following the continuation of Japan's aggressive policy of expansion and domination in north China which had begun in 1931, there occurred on 7 July 1937 the 'China Incident': a Japanese unit, ostensibly on night manœuvres near Peiping, clashed with Chinese troops and precipitated a full-scale invasion of China. Many regard this small incident as being the true flashpoint that started the Second World War. Facing the might of the 300,000-strong, well-equipped, and well-trained Japanese Army was Chiang Kai-shek's poorly-trained and ill-equipped 2,000,000-strong Chinese Nationalist Army; but in the face of invasion, Chu Teh grudgingly offered Chiang the support of his 150,000 guerrilla troops. The Communist general was given command of a newly-established Eighth Route Army, predominantly Communist-manned but ostensibly deployed under Chiang's strategic direction. The first major success achieved by Chu Teh's forces came on 25 September 1937 at the Battle of P'ingsinkuan, when the Japanese 5th Division under General Itagaki was ambushed and routed by the 115th Chinese (Communist) Division. The success of this battle gave rise to widespread propaganda sympathetic to the Communist cause throughout China, and greatly assisted Communist consolidation during the years 1937–40.

During the remainder of the Sino-Japanese War Chiang Kai-shek was forced to raise further Communist-based armies, and Chu Teh himself, while never far from the military planning scene, seems to have retired further into the realms of forward political projection. After the end of the war against Japan in 1945 he once again took up the assault on the Nationalists in the civil war which lasted from 1946 to 1949, and retained command of the Chinese People's Liberation Army until 1954. However, after the formation of the Communist People's Republic in 1949 his influence was more nominal than real. He had become a Politburo member in 1934, but was never regarded as any sort of contender for political power. There are few reliable analyses of Chu Teh's personality available in the West, but his military achievements suggest a remarkable influence skilfully wielded over the huge numbers of personnel who found themselves subject to his politico-military command. He died in June 1976.

Cid, El (Rodrigo Díaz de Bivar) (*c. 1043–99*)

'The lord who wins battles' – *El Cid Campeador* – was an extraordinary soldier who became a renowned figure in his own lifetime. His true historical significance, however, is obscured by his selection as a symbol of Christian Spain's eleventh-century struggle against the Moors. The status of Christian champion and national hero, falling not far short of saint, sits uneasily on Rodrigo's broad shoulders, and he would have been hard put to it to recognise himself under this encrustation of legend. It is in no way a denigration of his skill as a soldier to point out that his motives seem to have been those natural to all Frankish knights of the eleventh century – the lust for personal power and wealth. His long quarrel with his most able king, Alfonso VI, was probably based on motives far from noble, and kept him from Alfonso's side during the King's most significant campaigns against the Moors. Rodrigo's epic capture and holding of Valencia had little historical relevance; and he several times hired his services out to Moorish princes warring on Christian neighbours. Before the Almorávid invasion of 1086 the Moorish and Christian kingdoms of Spain were divided by no very marked barrier of religion; all played their part in the power-struggles of the day, making and breaking alliances according to personal advantage rather than religious solidarity.

Rodrigo was born at Bivar, near Burgos, in about 1043, the son of a minor Castilian noble named Diego Lainez. He was brought up at the court of Ferdinand I of Castile by the King's eldest son Sancho. Rodrigo seems to have distinguished himself young, in the later campaigns of Ferdinand's reign; when Sancho came to the throne on his father's death in 1065 the young knight of Bivar was appointed to the high office of standard-bearer. His reputation for skill in war grew rapidly during Sancho's reign; he had an effective combination of boldness, cunning, and painstaking patience. Sancho engaged in a dynastic war with his brother Alfonso of León, and it was largely thanks to the generalship of Rodrigo that he succeeded in seizing his brother's kingdom and sending him into exile. Rodrigo was therefore in a situation of some embarrassment when Sancho was killed in 1072 at the Siege of Zamora, and Alfonso came to the throne as Alfonso VI of Castile. Alfonso retained his services, however, and in 1074 he even married the King's niece Jimena, daughter of the Count of Oviedo. He served Alfonso in his campaigns for nearly ten years, steadily adding to his reputation. In 1079 he was with the army of Alfonso's Moorish tributary al-Motamid of Seville when he defeated an invading army under Abdullah of Granada. Alfonso's favourite, Count Garcia Ordoñez, happened to be fighting on the Granadian side, in the untidy manner of eleventh-century noblemen, and Alfonso was angry when *El Cid* took him prisoner. Two years later the feud between King and general came to a head when *El Cid* made an unauthorised attack on the Moorish kingdom of Toledo, over which Alfonso exercised a protectorate. In 1081 Alfonso sent him into exile.

Although his career might have seemed to be in ruins, Rodrigo still had his sword, his intelligence, and his reputation. He took service with the Moorish Emir of Saragossa, leading his armies against the Christian rulers of Barcelona in 1082, and of Aragon in 1084. His years among the Moors gave him a sophisticated understanding of Moorish attitudes and techniques which was to stand him in good stead. There was a temporary reconciliation with Alfonso in 1083, but it was not until 1087 that an emergency forced the King to invite Rodrigo to re-enter his service. Alfonso had been steadily pushing back the frontiers of Moorish power, and an emir had eventually appealed for help to Yusuf ibn Tashfin, leader of the Almorávids. (This loose federation of Berber clans had acquired cohesion and purpose through a religious reform movement in about 1040, and had spread its power over the whole Maghreb.) In 1086 the Almorávid army brought Christian expansion to a sharp halt at the Battle of Zallaca, near Badajoz, and the following year Alfonso recalled Rodrigo to rebuild the badly-mauled Castilian army. The arrangement did not last long, however, since *El Cid's* interests now lay entirely in the east. Armed with a written undertaking from Alfonso that he could keep for himself and his heirs any Moslem territory he could capture from the kingdom of Valencia, he went cheerfully into a second exile in 1089. While the Almorávids proceeded to swallow up all the Moorish kingdoms in Spain except Saragossa, revitalising the Moslem presence in the country, Rodrigo led a private army of Christians and Moors in the conquest of Valencia. First he made a tributary of the Moorish king, al-Kadir; then, when al-Kadir was murdered and replaced by the caid ibn Yenhaf, he besieged the city. The siege lasted until June 1094, when ibn Yenhaf capitulated on terms; this campaign had exercised Rodrigo's talent for undercover intrigue and playing off parties against one another, as much as his military skills. Several months after this *El Cid* broke the terms of the capitulation, had ibn Yenhaf killed in a most brutal way, and became virtually an independent monarch. He defeated Almorávid attempts on the city at Cuarte in 1094 and Bairen in 1097. His semi-royal status was underlined by the betrothals of his two daughters, one to the prince of Navarre and one to the count of Barcelona.

Rodrigo died at Valencia on 10 July 1099. His widow Jimena held out against the Almorávids for another three years, but the city could not exist in a constant state of siege, and she capitulated in 1102. The body of *El Cid* was taken to the monastery of San Pedro de Cardeña near Burgos, where the monks almost immediately began to foster a cult of reverence to his memory which bordered on the mystical. Rodrigo Díaz de Bivar has no need of such embellishments; he was a crafty, courageous, tenacious general who consistently defeated far larger forces, and raised himself from obscurity to the heights of fame and prosperity by his own resources.

Clausewitz, Karl von (*1780–1831*), *General*

Born at Burg, near Magdeburg, on 1 June 1780 of an impoverished professional middle-class family, Clausewitz entered the Prussian army in 1792, and in the following year, during the Rhine campaign, was commissioned at the age of thirteen. Thereafter while on garrison duty he devoted much of his time to self-education, which enabled him to gain entry to the Berlin War College in 1801.

He spent four years at the academy, being instructed in military science by Gerhard von Scharnhorst (q.v.) and developing original basic military strategic concepts. From these early associations it is clear that Scharnhorst recognised latent genius in the young man, introduced him at Court, and obtained for him an appointment as aide to Prince August. He then re-entered active service, but was captured at Prenzlau after the disastrous Jena campaign of 1806, being released and returned to Prussia two years later. During the period of Prussian military reform in 1809–11 Clausewitz was given senior administrative responsibility by Scharnhorst; however, when Napoleon (q.v.) invaded Russia in 1812 Clausewitz – like other Prussian patriot officers – resigned his commission and joined the Russian Army. As a staff officer in Russian service he distinguished himself on several occasions before returning to Prussia where, during the Waterloo campaign in 1815, he was appointed chief of staff of an army corps. In 1818 he was promoted general and elected head of the Berlin War College, an appointment that afforded him considerable opportunity during the next twelve years to pursue his military theories – his strategic studies were eventually published in the work *Vom Kriege* (*On War*). Before they could be completed, however, Clausewitz was assigned to the forces at Breslau, and was subsequently appointed as an observer of the Polish Revolution of 1830. He contracted cholera and died shortly after his return to Breslau on 16 November 1831. His military works were collected and published by his devoted widow, the Countess Marie von Brühl.

Clausewitz never had a command of his own, and his fame rests solely upon his extraordinarily influential strategic theories. Studying and analysing the military careers of Napoleon and Frederick the Great (q.v.), he isolated the factors that decided their respective successes. Demonstrating the almost infinite variety of tactical circumstances facing a military commander, he postulated the theory that in embarking on any successful military campaign it is necessary to aim broadly at three main targets: the enemy's forces, his resources, and his will to fight. He emphasised that rigid adherence to any other strategic concepts could be defeated by psychological and accidental factors; and while maintaining emphasis on the three main targets, he attempted to demonstrate military flexibility that would embrace all likely factors.

His most influential single strategic doctrine was contained in his thesis for the overall political direction of warfare, maintaining 'that warfare itself is nothing but an extension of political intercourse with the introduction of different means', and thereby denying that war is an end in itself. Further, he advocated warfare of a defensive nature, in that it is both militarily and politically stronger and more compact.

Such radical thought was wholly contrary to accepted Prussian militarism, and while the general tenets of Clausewitzian theories gradually came to be adopted in the military establishments of Sweden, Austria, Holland, and Switzerland, the political postulation was in fact reversed by Prussia in the mid-nineteenth century. The result was that Prussia, and later the German military confederation, abrogated to an increasing extent any political influence in military matters, so that by 1914 German military planning had become divested of almost all political motives

– a situation that reached its nadir when Ludendorff (q.v.) assumed virtual dictatorship in 1916.

Today Clausewitzian theories are read and held – in only slightly modified form – the world over. Communist theory on the nature of war, including the concept of 'the imperialist war', is derived from Clausewitz. In the context of long-range strategic missiles the theories may appear outdated; but since they are based on a fundamental understanding of the nature of war rather than upon any particular stage of technical advance, they are as relevant today as in Clausewitz's lifetime.

Clive, Robert, Baron Clive of Plassey (*1725–74*)

Clive was the British soldier-statesman who from relatively humble beginnings (but helped by influential family connections) rose to remedy many military and administrative shortcomings in British India. By an extraordinary combination of guerrilla and limited field commands he materially assisted in the creation of stable British power in the subcontinent.

Born on 29 September 1725 on the family estate at Styche, near Market Drayton in Shropshire, the son of a Member of Parliament, young Robert Clive was a high-spirited boy whose education proved difficult. At the age of eighteen he was sent to Madras as a 'writer' in the East India Company – more as a last resort than with any prospect of a career of substance.

Life in the Company depressed the young man; he fought a duel, and twice attempted to take his own life. But the outbreak of war with France in southern India led to the disclosure of young Clive's latent talent for action. At the capitulation of Madras in 1746 he was taken prisoner, but escaped to Fort St David, earning from the governor appointment as an ensign in the volunteers the following year. In 1751, in the war between France and England in support of rival Indian princes, Mohammed Ali was besieged in the fortress of Trichinopoly by the French candidate for the Carnatic nawabship, Chanda Sahib. Clive offered to lead a diversion against Arcot; with a force of only 500 he seized the city, and then successfully withstood a fifty-three-day siege by Chanda's brother. This achievement proved to be the turning point in the campaign against the French general Joseph Dupleix, and after the relief of Arcot Clive went on to demonstrate an extraordinary gift for the conduct of guerrilla warfare, which he successfully pursued for several months. In March 1753 he returned to England with something of a fortune.

After an unsuccessful bid to enter Parliament in 1755 Clive returned to India, reaching Madras in June the next year. He arrived in Bengal in the middle of a dispute between the British and the French-supported nawab Suraja Dowlah, who had captured the fort at Calcutta and imprisoned the surviving Europeans in a small underground cell – the so-called 'Black Hole'. In October Clive set out with a force of 900 Europeans and 1,500 Indians to relieve the city, the force being transported in a fleet commanded by Admiral Charles Watson. This object he achieved on 2 January 1757, and five weeks later he negotiated peace with the nawab. A four-month period of intrigue followed, during which it became clear to Clive that Suraja Dowlah was hostile to the newly-proposed ruler, Mir Jaffa. After seizing the French post at Chandernagore on 23 March (to secure his lines of communication with Calcutta) Clive therefore marched inland with 1,100 Europeans and 2,100 native troops to confront Suraja, and found him entrenched on the farther bank of the Bhagirathi river with a 50,000-strong army. Clive's small force crossed the river and took up positions in a mango swamp. As the nawab's army moved to encircle the British force a heavy rainstorm broke, soaking the powder of the nawab's artillery (which was under French command). But when a cavalry charge swept up to the British lines, Clive's guns – whose powder had been kept dry – opened up a murderous fire. A follow-up assault on the nawab's entrenchments swept the field, only the French gunners under General St Fraise fighting to the last. The Battle of Plassey, fought on 23 June 1757, decided the fate of Bengal. Suraja was assassinated a few days later, and was succeeded by Clive's nominee, Mir Jaffa. Clive, who had lost no more than twenty-three men killed in the battle, was now established as the virtual master of Bengal; he had secured the richest province of India over the heads of his political superiors.

Clive returned to England in February 1760, being raised to the Irish peerage as Baron Clive of Plassey in 1762, and knighted two years later. He now succeeded in entering Parliament as Member for Shrewsbury, and applied his considerable Indian wealth in the development of his estate. But jealous statesmen and retired East India Company members frustrated any desire he may have felt for power at home, and when, in 1764, troubles again threatened the Company in Calcutta, Clive eagerly accepted the governorship of Bengal once more, and set sail for the East. Arriving in Calcutta on 3 May 1765, he found that the situation had already been resolved at the decisive Battle of Buxar, but that there existed widespread confusion in the administrative assembly of Bengal, while a serious rift had opened between the military commands in Bengal and Delhi. The characteristic manner in which Clive resolved these problems firmly established the British Company as the scarcely-disputed administrator of India's two richest provinces, and marked the true beginning of the British Empire in India. He did much to eliminate fraudulent practices within the Company by the strict limitation of bribes – thereby establishing its increased prestige and reliability. His similar treatment of the Army, by realistic limitation of swollen allowances, brought much discontent, and for a time he was almost alone in his stand against corruption.

Later, however, it was the independent discovery of corruption among the Company's servants which led to an unjustified belief in British parliamentary circles that Clive had instigated frauds for his own benefit. He defended himself successfully against the charges in Parliament in 1772 and was entirely cleared of any guilt; but the ordeal had cost him his health, and in one of his fits of melancholy he took his own life at his home in London on 22 November 1774.

Clive paid the price for gaining the heights too soon. His meteoric rise in military command was due to a number of unusual circumstances – none of which equipped him to meet adversity or political intrigue when they were aimed at himself. He succeeded in matters of military importance when the East India Company itself was threatened – at a time when the Company was of more significance and possessed more power than the Army. In other words the Company owed its survival to the prowess of one of its own servants, and was in a position to advance that servant more effectively than it could have done had he been a senior regular officer. Clive's military achievements were scattered but wholly effective; their results were entirely disproportionate to his efforts or professional skill. His powers and acumen as an administrator were, on the other hand, unquestionable.

Condé, Louis II de Bourbon, Prince de (*1621–86*)

Le Grand Condé was one of France's great seventeenth-century commanders, the victor of Rocroi and Valenciennes, a master of battlefield adaptations, and an advocate of 'the first initiative'.

Born in Paris on 8 September 1621, the son of Henry II of Condé, the Duc d'Enghien (as he was called in his youth) was educated strictly by the Jesuits and at the Royal Academy in the French capital. He quickly grew into an arrogant young man, fully conscious of his royal blood, utterly disdainful of lesser princes, yet, when so inclined, a fairly consistent supporter of the French monarchy. By the age of seventeen he had been appointed Governor of Burgundy, and it was while in this post that he attracted the attention of Cardinal Richelieu and Louis XIII as a promising military commander.

At the age of twenty-two Enghien was given command of the French army at Amiens and, learning that the Spanish were attacking Rocroi to the east, led his forces without hesitation against the would-be invaders. Without giving the enemy the slightest chance to prepare any counter-move, Enghien opened the battle (on 19 May 1643) by attacking and defeating the Spanish cavalry wing, and turning his own cavalry on the rear of the Flemish infantry; in the confusion that followed he concentrated his main force to smash the core of veteran Spanish regiments. The encounter at Rocroi was France's greatest victory for a hundred years, and showed Enghien as a brilliant commander with a genius for direct, decided action.

The next three years brought continued warfare against Bavaria. Initially Enghien commanded a force of troops sent to reinforce the great Marshal Turenne (q.v.) in his campaign against Franz von Mercy. After his seizure of Philippsburg and occupation of the Rhineland as far as Mainz, Enghien kept strict control of his troops, thereby rendering military control of the inhabitants tolerable – an unusual phenomen for the age. In 1645 Turenne was again in difficulties on the borders of Bavaria, and Enghien once more arrived with reinforcements. The French victory at Nördlingen, on 3 August that year, further confirmed Enghien's military genius, and he was now sent to serve as second-in-command of the main French army campaigning in Flanders under the ineffectual Gaston, Duc d'Orléans. When Gaston was forced by gout to return to Paris, Enghien took control of, if he did not command, the army and set about recovering the important prize of Dunkirk – which he duly accomplished in two months.

Enghien's father died in December 1646 and he thus became the premier Prince of the Blood, the Prince de Condé. With power, glory, and military prestige, Condé now represented a dangerous element within France should his loyalty to the crown waver, and he was therefore closely watched by Cardinal Mazarin, the Queen Regent's chief minister. The following year Condé was sent to Catalonia to revitalise the defeated French army in an attempt to force Spain to accept peace on French terms. He invested Lérida, but when the garrison resisted, refusing to surrender, Condé realised his error in committing too much of his strength against a local threat, and abandoned the siege. Though his failure at Lérida was a blow to his prestige (a fact carefully exploited by Mazarin), he succeeded in holding Catalonia.

In 1648 Mazarin gave Condé command of the great French army in Flanders, but almost immediately split the army piecemeal on the advice of junior commanders. Thus at the battle of Lens on 20 August 1648 Condé's command was much depleted; yet he decisively defeated Archduke Leopold William's final attempt to invade France through Artois.

The following year civil war erupted in France following the rejection of demands on the royal family by the *Parlement*. Unrest and rioting in Paris forced the royal family and Mazarin to flee the city, while many nobles joined the rebels, among them Condé's younger brother, Prince Armand of Bourbon and Conti. Partly as the result of a feud with this brother, Condé himself led his army against Paris, captured the fortress of Charenton, and invested the capital. The *Parlement* sued for peace, Queen Anne declared a general amnesty, and Mazarin was reinstated. The Cardinal, however, pursuing his antagonism against Condé, ignored the soldier's part in restoring the monarchy – an attitude that so infuriated Condé that he commenced intriguing with the formerly rebellious nobles. In January 1650 Mazarin arrested and imprisoned the disgruntled Condé.

Following a further royal amnesty in 1651 Condé was released from prison and Mazarin fled to Germany. Once again civil war spread throughout France, but now Condé, allied with Spain, commanded the rebel forces. Trapped by an army under Turenne (who remained loyal to the crown) outside the walls of Paris, Condé only narrowly escaped capture, and made his way north to join Duke Charles IV of Lorraine. Thereafter, as a Spanish generalissimo, Condé pursued a fruitless campaign in Flanders, suffering a humiliating defeat at the great Battle of the Dunes (14 June 1658 – see under Turenne).

After the Peace of the Pyrenees, signed on 7 November 1659, which brought the cession of Flanders to France, the French monarchy found stability. Condé sought and obtained forgiveness from Louis XIV. Henceforth Condé and Turenne marched forth on campaign as colleagues, as of old. In the Dutch War, which opened in March 1672, Louis XIV personally led an army of 130,000 men in the great invasion of Holland with the two veteran commanders as his lieutenants; in a skirmish near Arnhem on 12 June Condé was wounded and his nephew killed. In 1674 he commanded the French army in the Netherlands, but was ordered to withdraw these forces as part of Louis's plan to concentrate against the Spanish Netherlands and to protect Alsace against invasion by the Germans. Condé succeeded in his disengagement and, seizing an opportunity to catch the Allied army on the march, boldly attacked with only part of his forces. In this, the Battle of Senef on 11 August 1674, Condé narrowly missed achieving a decisive victory; he was, however, engaged by William of Orange in a bitter fight, and when night fell both armies were glad to withdraw. Condé, now a sufferer from gout, had been in the saddle for more than twenty-four hours, had had three horses killed under him, and finally escaped from the battlefield wearing slippers and stockings.

On 31 July the following year, after the death of Turenne, Condé was ordered by the King to take command of the army in Alsace. At once the veteran commander set about restoring the army's flagging morale, manœuvring from one strong position to another until Raimondo Montecuccoli abandoned his ambitions in Alsace and the Rhineland, and withdrew across the river. This campaign was Condé's last. So crippled with gout that he was scarcely able to move, he ceased to attend the King's council, although he still enjoyed royal respect. He died at Fontainebleau on 11 December 1686.

Probably less gifted than Turenne, Condé was nevertheless a brilliant tactician, and was renowned for his ability to seize the fleeting opportunity. His victories, often achieved in the face of great odds, won for him considerable military prestige – although his periodic defeats deprived him of the much-sought legend of invincibility. Had he been less arrogant during his earlier years, his political *volte face* might have attracted less attention and his military career might have gone on unchecked; certainly his campaign partnerships with Turenne, when exploited to the full, rendered France unbeatable on the field in the mid-seventeenth century.

Constantine XI, formerly Dragases Palaeologus (*1404–53*), *Emperor of Byzantium*

Constantine's military fame rests on his participation in one of the most historically symbolic disasters of European history. He was the last Emperor of Byzantium, and with his death on the walls of the city more than 1,000 years of empire came to an end. It is no bar to his inclusion in this book that by the time he took up his unlucky inheritance the actual strength of Byzantium was a mere shadow of its former might; he was a courageous, intelligent, and energetic man who died the final defender of the heritage of Rome.

He was born the fourth son of the Emperor Manuel II and his Serbian wife Helen, of the Macedonian Dragas dynasty. Constantine spent his early adult years in the Peloponnese with his brothers Theodore and Thomas, completing the recovery of the modern Morea from the Franks, and subsequently governing the region. His brother John VIII died childless in 1448, and at Mistra in June 1449 he was proclaimed Emperor. Two years later the Ottoman Turkish sultan Murad II was succeeded by Mohammed II (q.v.). Mohammed determined to succeed where his predecessor had failed thirty years previously – before the walls of Constantinople. He led an army of more than 80,000 men and a siege battery of 100 heavy cannon to Constantinople, and in February 1453 invested the city. Constantine had only some 10,000 troops to defend the enormous length of its walls; for a time contact with the outside world was maintained by a Venetian fleet, but after this had been driven off by the Turkish galleys the blockade was complete. At the beginning of April the Turkish batteries west of the city began a heavy bombardment; they included a dozen super-heavy weapons which fired stone projectiles of enormous weight, and took hours to reload. Several breaches were made, both by artillery fire and by mining. These were sealed off by the building of interior walls, and the garrison repulsed several Turkish assaults by vigorous counter-attacks. Constantine showed great personal courage and qualities of leadership throughout the siege, ably backed by John Justiniani, commander of an effective contingent of Genoese mercenaries in the garrison. He attempted to persuade the Western nations to come to his aid, even offering to acknowledge the supremacy of the Church of Rome over the Greek Church, but without success. In the final days of May 1453 a last intensive Turkish bombardment levelled large stretches of the city defences, and the weakened garrison was unable to seal off the enormous breaches completely. For several hours Constantine led a gallant defence of the main breach and succeeded in holding off a very heavy Turkish assault; but another enemy force penetrated a section of the walls stripped of troops for the defence of the breach, and took the garrison in the flank and the rear. Constantine had a chance to save himself, but did not take it; and on 29 May 1453 he died fighting in the breach with the last of his men. The Turks then ravaged the city for three days, celebrating the capture, at long last, of the main European bastion against expansion from the East.

Coote, Sir Eyre (*1726–83*), *Lieutenant-General*

Born at Ash Hill, Co. Limerick, in 1726, Eyre Coote entered the Army at an early age (probably about 1741), and is said to have taken part in operations in Scotland during the 'Forty-Five Rebellion. In 1754, probably as a captain, he sailed to India with the 39th Regiment, and two years later accompanied the expedition sent from Madras to Bengal to punish Suraja Dowlah for the 'Black Hole of Calcutta' atrocity. He was present in 1757 at the capture of Calcutta, where he hoisted the English colours on Fort William, and at the capture of Chandernagore.

Under the command of Colonel Robert Clive (q.v.), the 39th Regiment then set out with 750 Europeans, 2,100 sepoys, and ten guns in pursuit of Suraja's army; finding it entrenched on the far side of the Bhagirathi river near the village of Plassey, Clive called his historic council of war. At first it was determined to delay the attack, but after representations by Coote, who suggested that such delay might result in French reinforcements reaching Suraja, it was decided to give battle without further ado. Against the small force under Clive's command was ranged an army of 50,000 with fifty-three guns. The Nawab's army moved to encircle the British, but a sudden rainstorm spoilt the powder of the Indian guns so that the supporting cannonade petered out. An Indian cavalry charge on the British (who had covered their powder) was repulsed with great loss. Clive then set about shelling the Indian entrenchments, and finally assaulted the Indian positions – in which only the French fought to the last. Coote, who commanded the 3rd Division of the 39th Regiment with outstanding skill at Plassey, was promoted by Clive, and as lieutenant-colonel took command of a newly-formed regiment, the 84th.

In 1758 a French expeditionary force arrived at Pondicherry under Baron Thomas Lally and laid siege to Madras, but failed to capture it – principally owing to the successful supplying of the garrison from the sea. At the Battle of Masulipatam on 25 January 1759 a relief force under Francis Forde defeated Lally and raised the Siege of Madras. As Lally moved away during the summer and threatened the important fortress of Trichinopoly, Coote marched out from Madras with 1,700 British troops and 3,000 sepoys to create a diversion. Moving with great speed, Coote captured the town of Wandewash and the fort of Carangooly, and these operations had the desired effect upon Lally, who abandoned his attack on Trichinopoly and at once moved against Coote's force, besieging it in Wandewash during December. Coote watched his besiegers carefully until, on 22 January 1760, he burst out of the town and utterly defeated the French – despite great odds. This victory, second only to that of Plassey, marked the beginning of the end of French military influence in India. Lally's prestige was gone, and Coote, after capturing Arcot, prepared to lay siege to Pondicherry – Lally's last refuge. With blockade assistance on the sea from Admiral Stevens, Coote reduced the garrison to starvation, and on 15 January 1761 Lally surrendered to the British forces.

In 1762 Coote returned home, to be generously rewarded by the British East India Company; he purchased the large West Park Estate in Hampshire, was promoted colonel, and was elected M.P. for Leicester in 1768. The following year he was again appointed Commander-in-Chief of the Madras forces, but no sooner had he arrived in his command that he found himself seriously at odds with the Governor, Josias Du Pré. Seeing no possibility of compromise, he forthwith resigned and returned home – a course of action wholly supported by the King and by the court of directors. Coote was invested with the K.B. on 31 August 1771, and promoted major-general four years later. In 1777 he was appointed Commander-in-Chief in India and promoted lieutenant-general, taking up his command at Calcutta on 25 March 1779. At once he found himself the victim of intrigues fostered by Warren Hastings, who attempted by bribery to gain Coote's support in the internecine squabbles between him and his council at Calcutta. Coote resisted these approaches – but at the same time managed to retain the support of Hastings in all military matters.

During Coote's absence from India, Hyder Ali, ruler of Mysore, had fought British forces to a standstill in the First Mysore War (1766–9), and then concluded a treaty with the East India Company. However, having received no assistance from the British during his own campaign against the Marathas, he sided with the French after the British declaration of war against France in 1778. He swept over the Carnatic, cutting to pieces a small British garrison force under Colonel Baillie at Parambakam on 10 September 1780. Warren Hastings at once despatched Coote with full powers and all the money he could spare to Madras, and ordered all available troops down the coast under the command of Colonel Pease. Coote reached Madras on 5 November, and two months later marched north with every soldier he could muster in an attempt to draw Hyder Ali after him. His march was successful, and he raised the siege of Wandewash. Hyder Ali, however, enticed him to move further north by threatening Cuddalore, and promptly interposed a large army across his vital lines of supply and communication with Madras. But duplicity by the French Admiral d'Orves left command of the sea to Sir Edward Hughes, and seaborne supplies continued to sustain Coote's force.

On 16 June Coote left Cuddalore, and two days later he attacked the pagoda of Chelambakam. He was forced to retire, and fell back on Porto Novo close to the sea, where he laid plans with Sir Edward Hughes to confront Hyder Ali. On 1 July 1781 the latter, encouraged by exaggerated reports of his success at Chelambakam, entrenched his forces, which numbered about 60,000, before Coote's 8,000 Europeans and sepoys. The British commander examined the position for an hour under heavy fire before ordering Major-General James Stuart to turn the enemy's right on the sandhills and attack in flank. Stuart attacked and was twice repulsed, but on the third attempt, aided by gunfire from the sea, he was successful. Coote then ordered his first line under Major-General Munro to advance, and Hyder Ali was utterly defeated.

Coote followed up his great victory by other successes. He joined Pease at Pulicat on 2 August; he captured Tripassoor three weeks later; and with an army now swelled to 12,000 he stormed Parambakam on 27 August. Further victories followed, but on 7 January 1782 illness forced Coote to return to Bengal – command of the troops passing to Stuart. A year in Calcutta partially restored his health, but during his return to Madras the ship in which he was travelling was chased by a French warship, and the resulting seasickness brought on a fever from which he died in Madras on 26 April 1783. His body was shipped home with great ceremony and buried at Rockbourne Church, Hampshire.

Coote's victory at Porto Novo saved Madras from Hyder Ali as surely as his victory at Wandewash saved it from Lally. He was not perhaps, a great tactician, but is remembered for his tactical maxim 'fight today, not tomorrow,' and his belief in carrying the assault to the enemy whenever possible, since (he thought) it was invariably wiser to attack in observed and known circumstances than to rely on possible advantages in the future.

Córdoba, Gonzalo de (*1453–1515*)

Córdoba's campaigns in Italy at the turn of the fifteenth and sixteenth centuries ushered in a hundred-year period during which Spanish infantry dominated European battlefields. His formative years coincided with technical developments in gunpowder weapons which increased their practical value in battle; Spain was in the forefront of firearm technology, and Córdoba's energetic and imaginative tactical applications of the new science raised him head and shoulders above contemporaries who clung to traditional medieval concepts.

He was the first general to devise tactics and formations specifically to exploit the potential of massed infantry armed with hand-held firearms.

Gonzalo de Córdoba was born in 1453, the second son of Pedro Fernández de Córdoba, Intendant of Andalusia, and Elvira de Herrera. In 1466 the boy was sent to Court and served as a page first to the pretender Alfonso, and later to Alfonso's sister Isabella. In 1474 Henry IV of Castile died and a war of succession broke out, in which Gonzalo served Isabella of Castile and her husband Ferdinand of Aragon. When their joint reign began, blessed by the Pope, Gonzalo saw much action and no little distinction in the final war against the Moslems of Granada, which lasted from 1482 to 1492 and was instrumental in the emergence of a fine Spanish standing army or 'constabulary'. Gonzalo showed great ingenuity in planning assaults on fortified positions – he was a true child of the age of gunpowder. He was chosen as one of the two commissioners who negotiated the final surrender of King Boabdil of Granada in 1492.

In 1495, high in the royal favour, Gonzalo de Córdoba was given command of an expedition sent to support the Aragonese King of Naples, Ferrante, against a French invasion. (The kingdom of Naples at that time comprised most of southern Italy.) Landing in Calabria on 26 May with some 2,100 men, he was defeated at Seminara by a French force under Aubigny on 28 June. It was to be the only defeat of his career; between 1495 and 1498 he slowly won back the captured territories from the French. He refused to meet them in major engagements, making use of his experience of campaigning against the Moors to exploit terrain and the enemy's long, vulnerable lines of communication. In the siege of Atella in July 1496 he captured the French Commander-in-Chief Montpensier; and his expulsion of the French garrison at Ostia at the request of the Pope earned him a Roman triumph and lavish rewards. During this period Córdoba was studying and adapting to his purpose the pike tactics of the Swiss, at that time the finest exponents of this type of fighting. In 1498 he returned to Spain, but two years later was back in Italy at the head of an army ordered to co-operate with the French against the Ottoman Turks; his brief also included keeping a close watch on French ambitions with regard to Naples. In December 1500, in collaboration with a Venetian force, his Spaniards captured the strongly-fortified island of Cephalonia. Following the removal of the Turkish threat, King Ferdinand and Louis XII of France concluded an agreement to share Naples between them; but the French violated the agreed demarcation line, and by the summer of 1502 Córdoba was fighting them once more. He had an army of some 4,000, opposed by 10,000 French under the Duke of Nemours, and initially he was forced to fall back before the enemy's advance. He held a perimeter round the port of Barletta, through which he was in due course reinforced to a strength of some 6,000. He marched out of Barletta and took up a hillside position; the French attacked it on 26 April 1503, and thus precipitated Cerignola, the first known battle in history to be won by the decisive use of hand-held firearms. Córdoba placed his arquebusiers and pikemen along a ditch and a staked palisade, where they were attacked by French heavy cavalry and Swiss pikemen supported by artillery; the Spanish artillery had earlier been rendered useless by the accidental explosion of the powder supply. Córdoba deliberately lured the enemy into rushing his apparently flimsy line, and slaughtered them by arquebus fire at close range. Only when they were thoroughly shaken, and the Duke of Nemours himself had been shot dead, did he launch his carefully-drilled mixed formations of pikes and arquebuses to drive them from the field.

On 13 May 1503 Córdoba took Naples, and he maintained a siege of Gaeta until the arrival of enemy reinforcements in October forced him to fall back on the Garigliano river. For two months the armies faced one another across the river in worsening weather; the Spanish made probing attacks, but no major operations were mounted against the much stronger enemy (some 23,000 to Córdoba's 15,000). But on 29 December, when the French were bored and inattentive, Córdoba launched a brilliant surprise attack on their extreme left by means of a sectioned pontoon bridge, prepared in secret and brought up to the river under cover of night and bad weather. The enemy army was rolled up from the left, and by tireless pursuit bundled back ten miles to the Formia Pass. There they rallied, but were broken in an hour and driven back to Gaeta with heavy casualties. This victory of the Garigliano was Córdoba's masterstroke, the crown to his career. It involved the use of picked units of fit men, moving fast under cover of the elements against an enemy whose natural adherence to more sluggish medieval custom had been deliberately encouraged. It hinged on efficient and imaginative engineering work, carried out miles from the site and assembled in secrecy by night. It was won by the relentless advance of carefully-trained mixed units of pikemen and arquebusiers, drilled to cover each others' weaknesses and shelter behind each others' strengths both in attack and defence, the whole preceded by fast-moving troops of skirmishers, both horse and foot.

On 1 January 1504 Gaeta fell to the Spanish, and before the year was out Córdoba – *El Gran Capitán* – was Viceroy of Naples. In 1504 the French were forced to abandon all claim to Naples by the Treaty of Blois. Córdoba never fought another campaign; he was recalled in 1507, due to the jealousy of the King at his enormous renown. He died in Granada, of malaria contracted in the Italian wars, on 1 December 1515. He was a true innovator, a revolutionary figure of his time who clearly deserves his title, 'The Great Captain'.

Cornwallis, Charles, 1st Marquess Cornwallis (*1738–1805*), *General*

One of the better English generals in America during the War of American Independence, Cornwallis later became Commander-in-Chief and Governor-General in India and Ireland.

Born the eldest son of the first Earl Cornwallis on 31 December 1738 in London, the young Charles entered the Army in 1757, serving on Granby's staff in Germany during the Seven Years' War in 1759. He commanded the 12th Regiment of Foot until 1762, when he returned to England on the death of his father. He became Member of Parliament for Eye, Suffolk, in 1760, and during the next few years displayed steadfast opposition to the taxation policies towards the American colonies. Notwithstanding this political stance in favour of the colonials, he temporarily abandoned his political career and sailed for America in 1776 as a major-general, to serve under Howe, and later under Clinton. His victories at Camden and Guilford Court House were, however, overshadowed by his defeat and capture at Yorktown. The severe strictures levelled against him by Fortescue were almost entirely unjustified, and his reputation at home remained unscathed; indeed, while still a prisoner-of-war in America he was offered the joint posts of Governor-General and Commander-in-Chief in India – offers he continued to decline until 1786.

After his arrival in India Cornwallis embarked on a sweeping reform of the administration, and though the majority of the individual changes he made were the result of work by his predecessors, it was by means of his own determination and tact that reform was effected. The aggression of Tipoo Sahib obliged Cornwallis to contravene his own policy of avoiding interference in the affairs of individual states, and in the Mysore War he conducted two successful campaigns in person. In 1791 he captured Bangalore, Nandydroog, and Sevendroog, and in the next year he decisively defeated Tipoo Sahib at Seringapatam and forced him to make peace. In these campaigns he displayed great skill and energy – so much so that news quickly reached Europe of his prowess, and even prompted Austria to seek his appointment as her Commander-in-Chief.

The general returned home in 1793 at a time when Pitt was heading the first coalition against revolutionary France, and it seemed likely that command of the troops in Flanders would pass to Cornwallis. Instead, however, he was appointed Master-General of Ordnance, created marquess, and given a seat in the Cabinet. In 1797 he was again appointed Governor-General of India, but this time the appointment was cancelled when trouble flared in Ireland, whither he was sent instead as Viceroy and Commander-in-Chief.

In 1798 he crushed the rebellion in Ireland, defeating and forcing the surrender of the French general Humbert. His administration was enlightened and tolerant; he actively advocated Catholic emancipation, and assisted

Castlereagh to pass the Act of Union. In 1801 he resigned and returned to London, being promptly set to work on negotiations leading to the Peace of Amiens. However, utterly unsuited to the world of European diplomacy and political intrigue, he soon showed himself to be no match for the subtle Talleyrand. In 1805 he reluctantly returned to India in his old dual post to replace Lord Wellesley; landing at Calcutta, he set out for the Upper Provinces, but was taken ill and died at Ghazipur on 5 October that year.

As a commander Cornwallis was a just, much-loved, and respected man. Unlike so many of his contemporaries, he never sought financial gain from his campaigns – thereby rendering himself impervious to military and political intrigue, although he was somewhat naïve in such matters. Despite his defeat at Yorktown his stature in the field recommended him for field command in India, while his experience of administration in Parliament provided the balance required by the governor-generalship. Only the division of his energies between military and political affairs prevented his ascent to a high place in British military annals.

Cortés, Hernán or Hernando (*1485–1547*)

Born at Medellín, Estremadura, Cortés inherited no military tradition, and having achieved no great success as a law student, settled as a farmer on the island of Santo Domingo. Such friends as he acquired, however, were of a restless nature, for ever sailing off on exciting expeditions, and it was not long before he determined to follow suit. In 1511 he joined Diego Velázquez on his expedition to Cuba, where he was appointed *alcalde* of the capital, Santiago.

On 19 February 1519 Cortés set out with eleven vessels and 500 soldiers for Mexico, and the following month landed at Tabasco, where he won over the natives before sailing on to another point on the coast, where he founded the city of Vera Cruz. At this spot he made himself captain general and chief justice – thereby repudiating Velázquez's authority. It was here that he ran his ships aground and dismantled them, 'burning his boats' in the grand tradition. He then set out for the interior, overcoming any opposition – occasionally by force, but always being careful to keep duress to the minimum – and invaded the independent state of Tlaxcala. At this time the Tlaxcalans were at war with Montezuma, ruler of the Mexican Aztec Empire, and at first they resisted Cortés bitterly; but after defeat by the Spaniards they became their staunch allies, and together the two forces moved against the Aztecs. By good fortune, when Cortés entered the Aztec capital Tenochtitlán (afterwards rebuilt as Mexico City) Montezuma was convinced that his conqueror was the reincarnation of an Aztec god and promptly received him with lavish honour. Realising the influence exerted by Montezuma, Cortés decided to seize the ruler and thereby lay hold over the entire nation. Achieving this end, Cortés released Montezuma and was henceforth recognised master of Mexico.

Meanwhile news of Cortés's conquests had reached Velázquez, who now dispatched a Spanish force under Pánfilo de Narváez to deprive him of command. Cortés at once led most of his men out to meet Narváez, defeated him, and enlisted Narváez's army into his own command. Returning to Tenochtitlán, he found dissident Indians laying siege to the city. Montezuma was killed during an attempt to arbitrate; Cortés decided to evacuate the capital, and retreated to the north end of Lake Texcoco. In the event he was forced to turn and fight, and on 7 July 1520 won the decisive – but costly – Battle of Otumba (Ottumwa). Although it transpired that this battle was to mark the end of the Aztec Empire, Cortés did not feel his forces strong enough to impose his will on the nation and he withdrew to Tlaxcala, where he set about reorganising his small army.

Having obtained reinforcements, Cortés now marched on Tenochtitlán, investing the capital on 26 May 1521. For almost three months the Aztecs resisted the Spanish assaults, but on 13 August the city fell and Cortés seized the new emperor, Cuauhtemoc, and through him once more imposed his control on the Aztec Empire. He razed Tenochtitlán and rebuilt the capital near by as Mexico City.

During the previous year Cortés had been

sending emissaries to the Emperor Charles V of Spain to convince him of the enormous wealth in Mexico, and in 1528 the conqueror returned home a hero, his past impetuosities forgiven. Created Marqués dell Valle de Oaxaca, he returned to Mexico to show himself a great administrator, pushing forward explorations and exploiting the natural resources of the land. But his jealous enemies eroded his authority and influence with the Emperor, with the result that he gradually lost credence with the colonial military. By 1542 the Emperor was unable to find suitable employment for his ablest commander, so discredited had the luckless Cortés become. He retired from Court and died on his estate of Castilleja de la Cuesta, near Seville, on 2 December 1547.

Craufurd, Robert (*1764–1812*), *Major-General*

Wellington's (q.v.) most brilliant subordinate in the Peninsula was 'Black Bob' Craufurd, a dark and bitter Scotsman whose career had been blighted by the fumblings of lesser men. He was as violent in his anger as Picton (q.v.), and as feared for his tongue; but while that Welsh juggernaut vented his rage in barrack-room oaths, Craufurd stripped his victims naked with his biting sarcasm. He was no mere county gentleman in a scarlet coat, but a serious student of military science, far ahead of most of his contemporaries both intellectually and professionally. Nevertheless his judgement was marred by the rancour which he felt at being passed over through no fault of his own, and by his urgent desire to make the best use of the command which eventually gave him a chance to shine.

Craufurd was born on 5 May 1764, the third son of Sir Alexander Craufurd, first Baronet, of Newark in Ayrshire. He entered the Army in 1779 as an ensign in the 25th Foot, was promoted lieutenant in 1781, and secured a captaincy in the 75th Foot in 1783. At this early stage he was already deeply immersed in military studies, an approach to his career which contrasted sharply with the professional indifference of all but a few of his contemporaries. He made himself an expert particularly on the career of Frederick the Great of Prussia (q.v.), and in later years was to produce English translations of German military treatises.

Craufurd first saw action in 1790–2 in the Indian war against Tipoo Sultan, the notorious 'Tiger of Mysore'. He later returned to Europe, and in 1794 his fluent German was instrumental in securing Major Craufurd an appointment as attaché to the Austrian headquarters in the Netherlands, and later on the Rhine. He accompanied the Austrian armies for three years, seeing too much of defeat at first hand. He returned to England and a lieutenant-colonelcy in 1797, and served on the staff in Ireland in 1798. In 1799 he travelled to the Continent once more, and this time was able to witness a Russian defeat, for he was attached to the Austro-Russian army in Switzerland (see under Masséna). His expert knowledge of stricken fields was further extended later in the same year, when he was posted to the staff of the mismanaged British expedition to the Helder. He seems to have discharged his duties with skill and diligence during these attachments; but Oman suggests that his constant reports of disaster to Allied arms, incisive and accurate though they were, may have given his name unpleasant associations in official quarters. He was certainly passed over for promotion in favour of many younger men of far less obvious professional insight.

In 1801, despairing of advancement, Craufurd went on half pay and became M.P. for East Retford. During the next five years his sound but savagely-worded criticisms of the military policies of successive administrations made him powerful enemies; but the 'Government of All the Talents' brought his friend William Windham the portfolio of War Secretary, and Craufurd a colonelcy, late in 1805. He subsequently received command of a 4,000-strong brigade of light troops destined for one of the more feather-brained British expeditions of that period, a projected invasion of Chile. Diverted while at sea to join another futile operation, Craufurd's brigade formed part of Whitelocke's force in the attack on Buenos Aires in 1807. In savage street fighting Craufurd handled his command most ably, but he was forced to capitulate when the expedition collapsed. He was embittered by this farcical conclusion of his first real chance to win distinction in the field; but the fact that he was held blameless was confirmed by his appointment to command a brigade in the Peninsula in 1808.

Craufurd missed Vimiero in August 1808, and Corunna in January 1809. Moore (q.v.) detached the light brigades for a separate and very punishing retreat to Vigo, during which Craufurd's determination and iron standards came to the fore. He returned to the Peninsula as a brigadier in 1809, but missed the battle of Talavera despite an epic forced march of forty-three miles in twenty-six hours, which has never been equalled since by a formed body of infantry. From 1809 to 1812 Craufurd commanded first the Light Brigade, and later the Light Division; and at last his talents were given scope. His was an *élite* command, consisting of the 95th Rifles, the 43rd and 52nd Light Infantry, and subsequently some fine Portuguese light units. Craufurd trained his men to a most exacting standard of skill and initiative. His mission was to operate on the flanks of the army, in the rear during withdrawals and far ahead during advances. His troops excelled at fast and flexible movement, at outpost work, and at maintaining sensitive contact with far superior enemy forces. Probably his finest achievement was the guarding of the north-west frontier of Portugal in 1810; with very small forces Craufurd constructed a magnificent screen of observation posts which proved impenetrable by enemy probes, and which was linked by a sophisticated communications network. In Oman's words, 'the whole web . . . quivered at the slightest touch', and the Light Division was trained to such a high standard of readiness that instant concentration at any threatened point could be achieved. Craufurd was a master of minutely-detailed contingency planning, and calculated every move, instead of reacting to enemy initiatives with whatever forces happened to be available, as did too many of his colleagues. The other side of the coin was that Wellington could not rely on him for the absolute obedience exhibited by Hill (q.v.); there was a streak of the *prima donna* in Craufurd, and on occasion his confidence in his splendid division led him to overreach himself unnecessarily, as on the Coa in July 1810.

Craufurd distinguished himself at Bussaco in September 1810, launching his troops from a concealed position on top of the ridge in a devastating counter-attack on Ney's (q.v.) infantry. So skilfully had the enemy been drawn on by his skirmishers, and so perfectly-timed was the blow, that his troops sustained only 177 casualties to Ney's 1,200. During the winter of 1810–11, while the army lay within the Lines of Torres Vedras, Craufurd returned to England on leave; during his absence his Division was led, most inadequately, by Erksine. Craufurd returned on the very day of Fuentes de Onoro, 5 May 1811; and at that battle the Light Division recorded another brilliant achievement. A French attack on the Allied right wing seriously shook the raw 7th Division, and Craufurd was sent to bring it to safety. He successfully covered the disengagement of the threatened formation, and then withdrew across two miles of open ground, menaced by five brigades of French cavalry and strong horse artillery. So perfectly did he co-ordinate the movements of his mobile squares that he suffered less than fifty casualties. On 4 June 1811 he was promoted major-general.

On 19 January 1812 Craufurd led his Division in the bloody assault on the Lesser Breach at Ciudad Rodrigo. He was shot through the body early in the action, and lingered in agony for five days before dying on 24 January. He was buried in the breach his men had taken at such cost. His Division continued to serve Wellington with distinction under commanders of less talent, a testimony to Craufurd's meticulous work; but the Commander-in-Chief often had cause to regret the loss of the short, fiery-tempered general. Despite his rigid code of discipline he was much respected by the rank and file of his Division, who recognised that much of their remarkable skill and *esprit de corps* was due to him.

Cromwell, Oliver (*1599–1658*), *Lord Protector of the Commonwealth*

Cromwell's importance in the constitutional history of Britain is immense, but for obvious reasons cannot be discussed here. From a purely military point of view his place in the history of British arms is sufficiently significant to assure him of immortality. Although the armies he commanded were small by later standards, and although he led them to victory entirely within the confines of the British Isles, the reforms over which he presided led to a revival of British military prestige overseas after nearly two centuries of relative obscurity.

Born of farming stock at Huntingdon on 25 April 1599, Cromwell was educated at a local school and at Sydney Sussex College, Cambridge. From his adolescence he was exposed to the Puritan religious philosophy, although he does not seem to have embraced it fully until his late twenties. He left the university in 1617 on his father's death. His early years as a small landowner were hard, but he slowly increased his security. He was elected to Parliament as Member for Huntingdon in 1628, but since Charles I ruled without a Parliament between 1629 and 1640 he had little opportunity to make his presence felt in the House. In 1636 he moved with his family to Ely in Cambridgeshire, and his prosperity grew from this moment. He lived simply, according to his station in life and his religious beliefs; his pleasures were those of a countryman. Cromwell was returned as Member for Cambridge when King Charles was forced to call a Parliament in 1640, and became associated with the leaders of the reforming faction, including Pym and Hampden. He spoke frequently, usually on religious questions. The political situation deteriorated rapidly, and in January 1642, after the King's unsuccessful attempt to arrest the 'Five Members', Cromwell was active in raising funds and stores and in organising levies in support of Parliament. He saw civil war as inevitable, and in August personally seized Cambridge Castle and secured the locality for the cause of Parliament. He raised a troop of cavalry from Cambridge and Huntingdon, and in September 1642 he was ordered to lead it to join the Parliamentary army in the midlands. His troop arrived late at the Battle of Edgehill on 23 October, and later retreated on London with the rest of the army. Captain Cromwell subsequently led his troop back to East Anglia. In February 1643 he was promoted colonel, and authorised to expand his command into a double regiment of horse, totalling fourteen troops. This was the force – later known as the 'Ironsides' – which he fashioned and trained according to standards spectacularly higher than those of his contemporaries.

He recruited very selectively, taking only men of proven good character and high motivation. Like any revolutionary army, the forces of Parliament were riven with factional quarrels and handicapped by a chaotic lack of organisation. Building on a firm basis of religious Puritanism and rigid discipline, Cromwell constructed a model regiment. He paid great attention to morale and discipline, to training, and to prompt payment and adequate provisioning. His methods were soon proved sound. Throughout 1643 his regiment was busy in the eastern counties, and they distinguished themselves at Grantham on 13 May, beating a Royalist force and inflicting some 145 casualties. Cromwell and his troopers were again prominent at Gainsborough on 28 July, and at Winceby on 11 October, when Cromwell had a horse killed under him while leading a charge. He attracted the favourable notice of Sir Thomas Fairfax (q.v.); and in 1644 was appointed lieutenant-general of the new army of the Eastern Association, under the inefficient Earl of Manchester. His troopers engaged in harassing operations between Oxford and Cambridge in the early months of 1644, and in May and June covered the sieges of Lincoln and York. On 2 July Cromwell commanded the cavalry on the left wing of the 27,000-strong Parliamentary army which inflicted the first major defeat on the Royalists at Marston Moor. The Parliamentary army attacked the 17,000-strong Royalists in the early evening. Fairfax, on the right, was repulsed and driven back by Goring's Royalist cavalry, and the opposing infantry in the centre became locked in a see-saw combat. The day seemed lost when Cromwell, on the appeal of Fairfax, faced his cavalry right and charged the Royalist horse, who were then occupying the original positions of the Parliamentary right. He defeated Goring's regiments decisively, and then turned on the centre and supported the infantry. The Royalist army was defeated, losing some 4,000 dead and many prisoners, as well as the bulk of its artillery.

The victory was not followed up, and Manchester retired to Lincoln. Cromwell was present at the Second Battle of Newbury, where despite odds of two to one in their favour, the Parliamentarians were unable to force a decisive result. Cromwell was frequently engaged in the fierce factional quarrels of this period which destroyed the advantage Parliament enjoyed over the Royalists in men and resources by preventing a unified command. In November 1644, disgusted by the failures and inefficiency of the command structure, Cromwell opened a campaign for reform by a fierce attack on the Earl of Manchester in the House of Commons. Eventually the raging political, religious, and military controversy was brought to an end with the adoption of a new system. All Members of Parliament – whose inexperience had caused many lost opportunities in the field – were to resign their commissions, and a New Model Army was to be formed under the command of Fairfax, free from political interference. The nucleus of the army was the Eastern Association force, and the planned strength was 6,600 horse, 1,000 dragoons, and 14,400 foot. (In practice these figures proved optimistic.) Cromwell's training methods were influential during the formation of the Army, and many sensible measures were adopted which brought order out of the previous chaos. Since most industry and trade was in Parliamentary hands, a serious attempt could be made to outfit and equip the regiments uniformly. (It is from the New Model Army of 1645 that the British soldier's traditional red coat takes its origin.) The discipline and training of the pike- and matchlock-equipped foot regiments were improved, but it was the cavalry arm which was to prove decisive, and which is always associated with Cromwell's name. The troopers were formed in regiments of about 600 men. They were protected by heavy buff-leather jackets, capable of turning a glancing sword-cut, and by breast- and back-plates. The usual headgear was a 'lobster-tail pot', a helmet of burgonet type with a long neck-guard, a peak, and a face-cage. Some troopers carried carbines, but the usual armament was a three-foot straight-bladed double-edged broadsword and a pair of wheel-lock or flint-lock holster pistols. In the attack the pistols were fired just before contact, and the troopers then charged home with the sword, which was the primary weapon. This emphasis on the shock action of the sword was in keeping with contemporary Continental developments under Gustavus Adolphus, Turenne, and Condé (qq.v.). The great advantage enjoyed by Parliamentary commanders from 1645 onwards was the superior discipline of their regiments, which could usually be relied upon to obey the recall ready for a second charge. The keynotes of the New Model Army were discipline, sound training, standardised equipment, and a sober professional approach to the business of warfare. The strong religious element in the indoctrination of the troops was a major factor in the maintenance of order.

Following the Self-Denying Ordinance, Cromwell (as a Member of Parliament) prepared to resign from active military life. But his gifts were obviously indispensable to the New Model Army, and shortly before the Battle of Naseby Fairfax summoned him to fill the post of lieutenant-general of cavalry and second-in-command. At Naseby on 14 June 1645 he commanded the cavalry on the right wing of the 14,000-strong Parliamentary army, facing Prince Rupert of the Rhine (q.v.) with 10,000 men. Rupert's attack on the Parliamentary left wing cavalry under Ireton was successful, but the Royalist horse then engaged in a wild pursuit. The rest of the Parliamentary line held; while Fairfax rallied the infantry in the centre, Cromwell fought off his immediate adversaries, regrouped, and charged to the support of the infantry. By the time Rupert's cavalry returned to the field after plundering the enemy baggage train, the battle was lost for the King. After the victory of Naseby Cromwell campaigned in the south-west, fighting at Langport in July, and in the Siege of Bristol. A notable incident occurred after his successful storming of a stubborn Royalist garrison at Basing House in Hampshire; some 300 survivors were put to the sword, as a deterrent to other near-by garrisons. This grim but strict interpretation of the traditional laws of warfare – which allowed the slaughter of a garrison which refused quarter and resisted after practicable breaches had been made – was not lost on the Royalists. Cromwell rejoined Fairfax at Exeter in October 1645 and wintered in the south-west. In 1646 he took part in the final stages of the Siege of Oxford which heralded the end of the first Civil War. He then returned to London and resumed an active career as a Member of Parliament.

There is no space in this book for a summary of the complex state of the country in the period between the Civil Wars. Cromwell worked hard for a constitutional settlement with the King, and also on behalf of the army, which had many legitimate grievances against Parliament. His record as a conciliator during this period is unchallengeable, and it was only the intransigence and deceit of the King and the extremism of his factional enemies which forced him, in his turn, to take up an extreme position. He became closely identified with the army party, and from the outbreak of renewed Royalist risings in early 1648 he was the King's implacable enemy. He was ordered to Wales by Fairfax, and in July took Pembroke Castle after a seven-week siege. A Scots army was advancing into the north of England to link up with Royalist forces in that area, and immediately after taking Pembroke Cromwell made a forced march north-eastwards. His infantry covered the 260 miles to Pontefract, Yorkshire, in just twenty-seven days. On 16 August, after being joined by another force under Lambert, but still at a considerable disadvantage numerically, Cromwell attacked the Royalists at Preston. He defeated them in an infantry fight, and then moved immediately against the Scots, who were near Warrington. He virtually destroyed them at Winwich Pass, taking some 6,000 prisoners; the Scots cavalry were subsequently forced to surrender to Lambert in Staffordshire on 25 August. This lightning campaign, in which superior forces were attacked and destroyed piecemeal, was probably Cromwell's masterpiece as an independent commander in the field.

Cromwell now cleared the north of England of Royalists and marched into Scotland, entering Edinburgh on 4 October. He agreed terms with the Marquis of Argyll, left troops to enforce them, and turned south once more. Returning to London in December 1648, he was instrumental in forcing the trial and execution of Charles I in January 1649. He took the not unreasonable view that Charles had proved himself incapable of keeping faith and that peace would never be achieved while he lived. Cromwell was then chosen as first chairman of the Council of State, the organisation by means of which the purged Parliament governed England as a republic.

Appointed to the post of Lieutenant-Governor of Ireland, Cromwell landed in that country in August 1649 to campaign against Royalist sympathisers who were then in arms. On 10–11 September he stormed the town of Drogheda after the garrison had refused terms. Once again he applied the strict laws of war with an implacability which shocked even his own men – the garrison was slaughtered. He justified this crime on the grounds both of religion and of military expediency, but although some near-by strongholds were cowed into early surrender, others (including Wexford and Waterford) held out to the end. Nevertheless by the end of 1649 most of the coast was in his hands, and he turned inland, taking a number of other fortified positions. In May 1650 a new threat arose in Scotland; the Covenanters had refused to recognise the English republic and had proclaimed Charles II king. Cromwell returned to England shortly after taking Clonmel.

An English army was sent to invade Scotland, and since Fairfax – a convinced opponent of the execution of Charles I – refused to command it, Cromwell was by Act of Parliament appointed Commander-in-Chief and captain-general. He led 10,500 foot and 5,500 horse into Scotland on 22 July 1650, but the Scots refused battle and he lost nearly half his men through disease and skirmishing casualties. Short of supplies, he eventually led his shrunken army towards the port of Dunbar to await reinforcement. The Scots, under Leslie, outmanœuvred him and took up positions on Doon Hill some two miles from Dunbar. They advanced from these positions on 2 September, and the next day Cromwell launched a daring dawn attack. His first wave was repulsed, but the perfectly-timed committal of his reserve turned the balance, and the Scots were decisively defeated; they lost some 3,000 dead and 10,000 captured.

A bout of malaria, contracted in Ireland, prevented Cromwell taking personal command of operations against the remaining Scots armies until well into 1651. In June he took the field again, outflanking the Scots and threatening their rear. They moved south into England as he had foreseen, and he followed. He caught up with them – and the young Prince Charles – at Worcester on 22 August. Splitting his force into two parts, and leading one personally across a bridge of boats on the Severn, he attacked the Scots from two directions at once. When they tried to break out eastwards he recrossed the river and personally led the decisive counter-charge. It was his final, and crowning victory. The future Charles II fled to the Continent, and from the end of 1651 Cromwell was *de facto* master of England. He still attempted to work within a constitutional framework, however, and it was not until 1653 that he stepped into the power vacuum and took over the control of the country as Lord Protector. Under his rule English arms – both at sea and on land – gained the respect of the Continental powers. He sent troops to fight the Spanish in the Low Countries under the terms of a French alliance, and the battle-hardened veterans of the New Model Army won several notable victories.

Cromwell died in London on 3 September – a mystic date in his life – of 1658. His remains were desecrated at the time of the Restoration two years later. His Calvinist religious principles, his rule by military law, and his occasional cruelties have tended to make his name a symbol of repression and dictatorship in the popular imagination. This is perhaps inevitable, but it is hardly just. Few men of his age were firmer or more patient supporters of Parliamentary principles against the genuinely arbitrary dictatorship of the monarchy. His New Model Army is usually considered to mark the beginning of the British professional Army of modern times. His own tactical skill – innate, and springing from only a few months of active experience in the field – was remarkable; and the buff-coated regiments which he raised and led remain one of the most impressive manifestations of English military prowess in our whole history.

Crook, George (*1829–90*), *Major-General*

George Crook was America's most accomplished Indian-fighting general, but in his attitude to the Indian he was a good deal more enlightened than the considerable majority of his countrymen. He was a frontier general who exercised command from his saddle in the desert rather than from a desk in headquarters; he understood the mentality of his enemy better than that of some of his civilian contemporaries, and it is quite possible that he preferred it.

Crook was born near Dayton, Ohio, on 23 September 1829. He entered West Point on 1 July 1848 and graduated in 1852. He was commissioned as a lieutenant of infantry, and served in the north-west until the outbreak of the Civil War. It was a hard apprenticeship which involved exploration of a huge and unknown country, as well as sporadic action against Indians who sought to resist the spread of white settlers. In September 1861 he was appointed colonel of volunteers, commanding the 36th Ohio Infantry Regiment. He served initially in West Virginia, and in May 1862 was raised to the regular rank of major, largely for his defeat at Lewisburg of a Confederate force under General Heth. The following August brought him the volunteer rank of brigadier-general. He led his brigade in the Kawanha Division of IX Corps in the Antietam campaign, fighting at South Mountain and Antietam and being promoted to regular lieutenant-colonel for his service in the latter battle. In 1863 he led a cavalry division in the Army of the Cumberland and took part in the Chickamauga campaign. He pursued the Confederate General Wheeler's cavalry corps and defeated it at Farmington, Tennessee on 7 October 1863, and this action brought him his regular colonelcy. He was sent to West Virginia once more in February 1864, and that spring carried out for Grant (q.v.) a most able programme of disruption of the Confederate rail communications. This service was rewarded by promotion to regular brigadier-general and the command of the Department of West Virginia. He led a corps in Sheridan's Army of the Shenandoah and fought at Winchester, Fisher's Hill, and Cedar Creek, subsequently being promoted to major-general of volunteers, and soon afterwards to regular major-general. Sheridan gave him a cavalry division for the final campaign of 1865, and he was present at Dinwiddie Court House, Sailor's Creek, Farmville, and Appomattox.

The standing army was cut to 39,000 in 1869, and again to 25,000 in 1873. Many officers had to revert to junior rank in order to stay in the Army at all, and career prospects were not glittering. The army on the frontier, which was faced with the task of policing the ever-expanding fringes of civilisation, and repairing the damage which was all too often inflicted by the venal and inefficient Bureau of Indian Affairs, never numbered more than 15,000 men – roughly one soldier for every fifty square miles. Pay was low, conditions were extremely harsh, and the standard of higher direction from Washington was miserable. The Indian tribesmen were the finest light cavalry in the world; trained from childhood to survive in country which was terribly punishing to white soldiers, masters of concealment and raiding, fast-moving and fearless, they were a formidable enemy.

The post-war reorganisation of the Army sent Crook to Boise, Idaho, as lieutenant-colonel of the 23rd Infantry. He served in a series of campaigns against the tribesmen in Idaho, southern Oregon, and northern California over the next three years, and met with considerable success. In 1871 Grant sent him to end the simmering trouble with the Apaches of northern Arizona. In two years Crook succeeded in bringing peace to the area, winning a measure of respect from his enemy as well as from his colleagues; and in 1873 he was promoted direct from lieutenant-colonel to brigadier-general, a most unusual distinction in that period of heartbreakingly slow promotions. Quiet, modest, and approachable, Crook was by all accounts a determined fighter but a magnanimous victor. He did not subscribe to the extreme views of many of his contemporaries, who regarded the Indian as vermin to be exterminated; indeed, he advocated the granting of full citizenship rights to all who accepted Government authority, a suggestion greeted by many with incredulous outrage. Crook was an honest enough man to understand and sympathise with the Indian's hopeless fight against white encroachment, while still performing his duty to his country to the best of his considerable ability.

In 1875 Crook was given command of the Department of the Platte in anticipation of trouble with the Sioux and Northern Cheyenne following the discovery of gold in the Black Hills of Dakota. He was prominent in the great Sioux War of 1876–7, in which Custer (q.v.) and his 7th Cavalry were annihilated on the Little Big Horn. In the bitterly-cold early spring of 1876 Crook led 800 men on a forced march into the Big Horn Mountains, and was initially successful against a far larger force of Sioux (under Crazy Horse of the Oglala tribe) on the Powder River on 17 March. Forced to withdraw, Crook caught up with Crazy Horse again on 17 June on the Rosebud; he fought a bitter but eventually indecisive battle against some 5,000 tribesmen, at odds of at least one to five. During this campaign Crook remained in the field with his troops for a year, suffering great hardships.

In 1882–6 Crook served with distinction against the Apaches in Arizona and New Mexico. He enjoyed fairly rapid success against those tribes which he had fought ten years earlier, but he had not previously encountered the Chiricahua Apaches. Their chieftain Geronimo led them into the Sierra Madre Mountains of northern Mexico, a desolate refuge not yet penetrated by either United States or Mexican forces, from which they raided north and south of the border. In 1883 Crook led a column of troops into the Sierra Madre and brought some 500 of the tribe back to the reservation. (Two years later Geronimo again headed for the mountains with about a quarter of the tribe, which was slowly worn down in a series of actions until the last two dozen surrendered to General Nelson Miles.) In 1886 Crook returned to the Department of the Platte, which he commanded until April 1888. He was then promoted major-general, and took over command of the Division of the Missouri, with headquarters at Chicago. He died there on 21 March 1890.

Cumberland, William Augustus, Duke of (*1721–65*)

Suppressor of the 'Forty-five Jacobite Rebellion, 'Butcher' Cumberland was successful in the bloody and decisive Battle of Culloden but, in opposition to the 'professional' French marshals Saxe (q.v.) and Estrées, proved to be as unlucky as he was incompetent and unimaginative.

Born on 15 April 1721, Cumberland was the second surviving son of George Augustus, Prince of Wales (later King George II of England). Created Duke of Cumberland at the age of five, he was educated for a naval career, but soon expressed a preference for the Army and, at the age of nineteen, was appointed colonel of the Coldstream Guards, being gazetted major-general two years later. In 1743 he fought with outstanding bravery at the Battle of Dettingen, and in 1745 was promoted captain-general of all British land forces and Commander-in-Chief of all British, Austrian, Dutch, and Hanoverian forces in Flanders – a post that had been in abeyance since the time of Marlborough – with Konigsegg as his adviser. In an attempt to relieve Tournai his forces were defeated in the Battle of Fontenoy on 11 May 1745 by the great Marshal de Saxe.

Towards the end of that year Cumberland was hurriedly recalled to England to oppose Prince Charles Edward's Jacobite rebellion and invasion, and followed the rebels in their subsequent retreat north from Derby to Carlisle. Cumberland returned to London in January 1746, but when news reached him of General Henry Hawley's failure to relieve Falkirk he returned north and set about preparing his army for a defensive battle against the Jacobites.

Prince Charles at first awaited the Hanoverians on Drummossie Moor in Inverness-shire but, hearing that Cumberland was celebrating his twenty-fifth birthday at Nairn, he determined to assault the British camp by marching his Highlanders over the moors and delivering a surprise attack. The long night march proved too much for the half-starved Scots, and dawn of 16 April 1746 found the 5,000 Jacobite Highlanders drawn up in two widely-separated lines opposite 9,000 Hanoverians, also in two lines. A preliminary artillery barrage caused terrible casualties among the Pretender's men, who sought to escape by charging forward against the disciplined ranks of Cumberland.

Cumberland's training had been methodical; he had taught his soldiers to engage 'diagonally right', thereby enabling each man to evade his opponent's sword arm and to thrust below it. On the left flank the Highlanders' attack was ineffective; on the right they succeeded in piercing the front rank but were taken in flank by supports. Seeing the flank attacks fail, the Jacobite centre faltered and gave way, under heavy fire from Cumberland's muskets, which did great execution. His dragoons meanwhile broke the Jacobite flanks and attacked the second line from the rear. About 1,000 Highlanders died in the Battle of Culloden – the last pitched battle between opposing armies to be fought on British soil – compared with no more than fifty killed in Cumberland's army. The Duke remained three months in Scotland to organise the systematic search for and pursuit of Jacobite refugees; these his men mercilessly slaughtered, earning for their commander his lasting nickname, 'the Butcher of Culloden'.

Returning to Flanders in 1747 Cumberland was again defeated by Marshal de Saxe, this time at the Battle of Lauffeld, near Maastricht, on 2 July. In this instance he was deprived of a possible victory by the failure of the Austrians. At the end of the war Cumberland retired from the Army to pursue a life of leisure at home, establishing his racehorse stables and, incidentally, founding the Ascot meeting.

On the outbreak of the Seven Years' War Cumberland was recalled to the colours and dispatched to defend Hanover against the French. At the Battle of Hastenbeck, near Hameln, on 26 July 1757 he was defeated by superior forces under Marshal d'Estrées, and his army was subsequently pinned between the North Sea and the River Elbe. On 8 September he acquiesced to the terms of the Convention of Klosterzeven, by which he undertook to evacuate Hanover – an agreement immediately overruled and repudiated by George II. Cumberland was recalled in disgrace, resigned all his military commands, and retired to live at Windsor.

During the early years of George III's reign he did much to recover his early prestige and popularity by contributing to the fall from power of the Earl of Bute and George Grenville, so that when he died at the early age of forty-four (on 31 October 1765) his military shortcomings had been almost entirely forgiven – if not forgotten.

Custer, George Armstrong (*1839–76*), *Major-General*

It is highly ironic that Custer's Last Stand – an affair involving a mere 250 dead, precipitated by the idiot conceit of one of the worst commanding officers ever to mishandle a regiment – should be as well known to the man in the street as Hastings, Waterloo, or the D-Day landings. To place the incident in its correct perspective, it might be as well to preface this entry by remarking that in January 1879 the Zulus wiped out 1,450 British and Natal infantry at Isandhlwana; that in March 1896 the Abyssinians killed some 5,300 Italian soldiers at Adua; and that as recently as July 1922 the Rif Berber tribesmen of Morocco killed General Silvestre and more than 12,000 of his Spanish troops at Anual. Even as a blunderer Custer does not rate very high.

Custer was born on 5 December 1839 near New Rumley, Ohio, and entered West Point in July 1857. His years at the Academy were remarkable for the number of his accumulated demerits, and for his hairbreadth escapes from dismissal. In a class much reduced in size by cadets heading south for a Confederate commission, he passed out thirty-fourth of thirty-four in June 1861, and was commissioned second-lieutenant in the 2nd U.S. Cavalry. He was present at the First Battle of Bull Run, but did not fight. Transferred in August to the 5th Cavalry, he was promoted first-lieutenant in July 1862; but since the previous month he had been serving as an aide to General McClellan, with the temporary rank of captain. He remained with McClellan until March 1863, and in June of that year was appointed brigadier-general of volunteers at the age of twenty-three. He distinguished himself in command of the Michigan Cavalry Brigade at Gettysburg two days later, leading a charge of the 7th Michigan which he later described in his report in these terms: 'I challenge the annals of war to produce a more brilliant charge of cavalry.' Custer served with the Army of the Potomac throughout the 1864 campaign, and gained further distinction in the Shenandoah Valley battles. He ended the Civil War as a major-general of volunteers leading a cavalry division, a flamboyant young showman with long red hair and a taste for velvet jackets with gold braid from wrist to elbow. He had an instinct for publicity, and adored the limelight; he had become much more widely known to the public than many officers of much greater age, experience, and skill. He was autocratic and conceited, and had no proper experience of that most essential school for an army officer – the undramatic routine of the junior company officer, learning about men, making mistakes, and taking orders.

Custer's first post-war command ended in the disbandment of the 3rd Michigan Cavalry after a mutiny caused, at least in part, by his heavy-handed discipline; it was a time when many volunteer regiments were pressing for demobilisation, and Custer had attempted to recall the troopers to his conception of their duty by the use of the lash. He mustered out of the volunteer service in February 1866 and reverted to his substantive army rank of captain – although he and his supporters tended to continue using the wartime title 'General Custer'. He negotiated briefly with President Juarez about the possibility of his taking the chief command of the Mexican cavalry; he was offered, but refused, the colonelcy of the negro 9th Cavalry; and in July 1866 he took up the lieutenant-colonelcy of the newly-organised 7th Cavalry. This was a *de facto* regimental command, as two successive colonels of the Regiment spent much of their time on detached duty.

In early 1867, at the conclusion of an indecisive expedition under General Hancock, Custer was ordered on a long scout with several companies of the 7th. His behaviour during this independent operation led to a court martial's finding him guilty later that year of absenting himself from his command without leave and using some of his troopers as an escort while on unauthorised personal business; abandoning two men reported killed on this march, failing to pursue the Indians responsible, and failing to recover the two bodies; ordering a party sent after deserters to shoot them dead, with the result that three were wounded and one killed; and unjustifiable cruelty to the three wounded men. He was suspended from rank and pay for a year. Lack of an experienced substitute led to his reinstatement before the sentence was completed, but ill-feeling divided the officers of the Regiment for several years thereafter. Over the next two years some parts of the Regiment saw minor skirmishing with Indian bands; Custer was not present on these occasions, although his published accounts greatly exaggerate the size and drama of the clashes. The 7th saw much hard riding, but had little contact with the Indians until the Battle of the Washita in November 1868. This was an attack on an encampment of Black Kettle's Southern Cheyenne in which Custer reported killing 103 warriors and 'some' women and children. The Cheyenne called him 'Squaw-Killer' from the Washita onwards, and one is tempted to doubt his arithmetic. Further bad feeling was caused within the Regiment during this action; a major, the sergeant-major, and fifteen other men were cut off and killed, apparently due to negligent orders. There was no further fighting for the 7th in this campaign, and in 1870 the Regiment was moved to the South to support the civil power at a time of unrest.

In the spring of 1873 the Regiment was ordered to Dakota, coming under command of Colonel D. S. Stanley at Fort Rice. Stanley was assembling a large escort for a party of Northern Pacific Railroad surveyors on an expedition to the Yellowstone river. In August there were some sharp skirmishes with the Indians, all of them successful for the Army. Custer was placed under arrest for insubordination during this operation, but Stanley was persuaded by mutual friends to take the matter no further. In 1874 the Government sent a 'scientific' expedition into the Black Hills country, which belonged both by right of occupation and by treaty to the Sioux and Northern Cheyenne Indians. Custer led an escort comprising ten companies of the 7th, two infantry companies, sixty scouts, and a detachment of Gatling guns. His orders specified a reconnaissance connected with the planned construction of a fort in the Black Hills to guard future railroad extension. The size of the 1,200-man force sent on this expedition suggests other motives. Custer is believed to have played a large part in spreading stories of gold finds during this operation, leading to a civilian gold-rush to the Black Hills. The Army column was far too strong for the Indians, but Crazy Horse's warriors harried the lawless bands of prospectors who flooded into their hills after the reports spread. In 1875 the Government tried without success to persuade the Indians to sell the Hills and withdraw to agencies. In 1876 the attempt at persuasion was abandoned, and preparations were made for a straightforward military campaign. The attacks on the prospectors – who had no right to intrude on Indian territory – were the excuse; the aim was to cow the Indians into agreeing to give up their lands. General Alfred Terry was given overall command of a strong mixed force, of which the main mobile element was to be the whole of Custer's 7th Cavalry, some 600 men strong.

Custer was allowed to lead his Regiment on this expedition only after the intercession of the popular and kindly Terry. He was in disgrace, having angered President Grant (q.v.), the Army Commander, General William Sherman (q.v.), and his division commander, Sheridan, by his behaviour in connection with a military-political scandal then rocking Washington. The issue was complex, but centred on alleged irregularities in the award of post traderships. Custer, available as ever to the Press, had talked a great deal about his inside knowledge of sharp practice in high places. Questioned under oath by a committee of inquiry, his testimony turned out to be nothing but lengthy and theatrically-phrased hearsay, most of which he later withdrew. He had embarrassed the Government and the Army, and did nothing to improve his standing with his superiors by leaving Washington against orders and returning to Fort Lincoln. After a stinging reprimand he was finally allowed to take the field.

Terry left Fort Lincoln on 17 May and proceeded to the Yellowstone river. Apart from the 7th, he had three infantry companies and a Gatling detachment, with numerous Arikara and white scouts. On the Yellowstone he met and took under command some 450 infantry and cavalry of another column under Colonel Gibbon. The Indian main force was thought to be somewhere to the south, but Terry's operations were complicated by the simultaneous manœuvres to the south of him of General Crook (q.v.). It was hoped that the two forces could converge and catch Crazy Horse's warriors between them, but Crook had been forced to abandon this plan by his severe handling at the Battle of the Rosebud.

The Yellowstone river runs roughly east and west at the point where the disaster occurred. Into it from the south run several tributaries: two of them, the Rosebud and the Big Horn (nearest to it on the west), run roughly parallel from the south-west. Some miles up the Big Horn the Little Big Horn flows into it from the south-east – the side nearer the Rosebud. The Indian camp was thought – correctly – to be in the valley of the Little Big Horn. On 22 June Terry split his forces. Custer was to ride south up the Rosebud, following the trail of a large force of Indians discovered by scouts; it was anticipated that at some point this trail would turn west over the hills towards the Little Big Horn. Gibbon was to move south up the Big Horn. Scout couriers were to keep Terry, Gibbon, and Custer in close touch. If the Indian trail turned towards the Little Big Horn, Custer was specifically ordered to *continue south* to guard against any possible break-out in that direction. A converging attack by the two columns would then be launched.

On 24 June Custer found that the trail did indeed turn towards the Little Big Horn. He chose not to follow orders, but to turn west along it. On the 25 June he halted on high ground from which he could see down into the Little Big Horn valley; many columns of campfire smoke could be seen, and a vast herd of grazing ponies. It was the encampment of the Sioux and Northern Cheyenne Indians, containing perhaps 5,000 warriors and 8,000 to 10,000 others. Custer's party was seen by Indian sentries, and he decided to attack immediately. He split his 600 men into three main parties and a small guard for the impedimenta. Major Reno was ordered to take three companies (112 men) into the attack at the southern end of the encampment, while Captain Benteen took three companies (125 men) off to the west. Custer himself planned to take five companies along the bluffs to the north, on the opposite side of the stream from the camp, and then swing down to attack the other end of the camp. He had no idea of the size of the enemy camp, or of whether there was a practical route down from the bluffs to the north. In fact the camp stretched unbroken for two miles, and Custer was never able to cross the stream. Reno's force attacked as ordered, but was soon driven back, and only with difficulty made its way back to high ground; this force was joined by Benteen, and managed to hold out on a hilltop until relieved by Terry the next day. Custer's men were driven back from the stream by Crazy Horse and his Oglala Sioux; they were trapped on a slope, cut to pieces, and overwhelmed by the Oglalas and a second group of the Hunkpapa Sioux led by Gall. In all, the Battle of the Little Big Horn cost the Army 250 dead and 44 wounded. The Sioux and Cheyenne Indians could probably have destroyed Terry and Gibbon as well, but they had suffered heavy casualties themselves and preferred to withdraw southwards, to celebrate and recover in the Big Horn Mountains.

There have been many attempts to explain Custer's behaviour in acceptable military terms, but none are convincing. He deliberately disobeyed orders, and then split his small command in the face of an enemy of unknown but obviously considerable strength. His previous career leads one to the conclusion that his motives were strongly coloured by a desire for personal glory. No doubt his luckless troopers sold their lives dearly – as do any soldiers worthy of the name when all hope is gone; but it takes a more-than-usually-impressive exercise of wilful blindness to historical fact to turn the Little Big Horn into anything but an unnecessary disaster.

Davout, Louis Nicolas, Duc d'Auerstädt, Prince d'Eckmühl (*1770–1823*), *Marshal of France*

Davout earned a place among the leaders of Napoleon's marshalate not by any single brilliant exploit, or by any blazing originality of character, but rather by stolid and reliable professional competence. A heavy-featured man whose manner has been described as dull to the point of surliness, he must have provided a sober contrast to such fiery personalities as Murat, Ney (qq.v.), and Lannes. Nevertheless, when it came to preparing and maintaining a command at peak efficiency, and employing them in the field in the most economical and effective manner, such dashing *sabreurs* were not in his class. Davout was also in a minority among the marshals in possessing the virtue of complete personal integrity.

He was born on 10 May 1770 at Annoux. His father was a junior cavalry officer, and Davout was destined for a military career from the start. He passed out of the Royal Military School at Paris in February 1788, and joined his father's regiment as a *sous-lieutenant*. An enthusiastic supporter of the Revolution, Davout led an eventful and sometimes uncomfortable life during the years 1789–91, but emerged as lieutenant-colonel commanding a battalion of volunteers raised in his home region. The years of the Revolutionary Wars were no less packed with incident.

Davout was at Neerwinden on 18 March 1793, when Dumouriez's French Army was defeated by the Austrians. Dumouriez was charged with treason shortly after and fled to the Allies, and Davout was involved in an armed attempt to arrest him. Promoted general of brigade that July, Davout spent a year in retirement during the height of the military and political confusion which accompanied the Terror, but returned to the colours as commander of a cavalry brigade in the Army of the Moselle late in 1794. He was captured at Mannheim in September 1795, but was paroled and subsequently exchanged in time to fight at Kehl in November 1796, and at Offenburg the following April. Initially selected for the Army of England, he was transferred to Bonaparte's headquarters for the Egyptian campaign, and fought in several engagements, including the Pyramids and Aboukir. While making his way back to France he was captured and briefly imprisoned by the British at Livorno. On his eventual return he was promoted general of division in July 1800.

Davout led the cavalry of the Army of Italy in the second half of 1800, and the following year was made Inspector-General of Cavalry and commander of the foot grenadiers of the Consular Guard. Raised to the rank of marshal in May 1804, he was made one of the four Colonel-Generals of the Imperial Guard shortly afterwards. In the 1805 campaign he led the III Corps of the *Grande Armée*, winning the Battle of Marienzell on 8 November. His corps played a major part at Austerlitz, after marching eighty miles from Vienna in fifty hours. The battle from which Davout took his ducal title, Auerstädt, was fought on 14 October 1806 – the day Napoleon routed another Prussian army at Jena. Davout led 27,000 men across the Prussian lines of communication, and fought a text-book defensive battle against 63,000 Prussians under the Duke of Brunswick. After six hours he had fought the enemy to a standstill; he then counter-attacked, and the Prussians' final collapse was accelerated by the belated arrival of Bernadotte's corps in their rear. It was the kind of engagement where Davout's competence and coolness shone – it is hard to imagine a temperament more foreign to the traditional idea of a Napoleonic cavalry officer, and one is forced to believe that Davout's original service in that branch was due to family tradition rather than any innate suitability. He was wounded at Eylau, where his steady pressure on the right wing was one of the few stable elements in a confused battle. In July 1807 he was made Governor-General of the Grand Duchy of Warsaw, and he received his dukedom, and considerable financial grants, the following March. He again commanded the III Corps in the 1809 campaign, distinguishing himself at Eckmühl on 22 April, when his command held firm in an isolated position until the arrival of support allowed him to deliver a telling attack. At Wagram on 5–6 July he had a horse killed under him while driving back the Austrian left wing. For two years after the conclusion of the campaign he held senior commands in Germany.

Davout led the I Corps of the *Grande Armée* into Russia in 1812; as always, his corps was a model of organisation, discipline, and planning. He immersed himself in detail, and issued a stream of far-sighted orders to meet every eventuality. At Borodino it was Davout who made a determined appeal to Napoleon that the main manœuvre should be a right flanking movement rather than a frontal assault; his advice was ignored, and both sides suffered enormous casualties without reaching a decisive result. Davout was wounded, and again had horses killed under him. During the early stages of the retreat from Moscow his corps formed the rearguard, and on 3 November it was cut off and badly mauled at Fëdorovskoye. Considering Davout to blame for this, Napoleon gave Ney the rearguard. By 27 November Davout's corps numbered about 3,000 men, out of the 72,000 he had led across the Niemen.

In 1813 Davout fought at Dresden and Lauenburg; and he subsequently defended Hamburg with such determination that it was not until he received news of the abdication in May 1814 that he evacuated the city. He lived in retirement until Napoleon's return from Elba, and was then made Minister of War, a post he held until 8 July 1815. He was not given a field command during the Hundred Days, but Napoleon entrusted the defence of Paris to this most reliable and self-sufficient of his marshals. At the time of the second abdication he still had some 100,000 men under command, and successfully checked Blücher when the Prussians appeared before the city on 30 June; but he was operating in a political vacuum, and the end was inevitable. He signed the Convention of Paris on 3 July, and shortly afterwards went into exile. Restored to his rank and titles in 1817, he lived quietly until his death from consumption in Paris on 1 June 1823.

Dayan, Moshe (*1915–81*), *Lieutenant-General*

Israel's best-known soldier, and one of the leading political personalities of the Israeli state, Gen. Dayan became the symbol for the world of immense professional skill and flexibility which gained for his beleaguered country the remarkable victories of 1956 and 1967.

He was born of Ukrainian Jewish parents on 4 May 1915 at Degania – the first child born in this important settlement, which has been termed 'the mother of kibbutzim'. His childhood was hard, owing to the physical demands of the life, the harassment of the settlers first by the Turks and later by the Arabs, and his father's frequent absences – Shmuel Dayan was extremely active in political life. In 1920 the family was among the first to move to the new settlement at Nahalal, near Nazareth. During his childhood and adolescence Moshe displayed qualities of independence and ambition, together with a precocious cultural sensitivity which he probably owed to his mother Dvorah, an educated woman of literary tastes. From 1930 onwards he, with his contemporaries, received rudimentary military training in the Haganah defence organisation, and he took part in minor skirmishes with Arab groups; it was a time of tension, but not of serious fighting, and casualties worse than bloody heads and bruises were rare. Between 1933 and 1939 the troubles intensified, however, and in the late 1930s the Jewish groups were losing many men killed and wounded every month. Moshe's marriage in 1935 brought him into contact with another influential section of the Jewish intelligentsia, and he made a six-month study trip to London. He was unhappy there, and the growing fury of the clashes between Arabs and Jews soon attracted him home. In 1936 a new policy led to the attachment of Haganah guides to British units policing Palestine. Moshe was attached for short periods to the King's Own Scottish Borderers and to the King's Own Yorkshire Light Infantry; he is reported to have been unimpressed by the rigid procedures of British military operations at that time. In 1937 Sergeant Dayan commanded one of the Mobile Guards of the Jewish Settlement Police. Impatient of discipline and routine, he was nevertheless keen to learn useful military skills. He became an instructor, and his manual on fieldcraft – in which he criticised 'bull', and stressed intelligent use of terrain, weapon training, and imaginative aggressiveness – was very influential. He saw considerable action against Arab terrorists and saboteurs, and met Orde Wingate (q.v.), who was at that time organising the Special Night Squads. Dayan and Wingate were kindred spirits in their tactical doctrine, and they fought together on several occasions.

In 1939 world-wide political considerations led the British Colonial Office to adopt a more pro-Arab policy, and to limit the activities of the Haganah. Friction between the Jews and the authorities intensified; and in October 1939, in an incident which reflects little credit on the British, Dayan and forty-two other Haganah men were taken into custody on arms charges. (It should be remembered, however, that the current activities of the murderous Jewish splinter organisation known as the Irgun had greatly contributed to the British attitude.) The group were interrogated with some brutality, tried, and sentenced to ten years' imprisonment. Dayan displayed great qualities of calm leadership while in custody. The sentences were widely criticised, in high British military and political circles as well as among international Jewry, and were later halved. In fact the Haganah men served only until February 1941, initially in the grim fortress of Acre and later at the Mazra'a work camp. A saner policy of co-operation was re-established; and a few months later Dayan was appointed commander of one of two Haganah companies which operated against the Vichy French in Syria alongside British and Commonwealth units. He distinguished himself in action on 7–8 June 1941 while accompanying an Australian patrol in the capture of a fortified French police post near Iskanderun. During the subsequent enemy counter-attack he manned a machine-gun on the exposed roof of the post. Then, while watching the enemy through field-glasses, he was severely wounded – a bullet struck the binoculars, driving glass and pieces of metal casing deep into his skull; his left eye was destroyed, as was much of the bone-structure of the socket. (The extensive bone damage – and not, as has sometimes been alleged, a taste for self-dramatisation – was the reason why Dayan wore his famous eye-patch; the socket would not support a glass eye.)

His injuries, and his slow adjustment to them, led to a fustrating period in Dayan's life. He fell behind his contemporaries in the Haganah, and despaired of ever again having an active military career. In 1941–2 he worked with a department which collaborated with British Intelligence in setting up clandestine projects involving Jewish personnel, but in 1943–7 he was in the doldrums. His political interests began to flower during this period, and he made many useful contacts as his fame spread after the Syrian episode. In 1947 he was appointed to a staff post in the Arab Affairs department of Haganah, combating the activities of Arab gangs inside Palestine and in Syria, and achieved some important successes. He had no combat command on the outbreak of the War of Independence, but on 18 May 1948 was appointed commander of the Jordan Valley sector in the face of serious attacks by superior Syrian forces. He successfully defended his birthplace of Degania with a tiny *ad hoc* force, handicapped by a tangled chain of command and chronic shortages of every necessity. A few days later Major Dayan was given command of the new 89th Battalion, and set about recruiting volunteers and acquiring vehicles in a thoroughly unorthodox manner, brazenly poaching men and machines from other units without the slightest authority. The 89th soon became known as a colourful and volatile unit, and distinguished itself in several actions during July 1948. Notable among these was an S.A.S.-type raid on Lydda and Ramle, in which Dayan's battalion, mounted in a small and motley assortment of home-made armoured vehicles, jeeps, half-tracks, and a single Jordanian armoured car personally salvaged under fire by the C.O., swept through the city and back again, firing everything they had at the startled Arab Legion garrison. At this period it is fair to say that Dayan's reputation was that of an extremely courageous, imaginative, and aggressive battalion commander – and a disobedient and professionally ill-educated subordinate.

In August 1948, promoted lieutenant-colonel, Dayan was appointed commander of the Etzioni Brigade in the Jerusalem sector; the formal command of the city came in November. This post, which he held until October 1949, marked the beginning of his serious career and his transition from a partisan leader into a soldier-statesman. He was deeply involved in high-level negotiations, both with the U.N.O. authorities and direct with the Jordanians, in the period which led up the cease-fire of July 1949. In October 1949 he was promoted major-general and appointed O.C. Southern Command, with headquarters at Beersheba. This period immediately following the end of the War was a difficult one for the Israeli army. A feeling that peace had been achieved led to a gradual revulsion against militarism throughout Israeli society. The transformation of a popular volunteer army fighting for national survival into a viable regular army was accompanied by severe birth-pangs. Psychologically hostile to the undemocratic traditions of conventional armies, Israelis tended to exercise to excess their individual rights when presented with military orders. Dayan himself was troubled both by this conflict, and by his own lack of formal training. He was a popular commander, informal and humorous, impatient of 'proper channels', and direct to a fault; but he was deeply conscious of his own shortcomings.

During 1950–1 a proper system of professional instruction was established for the officers of Zahal (the Israeli Defence Forces), and Dayan took advantage both of this and of a British command course. He brought a certain amount of criticism on his head by his insistence on practical and immediate action in any situation, at the expense of classic theory; this attitude was characterised by his all-out efforts to win manœuvres which his side were supposed to lose. In May 1952 he was appointed O.C. Northern Command, but only a few months later Ben Gurion personally nominated him for the post of Deputy Chief-of-Staff and Chief of Operations, a post he filled from December 1952 to December 1953. Following a period of tension between the Chief-of-Staff, Lieutenant-General Maklef, on the one hand, and Dayan and the Defence Minister, Lavon, on the other, Maklef was obliged to resign and Dayan took over the senior command.

During his time with Northern Command and as Deputy and later Chief-of-Staff, Dayan made his presence felt throughout Zahal. He played a central part in bolstering the morale of officers and men at this difficult time. He was a law unto himself, impatient of staff procedures, but tireless and direct in all matters of operational readiness. He ruthlessly pruned away 'dead wood' and promoted younger officers of ability. He established the principle that regular officers should be retired compulsorily at the age of forty, to keep the Zahal officer corps young and flexible. He made political pressures for economy in defence an ally rather than an enemy, using them as an excuse to cut the forces' administrative 'tail' in favour of the fighting units. Combat readiness was the goal, and Zahal had some way to go before satisfactory standards could be achieved. At this time the Israeli army was by no means the lean, superbly-trained force of today. It still had many internal problems, lacked experience and determination, and was too inclined to abandon objectives in the face of opposition. During the war of raids and retaliatory columns along Israel's borders, Dayan brought the realities of military life sharply into focus for officers and civilians alike when he preached the doctrine that any commander who abandoned his mission before suffering fifty per cent casualties should be dismissed. Demanding training, constant exercises, a no-nonsense approach to bureaucracy, and a merciless emphasis on the need to attack without counting the cost; these were the hall-marks of the Dayan period. He was instrumental in creating the *élite* paratroop branch, and did everything he could to encourage it. For a year it received all available combat assignments, and public praise in proportion; Dayan was attempting – successfully – to arouse the jealousy of the rest of the army to such a pitch that they would force themselves to match the combat standards of the paratroopers. He established as an absolute principle that under no circumstances whatever would wounded be abandoned to the enemy – a necessary step in view of Arab habits, and a strong factor in building *esprit de corps*. Dayan made great strides in changing the whole character of Zahal in only a few years. In doing so he made powerful political enemies who distrusted his unpredictable and secretive style, his impatience of political interference, and his informal personal mannerisms. Nevertheless, the relationship between Dayan and Ben Gurion was one of complete mutual respect and trust.

The Sinai Campaign of October–November 1956, which coincided with the Anglo-French invasion of the Suez Canal Zone, was a complete vindication of Dayan's philosophy that the whole of Zahal should be able to operate at the level of the commando units of other nations. Speed, surprise, and determination were the keynotes, together with a thorough understanding of the enemy. Dayan took risks which would have been unforgivable against a European enemy, but which paid off handsomely in practice. The campaign opened with a paratroop drop near the Mitla Pass, far behind Egyptian lines, on 29–30 October, coupled with widely-separated border crossings. The Egyptians were led to believe that these were isolated reprisal raids rather than a general invasion, and their response was sluggish and localised. The Israeli columns then achieved decisive breathroughs, racing through the enemy's rear areas, by-passing his strongest positions, and spreading out to disrupt the deployment of his reserves. Despite gloomy forecasts to the contrary, the isolated pockets of Egyptian resistance swiftly collapsed, as Dayan had predicted. The breakthroughs were achieved largely by armoured infantry, and there were no set-piece tank battles. By 5 November it was all over; Israeli units had reached the Canal Zone three days earlier, and on that date they took Sharm el-Sheikh at the mouth of the Gulf of Aqaba. In the aftermath of the war Dayan opposed the agreement of Israel to withdraw from captured territory, but he supported the Government loyally, and continued to serve as Chief-of-Staff until 1958, after which he entered political life in earnest. He became a member of the Mapai party, and entered the Knesset in 1960. He received the portfolio of Minister of Agriculture in 1959 and held it until 1964. The unusual complexity and ferocity of Israeli politics did not daunt him any more than the challenge of physical warfare. Nevertheless, difficulties with the Prime Minister, Levi Eshkol, obliged him to resign in November 1964, and he did not hold office again until he received the portfolio of Defence on 2 June 1967 – only days before the outbreak of the Six-Day War.

The emergency which led to the 1967 War blew up with bewildering speed and caused great public uneasiness. There seemed a real chance that the state might be destroyed, and the Eshkol Government did not enjoy widespread confidence. A political and public campaign for Dayan to be offered the portfolio of Defence gathered momentum; he expressed his willingness to take the Ministry, or to accept a field command, and toured Zahal units, familiarising himself with the situation while the political in-fighting proceeded in his absence. He was offered the portfolio on the night of 1 June, and immediately associated himself with the group which was pressing for a declaration of war. The armed forces greeted his return with joy and a burst of renewed self-confidence, which in retrospect was probably his greatest contribution to their astonishing victory. He rejected any idea of appeals for American military aid – not least because of his lively appreciation of the dangers of provoking Soviet involvement. While he injected a new and more ambitious strain into existing plans for a limited offensive, his overall influence on the campaign was one of moderation. He now displayed a wariness, a maturity, and a sensitivity of political judgement quite at odds with his old reputation as a hell-for-leather field commander. He only authorised the advance to the Suez Canal after the unexpected collapse of Egyptian resistance in Sinai made his original stop-line – the Mitla and Jidi Passes – militarily pointless. Despite fierce pressure in the Cabinet and the country, he resisted demands for an all-out attack on the Golan Heights until the situation in Sinai allowed major redeployment of ground and air forces.

After a period out of office Dayan took up the Defence portfolio again in 1969, and held it until 1974. During the lengthy debate over Israel's occupation of former Arab territories he was a voice of moderation, advocating liberal military government coupled with vigorous retaliation for guerrilla attacks.

The initial defeats suffered in October 1973 at the hands of the Syrians on the Golan and the Egyptians in Sinai caused an enormous shock to Israel's military and political establishment. Caught seriously unprepared and suffering unprecedented casualities and reverses, Zahal was faced after 48 hours by the real prospect of the imminent collapse of the northern defences. Informed of this, Dayan is reported to have requested from Prime Minister Golda Meir authority to activate Israel's nuclear capability – and to have received it. Mercifully, the tide was beginning to turn on the battlefield, and the last resort was avoided.

The fierce *post mortem* which followed Israel's eventual costly victory cleared Prime

Minister Meir and Defence Minister Dayan of blame, but both resigned in response to public feeling, as did several IDF senior officers criticised by the Agranat Commission. Dayan continued to lead an active political life, and was Minister of Foreign Affairs 1977–79; sitting as an independent M.P. from 1977, he founded his own party in 1981. He died on 16 October of that year.

Essentially a 'lone wolf', Dayan was always supremely unimpressed by orthodox conventions both in his public and his private life, and was a focus of controversy for much of his life. He guarded his freedom of speech and behaviour fiercely, and was not a good 'team man'. He did not suffer fools or bores easily, and quickly became bored with matters of detail unless they concerned combat effectiveness. A noted amateur archaeologist, he was a cultured man of great personal charm. He aroused either total loyalty or fierce loathing throughout his career.

De Gaulle, Charles André Joseph Marie (*1890–1970*), *General, President of the French Republic*

A controversial French military commander, possessed of only limited understanding of military strategy, de Gaulle assumed the leadership of the Free French movement after France's collapse in 1940. He became a postwar Prime Minister and eventually President of the Fifth Republic, and presided over the French withdrawal from Algeria.

Born at Lille, the son of a professor of philosophy, on 22 November 1890, Charles de Gaulle graduated with distinction at l'École Spéciale Militaire, St Cyr, and in 1913 was commissioned in the 33rd Infantry Regiment – then commanded by Colonel Philippe Pétain (q.v.). Actively engaged in numerous actions during the first eighteen months of the First World War, de Gaulle was wounded early in the Battle of Verdun in March 1916 and taken prisoner – which he remained until the Armistice.

After the war, de Gaulle lectured for almost two years at St Cyr, and went on to attend l'École Supérieure de Guerre, graduating in 1924 and joining Marshal Pétain's staff the next year. As a major in 1927 he served for two years at Trier with the Army of Occupation in the Rhineland, subsequently being posted to the Lebanon for two years, and returning to staff appointments in Paris in 1931.

Deeply affected by memories of the static warfare and the gigantic human sacrifices on the Western Front, de Gaulle's opinions on military planning conflicted with orthodox French thought during the late 1930s. Conventional wisdom based metropolitan defence on the concrete bastions of the Maginot Line. His strategy, if such it can be called, was based upon a highly professional mechanised army – itself a policy at odds with the French dependence upon conscripted personnel. In his recognition of the fatal inflexibility of dependence upon the Maginot Line de Gaulle was undoubtedly correct; but his proposals for armoured mobility were necessarily limited by wholly outdated equipment on the ground, supported by equally old-fashioned equipment in the air.

On the outbreak of the Second World War de Gaulle, then a colonel, was a tank brigade commander, but when the threat of defeat followed the *blitzkrieg* of May 1940 General Giraud appointed him divisional commander of the half-formed 4th Armoured Division. For reasons never satisfactorily explained, de Gaulle gained his first political appointment on 6 June, when Premier Paul Reynaud made him Under-Secretary of War – a post which brought the fifty-year-old colonel to London on the same day to meet Winston Churchill. Twelve days later, as France collapsed, de Gaulle broadcast his famous appeal to the French people to continue the fight against Nazi Germany, and placed himself at the head of the Free French movement.

Thereafter, though nominally Commander-in-Chief of the Free French forces, de Gaulle pursued a political career with occasional sallies into the realm of military affairs – usually to the accompaniment of complaints from professional Allied commanders who resented apparent interference from a man who, after all, had only just gained part-divisional command in the field – and then only for a few days. To be strictly accurate, however, de Gaulle's interference – if such it was – was usually well-conceived in the long-term interests of the French, who maintained a fragile and complex colonial administration in such areas as Algeria, Madagascar, and Indo-China. It was perhaps not unreasonable that, for the honour of France, French troops should be seen to participate in such operations as the 'Torch' landings, the Normandy landings, and the liberation of Paris. Yet the intransigence of de Gaulle and some of his commanders often led to bitterness among Allied leaders – and the frustrations and occasional humiliations which he suffered were never forgotten or forgiven by de Gaulle.

On 10 September 1944 he formed a provisional government in liberated France, but resigned as Prime Minister on 20 January 1946. In April 1947 he formed the Rassemblement du Peuple Française, a party which in effect exploited popular fear of communism; this was disbanded six years later.

Trouble in Algeria came to a head in 1958, when Army commanders acted in defiance of orders from Paris, leading to a serious threat of civil war in metropolitan France. De Gaulle, the traditional figurehead so beloved by France in times of strife, emerged from semi-retirement to form a new government on 1 June, pledged to reach a solution to the Algerian problems; three months later the French people approved a new constitution supported by de Gaulle who was, on 21 December, elected President of the Fifth Republic. A stormy and frequently controversial presidential career followed. Adopting tough, uncompromising postures over Algeria, the President was openly opposed by many factions in France and elsewhere; indeed, he probably survived more assassination attempts than any other European leader. Determined to see France independent of foreign influences – military, political and economic – de Gaulle withdrew his nation from the military organisation of N.A.T.O. in 1966, a step which gained the approval of many of his erstwhile leftist critics. He then committed what many international observers considered a gaffe in appearing to give support to Canadian separatists.

Always a believer in a flexible constitution in France, de Gaulle periodically sought sanction by referendum for alternatives or alterations to the administration; but by 1968 the left-wing elements had so gained ground (through student and industrial demonstration and open revolt) that when a constitutional referendum failed to support the aged President's uncompromising proposals, he retired to his home at Colombey-les-Deux-Eglises, where he died – widely mourned – on 9 November 1970.

Díaz de Bivar, Rodrigo, *see* Cid, El

Dietrich, Josef ('Sepp') (*1892–1966*), *SS Oberst-Gruppenführer und Generaloberst der Waffen-SS*

'Sepp' Dietrich, probably the best-known of Hitler's SS generals, rose to the highest rank in the *Waffen-SS* and to the command of what was probably Germany's best-equipped army in 1944, more by virtue of his connections than through the demonstration of any very remarkable military skills. Nevertheless it is less than just to characterise him simply as a political appointee. The higher ranks of the *Waffen-SS* contained some extraordinary figures, and after the enormous expansion of the service from 1943 onwards some of them found themselves holding senior frontline commands; in Dietrich's favour it can at least be said that he was a fighter by nature and that his men showed great loyalty to him even after suffering decimation under his leadership. He does not rank, however, among Germany's ablest Second World War leaders.

He was born in 1892 in Hawangen, Bavaria, of humble parents. He was originally destined for the trade of butcher, but in 1911 he enlisted in the Imperial army at the age of nineteen – a stocky, aggressive young man of rudimentary education. During the First World War he rose to the rank of sergeant-major, and in 1917 he was successful in a technical course. At the end of the war he was serving in one of Germany's first tank battalions, and in later years he was particularly proud to wear the combat badge of that branch, a rare distinction. He became involved in the embryo Nazi movement early in the post-war period, and was one of Hitler's closest associates in the days of street brawls and window-smashing. He was given command of the Nazi leader's picked bodyguard squad in 1928, but until the anti-SA purge of 1934 this fledgling SS remained in relative obscurity. The bodyguard was entitled *Leibstandarte SS 'Adolf Hitler'* in 1933, but it still comprised only 120 men. In June 1934 this squad, under Dietrich's personal leadership, played an important part in the summary arrests and executions without trial of the *Sturmabteilungen* leadership. After this the SA declined in power and the SS rose, and Dietrich rose with it. For the rest of the Nazi era he commanded the premier combat unit of the *Waffen-SS* in all its stages of expansion from a battalion to an army.

In September 1939 the *Leibstandarte* fought as a regiment in Poland under Reichenau's 10th Army; during the *blitzkreig* in the West it saw active service as a strong motorised regimental combat group, and was prominent in the Paris victory parade. It was expanded to brigade strength in August 1940; and in June 1941, after a period of sharp fighting in Yugoslavia and Greece, Dietrich led the *Leibstandarte* Division into Russia. The *Waffen-SS* was not taken too seriously as a military force at this time, but it still received the pick of recruits and new equipment. The disdain of the establishment generals was shaken by the undeniably impressive performance of SS units during the fierce Russian counter-attacks of the winter of 1941–2. The *Leibstandarte*, decimated in severe fighting, was withdrawn to France to re-equip as an armoured division in June 1942. In the face of the Russian breakthrough of January 1943 it returned to the Eastern Front, fighting most effectively on the Donetz and in Kharkov. In July 1943 the *Leibstandarte* was grouped with the other two newly-armoured SS divisions – *Das Reich* and *Totenkopf* – as the 1st SS Armoured Corps, which became Germany's *élite* combat formation. The *Leibstandarte* Division continued to serve in the Corps, which took that title and which Dietrich rose to command. He led the crack SS tank divisions in the Russian fighting of the second half of 1943, transferring to Belgium for refitting early in 1944. His divisions were heavily engaged in Normandy in 1944, suffering great losses; Dietrich was promoted to colonel-general, and awarded the Diamonds to the Knight's Cross. In October 1944 he was given command of the newly formed 6th Armoured Army, under which designation various units withdrawn from France were refitted and brought up to strength; although not exclusively a *Waffen-SS* formation, it was later termed the 6th SS Armoured Army as it contained a high proportion of crack SS units. Dietrich commanded this formation on the northern wing of the Ardennes counter-offensive in December 1944. Withdrawn to Germany in January 1945 after the advance was bogged down, it was quickly re-equipped once again – to an inadequate standard, but still enjoying a high priority – and sent into Hungary in late February. Dietrich was ordered to recapture Budapest, which had fallen early in February. This was a ridiculous mission for one of Germany's few armies still capable of realistic offensive action, at a time when every tank and gun was needed on the Oder front. Hoplelessly outnumbered and without sufficient artillery and air cover, the 6th SS Armoured Army was seriously mauled and forced to withdraw into Austria in March. Hitler was incensed by its failure to defend Vienna, and radioed a famous order to the effect that the personnel must remove their much-prized 'Adolf Hitler' divisional cuff-titles – to which Dietrich replied that he would rather shoot himself, and returned his medals to the *Führer*. The remnant of his forces took no further part in operations, and Dietrich concentrated on saving his men from Russian captivity – which in view of the activities of the *Waffen-SS* in Russia during more triumphant days, was liable to be perilous. He surrendered to American forces on 12 May 1945.

Dietrich was no strategist, but at divisional level he was an adequate general. His style of leadership was well suited to the men he led; he was a coarse, broken-nosed, foul-mouthed peasant who saw the war largely from the point of view of the man in the foxhole. His contribution to the very high morale and aggressive qualities of the *Waffen-SS* tank divisions was considerable. His promotion was opposed at every level by the establishment of *Wehrmacht* generals, who were appalled at the arrival of this gutter-brawler in the traditional officer corps; but he came to enjoy a measure of respect from the professional soldiers by virtue of his realistic view of military problems, and his willingness to commit his splendidly-equipped divisions to the fiercest fighting regardless of losses. He enjoyed a close relationship with Hitler, who sentimentalised over the old days of beer-hall comradeship and exaggerated his military prowess, but towards the end of the war the *Führer*'s increasingly unhinged behaviour led Dietrich to side with the professional soldiers.

After the war Dietrich was called to account for the atrocities committed by troops under his command. In 1946 he was sentenced to 25 years' imprisonment by an American court on charges connected with the murder of unarmed American prisoners at Malmédy during the Ardennes offensive; the atrocity was committed by a combat group of the *Leibstandarte* Division under Colonel Peiper, and as army commander Dietrich was held partly responsible. Like so many other condemned war criminals, he was released after only a few years' imprisonment in October 1955. In 1957 a German court sentenced him to eighteen months' imprisonment for his part in the killing of the SA leaders on 'The Night of the Long Knives' in 1934; he served this sentence and was released in due course. Russia also wished to try him for crimes committed by his troops in Kharkov, where he is known to have ordered that no prisoners be taken for a period of three days following the discovery of the bodies of German prisoners tortured to death by the Soviets. He never stood trial on these charges, and died at Ludwigsburg on 21 April 1966.

Du Guesclin, Bertrand (*c. 1323–80*), *Constable of France*

This outstanding French commander of the Hundred Years' War is not widely known in English-speaking countries, while the name of Edward (q.v.) the Black Prince, a far less remarkable military figure, has become a household word. It was due to Du Guesclin's shrewd ability to combine the dictates of chivalry with practical common sense that King Charles V (nicknamed 'the Wise') of France was able to recover much of what had been lost to the English under the previous two monarchs.

Du Guesclin was born in Brittany, at the castle of La-Motte-Broons near Dinan, about 1323. The details of his youth are obscure, but he first experienced warfare as an esquire in the War of the Breton Succession between Charles of Blois and Jean de Montfort. His military career began in earnest at the Siege of Rennes, which was invested by an English army under Henry of Lancaster between October 1356 and July 1357. Du Guesclin fought his way into the city and helped to defend it until the end of the siege; among his exploits was a famous single combat with Sir Thomas Canterbury. In December 1357 he received a pension from the Dauphin in recognition of his valour, and a few days later was named captain of Pontorson, for the Dauphin's uncle the Duc d'Orléans. He served the royal house of Valois from this time until his death more than twenty years later. In this period, in the aftermath of the Black Prince's victory at Poitiers, the French kept to their fortified strongholds while the English raided and plundered almost unchecked throughout France.

In 1359 Du Guesclin left Pontorson to take part in the Siege of Melun, but on his return journey was captured by Sir Robert Knollys – by no means the last time he was to go into captivity. He was freed after the signing of the Treaty of Bretigny in 1360, but was soon campaigning again, this time against the lawless mercenary bands which still ravaged the land in the aftermath of war. He was captured again, at Juigné. In 1363 he was named captain-general of the northern bailiwicks of Caen and Cotentin, and by the following year his power extended over the whole of Normandy and beyond. In 1364 Charles 'the Bad' – Charles II of Navarre – took advantage of the lawlessness and disorder in France to invade the southern provinces, and in April of that year Du Guesclin captured Mantes and Meulan from the Navarrese garrisons which held them. The following month he defeated them at Cocherel, capturing the famous Jean de Grailly. Later that year, on the accession of Charles V of France, the loyal Du Guesclin was appointed chamberlain; but the continued fighting over the Breton succession led to his recall to Brittany by Charles of Blois. At Auray on 29 September 1364 Charles of Blois was killed and Du Guesclin was captured – this time by Sir John Chandos – when they attempted to force an English army to raise the siege of the town; the English, naturally, were meddling vigorously in all France's internal and frontier troubles at this time. Du Guesclin was ransomed by King Charles for 40,000 gold francs, and in September 1365 he was given command of a mercenary army which Charles was sending into Spain to fight for Henry of Trastamara against his brother Pedro I, called 'the Cruel', of Castile. Pedro was toppled from the throne in 1366; but in 1367 Edward the Black Prince led another mercenary army across the Pyrenees from his lands in Aquitaine to support Pedro. At Nájera on 3 April 1367 Du Guesclin and his small force of French were isolated and forced to surrender when the English archers in Edward's army dispersed Henry's Castilian forces. Du Guesclin was ransomed with some difficulty and at enormous expense; but in 1369 he led a second Castilian expedition to victory at Montiel, where Henry personally killed Pedro and finally secured the throne. In 1370 Charles V named Du Guesclin as Constable, and for the next ten years he presided wisely over a marked recovery in French fortunes.

Du Guesclin had the best of reasons for appreciating the superiority of English long-bowmen; unlike some of his contemporaries, however, he saw no reason why his knightly code should oblige him to charge them head-on. He avoided attacks against prepared positions, and did not allow himself to be drawn into costly pitched battles during the various great English cavalry raids across France. He counselled a policy of containment and strategic withdrawal, coupled with a slow and determined campaign of siege warfare against English strongholds. One by one he recaptured the castles of Poitou and Saintonge, reducing English holdings to small areas around Bordeaux, Bayonne, Brest, Cherbourg, and Calais. He seized every opportunity to catch the enemy at a disadvantage, and excelled in night attacks. Apart from his success in the field, Du Guesclin laid the foundations for a new type of military organisation in France by creating some permanent standing units and a permanent 'staff'; regular artillery companies also appeared under his guidance. He refortified Paris, and gave some attention to the French fleet. Although the wars continued to ravage the countryside of France, his victories doomed any permanent British political presence on French soil.

In 1380 the Constable was sent into Languedoc to suppress an outburst of disorder. While laying siege to the castle of Châteauneuf-de-Randon, he died on 13 July.

Edward I, called 'Malleus Scotorum' (*1239–1307*), *King of England*

Duke of Gascony and Ireland before his accession to the English throne on the death of Henry III, Edward I was an unpopular intriguer and an inept general in his youth, but on maturity became shrewdly realistic, and a powerful administrator. His determination upon an autocratic militarism engendered constant conflict in Wales and Scotland, which in turn led the King to pursue the constant and successful development of his forces.

Born at Westminster on 17 June 1239, Edward was granted numerous duchies by his father, Henry III, on reaching the age of fifteen, including much of the King's land in Wales. His first campaign was fought only one year later when, opposed by Llewelyn ap Gruffydd, he marched to support the application of the English shire and hundred system to his lands in Wales. Without the support of his father, Edward was defeated ignominiously.

In the Civil War of 1263–5 Edward fought beside his father in the Battle of Lewes (14 May 1264), in which the King's army was decisively beaten by Simon de Montfort – a defeat that was largely brought about by Edward's impulsive pursuit of part of the baronial army from the field. Thereafter Simon made an alliance with Llewelyn, and Edward (in the absence of Henry III, who had been taken prisoner after Lewes) began a campaign to bring the rebel forces to decisive battle. After a number of minor engagements in which the baronial army attempted to evade the royalist forces and break into the Midlands, Edward defeated Earl Simon at the Battle of Newport on 8 July 1265.

Learning of his father's defeat at Newport, Simon de Montfort the younger assembled an army of about 30,000 men and marched west to the rescue. Caught between two armies of a combined strength double his own, Edward determined to strike before they could join forces; he surprised the younger Simon's army and almost annihilated it at the Battle of Kenilworth on 2 August. Earl Simon's army now crossed the Severn but, with only 7,000 men, was trapped in a bend of the river Avon; it was in its turn destroyed in the decisive Battle of Evesham by Edward on 4 August. These two battles confirmed the young Edward as a brilliant tactician and, without doubt, one of the greatest commanders of his age. Although he had displayed almost Napoleonic ruthlessness in driving his forces to exhaustion, his movements had been vigorous and decisive, and his personal example provided much of the stimulus needed to defeat the powerful rebel armies. The Evesham campaign crushed the rebellion and left Edward ruler of England in all but name.

In 1270 the Prince left England for the Holy Land, but on the death of his father on 16 November 1272, returned for his coronation in August 1274. While commanding the loyalty of the great majority of English barons, Edward was from the outset of his reign constantly beset with troubles in Wales which, under Llewelyn, had gained a degree of independence by the Treaty of Shrewsbury in 1267. Determined to reduce Llewelyn, and using his constant evasion of dues as a pretext for attack, Edward invaded Wales in 1277 with three co-ordinated armies with naval support. He isolated Llewelyn in Snowdonia, starved him into submission, and deprived him of all territorial gains made since 1247. He then built a great ring of fortresses round Gwynedd, and divided the land into shires and hundreds – a system of administration hated in Wales, and one that was bound to provoke further rebellion. When this broke out in 1282 Edward again moved against the principality, and decisively beat Llewelyn, who was killed at Radnor in that year; his brother David was executed shortly after. Introducing the Statute of Rhuddlan, Edward finally reorganised the principality on English lines, and a last rebellion in 1294 was ruthlessly crushed, leaving Wales supine for more than a hundred years.

In the north Edward's forceful administration was provoking trouble with the Scots as well, and in 1295 border disputes led to war. Edward invaded Scotland, stormed Berwick, and defeated his former protégé King John Baliol at the Battle of Dunbar on 27 April 1296. Open rebellion then flared up under Sir William Wallace (q.v.) who was decisively defeated by Edward at the Battle of Falkirk. In this battle, involving about 30,000 Scots and 23,000 English, Edward's infantry made the first substantial use of the long-bow by an English army: the Scots infantry was deployed in four large circular phalanxes which repulsed all charges by the English, but which were greatly weakened when Edward ordered his bowmen to shower arrows into them. The victory at Falkirk enabled Edward to annex Scotland – a step which was to inaugurate 250 years of bitterness and savage warfare. Further operations led to the capture of Sir William Wallace, whom Edward executed as a rebel in 1305.

Further unrest in Scotland led to the rebellion under Robert the Bruce (q.v.), later to become King Robert I of Scotland. Bruce beat the Earl of Pembroke at Loudon Hill in May 1307, and began to clear the English out of Scotland. Edward, on his way north to deal with the Scottish uprising, died suddenly from an attack of dysentery while at Burgh-on-Sands, three miles south of the border, on 6 July.

Edward I had shown himself to be a gifted commander in the field and a brilliant tactician. His constant study of warfare had enabled him to bring about substantial improvements in the royal armies, greatly increasing the centralised control of the English militia. The many frontier castles which he built represented a whole new chapter in military engineering; and his campaigning in Wales convinced him of the value of the long-range fire-power of the long-bow, a weapon he did much to introduce into the English Army in very large numbers. But the repeated wars in Wales and in Scotland proved so expensive that by the turn of the century his treasury was virtually exhausted, with the result that many of the King's military responsibilities across the Channel could not be met. Philip III and Philip IV of France both nibbled at his Gascon borders with impunity, at the same time severely undermining the English King's authority in the province. Unable to find sufficient funds to support a military expedition until 1297, Edward eventually sailed to attack France from Flanders in that year, but his barons refused to accompany him and the Scottish uprising forced him to return to England. There is no doubt that, brilliant soldier though he was, Edward's strong military preoccupation provoked civil strife that was to remain with his successors for several hundred years.

Edward III (*1312–77*), *King of England*

A constant campaigner in Scotland and France who sought to restore peace in and authority over the lands originally pacified by Edward I (q.v.), Edward III was the victor in the naval battle of Sluys, and at Crécy, where many of the great continental commanders learnt to their cost of the constant development of English arms and tactics that had followed in the wake of fierce campaigns in Scotland and Wales in the previous hundred years.

Born the eldest son of King Edward II and Isabella of France at Windsor on 13 November 1312, Edward of Windsor never received the title Prince of Wales, but was summoned to Parliament as Earl of Chester at the age of eight, and created Duke of Aquitaine at the age of eleven. He was proclaimed 'keeper of the realm' on 26 October 1326 at the time of his father's flight with the Despensers, and after Edward II's abdication was accepted as King, being crowned on 29 January 1327 at the age of fourteen. Before his fifteenth birthday he had participated in an abortive military campaign against the Scots, but in reality his mother and Roger Mortimer (a previously exiled baron) ruled in his name. At the age of fifteen he married Philippa, daughter of William I, Count of Hainault; and his eldest child, Prince Edward (q.v.), later known as the Black Prince, was born on 15 June 1330. It was at this time that the young, headstrong king determined to throw off the degrading influence of Mortimer, seized him at Nottingham, and procured his execution. His mother, whom he treated with respect, exercised no further political influence.

Edward now began to exercise his determination to restore England's power to what it had been during his mighty grandfather's reign. Robert the Bruce (q.v.) had died in 1329; following this the Scottish barons who had been exiled by Robert for their support of the English invaded Scotland in 1332 under the leadership of Edward Baliol, and he, after a victory at Dupplin Moor, ascended the Scottish throne for a short time. He was defeated by a Scottish coalition under the boy-king David II. Edward III, in support of Baliol, marched north and defeated the Scots at Halidon Hill in July 1332. Despite his youth, the English King learnt in this battle the importance of controlled infantry tactics – a lesson he applied with great thoroughness during later campaigns.

In the course of the next decade relations between England and France deteriorated steadily for two principal reasons: Edward's undisguised alliance with the Flemish, and his own claim to the French Crown, first made in 1328. In 1337 Philip IV of France declared Gascony forfeit, and Edward negotiated an alliance with the Emperor Louis of Bavaria and twice attempted to invade France from the north; he achieved nothing save his own reduction to near-bankruptcy. Despite his failures he assumed the title of King of France in 1340. As his pretensions to the French crown became more persistent any likelihood of stable peace disappeared; the struggle that started in this way grew into the Hundred Years' War. It was during the early stages of this war that, having withdrawn his army from France on the approach of a much stronger force under Philip, Edward, with an English fleet of about 140 vessels, annihilated a French fleet of 190 ships at the entrance of Sluys harbour. By bringing to bear large numbers of archers and fire machines in his own vessels, Edward captured or destroyed 166 enemy ships in the famous battle fought on 24 June 1340.

When in 1346 a French army under Duke John of Normandy invaded Gascony, Edward assembled a small army and sailed from Portsmouth, to land near Cherbourg in July. With about 20,000 men (of whom some 12,000 were archers) he captured Caen before the end of the month, and marched down the Seine towards Paris, where he knew that Philip was amassing a large army. At Crécy-en-Ponthieu, near Abbeville, he discovered a suitable field of battle and set about preparing his positions. In front he placed two divisions: that on the right was commanded nominally by the sixteen-year-old Black Prince, but in practice by the Earl Marshal, Warwick; and that on the left was commanded by the Earls of Arundel and Northampton. A few hundred yards to the rear was located the third division, under Edward's personal command. The archers of each division, probably about 4,000 men, were echeloned forward on the flanks of each division, in the centre of which was a phalanx of about 1,000 dismounted men-at-arms deployed six deep on a front of about 250 yards. Behind each division was a small reserve of heavy cavalry, positioned to deal with any break-through by the enemy.

The French army, numbering about 60,000 men (including about 12,000 heavy cavalry) but without any scouting screen ahead, stumbled on the British Army in the late afternoon of 26 August. Philip managed to halt his straggling masses and, as a thunderstorm broke over the field, ordered his Genoese crossbowmen to advance; at about 150 yards they fired their bolts, but most of their missiles fell short. As they moved on, the English bowmen launched a shower of arrows which decimated the ranks of the Genoese, who reeled away. Impatient to enter the fight, the French cavalry charged through the ranks of the panic-stricken Genoese; but the rain-soaked grass proved too slippery for quick approach, and within minutes the ground in front of the English line was heaped with dead and churned up by the milling throng of men and horses. All the while the English kept up their devastating hail of arrows. Some French cavalry reached the English line, but were immediately routed by Edward's heavy horse. The French launched more than a dozen separate waves of infantry and cavalry, and the terrible slaughter lasted well into the night; but the English stood steadfastly in their magnificently-sited lines until dawn, by which time the French army had abandoned the battle, leaving about 15,000 dead or dying, including 1,500 lords and knights. The French King himself was wounded, and the blind King John of Bohemia was among those killed. The English losses amounted to two knights killed and no more that 200 dead and wounded all told.

Europe was staggered by Edward's stupendous victory at Crécy. Only vaguely aware that English armies had been engaged in campaigns within their own islands, continental leaders had no inkling that these forces had been so highly developed both in command discipline and in tactics. In this single battle the thousand-year-old doctrine of the universal superiority of heavy cavalry was destroyed; from this time – and in a sense almost ever since – the fire-power of disciplined infantry ruled the battle-field.

After the destruction of the French army Edward resumed his leisurely march through northern France; he laid siege to Calais and forced its surrender in August 1347, ejecting most of the French inhabitants and colonising the town with Englishmen to provide a base from which to conduct further invasions of France. However, the lengthy campaign in France and the ravages of the Black Death which swept England in 1348 and 1349 exhausted his army, and again almost brought about bankruptcy. Henceforth Edward's military exploits were largely overshadowed by the initial achievements of his son, the Black Prince, who forced the French to accept a new peace after his capture of the French King John II at the Battle of Poitiers on 19 September 1356. When the French repudiated the terms demanded in the Treaty of London in 1359, Edward again invaded France and laid siege to Rheims, where he planned to have himself crowned king; he was frustrated by the resistance of the inhabitants and was forced, by the Treaty of Calais in 1360, to renounce his claim to France.

The last seventeen years of Edward's life were marked by a decline into dotage, with court intrigues sapping his energy and will to impose his authority. The Black Prince and his brother John of Gaunt headed opposing factions, and when the former died on 8 June 1376 the commons were robbed of their strongest support. Saddened by the death of his famous son, Edward himself died peacefully at Richmond (Sheen) on 21 June the following year. He had been an able ruler and a brilliant military commander, entirely suited to head the most illustrious court in contemporary Europe. His ability to humble his neighbours on the field of battle gave force to his constant demands for money and military resources, and in return he worked hard to strengthen Parliament. Like his grandfather, however, his ambitions far exceeded the scope of his resources.

Edward Plantagenet, called the Black Prince (*1330–76*), *Prince of Wales, Prince of Aquitaine*

The son of Edward III (q.v.) of England and Philippa of Hainault, Edward of Woodstock was born on 15 June 1330. He was to become the best-known English commander of the Hundred Years' War, although his strictly military reputation rests largely on a single victory at Poitiers. An energetic and personally courageous leader, he was neither a great strategist nor a far-sighted ruler; by the time of his relatively early death he was a discredited man, a tragic example of a prince who failed to fulfil his early promise.

The young Prince of Wales (as he was created in 1343) saw his first campaign in 1346 during his father's expedition to France. At Crécy on 26 August 1346 he was placed in nominal command of the right-hand division of the English army, consisting of about 1,000 dismounted men-at-arms ranged six ranks deep over a front of some 250 yards, with several thousand archers ranged on each flank. (Actual command was in the hands of the veteran Earl of Warwick.) The sixteen-year-old Prince acquitted himself well: as well as his knightly spurs he won the badge of the three feathers – associated with the title Prince of Wales to this day – taken as a mark of respect for the blind King John of Bohemia, who fell among the French knights.

During the late 1340s and early 1350s both England and France were devastated by the Black Death, and it was not until 1355 that the war was resumed. Edward III, the Black Prince, and his younger brother John of Gaunt led independent commands in a series of devastating raids across northern France. (The Prince of Wales is popularly supposed to have taken his nickname from his habit of wearing black armour; while this has a fine romantic ring to it, the fact is that the earliest known reference to this *nom-de-guerre* is in Grafton's *Chronicle of England* in 1569.) In early August 1356 Prince Edward started out from Bergerac with about 4,000 heavy and 4,000 light cavalry and some 1,000 infantry, all largely professional indentured troops raised in England's Aquitainian possessions, and some 3,000 English archers. Concentrating on relatively swift movement, the Prince cut a swathe of destruction through France and had reached Tours by 3 September. When word reached him that John II of France was leading a large army to cut him off he withdrew southwards, but was hampered by a large convoy of plunder. The French moved faster, but neither side was prepared for the eventual meeting, which took place unexpectedly near Poitiers on 19 September. Edward, knowing that all hope of evasion had gone, prepared a good defensive position on a gentle slope facing north. One flank was guarded by a marshy stream; the other he anchored on a wagon-park garrisoned by archers. His front was covered by hedges across the slope, with only a few sunken lanes giving direct access to his line. Archers were placed behind the hedges and along the lanes, and manned the marshes well forward on the flank of the main line. All but a small mounted reserve dismounted and formed three main 'battles'; that on the left was led by Salisbury, that on the right by Warwick, and the Prince kept a third under his own hand in a second line behind Warwick. The battle was joined on 19 September, when the French (some 16,000 cavalry and between 15,000 and 20,000 infantry) advanced frontally in four successive waves. The first was led by some 3,000 mounted men, but all the remaining French men-at-arms were ordered to fight on foot – John thought to avoid a repetition of Crécy in this way, but failed to understand that it was not merely the fact that they were mounted which made the French knights vulnerable to English archers. The first attack was repulsed in classic manner by the longbowmen, only a few of the enemy reaching the main line. The second wave, led by the Dauphin, closed and fought with determination, and Edward was obliged to move his third division up into the front line, holding out only some 400 reserves. The losses were severe on both sides, but the French eventually fell back. The third wave lost heart at the sight, and did not close. The fourth and largest division, led by King John, now approached the thinned and shaken English ranks. Edward seized the initiative; sending his small cavalry reserve in a hook to strike the French left rear, he put his remaining fresh men in the front rank and ordered a general advance. After bitter fighting the French broke, and thousands were captured, including the King himself. Losses were probably about 2,500 French dead and a similar number captured, and more than 1,000 English dead and a similar number wounded. While John II waited for the payment of his ransom of three million gold francs he had leisure to reflect on his mistakes. He had robbed his knights of their only real advantage – mobility and impact – and had fed them on a narrow front into the killing-ground swept by English arrows, instead of trying to envelop the weak right of the heavily-outnumbered enemy. In 1360 the Treaties of Bretigny and Calais recognised England's rule over Aquitaine and English possessions at Calais and Ponthieu; and three years later Edward, now Prince of Aquitaine, left England to rule his new domains.

His rule was a disaster. He kept a lavish court, and taxed all classes heavily. He made no attempt to gain the goodwill of either nobility or clergy, but treated his vassals as if he were a foreign conqueror rather than a responsible and legitimate feudal lord. His arrogance and profligacy led to a strong spirit of revolt, which increased during his absence in 1367 on campaign in Spain. He had allied himself to the deposed Pedro the Cruel of Castile, whose brother Henry had taken power with French support. Edward made a skilful advance through the Pyrenees, evading a Franco-Castilian army led by Henry and the French commander Du Guesclin (q.v.) and forcing them to fall back south of the Ebro. When the two armies met at Nájera on 3 April 1367 the Prince won an impressive victory; with some 20,000 men, of whom only about 5,000 were English bowmen, he broke and dispersed a Castilian army of 11,500 cavalry and up to 25,000 infantry. Although the French contingent of some 2,000 knights fought stubbornly, they were eventually forced to surrender by the defeat of the Castilians, who lost some 12,600, as against 400 French and less than 200 English dead. Despite this remarkable achievement Edward returned from Spain with his health and his exchequer broken, to find his vassals in revolt. The clergy and nobility appealed for help to Charles V of France, and so thoroughly had the Prince alienated all classes that by March 1369 some 900 towns and castles had declared against him. His mercenary army dwindled in proportion to his treasury, and such savage episodes as the terrible sack of Limoges in October 1370 merely increased the hatred of his enemies. In 1371, sick and ruined, he returned to England; the following year he formally gave up to his father the Princedom of Aquitaine. His last five years are shrouded in obscurity, and he died at Westminster on 8 June 1376. He is buried at Canterbury, where his tomb and his accoutrements are still to be seen.

Like all his family he had great physical presence; he was literate and fairly pious, and had a great love of fine possessions and artistic treasures, particularly jewels. He consistently lived beyond his means, but despite his self-indulgence and his capacity for cruelty in warfare, he was capable of very generous gestures to subjects whom he considered deserving.

Eisenhower, Dwight David (*1890–1969*), *General of the Army, 34th President of the United States of America*

Dwight Eisenhower's rise to prominence was as spectacularly swift as that of any leader in military history. At the beginning of 1942 he was a promising lieutenant-colonel who had never heard a shot fired in anger. Yet at the beginning of 1944 he was appointed supreme commander of the greatest invasion force the world has ever seen – a force which, with its air and sea support and its logistic 'tail', totalled nearly 3,000,000 men of a dozen nations. If the invasion had failed the Allies would very probably have lost the Second World War. That it did not fail was due, to an incalculable degree, to Dwight Eisenhower.

He was born in Denison, Texas, on 14 October 1890. His parents later moved to Abilene, Kansas, where they brought up their large family of boys in the mid-western tradition of hard work and piety. Dwight drifted for a year after leaving school, unsure of his course; his eventual decision to enter the competitive examinations for Annapolis and West Point was not due to any burning desire for a military career: he only wished to better himself by taking advantage of the general educational facilities offered by the academies. He passed the examinations in first place, and opted to follow a friend to Annapolis; but it transpired that he was slightly over-age for the naval academy, so in 1911 he entered West Point. He graduated in 1915, about half-way down a class which was to provide nearly sixty general officers, including Omar Bradley (q.v.). As a lieutenant he served with the 19th Infantry between 1915 and 1917 at Camp Sam Houston, Texas. In 1918 he organised training courses for tank troops at Camp Colt, and was promoted captain. He reached the rank of major in 1920 – a meteoric start for an ambitious officer – but the peacetime army offered no more quick promotions, and he was to serve in the same rank for no less than sixteen years. His career proceeded through a series of staff appointments. In 1926 he graduated first out of 275 from the Command and General Staff School at Fort Leavenworth, Kansas; and in 1928 he graduated from the Army War College, Washington. In 1929 he was sent to Europe to prepare a detailed study of all aspects of American involvement in the final battles of the First World War, and he made himself a leading expert on the subject. In 1932 he graduated from the Army Industrial College, an institution he had helped set up; he became an expert on the subject of military-industrial co-operation in the event of war. In 1933 he was named special assistant to General Douglas MacArthur (q.v.), and between 1935 and 1939 served under MacArthur as assistant military adviser in the Philippine Commonwealth. Eisenhower earned his chief's praise for his efficiency, but the two men did not enjoy a particularly cordial personal relationship. In 1936 Eisenhower was promoted lieutenant-colonel at last. He returned to America on the outbreak of war in Europe; and during the 1941 summer manœuvres, when he was serving as Chief of Staff of the 3rd Army, he attracted the attention of U.S. Army Chief of Staff General George C. Marshall (q.v.). In February 1942 Eisenhower was named chief of the War Plans Division on Marshall's staff, and subsequently became chief of the Operations Division. His promotion now was rapid – too rapid to please some officers, passed over themselves.

In June 1942 this extremely able staff officer, with his capacity for absorbing a mass of detail and instantly extracting the essentials, was given an independent command – that of the American troops in Europe, which in practice meant in Britain. It was a sensitive post, carrying the responsibility for avoiding friction at all levels between allies of sharply differing attitudes. It was natural that Americans – fresh, loudly confident, and inexperienced – should have mixed feelings about the British, who were still dogged after three years' hard and usually unsuccessful fighting, but who tended to resent advice. At this time, and throughout his later career, 'Ike' earned great popularity and respect for the way in which he managed to reconcile national temperaments and aspirations. He was a man of enormous personal warmth and charm, with an unaffected manner and a natural friendliness and humility which disarmed all opposition. He made no secret of his determination that the Allied cause should not be jeopardised by displays of chauvinism on either side; and by a combination of ruthlessness towards those who flouted his instructions, and flexibility in negotiation, he worked wonders.

In November 1942 Lieutenant-General Eisenhower commanded the American landings in North Africa: his first field command. The landings were opposed at first by French garrisons, and a cease-fire was essential if the Americans were to advance to Tunis in time to prevent its reinforcement by the Axis and a link-up with Rommel's (q.v.) *Afrika Korps*, which was retreating westwards pursued by Britain's 8th Army. Eisenhower concluded a cease-fire agreement through Admiral Darlan, but his apparent endorsement of this prominent Vichy officer caused a storm of protest in American, British, and Free French circles. As a military necessity the move was understandable, but politically it was dynamite. This political distraction, together with appalling weather and inadequate logistic organisation, prevented a swift advance to Tunis, and led indirectly to defeat at Kasserine Pass in February 1943. Rommel linked up with V Panzer Army in Tunisia, and from his front on the Mareth Line facing the 8th Army launched a daring attack northwards which badly mauled certain American formations. Eisenhower, who had himself exercised inadequate command supervision, was much sobered by his first sight of the scene of a major American defeat. He made sweeping changes in his command structure, and those among his subordinates who had failed the test of action were replaced at once. Eisenhower preferred an informal style of command, and was usually at pains to gain a consensus of opinion for any decision which he took; but as Tunisia showed, he could adopt a harsher method if circumstances demanded it.

Leaving the Mediterranean theatre late in 1943, Eisenhower was named as the supreme commander of the Allied Expeditionary Force for the invasion of Europe in January 1944. The preparations for this vast, multi-national undertaking demanded all his considerable reserves of calmness, moderation, patience, good humour, and capacity for sheer hard staff-work. It was the perfect appointment for him, and he succeeded brilliantly. When he finally took the decision to launch the invasion on 6 June 1944, the Force was in as good a state of readiness as it would ever be. In the event, the invasion proved successful; and from then on Eisenhower's problems were to be concerned more with the methods than with the outcome of the campaign. His main personal problem was Sir Bernard Montgomery (q.v.), who favoured a narrow thrust into the heart of Germany, led of course by himself. Eisenhower would have had to halt the Allied advance on all other fronts to provide sufficient logistic support for such a thrust, and considered the risk unjustifiable. Considering the performance of the German armies in the Ardennes at the turn of 1944–5, it is arguable that such a manœuvre could have led to a tragedy of major proportions for Allied arms.

Shortly after his promotion to General of the Army in December 1944 Eisenhower was faced with this dangerous German counter-offensive in the Ardennes – 'The Battle of the Bulge'. Montgomery was given temporary command of certain American formations to the north of the threatened sector, and in due course a combined effort by many American and some British divisions averted the danger. While the British contribution had not been negligible, Montgomery caused damage to inter-Allied relations and to morale as a whole by his reported remarks regarding the performance of the Americans. Eisenhower was so incensed that a signal to London was actually prepared, demanding Montgomery's replacement by Alexander (q.v.) if Eisenhower was not to resign. Luckily the threat was averted by the common sense of the respective chiefs of staff, and Montgomery sent a generous and loyal message. It was yet another reminder that beneath Eisenhower's surface of unflappable good humour there was a determined commander who was not afraid to face the implications of his decisions.

After the surrender of Germany in May 1945 Eisenhower served briefly as commander of the U.S. occupation forces; he replaced Marshall as Chief of Staff that November. In 1948 he retired from military life to take up an appointment as President of Columbia University, but returned in 1951 as Supreme Commander, Allied Powers Europe. He had long resisted the advice of friends and associates to turn to politics; but in 1952 he resigned his command to campaign for the Presidency. He became the Republican candidate, and was voted into the White House with the biggest popular vote in American history. He was inaugurated as President on 20 January 1953. His activities in that office have no place in this book. He will be remembered as a soldier, one of the handful of truly great commanders to emerge from America during her short history, and a man whose particular talents were representative of enormous changes in the military art over a few generations. After a series of heart attacks, Dwight Eisenhower died in the Walter Reed Army Hospital, Washington, on 28 March 1969.

Eugène, Prince (properly François Eugène de Savoie-Carignon) (*1663–1736*), *Field-Marshal*

Son of the Comte de Soissons, Eugène entered the service of Austria after humiliation in France and the Netherlands, to become the greatest soldier of his generation: the victor at Zenta over the Turks, the ally of Marlborough (q.v.) at Blenheim, and a master of the art of military movement.

Born in Paris the younger son of Eugène Maurice of Savoie-Carignon, Comte de Soissons, on 18 October 1663, Eugène was left in the care of his grandmother when his father was compromised as a result of the scandal of the 'affair of the poisons' in 1680 and was forced to flee to Brussels. Of poor physique and ugly in the extreme, the young man was neglected by his relatives in Paris, and was steered in the direction of an ecclesiastical career. Humiliated at court, he applied to Louis XIV for permission to enter the French army, but was refused on the ground of his family's disgrace. Further humiliations followed, and the young man abandoned Paris and made his way first to the Austrian Netherlands, and thence to the Holy Roman Emperor Leopold I, hoping to be granted command of the regiment of his elder brother, who had been killed at Petronell on 13 July 1683. As this command had passed elsewhere Eugène had to content himself with the Kufstein dragoon regiment, with which he displayed outstanding bravery during the reconquest of Hungary between 1684 and 1688, being promoted major-general in 1685 and lieutenant-field-marshal in 1688. He also acquired Spanish grandeeship and became a Knight of the Golden Fleece.

After the outbreak of the War of the Grand Alliance Eugène served for three months on the Rhine and was then sent as general of cavalry to Italy, where he was appointed field-marshal in 1693, rising to the post of imperial commander-in-chief in Italy in 1694 – extraordinary seniority for a thirty-one-year-old foreigner in the imperial service. It was at the request of Rüdiger von Starhemburg, President of the Austrian War Council, that Eugène was appointed assistant commander-in-chief in Hungary in 1697, and when Frederick Augustus became King of Poland, Eugène succeeded him as commander-in-chief. In the Turko-Hungarian War of 1688–99 Eugène, having wholly reorganised the dilapidated army of Hungary, marched to oppose a major Turkish invasion in 1697; at the Battle of Zenta, on 11 September, he attacked and almost annihilated the main Turkish army under Sultan Mustafa II, effectively ending the Turkish threat to Hungary. The victory at Zenta brought universal fame to Eugène.

Probably much influenced by his own dislike of the French and his past humiliation, Eugène participated in intrigue, to bring about the War of the Spanish Succession, although he must have known that the Empire was as yet militarily unprepared for a major war. In 1701 he nevertheless took a small *élite* army across the Tirolean Alps into northern Italy against a French army under Catinat. Winning a battle at Carpi on 9 July, he pursued Catinat back to the Oglio river and defeated his successor, Villeroi, in a number of engagements, eventually taking him prisoner in a night attack on Cremona in February 1702. His small army was not equal to the task of meeting Vendôme (q.v.) and was defeated at the Battle of Luzzara in August that year, after which Eugène returned to Vienna to draw attention to the Empire's weak conduct of the war.

Soon afterwards Vienna itself was seriously threatened by the approach of French forces, which had been joined by the Bavarians and a rebellious force of Hungarians under Rakoczy, and Eugène was nominated President of the Imperial War Council and commissioned to purge the imperial administration. With the approach of Marlborough in southern Germany Eugène mobilised all available forces to join with the English commander and, by holding the superior forces under Marsin and Maximilian Emmanuel at the Battle of Blenheim on 13 August 1704, greatly assisted in the preparation of the final and victorious offensive against the French. Thereafter the two great commanders became close friends and formed an indomitable partnership.

Eugène then returned to Italy and claimed a victory at Cassano which, in fact, proved inconclusive. In 1706, however, he outgeneralled and outflanked the French force besieging Turin, and on 7 September dealt the French army a crushing blow, sweeping the investing force from the field. The French commander, the Duke of Orléans, was wounded, and Marshal de Marsin killed. In the course of the next three months Eugène drove the French out of north Italy.

Ordered to the Netherlands in 1708, Eugène joined forces again with Marlborough, and was present at the victory of Oudenarde on 11 July, which led to the capture of Ghent and Bruges; while the reduction of the fortress at Lille in December broke Vauban's (q.v.) hitherto impregnable defence system.

In the Turkish war of 1715–18 Eugène, previously the sworn enemy of France, advocated an alliance with the French Regent, the Duke of Orléans, thereby leaving the Empire free to assist Venice against the Turks. By his victory at Peterwardein in August 1716 he drove the Turks back, and he captured Belgrade a year later. The war was concluded by the Peace of Passarowitz in 1718, by which Austria gained Belgrade, Little Walachia, and part of Serbia; Eugène brilliantly administered these territories by bringing in German settlers to develop the hitherto sterile lands and provide a profitable buffer and stronghold against the Turks. He retired from active military campaigning and settled down to become the influential adviser to the Emperor Charles VI (as he had been to the Emperor Joseph I). When in 1725 England entered into alliance with France and Prussia, Eugène opted for an alliance with Russia and Prussia which he felt placed Austria in an unassailable position of leadership in Europe. However, Austria did not follow Eugène's advice to strengthen the imperial army, so that when, in 1733, the War of the Polish Succession broke out, Eugène was almost powerless against the French in Italy and on the Rhine, though he directed a masterly defence of southern Germany. He died in Vienna during the night of 20–21 April 1736.

Despite his early hatred of France and the French, Eugène was a passionate European devoted to the development of European culture. His fate and his career were linked to Germany and Austria, though he admired the English and Italians. As a military commander he was a master of movement and surprise attack, as displayed at Cremona and the crossing of the Alps; he excelled in the frontal breach, as demonstrated at Blenheim and Malplaquet. By personal bravery he could always inspire his troops, even when they were exhausted or badly-equipped. His campaigns provided classic examples studied closely by Maurice de Saxe and Frederick the Great (qq.v.). Above all, he was the instrument by which Austria displayed her leadership of the German states and rose to become a leading European power in the early eighteenth century.

Fairfax, Thomas, 3rd Baron Fairfax of Cameron (*1612–71*)

Parliamentary army commander-in-chief during the English Civil War, and later commander of the Commonwealth army, Fairfax was the true organiser of the New Model Army, and the victor, with Cromwell (q.v.), in numerous battles and engagements, including Winceby, Selby, and Marston Moor.

Born at Denton, Yorkshire, on 17 January 1612, the son of a Member of Parliament, Thomas Fairfax was educated at Cambridge, and fought in the Netherlands from 1629 to 1631. In the first of the Bishops' Wars (1639) Fairfax raised a force of dragoons, dubbed the Yorkshire Redcaps; for his services in this company he was knighted; and he went on to fight in the second campaign until the Treaty of Ripon brought hostilities to an end.

His father had from time to time opposed royal policies in Parliament, and so Sir Thomas Fairfax found himself on the Parliamentary side on the outbreak of the First Civil War in 1642. In September that year he was appointed to command the Parliamentary cavalry in Yorkshire. He occupied Leeds in January 1643, and forced the Royalists to evacuate Wakefield, but was defeated at Seaforth Moor, near Leeds, on 2 April. After a confused struggle he and his father were again beaten at Adwalton Moor, near Bradford, in June. Notwithstanding these setbacks Fairfax managed to execute a skilful withdrawal to Hull where, on 26 September, he met Oliver Cromwell for the first time. Together they fought and won the Battle of Winceby (in Lincolnshire), thereby relieving Royalist pressure in the north-east.

In January 1644 Fairfax was ordered to lead Parliamentary reinforcements to raise the Siege of Nantwich, which was invested by Irish Royalists and local forces: he routed the Royalists, almost half of whom defected to the victorious Parliamentarians. Returning to Yorkshire, Fairfax besieged Royalist forces under William Cavendish, Duke of Newcastle, at York, but on the approach of Prince Rupert (q.v.) and Royalist reinforcements he abandoned the siege and marched to meet them in the vicinity of Long Marston. In the battle that followed at Marston Moor on 2 July, both armies fielded about 7,000 cavalry – but the Parliamentarians deployed 20,000 infantry to 11,000 on the Royalist side. On the right of the Parliamentary line Fairfax's men suffered severely from attacks by Lord George Goring's cavalry; himself wounded, Sir Thomas rode over to the other flank to seek assistance from Cromwell, who then swung his highly-disciplined 'Ironsides' to the right, routed Goring, and helped to crush the Royalist centre. This victory led to the surrender of several northern Royalist strongholds, including York on 16 July, and Newcastle on 16 October.

In February 1645 Fairfax, at the age of thirty-three, was appointed Captain-general of the New Model Army, and it was he (not Cromwell as is often suggested) who organised its formation and training. Cromwell urged Parliament to adopt proposals to sustain a standing army of 22,000 men; but it was Fairfax who organised the impressment of manpower and its deployment into twelve infantry and eleven cavalry regiments, and one regiment of dragoons. Modelled superficially upon Cromwell's Ironsides, the New Model Army adopted for the first time uniform scarlet clothing, and (of far greater significance) overcame the traditional reluctance of local militias to campaign far from their home territory by the creation of a mobile, professional force.

The New Model Army was first ordered to relieve Taunton, but was recalled to blockade Oxford. When King Charles left the city on 5 May 1645 Parliament ordered Fairfax to 'march to defend the association'; reinforced by a cavalry force under Cromwell, he surprised Charles with about 7,500 Royalists at Naseby. After Prince Rupert's cavalry had routed the Parliamentary left flank and left the battlefield in reckless pursuit, Cromwell was able to smash the Royalist infantry; thereafter the Royalist collapse was swift. In the south-west Fairfax and Cromwell defeated Lord George Goring at Langport on 10 July, and Royalist strongholds at Bridgwater, Bristol, Leicester, and Winchester surrendered soon after. King Charles himself surrendered to the Scots on 5 May 1646.

Sir Thomas Fairfax's role in the Second Civil War and afterwards was to some extent placatory, as he was fearless in condemning the more excessive policies of the Parliamentary factions. He strongly disapproved of Colonel Thomas Pride's 'purge' of Parliament on 6–7 December 1648, and also of the King's trial and execution; but he agreed to continue to serve as a member of the Commonwealth's council of state. He remained Commander-in-Chief until 1650, when he refused to lead an expedition against the Scottish uprising of that year on the grounds that this would lead to an aggressive war. He resigned from the Army and retired to Nunappleton to 'write bad verse, grow roses and patronise historical research'. He died at Nunappleton on 12 November 1671.

Farnese, Alessandro, Duke of Parma (*1545–92*)

Alessandro Farnese was Governor of the Spanish Netherlands for fourteen critical years during their struggle for independence. He was also one of the leading generals of his day, in an age when the superiority of Spanish infantry in the open field was not seriously challenged anywhere in Europe.

He was born in Rome on 27 August 1545, the son of Ottavio Farnese, Duke of Parma, and Margaret of Austria, natural daughter of the Holy Roman Emperor Charles V. At the age of eleven the boy was sent to the court of Philip II of Spain; his status, although comfortable and honoured, was one of hostage. The court was at Brussels at this time; the greater part of the modern Netherlands, Belgium, Luxemburg, and the French provinces of Artois and Flanders had been under Spanish rule since the days of Charles the Bold of Burgundy. Alessandro accompanied Philip II on his journey to England, and was taken back to Spain in 1559. His mother Margaret was made Regent of the Netherlands in that same year, and in 1565 Alessandro returned to Brussels for his marriage to the Infanta Maria of Portugal. After this he was allowed to return home to Parma, but he soon became bored. In Spain he had met Philip's soldier half-brother Don John of Austria, and now he obtained permission to serve under him. He fought under John's command at

Lepanto in 1571 and in Italy until 1574; in 1577 he was summoned to the Netherlands, to reinforce John in his efforts to suppress the revolt. This had been raging off and on since 1566, when rioting had brought the Duke of Alva (q.v.) and a Spanish army to restore order. When Farnese arrived a temporary truce had just broken down after the failure of complicated negotiations, and Don John was preparing to reconquer the provinces.

On 31 January 1578 Farnese made an important contribution to Don John's crushing defeat of a Dutch army at Gembloux: he overcame their rearguard, and led his cavalry into their retreating infantry columns, causing great execution. Between February and September Don John was partly successful in reimposing Spanish rule in the southern provinces; the country was open, and suitable for his troops. The campaign then ground to a halt because of internal problems on both sides. In October Don John died, and Farnese was appointed Regent in his place. His first major success was the conclusion in May 1579 of the Treaty of Arras, by which the Catholics of Douai, Artois, and Hainault gave up the rebellion and reaffirmed their loyalty to the Spanish throne. Having thus weakened his enemies, Farnese prepared to move against the rebellious Union of Utrecht, a confederation recently formed among the northern provinces which acknowledged William the Silent, Prince of Orange, as Stadtholder. The political background to the next ten years' fighting was extremely complex, involving not only religious but also dynastic implications for all the ruling houses of Europe; Farnese sturdily ignored these factors and concentrated on the military reduction of the Netherlands. In March 1579 he took Maastricht, which was sacked with great ferocity, and then proceeded to take city after city. Tournai fell to him in 1581, but in 1582–3 he was distracted by the necessity of campaigning against François of Anjou, the French pretender. In 1584 he returned to his central task, and in that year the patriots lost Ypres, Bruges, and Ghent. Late in the same year Farnese began his great siege of Antwerp; he had a bridge built across the Scheldt to prevent relief by sea – Dutch mastery of the coastal waters had long been a thorn in the Spanish side – and on 17 August 1585 the city fell to him. The southern provinces of the Netherlands were now firmly in Spanish hands; and William the Silent, the spirit behind the revolt, had been assassinated in July 1584. (Farnese had condoned several previous attempts on the life of the much-respected Prince of Orange; this, with his apparent indifference to the excesses committed by his troops among the civilian population, must remain a shadow on his name.) On the death of his father in 1586 he became Duke of Parma, but was retained as Regent of the Netherlands.

An English expedition to help the Dutch patriots arrived in 1585 under the Earl of Leicester, and operated in the Netherlands until 1587, but its contribution was negligible; notable were Leicester's failure at Zutphen in 1586, and Parma's capture of Sluys in August of the following year. Parma was now prevented from pursuing the final destruction of the patriots by Philip II's determination to invade England. Parma was forced to assemble a force of 25,000 men, badly needed elsewhere, for the proposed invasion, and to gather a fleet of transports on the coast. Even when the scattering of the Armada in 1588 doomed the invasion attempt, Parma was not able to concentrate on the Dutch, but was forced to become involved in French affairs. In 1589 the Protestant Henry of Navarre had come to the French throne as Henry IV (q.v.), but civil war had continued, and Parma was ordered to support the Catholic League against Henry. His troubles were increased by the weakness of the Spanish treasury at this time; Parma was forced to pledge his personal jewels in order to pay his army, which was disturbed by mutinies over withheld pay in 1590. In August of that year Parma led 15,000 men into France to relieve Catholic Paris, besieged by Henry IV. For the next two years he and Henry fought a skilful war of manœuvre in northern France, but there were few serious engagements. Simultaneously Maurice (q.v.) of Nassau, the son of William the Silent, took advantage of Parma's absence to win a series of startling victories in the Netherlands. Denied strategic concentration on one task at a time, Parma was moved backwards and forwards, unable to achieve decisive results anywhere. It was in this period of frustration that he received the wounds from which he died at the Abbey of St Vaast, near Arras, on the night of 2–3 December 1592. He was given a magnificent funeral in Brussels, and was buried in Parma beneath a monument bearing an epitaph of massive simplicity and arrogance: 'Alessandro'.

Parma was one of the most able generals of his day. He was a good strategist who was denied by his King the opportunity and the money to carry out his strategy. His troops were the finest infantry in the world at that time – a delicately balanced combination of pikemen, heavy musketeers, and light arquebusiers on whose best employment he was an expert; but their formations were those of the open field battle, and Parma's army fought amid the foggy streams and mud-flats of Holland. He was also accomplished in the employment of his superb Spanish engineers, and on more than one occasion used them with great imagination to solve a battlefield problem, as well as directing them in the more conventional exercises of siege warfare.

Foch, Ferdinand (*1851–1929*), *Marshal of France*

Allied generalissimo on the Western Front in 1918, and architect of the plan to halt the final great German offensive of the First World War, Foch was also the French commander whose parsimonious release of reserves earlier in the war was one of the main reasons for the enormous loss of life on the Yser and at Ypres.

Born the son of a civil servant at Tarbes (Hautes-Pyrénées) on 2 October 1851, Ferdinand Foch received what was then regarded as a conventional Jesuit upbringing. He enlisted in the French infantry in 1870 but saw no active service during the Franco-Prussian War. He entered *L'École Polytechnique* at Nancy during the Prussian occupation, and gained a commission with the 24th Artillery Regiment in 1873. Foch quickly distinguished himself as a horseman, but while others gained honours in the colonies the world over, Foch's progress and promotion were slow by comparison. The result was a marked tendency towards an academic approach to tactical problems, with little real understanding of field conditions. He was connected with *L'École Supérieure de la Guerre* three times: as a student in 1885, as a lecturer in military history ten years later, and as director in 1908; he was also an avowed disciple of Prussian military doctrines, particularly those of Clausewitz (q.v.). Nevertheless, realising the inevitability of war with Germany, it was much to Foch's credit that he, probably more than any other Allied commander, made real efforts to study the military strategies of the Kaiser's generals.

His own special contribution was a staunch belief in the importance of the will to win, and he it was – perhaps more than any other French general – who nurtured the offensive spirit, the famous *élan*, among his troops. His conviction, developed before the war, that superiority of support fire-power was of vital importance in any victory was later perverted into a blind doctrine of *l'offensive à l'outrance* – attack at all costs, at any cost, and without consideration of alternatives. But it must be said that the responsibility for this perversion lies with Foch's subordinates.

In 1913, at the age of sixty-one, Foch was appointed to command the crack XX Corps on the frontier with Germany, with his headquarters at Nancy. At the beginning of the war the Germans withdrew on the XX Corps' front before turning to inflict a severe defeat on the French army at Morhange. The XX Corps displayed the greatest *élan*, but in the face of the Germans' disciplined control of counterfire, the French suffered huge losses. Foch, far from losing the support of his superiors, was withdrawn and given command of the new French Ninth Army. He soon found himself heavily engaged in the Battle of the Marne, situated on the right flank of the Allied armies which stood before Paris. A dangerous gap existed between the French Ninth and Fourth Armies, and it was on the edge of this gap that the Germans launched a colossal four-division bayonet attack on the night of 8 September 1914. Although much of his Army was routed and thrown into confusion, Foch ordered an immediate renewal of his own attack, and in spite of casualties in excess of 35,000 men, succeeded in halting the German penetration. Meanwhile the French Fifth Army was launching successful counter-attacks, with the British Expeditionary Force, on the left flank, and the German offensive was turned. Although Foch's own contribution had lain in withstanding enemy attacks rather than in inflicting any specific defeat, his conduct so impressed Joffre (q.v.) that he was appointed to co-ordinate the Allied Armies of the North – an extraordinary task, since no authority had been given or discussed for the overall command of British and Belgian armies by the French. Foch's task would have been impossible had he not exercised sole control over French reserves, and the parsimony he displayed in their use, both then and later, while probably correct in the long term, undoubtedly endangered the very survival of the northern armies, infuriated field commanders, and cost the lives of tens of thousands of French, British, and Belgian soldiers. In the second Battle of Ypres during April and May 1915, following the German use of poison gas, a large gap was torn in the Allied line, and Foch committed inadequate reserves to bolster his two terrorised French divisions; only the lack of German preparation to exploit the breakthrough saved the Allied line from catastrophe. British losses alone were 60,000 dead or seriously wounded.

At the end of 1916 the failure of the Battle of the Somme to achieve any worthwhile results brought about the replacement of Joffre by Nivelle. As army group commander, Foch followed his chief. His next responsibility was to plan the supply of military aid to Italy in the event of Germany's materially reinforcing the Austrians in that theatre.

On the Western Front Nivelle's ambitious offensive failed, and its architect was dismissed in May 1917, to be replaced by General Philippe Pétain (q.v.), the 'saviour of Verdun'; Foch replaced the latter as chief of the general staff. Meanwhile Russia had collapsed, and by the end of the year German divisions disengaged from the Eastern Front were streaming westwards for the anticipated ultimate offensive. Recognising the danger, Foch strove to create an international military reserve for the Supreme War Council, but neither French nor British commanders would yield a single man to be controlled and commanded by such a committee.

The German offensive opened on 21 March 1918, in heavy fog, on a sixty-mile front between Arras and La Fère; its object was to split the Allied front by driving in a deep wedge on the Somme, and following this by a massive thrust towards Paris. It was expected that the British would be crushed against the Channel coast, and that Pétain would be destroyed while falling back on Paris. Largely as a result of Haig's (q.v.) initiative and support, Foch was appointed Allied generalissimo at Doullens. He immediately set about a reorganisation of the crumbling front – a reorganisation which must stand as his greatest triumph. Once again his cautious application of reserves – though now French manpower had become so depleted by four years of exhausting warfare that few effective reserves remained – angered the British command, and the German offensive was not immediately held. But the growing resistance, to some extent aided by the arrival of American troops in France, gradually exhausted the German effort. The American General Pershing's (q.v.) insistence upon recognition of an independent American sector seriously upset Foch's plan to operate a co-ordinated counter-offensive under a single unified command, and led to continued acrimony. However, the Allied commanders eventually compromised and the co-ordination painfully agreed between Pershing and Foch was a vital factor in Foch's final triumphant counterblow. It should be said, however, that Foch's influence in the wider sphere of command was never as great in offensive as it had been on the defensive. His choice of objectives was frequently questioned by Haig and Pershing – who usually got their way. Haig, however, could never be considered disloyal; on the contrary, he followed Foch's vital war directive (issued on 3 September 1918) to the letter.

In August 1918 Foch received the baton of a marshal of the French army, and on 19 July the following year he was gazetted as a British field-marshal. From January 1920 he was appointed President of the Allied Military Committee at Versailles to administer the terms of the Armistice – a period of his career marred by factional quarrels in the French High Command, and one not simplified by the suspicions constantly voiced by British politicians determined to wring the last ounce of reparations from the defeated and potentially anarchist Germany.

Foch died in Paris on 20 March 1929, and his body was placed under the dome of Les Invalides with those of Napoleon and Turenne. Undoubtedly his military strategy was vindicated by final triumph; yet he must remain responsible for his near-oriental wastage of human life, born of his support for the mystique of French *élan* and the doctrine of *l'offensive à l'outrance*. Yet for all the sacrifices imposed on France, only a man of Foch's military stature was able to turn the supposedly inevitable defeat at the hands of Ludendorff (q.v.) in 1918 into final victory. It was the very magnitude of the bloodbath of Flanders and Verdun that impressed it upon the commanders of the Second World War that such static warfare must be avoided at all costs.

Frederick I, called Barbarossa

(*c. 1123–90*), *Emperor of the Holy Roman Empire*

Nephew of Conrad III, the first Holy Roman Emperor of the Hohenstaufen dynasty, the gifted Frederick accompanied him on the disastrous Second Crusade (1146–8). In 1152 Frederick himself succeeded as Emperor, and was crowned at Aachen on 9 March. He pursued sporadic wars against the Welfs under Henry the Proud, Duke of Saxony, and was victorious in a number of campaigns in Poland, Bohemia, and Hungary, in addition to various campaigns in Italy.

Germany during the twelfth century was experiencing severe internal religious troubles brought about by the hundred-years-old papal insistence upon Rome's right to appoint bishops – a right which threatened imperial control over the clergy in Germany. These troubles came to a head in the mid-twelfth century, and prompted Frederick to move against Rome. Having secured peace at home and obtained homage from King Svend III of Denmark, Frederick embarked on the first of six expeditions to Italy which were to engage his main energies for thirty years. Setting out in 1154, he seized and hanged Arnold of Brescia, who held sway in Rome. Sectarian troubles at home demanded his return to Germany almost immediately afterwards. Further expeditions followed in 1158 and 1163. During the first of these Frederick put down a rebellion in Milan and razed the city to the ground.

In the fourth expedition (1166–8) Frederick succeeded in capturing Rome but was forced to return home once more when plague decimated his forces. Militarily his fifth expedition was the most significant, though again he achieved very little. It ended disastrously at the great Battle of Legnano on 29 May 1176 when Frederick, believing he could defeat a superior force of infantry using only cavalry, was routed by the Army of the Lombard League. In the event his cavalry charges were repulsed by the steadfastness of the Lombard pikemen, and his forces were then enveloped by the opposing cavalry. (It was not, as is often stated, a victory of infantry over cavalry, but rather a victory obtained by the co-ordination of infantry and cavalry.) Relations with the papacy continued strained, despite periodic reconciliations; after Pope Urban III had contrived to support a revolt by Philip of Heinsberg, Archbishop of Cologne, further warfare might have broken out had not the Third Crusade intervened.

As preparations were made for the forthcoming Crusade, Frederick took up the Cross in late 1188, and in May the following year set out from Regensburg with a splendid army. Overcoming the hostility of the Byzantine Emperor Isaac II Angelus, he marched into Asia Minor, but on 10 June 1190, while crossing the Saleph, Frederick was drowned. He left five sons by his wife Beatrix, of whom the second later reigned as the Emperor Henry IV.

Frederick's military stature rested solely upon his earnest and commanding personality. His military status was maintained on the pretext of championing the German Church and nobility – who in return afforded him active support. Nevertheless much of this support was squandered by inept generalship, and he was frequently headstrong to the point of recklessness. He was, however, a ruler respected for the care with which he exacted his feudal dues. In appearance he is said to have been well-proportioned, of medium height, with flowing yellow hair and a reddish beard – hence his Italian surname.

Frederick II, called the Great (*1713–86*), *King of Prussia*

The outstanding general of the eighteenth century, Frederick II's military developments rendered the Prussian army technically superior to any contemporary neighbour for fifty years.

Grandson of King George I of England and elder surviving son (among fourteen children) of Sophia of Hanover and King Frederick William of Prussia, the Crown Prince Karl Frederick was born on 24 January 1713. He was a weakling in his childhood but was destined, as were all male members of the Prussian royal family and nobility, for a military life. In spite of a singular lack of academic aptitude (he spoke and wrote German indifferently and used French as a mother tongue), he was placed in command of one hundred cadets at the age of nine. Before long, however, young Prince Frederick outraged his father by developing a preference for the 'soft' life of music and literature, and a profound distaste for the army. After a public admonition from the King, Prince Frederick was made a major in the Potsdam Guards at the age of fifteen, but four years later he attempted to desert and leave the country. He was arrested; his accomplice (a fellow-officer) was condemned and executed; and he himself was incarcerated at Küstrin, under sentence of death by strangulation. However, after intercession by several foreign diplomats, and having declared himself penitent, he was released by his father, who restored him to his former rank and regiment.

Thereafter Prince Frederick appears to have been reconciled with King Frederick William, and certainly applied himself to his military duties. At the age of twenty he was appointed to the colonelcy of the Von der Goltz infantry regiment, a post he filled with apparent enthusiasm and ability, with the result that the King allowed him to volunteer for service with the Prussian contingent fighting the French on the Rhine, where he served for several months under the great Prince Eugène (q.v.).

On 31 May 1740 King Frederick William died and Prince Frederick was proclaimed king. He inherited the strongest *per capita* army in Europe, built up by his father to 80,000 men recruited from the dregs of society, but officered – reluctantly – by members of the nobility. By the use of iron discipline and the training genius of Prince Louis of Anhalt-Dessau – 'the Old Dessauer' – King Frederick William had transformed this unpromising material into a highly effective force.

Frederick inherited other leanings from his father: notwithstanding his culture and sensitivity in artistic matters, he was a coldhearted, ruthless disciplinarian – personally a man of honour, but as a ruler treacherous, untrustworthy, and as sly as a fox. Such a complex personality and such genius defy analysis; but his military achievements command admiration.

When the Emperor Charles VI of Austria died on 20 October 1740, Maria Theresa inherited the Hapsburg Empire in accordance with the Pragmatic Sanction, but her succession was disputed by three foreign claimants. Frederick II offered support to Maria Theresa against these claimants, but in return announced his intention to occupy Silesia, pending the settlement of an old claim for cession of the Brandenburg province to Prussia. When Maria Theresa refused to accept this arrangement Frederick promptly invaded Silesia with 28,000 men.

The First Silesian War started with a near-walkover for Frederick, whose invasion, early in 1741, surprised the scanty Austrian garrisons and shut them up in the towns, including Neisse and Glogau. Meanwhile Count Niepperg set about raising a formidable Austrian army in Bohemia, which in March and April surprised Frederick in turn and regained much of the lost territory. Confrontation occurred at the Battle of Mollwitz on 10 April, and Frederick was obliged to flee the battlefield when his flank cavalry was badly defeated by superior Austrian cavalry; but the Prussian infantry – commanded by Count von Schwerin, a veteran of Blenheim – stood firm, counter-attacked, and eventually drove the Austrians from the field. This was the only time that Frederick ever left a still-disputed battlefield, but – far more significant – it stiffened his resolve to reorganise and greatly improve the Prussian cavalry, a process that occupied much of his energy during the following twelve years.

The other important battle of the First Silesian War was fought just over a year later near the village of Chotusitz, on 17 May 1742, and was decisive. The early stages of the battle were almost exactly the reverse of those at Mollwitz. On this occasion it was the Prussian flank cavalry – already displaying the benefits of Frederick's attention – that rode down the Austrian horse, while the Austrian infantry marched into the confined spaces of the village where the opposing cavalry found it difficult to engage. A hard-fought infantry battle followed, but it was the efforts of the Prussian cavalry that carried the day, and the Austrians were decisively beaten and suffered heavy casualties. This defeat prompted Maria Theresa to sue for peace with Frederick by the Treaty of Breslau, which ceded Silesia to Prussia; it also sowed widespread belief in the invincibility of the Prussian army.

The danger from Austria still loomed over Prussia, however, and during 1743–4 French reverses, particularly at the Battle of Dettingen on 27 June 1743, threatened to leave Frederick alone to face the power of Austria. He also suspected that Maria Theresa was planning to reoccupy Silesia. He accordingly allied himself once more with France and Bavaria, and when Austria invaded Alsace, marched an army of 80,000 men in three columns through Saxony, Lusatia, and Silesia into Bohemia, capturing Prague on 2 September 1744.

Following the Treaty of Füssen of 22 April 1745, by which Austria contrived a reconciliation with Bavaria (after her army had caught the Bavarian army in winter quarters and defeated it at the Battle of Amberg on 7 January), Frederick found himself isolated from France, his only effective ally. His army in Silesia was constantly harassed by Austrian and Hungarian irregulars, but by the cunning use of spies and by providing the enemy with false intelligence he concealed the concentration of his army of 60,000 men, and then quickly and secretly moved it to confront the main Austrian army of 80,000 men under the Emperor Charles VII at Hohenfriedberg during the night of 3–4 June. At dawn the following day Frederick struck, and within two hours he had won a battle which he claimed was as decisive as any since Blenheim. The Prussian infantry simply overwhelmed the Austrians with their superior discipline and musketry, while the imperial horse was unwilling to close with Frederick's new model cavalry. Austrian losses amounted to 9,000 dead and wounded, 7,000 prisoners, and 8,000 deserters. Frederick's losses were barely 1,000 men.

Now, however, Frederick's army began to feel the full effects of its isolation and the aggravation of harassment by the guerillas; the Prussian leader therefore determined to break back to Silesia. He halted at the village of Sohr with 18,000 men on 30 September, but soon found that he had been outmanœuvred by Charles's army, which had seized high ground overlooking his right rear. Believing that Frederick would not dare assault this superior position, the Austrians opened heavy artillery fire on the Prussians below them, but, quite undaunted, Frederick wheeled his entire army and threw his whole weight in echelon on the Austrian left wing, smashing all resistance. The dazed Austrians retreated to the north-west, leaving 8,000 dead and wounded, twenty-two guns, and the passes open for Frederick's unrestricted withdrawal into Silesia.

Both Charles and Frederick each believed the other would winter his army. Frederick indeed wintered his, but Charles reinforced his columns and invaded Prussia in concert with a Saxon army under Marshal Rutkowski; they converged on Berlin. But Frederick's spies quickly discovered the Austrian plan, and Prussian elements of light horse, cuirassiers, and foot struck the Saxon marching columns at Hennersdorf on 23 November and inflicted heavy casualties. Having lost the element of surprise, Charles turned about and retreated into Bohemia. A further defeat for the Allies at Kesselsdorf on 15 December, magnificently fought and won by Prussian infantry, supported by the devastating Prussian cavalry, advancing uphill in snow under fire from 9,000 muskets, spelt the end of the Second Silesian War. The Treaty of Dresden was ratified ten days later.

Prussia had repeatedly faced defeat in both wars, but had been saved by the invincibility of its infantry and cavalry, both brilliantly led, both superbly handled in conjunction by Frederick, whose audacity and tactics invariably surprised and overwhelmed his opponents.

The Silesian Wars introduced new tactics which, despite its successes, found Frederick's army lacking in several important respects. In general his infantry was adequately established, although it was progressively increased in strength. The line cavalry, which had failed at Mollwitz, was, however, inadequate to meet the harassing tactics of the irregular Magyar hussars and Croat light horse; Frederick therefore formed a new guards regiment (the *Garde-du-Corps*), five new dragoon regiments, and, during the period 1742–73, no fewer than eighty-one German and nine Bosnian hussar squadrons. In 1740 he had inherited artillery from his father amounting to no more than six field companies, but the following year he doubled this establishment; within twenty years his artillery had increased to thirty companies, organised in three distinct branches – Fortress, Regimental, and Battery Artillery. In 1759 he also introduced fast horse-drawn field artillery into the Prussian army and, as he had seen it used successfully by the Russians, applied it in battle during the Seven Years' War in support of his already-formidable cavalry.

For all his military ascendancy, Frederick pursued an arrogant diplomacy against his neighbours after the two Silesian Wars, thereby alienating the French king and the Russian empress; and an inept conspiracy to disrupt the Anglo-Russian coalition resulted in a reactionary alliance being made between France, Russia, and Austria against Prussia, which retained a somewhat incoherent coalition with England. Frederick did little to foster this coalition when he pre-emptively attacked and invaded Saxony on 29 August 1756 without consultation with England, his pretext being that he anticipated an Austro-Russian attack on himself. Leaving 80,000 men to protect his east and west frontiers, he struck deep into Saxon territory with 70,000 men, occupying Dresden on 10 September, and thereby precipitating the Seven Years' War – a foolhardy conflict that was to kill a quarter-of-a-million men, and to bring exhaustion and near-bankruptcy to the principal participants, and little territorial benefit to anyone.

Frederick's first success in the war came about when, having shut up and invested the Saxon army at Pirna and beaten off all attempts at its relief, he finally starved it into surrender in October; he thereupon dismissed the Saxon officers and incorporated the entire fighting force into his own army. The following spring he invaded Bohemia with an army of 64,000 and moved on Prague, where Prince Charles of Lorraine had assembled about 70,000 troops. On 6 May 1757, after a short, savage battle, Frederick routed the Austrians, although the fight was marked by several critical mistakes in the conduct of Frederick's attack. He had inadequately reconnoitred the ground over which he intended to move, and when his infantry moved forward they found themselves floundering in deep muddy ponds at the mercy of Austrian artillery at 400 paces' range. The veteran Schwerin was killed while leading the first assault. Frederick, seeing his infantry hopelessly tied down, then sent his cavalry to envelop the Austrian right flank, and thereby opened a gap in the opposing centre. Penetrating this gap, he threw the Austrians back into Prague which he then proceeded to invest. Losses in this battle were severe, each side losing about 14,000 men.

These actions in the south were but a prelude to a massive movement against Prussia from all sides. In July that year 100,000 French under Marshal d'Estrées invaded Hanover, while 24,000 French and 60,000 Austrians under Charles, Duke of Soubise, moved up from Franconia in an attempt to join forces with Estrées. The Austrian army, now 110,000-strong under Marshal Daun, was moving against Prussia from Bohemia, while 100,000 Russians under Marshal Apraskin were invading East Prussia – albeit at a snail's pace. 16,000 Swedes also landed in Pomerania.

Such an overwhelming attack would have daunted any lesser man, but Frederick now embarked on an amazing series of lightning campaigns aimed at defeating the invaders before they could join forces and thus present an unbeatable single threat. He first moved west by forced marches with only 23,000 men to meet the invaders from France; but neither the French nor the Austrian army would give battle, and the Austrians fell back before him to Eisenach. Frederick then hurried back to block the Austrian invasion from Bohemia and the lethargic Russian advance on Berlin. Leaving token screening forces to delay these attacks from east and south, Frederick was back in the west by the end of November facing 64,000 imperial troops who had entrenched on commanding ground at Rossbach. There followed one of the classic battles of the period, in which the skilful use of disciplined men defeated clumsy tactics by a numerically superior force. The French detached 40,000 men to envelop Frederick's left flank, whereupon he pretended to withdraw his entire army, but instead stealthily wheeled it in echelon to the rear to face the envelopment, so that when the imperials attacked in three columns they were suddenly faced at short range with murderous fire from Frederick's artillery. Before they could recover from the shock the Prussian cavalry was unleashed (under General von Seydlitz, q.v.) and in less than two hours the imperial army was routed with the loss of 8,000 men, the fugitives being joined in flight by the 20,000-odd troops that had not been fully committed. Prussian casualties were no more than 500.

Hearing that Prince Charles's 80,000-strong Austrian army was moving towards Breslau in Silesia, Frederick then hastened east with 36,000 troops to meet the invaders at the Battle of Leuthen on 6 December 1767. The Austrians, drawn up and partly entrenched on a five-mile front behind a line of low hills, and anticipating that Frederick would employ his favourite enveloping tactic, retained their main reserves behind their left wing, while the right wing was positioned behind marshy ground. Frederick moved against the Austrian right but, under cover of the screening hills, changed direction diagonally to the right, at the same time leaving his cavalry to give a demonstration against the Austrian right. Believing that the Prussian attack would fall on this sector of his line, Charles moved his reserves to this flank, but almost simultaneously Frederick's infantry fell upon the Austrian left with devastating artillery support. The Austrians tried desperately to re-form left to face the Prussian attack in echelon, and in an attempt to divert Prussian weight launched a cavalry assault on the right wing against the Prussian horse. A delayed counter-attack by the Prussian rear cavalry formations caught the Austrians off balance, and a bloody rout ensued. Only darkness enabled Charles to retire across the Schweidnitz river, his army shattered by the loss of 20,000 prisoners, nearly 7,000 dead and wounded, 116 guns, and fifty-one colours. The battle of Leuthen was probably the greatest battle ever fought by one of history's greatest tacticians.

Frederick's victories came thick and fast during the next three years. At the bloody battle of Zorndorf on 25 August 1758 he brought his brilliant tactics to bear upon the Russian army under Fermor; a costly stalemate threatened until, at the critical moment, Seydlitz led his cavalry into the milling Russian infantry, and sheer butchery was only ended by the coming of night. Fermor's army lost nearly 20,000 men, but the Prussians, utterly exhausted and with 14,000 casualties, could not pursue. The first phase of the war ended at the close of 1758: Frederick was in control of Saxony and Silesia and had an unbeaten army of 150,000 troops in the field. But the great Prussian army was by this time in decline: during the previous eighteen months no fewer than 100,000 of the superbly-trained veterans who had been the backbone of Frederick's astonishing campaigns had fallen.

In the summer of 1759 the next stage of the war opened with invasions by the Russians under Count Soltikov and the Austrians under Laudon. Frederick, gathering all available forces, met the invaders at the disastrous Battle of Kunersdorf on 12 August; in this the Prussians attempted a double envelopment but, losing their way in the intervening woods, were thrown back with heavy losses. The defeat cost the Prussian army about 20,000 men and 178 guns, but the victors were too exhausted to follow up their success. Such was the effect of this disaster upon Frederick that he seriously considered abdication, but realising that the battle had not in fact been entirely one-sided in result he decided to continue in command.

Astonishingly, he quickly recovered the initiative, but his subsequent successes (the Battle of Liegnitz on 15 August 1760 and the

Battle of Torgau on 3 November 1760) were largely victories of survival, and at Bruzenitz in August 1761 Federick only escaped total annihilation by overwhelming Russian and Austrian forces as the result of vacillation on the part of the Allied commanders. In 1762 the death of the Empress Elizabeth of Russia was followed by the succession of Peter III, an admirer of Frederick, with the result that the Treaty of St Petersburg ended hostilities between Russia and Prussia – indeed Frederick even received Russian reinforcements from Peter. Events now moved fortuitously in the Prussians' favour, and when the Treaty of Hamburg ended hostilities with Sweden, Frederick was able to concentrate his efforts against Austria. After significant victories at Burkersdorf (on 21 July 1762, when Seydlitz's cavalry once again won the day) and at Freiberg (on 29 October 1762) an armistice was agreed between Prussia, Austria, France, and England in November 1762; the Seven Years' War was formally ended by the Treaty of Hubertusberg, by which Frederick was, after all, permitted to retain Silesia.

Despite the eclipse of Prussian brilliance in the closing stages of the war, Frederick had acquired and retained widespread recognition as the leading military strategist, tactician, and administrator of the age. His last twenty-four years of rule were spent in giving further polish to his army, but more and more he retired into the loneliness of old age. His long-deferred inclination towards the arts now reasserted itself, but he remained concerned at the inevitability of the succession of his undistinguished nephew, Frederick William II. At the age of seventy-three *Der alte Fritz* died at Sans Souci on 17 August 1786.

It is not the purpose of this work to dwell at any length on activities and attributes other than those of a military nature, but an unbalanced impression of Frederick's reign will be given unless it is said that he was as great a ruler as any in the eighteenth century. In Prussia he created a state of 'enlightened despotism', in itself a contradiction, for anomalous with his country's expansionist drive was his tendency towards a humanitarian administration. His own personality represented an unresolved tension between his artistic philosophies and a ruthless brilliance on the field of battle. It has been suggested that it was his creative capacity which spurred his passion to improve 'his' army. Whatever the inner motivations of this extraordinary, contradictory personality, Frederick is rightly honoured as one of the half-dozen 'great captains' of history – a worthy rival to such names as Gustavus Adolphus and Napoleon (qq.v.).

Gamelin, Maurice Gustav (*1872–1958*), *General*

Gamelin was a brilliant French divisional commander in the First World War, but his command policies and outdated strategies were largely responsible for the French collapse in the west in 1940, and he was eventually brought to trial for his part in the defeat of France.

Born in Paris on 20 September 1872, Maurice Gamelin later entered St Cyr, and graduated in 1893 with a commission in the Algerian *Tirailleurs*. By the outbreak of the First World War he had risen to colonel, and as a senior staff officer on General Joffre's (q.v.) staff he was personally responsible for drawing up command orders which resulted in victory on the Marne in 1914. Promoted major-general the following year, he was given command of a division – at the age of forty-three he was probably one of the youngest French divisional commanders, and by the end of the war proved himself to be one of the best.

Rising to general in 1931, Gamelin was appointed chief of the French general staff, and by the mid-1930s should have been energetic in forcing through substantial rearmament plans in the knowledge of German military resurgence. As it was, Gamelin's strategy was largely one of passive defence based upon the concrete of the Maginot Line, while his tank force – to all outward appearances one of formidable numbers – was almost entirely outdated. When the Second World War broke out Gamelin, now nearly sixty-seven years of age, was still in the Army, having just previously been appointed its Commander-in-Chief. Realising that the safety of France behind the Maginot Line depended solely upon frontal defence, he issued instructions as early as 15 November 1939 for the wheeling forward of Allied forces on the left flank in the event of Germany's attempting to outflank the Maginot Line by striking through Belgium. When the blow fell the movement forward was delayed, with disastrous results, principally because of reservations voiced by the Belgian government; Gamelin later professed that he had known nothing of any such reservations beforehand, and stated his conviction that had the Allies not been thrown forward, the Belgian Army in any case would not have offered any resistance to the German invasion.

As his front line crumbled and the French armies retreated in the face of the German armoured thrusts, Gamelin was relieved of his command and replaced by General Weygand – whose efforts to stem the flood proved no more effective. Gamelin was afterwards arrested and placed on trial for his defeat, but was able to make a strong case for his actions, with the result that the trial was abandoned. In 1943 he was, on the insistence of the Nazis, interned in Germany, where he remained until the end of the war. He died in Paris on 18 April 1958.

Apart from his divisional command, the career of Gamelin in the broad perspective of world war was of little or no consequence. Suspicion has been voiced that he planned to attempt a second battle on the Marne in May-June 1940; if such was his intention, this would have provided all the evidence needed to support the view that his whole understanding of land warfare was thirty years out of date. In the age of fast armour and dive-bombers one shudders to think of the result of any 1914-pattern stand before Paris.

Genghiz Khan (*1167?–1227*)

Temujin, later Genghiz Khan of the Mongols, was one of the half-dozen great captains and conquerors of recorded human history. (The spelling used here is conventional, although 'Chinggis' is thought to be the nearest transliteration to the original. Dates are also confusing, due to the difficulties of exactly reconciling different calendars. Some sources maintain that Genghiz Khan was born in 1155 or 1162; here the sequence used in the *Encyclopedia Britannica* is followed.) His military record seems to be without flaw; his impact on the world of his times was revolutionary, and the fact that his name is still in household use as a synonym for an invincible conqueror speaks for itself.

So complex and ephemeral were the exact relationships between the peoples and nations of central Asia in the twelfth and thirteenth centuries that no detailed background description can be attempted; in any event, the Khan's importance lies in his tactical and organising brilliance, not in the structure of the enemies he overthrew. Briefly, the mid-twelfth century found a great variety of nomadic peoples of Turkish and Mongol stock spread across Asia from Turkestan to the Sea of Japan, and from Siberia to the Himalayas. The Chin Empire was centred on Peking and comprised, roughly, Manchuria and northern China. The Sung Empire, driven from these regions by the Chin, held sway to the south on the Yangtse. A semi-independent buffer state, the Hsi-Hsia, existed on the north-western frontiers of China. The Tartar nomads and the various Mongol clans occupied the northern and eastern regions of central Asia, and people of Turkish stock the southern and western areas. In the west lay the Empire of Khwarezm, inhabited by Turco-Iranian peoples and bounded roughly by the Persian Gulf to the south-west, the Caspian Sea to the west, the Kipchak Steppes north of the Caspian and Aral Sea to the north, the river Syr-Daria (Jaxartes) to the north-east, and the Hindu Kush Mountains to the south-east; its great centres were Samarkand and Bokhara. The nomad tribes of the grassy steppes and the Gobi Desert were as physically tough, and as ruthless, as any people on earth. They lived in tents, and their wealth was based on horses and draft animals. Those who lived to be adults were inured to hardship and privation, extremes of weather, and a constant state of bloody warfare. They were expert horsemen, archers, and javelin fighters. By this stage in their history many of their leaders had also acquired a sophisticated insight into power politics. Not only were inter-tribal feuds a way of life, but their static neighbours had long engaged in a tangled policy of alliance, treachery, and manipulation among the tribes in order to safeguard their frontiers.

In about 1167 a Mongol chieftain named Yesugei had a son whom he named Temujin, in honour of a courageous Tartar chief whom he had recently killed. When he was eight the boy's father was killed by Tartar enemies, and for the next few years the fatherless family lived a precarious life, exploited and abused by other relatives and without a man to defend them. At an early age Temujin learnt self-sufficiency, stoicism, and cunning. Achieving warrior status in his mid-teens, he claimed the bride to whom he had been betrothed in childhood and used her dowry to cement relations with a powerful ally. For the next thirty years he devoted himself to a single-minded process of building and extending a power-base among the Mongol clans. By judicious alliances, raids, treacheries, and negotiations he spread his domination from his original group of young followers to a unified nation of more than thirty tribes. During the years of inter-clan warfare he did not automatically massacre defeated rivals, but tried to bind able men to his cause. He successfully eliminated his opponents one by one, and then turned on the Tartars. By 1194 he was secure enough to assume the title Genghiz Khan; its meaning has been the subject of endless controversy, but the most likely translation is something like 'universal lord', or alternatively 'rightful lord'. Another twelve years of campaigning and consolidation passed, however, before he was finally confirmed in the title by the election of the entire Mongol federation which he had welded together. Considering the environment in which he lived and worked, his vision and strength of character are awe-inspiring. He was soon in a position to turn the weapon he had forged on the outside world. The aim behind his policy of conquest was to found an empire in which a nomad *élite* lived and fought (the two were inseparable) supported by tribute from static subject states.

Each tribal chief in the confederation was responsible for answering the call to muster instantly, with his riders in full readiness for a campaign. The army was entirely composed of cavalry, divided into groups of 10,000 men – *toumans* – and subdivided into regiments, companies, and sections according to the decimal system, down to squads of ten riders. Each man had a spare pony, and carried the basic equipment for maintaining himself and his mounts in the field; thus there was no need for any but the most rudimentary supply train, and the army lived off the country through which it was passing. Many of the horses were mares, and the Mongol taste for mares' milk gave the warriors an unusually convenient form of campaign ration. About forty per cent of the army were heavy cavalry, whose main weapon was the lance; they wore helmets and leather or captured mail armour. The remainder were light cavalry, unprotected except for helmets, perhaps. They carried bows and quivers slung to their saddles, or light javelins. Every warrior had a battleaxe or curved sabre slung to his belt; and the use of the lariat in combat was not unknown. In battle the usual formation was two ranks of heavy troopers followed by three of light cavalry. As the distance closed the light troops passed through the gaps between the heavy troops, and sent a withering fire of arrows and javelins into the enemy ranks. They then held back while the heavy riders passed through them, to break the shaken enemy lines by shock action. The light riders then joined in the mêlée, and a wild sword- and axe-fight decided the day; it was almost invariably a Mongol victory. It is supposed by many that the Mongols' series of devastating victories was due to weight of numbers, and the uncontrollable savagery of the barbarians of the steppes. Nothing could be further from the truth. Genghiz Khan's genius lay in forging as his weapon of conquest the most disciplined army in the world, and wielding it with sophisticated precision in the first example of *blitzkrieg* warfare known to history.

In peacetime manœuvres every Mongol unit, from squad to division, was trained and drilled and practised in a series of perfectly co-ordinated battlefield movements, following a system of signals transmitted by flags, whistling arrows, and fire arrows. Central to Genghiz Khan's method was a small group of perfectly reliable generals – among them his sons and several of his earliest companions – who could be trusted to use their initiative in command of dispersed formations, while sticking rigidly to an overall strategy laid down by the Khan. Through the use of a vast network of spies and a large screen of forward scouts when on campaign, the Khan was kept informed of every development – the enemy's strength and disposition, his speed and direction of march, the exact progress of the various Mongol formations – and the corps of fast couriers attached to his headquarters made certain that each *touman* commander in his turn was kept informed. Thus Genghiz Khan's army was probably handled in the field according to more complete and up-to-date information than any other army in history before the invention of the radio. To these strengths – perfectly trained and disciplined units, self-sufficient and physically formidable, reliably led and handled according to a constant stream of intelligence reports – were added the Khan's two greatest weapons: mobility, and psychology. No land forces on earth have equalled the mobility of the Mongol armies. At the outset of every campaign Genghiz Khan seized the initiative, and kept it until the final destruction of the enemy. The typical offensive consisted of a very rapid advance over a broad front by several strong, widely dispersed columns connected only by couriers. The Khan laid down general directives, and the *orloks* (generals) followed these, but were free to react to local conditions as they thought best, provided that the Khan was kept informed. This swift advance would continue until the enemy was encountered in strength; every man, woman, child, and beast in the path of the Mongol army was butchered, and every building burnt down, but nothing was allowed to interfere with the speed of the advance. If one of the columns ran up against an enemy force which it could handle unaided the columns on either side pressed on fast, outflanking the enemy and usually converging behind him to destroy his lines of communication. The engaged Mongol column would overwhelm the enemy in its path, and then press on to catch up with the others. If the enemy was too formidable for a single column, those on either side would close in and launch perfectly co-ordinated attacks from two, three, or four sides. The enemy never knew the strength of the opposition he would have to face, nor which direction it would come from – the Mongols were masters of ruse and deception, simulated flight and stealthy infiltration. Diversionary attacks might be made by corps separated from the main Mongol host by hundreds of miles, to draw the enemy forces into an unfavourable position which could be exploited by the other columns. Complete control and maximum flexibility, based on accurate intelligence – it was the perfect example of the maxim 'march divided, fight united'.

The Khan was well aware of the value of seizing the psychological advantage, and while the savagery of his race was a natural attribute, he deliberately harnessed it to further his campaigns. The speed and devastation of a Mongol attack invariably made his enemies believe that they were outnumbered, while in fact the greatest army the Khan ever put into the field numbered less than 240,000, and was opposed by twice that number. Added to the fear of envelopment and the confusion engendered by the widely dispersed but co-ordinated advance was the sheer terror which gripped any nation overrun by the Mongols. They took no prisoners during the advance, and massacred defeated armies and helpless civilians with equal enthusiasm. Their cruelty was bestial, their blood-lust insatiable; and Genghiz Khan's victims must be numbered in tens – perhaps scores – of millions. If a fortified city defied him, the eventual storming was followed by a bloodbath. If the territory in his path instantly threw itself on his mercy and acknowledged his suzerainty, it might be spared all but casual butchery, but many cities and provinces were not saved even by immediate capitulation. Genghiz Khan deliberately spread the terror of his coming ahead of his armies, to demoralise the enemy; and so fearful was the reputation of the Mongols that it was widely believed in far-off Europe that they were a demoniac horde sent by the Powers of Darkness to scourge the world.

In his first conquests Genghiz Khan was sometimes seriously frustrated by strong static defences, particularly in northern China. He therefore added 'flying artillery' to his army, in the form of siege engines carried by pack animals in sections which could be assembled rapidly in the field. Such military skills as his people could not provide were acquired by the simple process of conscripting captured experts, such as Chinese engineers and surgeons.

Once his position in central Asia had been secured, and the Tartars defeated and scattered, the Khan began his external conquests with a campaign against Hsi-Hsia in 1206–9; the ruler was allowed to retain reduced territories after acknowledging Mongol suzerainty. In 1208 the Khan defeated his last local enemy, Kushluk of the Naiman tribe, at Irtysh, although Kushluk himself escaped westwards. In 1211 the Mongols moved against the Chin Empire. Triumphant in the field, the Khan was held up by static fortifications for several successive campaigning seasons, but returned each spring to capture more and more towns until he finally captured and sacked Peking in 1215. The Chin Emperor bought his life by complete abasement to the Mongols, and Genghiz Khan returned to his capital at Karakorum with huge tribute and many useful prisoners, leaving northern China a wilderness. While this major campaign had been occupying the Khan his enemy Kushluk had gained the patronage of the apprehensive Emperor of Khwarezm, and with his help seized Kara-Khitai, a buffer state in modern Turkestan which lay between Khwarezm and the 1,000-mile Empire of the Mongols. In 1217 Genghiz Khan sent two *toumans* only against Kara-Khitai, led by a brilliant young general named Jebei; Kushluk was defeated west of Kashgar and executed, and Kara-Khitai joined the Mongol Empire.

War with Khwarezm was the logical next step; but this huge territory presented a serious obstacle, and the ruler, Ala-ud-Din Mohammed Shah, had forces totalling some 500,000 men. In 1218 Genghiz Khan mounted his offensive with an army of about 240,000, divided into four main columns commanded by himself and his sons Juchi, Jagatai, and Ogatai. While these forces penetrated the north-eastern frontier, Jebei led a smaller force up from Kara-Khitai. In 1219 Juchi and Jebei, with a force of no more than 30,000, engaged some 200,000 enemy troops in the Ferghana Valley and fought them to a draw at the Battle of Jand. Reinforced to about 50,000, Juchi advanced down the Ferghana Valley and besieged Khojend. Jagatai and Ogatai, with two columns of similar strength, invested Otrar: Genghiz Khan himself took another army in a wide swing to the north of Lake Balkash, and then moved south again, to encircle the Khwarezmian armies. Jebei moved south through the Pamirs with 20,000 men and followed the Oxus river into the heartland of Transoxiana. In 1220 the Shah reacted by moving some 100,000 men south to prevent Jebei cutting him off around Samarkand from his centres at Khorasan and Ghazni. Genghiz Khan then appeared, moving against Bokhara from the west – a totally unexpected direction. Panic-stricken by this pincer movement, the Shah put most of his 500,000 men into Bokhara, Samarkand, and other cities, and fled south with a small personal guard, convinced that he was heavily outnumbered. Genghiz Khan took Bokhara, and his sons sacked Khojend and Otrar; Jebei destroyed the enemy facing him and moved on to Samarkand, devastating everything in his path. The Mongol armies united outside Samarkand, which had a garrison of 100,000; it was quickly stormed and sacked, with enormous loss of life. Jebei and Subatai (q.v.) took three divisions and relentlessly pursued the Shah; in February 1221 he died, possibly by his own hand, on a small island in the Caspian Sea. Throughout that year Genghiz Khan supervised the progressive destruction of various centres of Moslem resistance by Mongol corps led by his sons. Jellal-ud-Din, the dead Shah's able son, raised a new army in what is now Afghanistan; while Genghiz Khan assembled his sons' armies for a new campaign this Moslem army of some 120,000 men defeated an advanced force of 30,000 Mongols in the Hindu Kush Mountains at Pirvan. Genghiz Khan led some 50,000 men to avenge the defeat, and the Prince was promptly deserted by much of his army. He retreated into the Punjab with about 30,000 men, and was routed on the banks of the Indus; he himself escaped by leaping his horse over a cliff into the river, a feat which impressed the Khan. The pursuit and pillaging which followed devastated Lahore, Peshawar, and Multan, but when the Sultan of Delhi refused Jellal-ud-Din asylum the Mongols withdrew. Between 1222 and 1224 Genghiz Khan consolidated his conquest of Ghazni. Meanwhile Jebei and Subatai led a detached force completely round the Caspian Sea (see under Subatai).

While the Mongol conquest of Persia had

been progressing, unrest had been increasing in the east. The Hsia and Chin Empires allied themselves in an attempt to throw off the Mongol yoke in about 1225. Genghiz now named Ogatai as his successor, studied the reports of his spies, and raised an army of 180,000 for his final campaign. In 1226 he launched a winter offensive, and destroyed an army of some 300,000 on the Yellow river. The Hsia Emperor was pursued and killed, and after a siege of the strong capital, Ninghsia, his heir submitted to Mongol vassalage. Genghiz Khan was determined to destroy the Chin state once and for all, however, and refused all peace offers. At this point (1227) the sixty-year-old Khan apparently felt his death upon him. He started back to Mongolia, but died on the journey. On his deathbed he instructed his youngest son Tuli on the strategy to be adopted against the Chin.

The Empire which Genghis Khan founded was extended by his sons and grandsons until the Mongols held sway from the borders of Poland to Korea, and from Siberia to Vietnam. Advanced units of the dreaded *toumans* reached Vienna and Northern Italy at one point; yet despite their apparent invincibility, and the considerable thought which Genghiz Khan had given to the internal structure of his Empire, the period of Mongol supremacy was to last little more than a century and a half after the great Khan's death. His Empire fragmented, and though the consequent dynasties ruled large areas of Russia and Asia the dynamic high-point was past. It is as a captain of war the Temujin is remembered, rather than as the father of a viable politico-economic empire; he ranks with Alexander, Hannibal, Bonaparte, and not more than two or three others as one of the greatest military geniuses of recorded history.

Geronimo (*c. 1829–1909*)

As far as can be ascertained Geronimo was born in June 1829, and it is said that he was born in No-doyohn canyon, Arizona. He never rose to be a chieftain in the Apache tribe, but became leader of a faction of braves within the tribe. He was given a surprisingly thorough education by his mother, and also quickly learned the Indian arts of stealing, hunting, and tracking. When his mother, wife, and children were massacred by the Mexicans in 1858 he assumed the leadership of an avenging group of Apaches, displaying his extraordinary courage, skill, and determination in numerous raids upon Mexican settlements.

Devastating Apache raids and massacres in Arizona and New Mexico during the 1860s led to action by the U.S. Army under command of General George Crook (q.v.), as a result of which offenders, including Geronimo, were placed on reservations. In 1876 Geronimo fled to Mexico. During the following decade he led outlaw bands in intermittent raids against American settlers, but in 1886 General Crook succeeded in bringing Geronimo to a meeting at which he and his Apache warriors agreed to surrender if they were taken to Florida where their families were being held. The terms were agreed, but on the way the Indians escaped. General Crook was replaced by General Nelson A. Miles who, after months of pursuit, finally secured another conditional surrender of the elusive Geronimo and his followers, who had been operating from refuges in the harsh Sierra Madre.

During this final campaign, which lasted eighteen months, no fewer than 5,000 troops and 500 Indian auxiliaries were employed in apprehending a band of Apaches which numbered only thirty-five men, eight boys, and 101 women operating in two countries without bases of supply. U.S. Army and civilian casualties totalled ninety-five, while Mexican losses were much higher. Geronimo's losses were thirteen killed, but none from direct U.S. Army action.

Under orders from President Grover Cleveland, and contrary to terms agreed upon by Miles, Geronimo and fourteen companions were placed under military confinement at Fort Sill, Oklahoma. There Geronimo was allowed to carry on stock-raising and farming, and died on 17 February 1909. He represented the last warrior generation of Indians who, faced with the rapidly rising tide of white soldiers and settlers, had pursued a long and cunning military campaign, and he displayed all the stealth and inbred tactical inventiveness of his race.

Giap, Vo Ngyuen (*b. 1909*), *General*

Vo Ngyuen Giap was born on 1 September 1909 in the village of An-Xa in the Quang Tri province. A student at Hue involved in left-wing subversive activities he was gaoled by the French from 1928 to 1931. Upon his release Giap obtained a degree in economics and politics at Hanoi and taught for the following eight years. He took refuge from the French police in the Communist zone of China in 1939 and served his apprenticeship as a guerrilla theorist and tactician with Mao's guerrillas during the Second World War.

Returning to Indo-China towards the end of the war Giap, leading the Indo-Chinese Nationalists, and the Communists under Ho Chi-Minh combined forces to mount guerrilla activities in Indo-China against the occupying Japanese forces. After the collapse of Japan the Communist-Nationalist faction determined to eliminate French imperial influence in Indo-China and widespread guerrilla warfare broke out towards the end of 1946.

Sporadic guerrilla activity continued against the French, who were able to make little significant headway. In 1950 the Viet Minh gained formal recognition by Communist China and the U.S.S.R., which countries progressively increased their military aid to the guerrillas and, at the instigation of Giap, provided training facilities in southern China. The anti-Communist posture adopted by America – nurtured by the Korean War – resulted in much U.S. aid being given to the French in Indo-China, and it became clear that open confrontation and a military show-down were inevitable. Despite the assistance thus forthcoming, the French were generally unable to master the specialised nature of jungle warfare; and the infiltration tactics of Giap's forces, well-trained and resourceful, continued to demonstrate the futility of pursuing conventional methods of warfare. The first major French defeat occurred in October 1950 when, during the course of a guerrilla offensive in the north, French forces were badly beaten on the Cao-Bang Ridge by Viet Minh troops under Giap's command.

This defeat and the French loss of territorial control prompted France to send her leading soldier, General de Lattre de Tassigny (q.v.), to Indo-China to restore the situation. General de Lattre apparently accomplished his mission, as well as restoring sagging morale; but within a year, following widespread Communist infiltration throughout the country, guerrilla activity exploded once more, this time accompanied by urban terrorism against the French in the cities. A fine military commander, Lattre de Tassigny was no politician; impotent against these terrorist activities, and discouraged by local corruption and indecision at home, the French general returned to France a sick, disillusioned man, and died some months later.

General de Lattre's position was filled by General Raoul Salan (q.v.) who pursued an unjustifiably leisurely campaign against Giap, although early in 1953 the French did seize the rebel base at Quinhon, and destroyed several arms factories concealed in the jungle. Propaganda suggested that real progress was being made again the Viet Minh, which encouraged America to step up the military aid being sent to the French in 1953. But Giap's continued activity was undeniable, and was not lost upon the French High Command; Salan was replaced by the equally unimaginative General Henri-Eugène Navarre.

The French command in Indo-China was simply unable to grasp the fact that, although its enemy operated in small, mobile, and apparently isolated cells of guerrillas, the French faced a highly integrated network of well-trained, well-armed, dedicated units. The very fact that those units did not follow the accepted pattern of Western military organisation and operation seemed to the French to recall colonial warfare of a previous century; their methods of dealing with the problem were therefore quite outdated. While France continued to suffer under this delusion, America followed suit.

Thus it was that in November 1953 Navarre decided to present the guerrillas with a 'target' garrison in the hope of attracting a substantial part of Giap's army to one point where it might be brought to decisive battle and wiped out. Accordingly a force of 15,000 men under Colonel Christian de Castries, spearheaded by air-dropped paratroop units, occupied the village and airstrip at Dien Bien Phu, 220 miles west of Hanoi, far behind enemy lines near the Laotian border. This force, consisting of French paratroops, Foreign Legion units, North Africans, and indigenous troops, was provided with a squadron of light tanks, and twenty-eight guns of medium calibre. It was to be wholly supplied and supported by air from Hanoi.

Giap, accepting the challenge, moved up four newly-trained regular divisions – between 70,000 and 80,000 men – and more than 200 guns, including Chinese anti-aircraft guns and Russian rocket launchers. The French command was wholly unprepared for a fire-power confrontation on such a scale. Assault after assault was mounted against the French pocket; one by one the perimeter strongpoints were overrun, despite great gallantry by the defenders. The volume of air supplies that was flown in (or rather, that survived the Viet Minh anti-aircraft fire) was but a fraction of what was needed, and on 27 March 1954 Giap's forces overran the airstrip. A subsequent air drop of supplies to the garrison failed; of more than 400 French aircraft available, the siege cost sixty-two destroyed and over 100 damaged. In the last stages several hundred untrained volunteers parachuted into the pocket by night.

On 7 May the garrison was forced by starvation, exhaustion, and lack of ammunition to surrender. 5,000 French had died; 10,000 (about half of them wounded) were taken prisoner; only 73 men escaped. It was estimated that the capture of Dien Bien Phu cost Giap 25,000 casualties.

The loss of this garrison was the beginning of the end of French influence in Indo-China. The prestige reflected upon the Communist forces in the area was fully exploited, and the strengthening of Giap's guerrilla armies was easily accelerated. His successful tactics, suitably updated, provided the basis of those employed by the Viet Cong after the further escalation of the war during the disastrous involvement of America in the 1960s. As commander-in-chief of the North Vietnamese army Giap is a revered figure, and may sometimes attract a degree of admiration based more upon political expediency than historical fact, but his central achievement is unchallengeable. While maintaining a carefully-paced guerrilla war against the occupying forces – a war in which he seized the initiative early on, and held it – he presided over the building up of a modern field army fully capable of meeting a European army in a conventional battle. Supported by an enormous number of coollies who carried his disassembled guns through jungle and over mountains on their backs, and who later braved French tactical air power to supply and feed the siege force, Giap quickly established an iron ring around Dien Bien Phu. His anti-aircraft artillery largely negated the French air effort, and his artillery was more than a match for the fortress. He denied the garrison freedom of movement, and throttled it by a classic siege campaign. His clear vision, decisiveness, determination, and realism were in marked contrast to the confused performance of the French generals, although at battalion level the forces were evenly matched.

Glyn Dŵr, Owain (*c. 1354–c. 1416*), *Prince of Wales*

Immortalised by Shakespeare as Owen Glendower, this last independent Prince of Wales is hard to assess as a straightforward general – as is so often the case in this period. He was an irregular leader of great ability, and in the centuries which have passed since his death he has been elevated into a national hero: hence his inclusion in this book.

He was born in about 1354, son of Gruffydd Vychan (who was connected with the royal line of Powys) and of Helen (a descendant of the royal family of Deheubarth). Owain was thus descended from rulers of two of the three old kingdoms of the province. His early life is shrouded in uncertainty, but he was educated for a time at the Inns of Court in London: this lord of Glyn Dyfrdwy and Sycharth was by no means a simple hill-country chieftain. Generally he seems to have favoured the opponents of the weak English king Richard II, a tendency which may have been reinforced by first-hand experience of Richard's bungling during the futile campaign against Scotland in 1385. Owain's name has been associated with both Richard, Earl of Arundel, and Henry of Lancaster, later King Henry IV.

As with *El Cid* (q.v.), the posthumous elevation of Glyn Dŵr into a national cult-figure makes it almost impossible to establish his original motivation. He first went to war against the English in September 1400, but there is a suggestion that his actions were initially purely personal, and that it was only by chance that he found himself at the head of a spontaneous Welsh rising against English oppression sparked off by the example of his attack on Ruthyn. For years Reynold, Lord Grey of Ruthyn had been trying to add Glyn Dŵr's lands to his estates, and in September 1400 the feud boiled over. Owain ravaged and burnt the town of Ruthyn, although Grey held out in his castle. In the euphoria induced by this successful raid Owain's followers hailed him as Prince of Wales; but only days later, on 24 September at Welshpool, they were beaten and dispersed. It seemed that the rising was over; but although it had been put down in the north it now flamed up in the south, and gathered impetus. Owain proved to be a resourceful leader, expert at limited operations of ambush and raid. Henry IV was at this time distracted by baronial unrest at home; English campaigns of suppression in late 1401, 1402, and 1403 all proved ineffective, and English garrisons continued to fall to the patriots. In April 1402 Owain had the great satisfaction of capturing Lord Grey after defeating an English force at Vyrnwy; he imprisoned his old enemy in Dolbadarn Castle and extracted a ransom of 10,000 marks.

In June 1402 Owain achieved another and most significant victory, ambushing an English force in the defile at Bryn Glas near Pilleth, Radnorshire, and capturing Sir Edmund Mortimer, the powerful Marcher Lord. After being held prisoner until November 1402 Mortimer came to terms with Owain, and married his daughter Catherine to seal the bargain. Through Mortimer Owain formed a loose alliance with another turbulent nest of Henry IV's subjects – the Percys, Earls of Northumberland and for centuries virtually a law unto themselves. In the spring of 1403 the Earl and his famous son Harry Hotspur advanced south-westwards into England with some 4,000 men, planning to effect a junction with Owain's Welshmen. By this time Owain was widely supported throughout the province, a score of castles had fallen to him, and he was credited with powers verging on the magical. His prestige was sufficient to survive even the disaster of Shrewsbury in July 1403, when Henry IV placed his army between those of Percy and Glyn Dŵr and managed to rout the Northumbrians and kill Hotspur before the Welsh could come up. Owain himself had been defeated by an English force near Carmarthen shortly before Shrewsbury, and that spring Prince Hal – later Henry V (q.v.) – had ravaged his lands at Glyn Dydrdwy and Sycharth unopposed. It is a measure of Owain's strength of will and the loyalty he commanded that 1404 was to be his greatest year. He captured the strategic castles of Aberystwyth and – after a long and grim siege – Harlech. He treated successfully with the French, and a French fleet in the Irish Sea prevented the English from relieving their beleaguered coastal garrisons. Owain made mighty Harlech Castle his headquarters; from there, in the summer of 1404, he ruled Wales from Anglesey to Glamorgan. He summoned at least one parliament to meet at Machynlleth, and concluded a formal agreement with Charles VI of France. It was in 1404 that Owain began to style himself 'Prince of Wales', back-dating his reign to 1400.

In 1405 the tide began to turn against him. The Welsh underestimated the slow-burning determination of Henry IV to enforce his grip over the kingdom he had won by guile and ruthless ambition. In March 1405 Owain was defeated at Grosmont; in May he was beaten again at Pwll Melyn, near Usk, and his eldest son Gruffydd was captured. That summer the Earl of Northumberland was forced to flee to Scotland, and any hope of assistance from the Percys died. Only foreign intervention could bolster up the Welsh resistance to grim Henry and his fiery son, and in August 1405 a few thousand French soldiers landed near Milford Haven. Owain carried the war to the English again, advancing far enough to threaten Worcester, but the expedition accomplished little and the French withdrew from Wales in 1406. Gradually Henry beat down his rebellious barons, and gradually the forces available to Prince Hal for the re-conquest of Wales increased. The number of castles in Welsh hands began to decrease as the English retook them in a hundred small, savage, and largely unrecorded battles. In 1407 Prince Hal recaptured Aberystwyth;

Glyn Dŵr took it back briefly, but in 1408 it was retaken by the English for good. In 1409 Owain was forced to flee from Harlech when 1,000 men under John Talbot attacked and captured it, and with it the Welsh prince's wife, daughter, and grandchildren. With much of his power-base destroyed, the man who only three years before had negotiated with the Pope for religious independence for Wales was forced to revert to the campaign methods of raid and petty ambush with which he had started his revolt. The bloody pacification of the country went on, and Owain's supporters dwindled. By 1410 he had to go underground completely, moving from one refuge to another as the hunt came near. He was active for another year or perhaps two, and then he simply disappeared. In 1413 Prince Hal came to the throne as Henry V; he proclaimed generous terms if Glyn Dŵr would give himself up, but nothing was ever heard of the last Welsh Prince of Wales again. It is supposed that he lived out his last years sheltered by his son-in-law, Sir John Scudamore of Bacton in Hertfordshire, and that he died about 1416.

Gneisenau, August Wilhelm Anton, Graf Neidhardt von (*1760–1831*), *Field-Marshal*

Gneisenau was the brilliant staff officer best remembered for his defence of Kolberg in 1807 and for the part he played as Blücher's (q.v.) chief of staff in the 1815 campaign. He was one of those rare officers who combined the energy of a field commander with the organising ability of a born staff officer and the vision of a military intellectual. With Scharnhorst (q.v.) he played a major part in the revival of Prussian military standards in the first quarter of the nineteenth century.

Gneisenau was born at Schilda, near Torgau, on 27 October 1760, the son of a noble but impoverished Upper Austrian family. His father was an artillery officer in the Saxon forces; his mother died while following his father on campaign during the Seven Years' War. August was educated at the expense of his maternal grandfather at Würzburg and at Erfurt University, and joined the Austrian army as a cavalry subaltern. Transferring to the service of Anspach-Bayreuth, he was shipped to America in 1782 to serve with the British Crown forces against the American colonists, but arrived too late to see any action. He returned to Europe in 1783, and three years later joined the Prussian army as a staff captain. He served in various Silesian garrisons for the next twenty years. In the 1806 campaign he fought at Saalfeld and at Jena, rising from a company to a battalion command. In 1807 his heroic and tenacious defence of Kolberg brought him fame; this became one of the best-known Prussian military epics, and earned him the *Pour le Mérite*. Relinquishing his commission in 1808, he travelled on clandestine missions to England and Russia to discuss the organisation of resistance to Napoleon (q.v.). It is certain that he felt a strong dislike of the Russians, and was extremely suspicious of British good faith throughout his later career – whether these feelings stemmed from this period in his life is not known for certain.

In 1813 Gneisenau joined Scharnhorst's staff as 'first general staff officer'; he deserves considerable credit for the strategic planning of the Russian and Prussian operations in the spring of 1813, for the Leipzig campaign, and for the winter fighting of 1813–14. After Scharnhorst's death in June 1813 he became Prussian chief of staff. He was a visionary planner, in the sense that he appreciated that the old mercenary army directed by an aristocratic hierarchy headed by the king would have to disappear, to be replaced by a well-organised citizen army directed by a staff of well-educated military professionals. This reform of the whole approach to soldiering in Prussia ran counter to prevailing attitudes in high places, and was inseparable from a general liberalisation in many areas. Gneisenau was strongly supported by several of his able contemporaries, but made powerful enemies.

Although their approaches were completely dissimilar, Gneisenau and Blücher worked well together when serving as chief of staff and Commander-in-Chief respectively of the Prussian Army of the Lower Rhine in 1815. Blücher provided the traditional fire and determination, Gneisenau the technical expertise – yet neither seems to have resented the other, as their skills were complementary. Yet Gneisenau was by no means the cold technocrat – he was an impressive figure physically, and it was said of him that no portrait ever did him justice, because the main impression he left on those who met him was of intense animation and mental energy. In the Ligny-Waterloo campaign he handled the practical details of the army's movements with consummate skill. It seems likely that his distrust of the British led him to behave in a way which might have been seriously damaging to the Allied cause, had Blücher not insisted on making an all-out effort to reach Waterloo and support Wellington (q.v.). Nevertheless it was almost entirely due to Gneisenau that there was a Prussian army left at all after Ligny. While Blücher was near to getting himself killed leading the ulans in a forlorn hope at the end of the battle, Gneisenau was coolly taking a grip on the situation and presiding over a difficult and risky disengagement, which proved wholly successful.

The reactionary policies followed by the Prussian Government led to Gneisenau's resignation in 1816, and it was to be 1825 before he received his well-deserved promotion to field-marshal. In 1831, at the time of the Polish rising, he returned to the colours to command the forces guarding Prussia's eastern frontier. He died of cholera at his Posen headquarters on 23 August of that year.

Gordon, Charles George (*1833–85*), *Major-General*

Gordon's death, faced with great courage and devotion to a very personal sense of duty, has become so famous that it has tended to overshadow the other episodes in a remarkably active career. While the epic of Khartoum stands unchallenged, it should be remembered that Gordon was already a national hero before his final appointment in the Sudan, and that it was upon his campaigns in China and against African slavers that his contemporary reputation was based.

Born at Woolwich on 28 January 1833, Charles Gordon was the fourth son of Lieutenant-General Henry William Gordon, R.A. In 1848 he entered the Royal Military Academy, and was commissioned in the Royal Engineers in June 1852. Ordered to the Crimea in December 1854, he quickly became conspicuous for his daring and his aptitude for detecting enemy movements. In June the following year he was slightly wounded, but was sufficiently recovered to participate in the attack on the Redan. For his services during the Crimean War Gordon received the British Crimean Medal with clasps and the Turkish War Medal, and was elected to the French Legion of Honour.

Gordon was promoted to captain in April 1859, and the following year joined the forces under Sir James Hope Grant operating with the French against the Chinese. He was present at the capture of Peking and at the destruction of the Emperor's Summer Palace. He received the British War Medal with clasp (Peking) and a brevet-majority in 1862. While the British forces remained at Tientsin, Gordon accompanied several expeditions into the interior and explored part of the Great Wall. In April 1862 he commanded the Royal Engineers during the storming of Sing-Poo and other fortified towns in efforts to clear the T'ai P'ing rebels from a thirty-mile radius around Tientsin.

After these operations against the T'ai P'ing rebels had continued for some weeks under different commanders Li Hung-chang, Governor-General of the Kiang province, asked the British Commander-in-Chief for the services of an English officer, and Gordon was selected. He entered the Chinese service as a mandarin and lieutenant-colonel on 24 March 1863, his force of 4,000 Chinese being officered by 150 Europeans – a heterogeneous collection forged into a well-disciplined body by the sheer indomitable will of its commander. It fought no fewer than thirty-three engagements with the T'ai P'ings in under two years, and this campaign resulted in the outright suppression of the rebellion. Gordon was deeply respected by his men, who attached almost superhuman powers to this white leader who fought at their head, carrying nothing but his officer's cane. After the end of the rebellion Governor Li broke his promise to Gordon, having given an undertaking not to harm the rebel leaders after capture. When they were murdered the infuriated Gordon refused to serve Li any further, and declined all rewards offered by the Emperor. There was a short-lived recurrence of the T'ai P'ing uprising, and Gordon reluctantly returned to his previous command. His victory at Chan-Chu-Fu ended the campaign, but again Gordon refused all honours. In 1865 he returned home and was appointed a C.B.

The next six years were uneventful – 'Chinese' Gordon was appointed to command Royal Engineers at Gravesend. During 1872 he met Nubar Pasha, who sounded him out as to his succeeding Sir Samuel Baker in the Sudan. When Baker resigned Gordon was appointed Governor of the equatorial provinces of Central Africa. He travelled to Egypt in 1874 and reached Khartoum in March, where he stopped only long enough to take on supplies before continuing to Gondokoro. Within a year he had gained the confidence of the natives, and the country was made secure; he had also enforced the Government monopoly in ivory.

Gordon next transferred the seat of government from Gondokoro (a most unhealthy place) to Laido, and by the end of 1875 these two places had been linked by a line of fortified posts a day's march apart. He had also dispersed all the slave dealers who had operated in the area. (It was by his personal observations that it was established for the first time that the Victoria Nile flowed into Lake Albert.) Notwithstanding his early success against the slave trade, Gordon soon realised that a permanent solution was impossible as long as the government of the Sudan remained separated from that of the equatorial provinces. In frustration he resigned and returned to England in 1876.

Under strong pressure from the Khedive to return, Gordon did so in January 1877 with an appointment as Governor-General of the Sudan, Darfour, the equatorial provinces, and the Red Sea littoral. He was charged with the responsibility of improving communications and the final suppression of the slave trade. Almost immediately news reached him that 6,000 armed men, led by Suleiman, had moved from their stronghold at Shaka towards Dara. Gordon himself rode out and alone entered Suleiman's camp; there he solemnly upbraided the rebel leader for his disloyalty. Gordon returned to Khartoum in October. Some months later news arrived that Suleiman was leading the slavers in another rebellion in the south, whereupon Gordon sent Captain Romulus Gessi, a much-trusted Italian, to deal with the trouble while he himself put down a rising in the western districts round Darfour and set free thousands of slaves. Meanwhile Gessi succeeded in catching Suleiman and other traders, and after a year's campaigning the ringleaders were caught, tried as rebels, and shot. By July 1879 the slave trade had been virtually stamped out.

During this period of campaigning Gordon travelled extensively throughout the difficult terrain of north-east Africa, on one occasion being captured by irregulars under King John of Abyssinia and left to make his own way to the shores of the Red Sea across snow-covered mountains. Ill health coupled with such continous exertions, as well as differences of opinion with the Khedive, prompted Gordon to tender his resignation, and he returned to England in January 1880. He had been appointed to the First and Second Class of the Order of Medjidieh while in the Khedive's service.

There followed a long period of frustration for this unusually gifted and energetic man. After an enforced convalescence in Switzerland he accepted the post of Private Secretary to the Marquis of Ripon, the new Viceroy of India, but soon resigned. He then visited China at the invitation of Sir Robert Hart, Commissioner of Customs at Peking, to advise the Chinese Government on matters relating to the strained atmosphere that existed with Russia. His wise counsel in favour of peace prevailed, and with his mission successfully completed he again returned home. In 1882 he accepted a post in Cape Town to assist in bringing to an end the war in Basutoland, and shortly afterwards took command of the colonial forces – much against his own inclinations. Finding himself at the centre of intrigue, Gordon once again resigned and returned home – now a major-general on half pay.

An abortive mission to the Belgian Congo at the instigation of the King of the Belgians (which was frustrated by lack of support from the Belgian Government), and a year's sojourn in the Holy Land decided Gordon to renounce his commission and pension in 1884. He was only just prevented from so doing by a summons to the War Office from Wolseley (q.v.).

Here he learnt that following the success of the Mahdi in the Sudan and the catastrophe to Hicks Pasha the previous November, the British Government had decided to withdraw its military presence in the Sudan and to insist on abandonment by the Khedive. This in turn would involve a possibly dangerous disengagement by forces scattered among the garrisons throughout the Sudan, and Gordon was instructed to take charge of a mission to organise this delicate task.

Gordon's arrival in Cairo early in 1884 was followed by his appointment as Governor-General of the Sudan, and an examination of his mission convinced him of the difficulty of his task. Not only was he to organise the evacuation of the garrison forces, but also to leave behind a coherent form of government and secure lines of communication between Cairo and the Sudan. Reaching Khartoum on 18 February amidst a great ovation, Gordon straightway proclaimed the Sudan's independence and allowed the retention of slaves. He sent word to Cairo that Zebehr should be sent to Khartoum as the only person whose influence was capable of competing with that of the Mahdi, and asked that Turkish troops should be sent to his assistance. He made numerous other requests in the furtherance of his mission: that the lines of communication between Suakin and Berber should be maintained; that Indian Moslem troops should be sent to Wadi Halfa. He asked permission to confer personally with the Mahdi and, if necessary, to take action south of Khartoum. Not only was each of these requests refused, but when Sir Gerald Graham, after his success on the first Suakin expedition, suggested that he reinforce Gordon through Berber, this was also turned down.

During the first month Gordon managed to dispatch about 2,500 people down the Nile in safety, but he then found himself completely isolated by the Mahdi. His last telegram, sent from Khartoum before his lines were cut, contained a bitter complaint at the shameful neglect demonstrated by the British government.

The Mahdi's forces opened their attacks on Khartoum in March, and these continued over a period of ten months. With only about half a dozen European assistants, Gordon displayed great skill and ingenuity in his defence. He converted his river steamers into 'ironclads' and even managed to construct others; he laid rudimentary minefields and wire entanglements, and made frequent sorties against the surrounding forces. Meanwhile news had of course reached England of Gordon's desperate predicament, and by May public feeling was running so high against the Government for its apathy that £300,000 was voted to support a relief expedition under Wolseley. But hysteria and indecision delayed the departure of this force until September – by which time it was too late.

In an attempt to re-establish his communications, Gordon sent Colonel Stewart (and Frank Power, *The Times* correspondent) down the Nile towards Dongola, but when their steamer struck a rock they were captured and murdered out of hand. Thus Gordon was left as the only Englishman in Khartoum.

By December Wolseley's relief expedition had set out for Khartoum, and on the 30 December an advance expedition was sent forward across the desert to Korti under Sir Herbert Stewart; this made slow progress and was in action seven times. Stewart was killed, and Sir Charles Wilson assumed command before reaching Korti on 20 January 1885. Learning of the approach of the relief expedition, Gordon unselfishly provisioned five of his precious river steamers and sent them down river to meet Wilson; one of these vessels contained the last dispatches prepared by the British commander.

When Wilson reached Khartoum he found the town in the Mahdi's hands – after a siege that had lasted 317 days. Gordon had, by all accounts, continued to offer resistance to the last against tremendous odds, but learning that relief was near at hand, the Mahdi determined to overwhelm the garrison before its arrival. Although accounts have never been fully authenticated, it seems that Khartoum fell only two days before Wilson's arrival, and that Gordon was killed near the gate of the palace, his head being taken to the Mahdi's camp.

News of Gordon's death spread quickly to England and throughout the Empire; grief was mingled with fury at the War Office's incompetence and the Government's apathy.

Not unnaturally General Gordon was acclaimed a hero worldwide. Much more significantly, the tragic episode provided vital fuel for the determined arguments in support of total repeal of many immoral practices in the Army's administration.

Graham, James, *see* Montrose

Graham, Thomas, 1st Baron Lynedoch (*1748–1843*), *General*

A British soldier who fought the French throughout the Revolutionary and Napoleonic Wars, Graham rose to become Wellington's (q.v.) second-in-command during the Peninsular War.

Born the son of the Laird of Balgowan, in Perth, on 19 October 1748, Thomas Graham might have lived out his long life as an obscure country gentleman, occasionally touring abroad for the benefit of his wife's health. But that remarkably beautiful woman (the subject of a famous Gainsborough portrait) died at sea off the French coast in 1792; and as her body was being shipped home for burial, a revolutionary French rabble broke open the coffin in a search for hidden arms. This act of desecration so incensed the bereaved Graham that he forthwith devoted his life to active military service against the French nation, his inveterate hostility leading him in time to the highest ranks of the British Army. In 1793 he personally raised the 90th Regiment of Foot, serving with it at Quiberon and Ile d'Yeu in 1795. The following year he was appointed British military commissioner with the Austrians in Italy; it was during this appointment that he made a daring escape from Mantua to bring news to the relief forces of the critical state of its garrison. He distinguished himself during the capture of Minorca in 1798, and conducted the Siege of Valetta, Malta, in 1799. Then he returned to England to take his seat as Member of Parliament for Perthshire.

On the renewal of the war with France Graham rejoined the Army with the rank of colonel, serving on the staff of Sir John Moore (q.v.) in Sweden in 1808 and in the Corunna campaign of 1808–9; he was present at Sir John's death and his burial on the ramparts of Corunna in January 1809. Returning once more to England, he was at once ordered to accompany the Earl of Chatham's expedition to Walcheren, sent in July to divert Napoleon's attention from central Europe. The project was a dismal failure, the troops' lives being frittered away in the pointless capture of Flushing. When Chatham left to return home, Graham accompanied him.

Promoted major-general in 1810, Graham was given command of a force in southern Spain with which he won a brilliant victory against odds at Barossa, near Cadiz, the next year. Joining Wellington's army in command of the First Division he distinguished himself at the Siege of Ciudad Rodrigo, and was present at the Battles of Badajoz and Salamanca in 1812. He commanded the left flank of the army at the decisive Battle of Vittoria on 21 June 1813, and he went on to conduct the successful Siege of San Sebastian. In 1814 he was transferred to command British forces in Holland, winning a minor victory at Merxem, but failing in an attempt to storm Bergen-op-Zoom in a surprise attack.

This was the last active episode in the sixty-six-year-old lieutenant-general's military career. He was created Baron Lynedoch of Balgowan in May 1814, and promoted full general in 1821; he died at the age of ninety-five in London on 18 December 1843. By no means an outstanding tactician, Graham has nevertheless earned respect and some fame for his persistent determination to defeat the French whenever and wherever he was ordered. Much respected by his men, he was typical of the abler commanders who served Wellington in the closing stages of the Peninsular War. Apart from Hill (q.v.), Graham was the only subordinate whom Wellington trusted to handle a degree of independence. His personal character was apparently endearing; his grim dedication to the war against France did not prevent him treating captured Frenchmen with an old-world courtesy and humanity.

Grant, Ulysses Simpson (*1822–85*), *General in Chief of the Army, 18th President of the United States of America*

Grant's claim to immortality as one of America's greatest generals is sometimes contested on the grounds that he was a mere butcher, buying victory by lavish expenditure of his men's lives. He was by no means consistently victorious, and he was quite ready to accept heavy casualties – but the same could be said with equal justice of many other generals, starting with Napoleon. More significant is Grant's positive role in bringing the American Civil War to a close. His strategy was the first unified plan of action for all the Federal armies since the outbreak of war. Soundly based on recognition of the necessity of cutting the Southern homeland into fragments and of engaging all her armies simultaneously to avoid their mutual support, it envisaged a war of attrition in which the Confederate forces could be worn down by unrelieved pounding. It worked – and there is really very little else which needs to be said about any military strategy.

Grant was born on 27 April 1822 at Point Pleasant, Ohio, the son of poor parents. He was raised austerely by his hard-working father and his pious, undemonstrative mother; he attended school from the age of six until the age of seventeen, but did not escape a heavy share of the family chores. He spent much of his time in the open air working with animals, and was a keen and skilled horseman at an early age. In 1839 he secured a place at West Point, and this was instrumental in changing his name. He was christened Hiram Ulysses, but did not wish to have the potentially humorous initials HUG on record at the academy, so reported himself as Ulysses Hiram. Confusion was compounded when the congressman through whom his application had been handled wrongly gave his name as Ulysses Simpson. The consequent forest of red tape was so daunting that Grant simply accepted the congressman's version and used it for the rest of his life. He did not at this stage envisage a career in the army; although the best rider in his class, he was academically placed in the worthy but undistinguished middle of the list. When he graduated in 1843 there were no cavalry vacancies available, so he joined the 4th Infantry in Missouri. After service in Louisiana his Regiment was sent to Corpus Christi, Texas, to join the army of Zachary Taylor (q.v.) guarding the Mexican border. When the war with Mexico began Grant saw action in Taylor's early engagements, and greatly distinguished himself in combat at Monterey in September 1846. His unit was transferred to Winfield Scott's (q.v.) army in due course, and Grant took part in the gruelling fighting advance from Vera Cruz to Mexico City as the quartermaster of his regiment. He saw a good deal of fighting and was mentioned in dispatches, ending the war as a first-lieutenant and brevet captain. He served in Mississippi after the war, and then on the Pacific coast, making a nightmare journey through disease-ridden Panama in 1852. He was unable to take his wife and child to California, and although he was promoted captain in September 1853 his pay was still insufficient to keep them in any comfort. He was made commander of a tiny and inexpressibly dreary frontier post in California, and in his loneliness and frustration took to heavy drinking. When he was warned about this in July 1854 he immediately resigned his commission.

The next six years saw many moves from town to town, many attempts to establish himself in businesses of one kind or another, and many failures. In 1861, after a frustrating series of attempts to find himself a military position worthy of his experience, he finally gained the colonelcy of the 21st Illinois Volunteer Infantry Regiment. Only two months later, to his considerable surprise, he was made a brigadier-general of volunteers. His first serious action of the Civil War was a confused attack on the Confederate position at Belmont; Grant's raw troops won, but then got out of hand and scattered to loot, narrowly escaping disaster at the hands of Confederate reinforcements. This escapade, in November 1861, was followed by Grant's capture of Forts Henry and Donelson, which brought him much credit. These two Confederate positions on the Tennessee and Cumberland rivers, which are at that point parallel and close together, were keys to the Confederate cordon which had been established across Kentucky to prevent Federal navigation or land advance up these two rivers and the Mississippi. With some 17,000 men and a squadron of ironclad gunboats Grant took Fort Henry on 6 February 1862, marching overland to Fort Donelson while the gunboats followed by the long route.

Grant maintained his encirclement despite a strong counter-attack, and the position fell on 25 February; subsequently he was promoted major-general of volunteers. During March and April 1862 he assembled a force for the planned move against Corinth, but took inadequate precautions against surprise attack; at dawn on 6 April the Confederate General A. S. Johnston fell upon him near Shiloh Church. The Union army was forced back in bloody fighting but managed to hold out until the arrival of reinforcements from General Buell; on the following day they counter-attacked and pushed the now heavily-outnumbered Confederates back towards Corinth. It was a battle marked by unimaginative generalship and great heroism on the part of the troops; some 13,000 Union and 10,600 Confederate casualties were recorded, of the 62,700 and 40,000 men engaged. Grant was much criticised, but Lincoln refused to dismiss him, making the famous remark: 'I can't spare this man – he fights.'

The autumn of 1862 was spent at Corinth, planning moves against the Confederate bastion of Vicksburg on the Mississippi. In November Grant moved south with his Army of the Tennessee from Grand Junction towards Vicksburg, while Sherman (q.v.) made an amphibious expedition down the river itself. This latter attack failed, and Grant was seriously hampered by raids on the railway system and by the destruction of his supply depot. The campaign was abandoned, and Grant went back to probing the Vicksburg defences. In April 1863 his preparations were ready; he moved to a new concentration point at Hard Times, on the Mississippi south of Vicksburg, and then ferried his army across to the enemy's eastern bank, supported by the gunboats of Rear-Admiral S. S. Porter. Grant now led his 41,000 men deep into enemy country, abandoning his communications and placing himself between two Confederate armies. He disposed of J. E. Johnston's 9,000 men at Jackson on 14 May; then he turned west towards Vicksburg and beat Pemberton's 22,000 men at Champion's Hill on 16 May. Pemberton's survivors retreated into Vicksburg; Grant settled down to besiege them, with Sherman covering his rear. Vicksburg fell on 4 July 1863. This 'Big Black River Campaign' was a classic example of audacious generalship, and the eventual fall of Vicksburg placed the Mississippi in Union hands from source to mouth and cut the Confederacy in two. Grant was made a major-general on the regular list, and in October he was given command of all territory from the Mississippi to the Alleghenies.

He went immediately to Chattanooga, where Confederate General Braxton Bragg had locked up General Rosecrans since the Union defeat at Chickamauga. Within five days of his arrival Grant had opened communications with his base at Nashville, and supplies were reaching the besieged Army of the Cumberland. He then co-ordinated plans for driving Bragg off the commanding heights of Lookout Mountain and Missionary Ridge. These came to fruition on 24 and 25 November; it was the Army of the Cumberland which sealed the Confederate defeat on the latter feature with a wild, spontaneous charge after the assigned objective had been secured. Grant was promoted lieutenant-general, and subsequently named General-in-Chief of the Union armies. In the spring of 1864 he presented an overall strategy for exhausting the Confederacy's power to continue the war. Lincoln liked him and trusted him, and gave him a completely free hand.

On 4 May 1864 some 253,000 Union soldiers began to move. Meade's Army of the Potomac was to move directly against Lee's (q.v.) Army of Northern Virginia, and destroy it, no matter how far they had to follow it; Grant himself accompanied this army. Butler's Army of the James was to move against Lee's lines of communication, and Richmond; and Sherman's army group was to drive towards Atlanta, against the Confederate Army of Tennessee. In early May there was bloody fighting between Grant and Lee in the Wilderness, but Lee identified the nature of the threat facing him brilliantly. Grant marched around Lee's right flank in an attempt to chop it loose from Richmond, only to find that Lee had anticipated this and blocked him at Spotsylvania, where another major battle ensued. Again Grant moved around Lee's right, and again he found himself blocked, this time on the North Anna river. Lee's final blocking position was at Cold Harbor, only ten miles from the centre of Richmond, and Grant lost yet more men before giving up the idea of carrying it by assault. Casualties throughout the campaign had been very heavy; Lee had not once been beaten, and had outmanœuvred Grant all the way from the Wilderness. Yet still Grant had forced Lee to react to his initiative, and Confederate reserves were far skimpier than those of the Union. Grant now carried out a very skilled shift of axis, pulling back out of touch with Lee and swinging south across the James River. Subordinates bungled an attempt on Petersburg, Virginia, and Lee was able to reinforce it before Grant arrived in strength. Nevertheless, Grant was in a good position. Lee was forced to hold a line of thirty-seven miles between Petersburg and Richmond with 60,000 exhausted and starving men; Grant, who eventually assembled 122,000, maintained his pressure by constant probing attacks up and down the line, while Confederate resources drained. The Siege of Petersburg lasted from 18 June 1864 until 2 April 1865. The Battle of Five Forks on 1 April 1865 compromised the entire right wing of Lee's position, and the next day he withdrew from Richmond and Petersburg and marched west. Grant paralleled him to the south, and threw Sheridan's cavalry across his path long enough to trap him. Lee accepted Grant's generous surrender terms at Appomattox Court House on 9 April 1865.

Grant's later career as eighteenth President of the United States has no place here, but the immediate post-war years were eventful enough. The position of General-in-Chief of the Army was ill-defined, and Grant found himself involved in a lively three-cornered struggle with Congress and President Andrew Johnson; for a while in 1867 he was his own Secretary of War. The rank of full general, unused since 1799, was revived for Grant in 1866 by Congress. In 1868 he was elected President, and served two terms, leaving office in 1877. In his later years he suffered some financial distress, but he had many loyal friends to the end. Physically he is described as rather short and stocky, round-shouldered, and untidy as to his uniform. He had the inscrutable face which was no doubt the legacy of his lonely youth, a heavy jaw partially concealed by a rough, light-brown beard and moustache, and clear blue eyes. He died of cancer of the throat at Mount McGregor on 23 July 1885.

Graziani, Rodolfo, Marchese di Neghelli (*1882–1955*), *Marshal*

Born at Filettino, near Frosinone, Italy, on 11 August 1882, Graziani entered the Italian Army before the end of the nineteenth century, serving in Eritrea and Libya before the First World War, and rising steadily through the lower commissioned ranks of Italian colonial forces in North Africa. By 1919 he had assumed command of the Italian garrison forces in Macedonia. His next command took him back to North Africa, this time to Tripolitania, where he remained until in 1927 he assumed command of the Italian forces in Libya. Here, as in other North African countries which seethed and suffered in the backwash of the world economic depression, there were constant civil disorders, and between 1930 and 1934 Graziani earned himself the reputation of being a harsh administrator and a malevolent military governor, a posture tolerated if not actively supported by Mussolini's totalitarian Fascist régime.

In 1935 Graziani was appointed Military Governor of Italian Somaliland during the period of the League of Nations' investigation of border clashes between Italian and Ethiopian troops. Though not personally involved in the Ethiopian–Italian War of 1935–6 (see under Badoglio), Graziani was appointed Viceroy of Ethiopia in 1936 after Marshal Badoglio had voluntarily relinquished the post, and in 1938 he was made Honorary Military Governor of Italian East Africa.

With very little experience of senior command in the field, Graziani was given command of the 200,000-strong Italian army in Libya after the death of Marshal Italo Balbo in an air crash in June 1940. Against him was deployed General Wavell's (q.v.) force of 35,000 men (of whom roughly half were administrative or local levies). For three months Graziani conducted his ponderous preparations for an invasion of Egypt, knowing full well the weakness of Wavell's forces. Then on 13 September 1940 he moved, pushing forward five divisions on a narrow coastal front, while light British covering forces fell back before him; on reaching Sidi Barrani he halted, and then set about an extraordinary consolidation of fortified camps over a fifty-mile area, the real purpose of which has never been logically explained. Wavell, meanwhile, had left only light screening patrols forward, preferring to retain his main force, amounting to no more than two weak divisions, seventy-five miles to the east.

Graziani's military ineptitude was demonstrated to the world when, on 9 December 1940 in a surprise night attack, his forward positions were ripped apart by British forces under General O'Connor. Attacked by no more than 30,000 men and 275 tanks, Graziani's entire fortified area was overrun within four days, each fortified position being separately winkled out without the slightest possibility of support from any other. Inside a week Graziani's command suffered the loss of more than 12,000 dead and wounded and a further 38,000 prisoners, as well as almost all the guns, vehicles, and tanks of five divisions. This defeat must stand as one of the most disgraceful displays of military incompetence by an army in the Second World War. Further disasters followed in quick succession as Wavell's tiny army advanced 400 miles into Libya, capturing Benghazi on 7 February 1941. In the course of no more than two months Graziani suffered the loss of nearly 150,000 men (three-quarters of his original army), of whom 138,000 were taken prisoner, 400 tanks, and more than 1,200 guns. British casualties amounted to the extraordinary total of 500 killed and 1,373 wounded.

Not surprisingly, Graziani resigned from his command and returned to Italy, where he took no further interest in active operations. In 1943, after the Italian armistice, he took the side of the pro-German Fascists and became defence minister in the Italian Republican Government. At the end of the war he surrendered to the American forces, who imprisoned him until 1946, when he was handed over to the Italian Government for trial. He was confined to Rome for four years, and in the spring of 1950 was sentenced to nineteen years' imprisonment. So-called humanitarian reasoning and political intrigue secured his release in August that year, however, whereupon he promptly became a member, and ultimately President, of the neo-Fascist Italian Socialist Movement. He died in Rome on 11 January 1955.

Graziani represents the disagreeable archetype of a totalitarian military commander. His reputation was built upon a repressive colonial career, and he met his fate at the hands of highly professional and dedicated military opponents.

Gribeauval, Jean Baptiste de Vaquette
(1715–89), Lieutenant-General

This great French artillery innovator was born in Amiens, the son of a magistrate, on 15 September 1715. In 1732 he entered the artillery as a volunteer, being commissioned in 1735. For nearly twenty years he occupied himself with regimental duty and with scientific studies; the artillery provided technicians of various types, and in 1752 it is known that he was a captain commanding a sapper company. A few years later he served on a military mission to Prussia. He was a lieutenant-colonel in 1757 when, on the outbreak of the Seven Years' War, he was seconded to the Austrian service. He served as a general of artillery, seeing action at the Siege of Glatz (Klodzko) in June-July 1760, and in the defence of Schweidnitz. He was raised to the Austrian rank of lieutenant field-marshal, and awarded the Cross of the Maria-Theresa Order. On returning to France he was made *maréchal de camp*; he was appointed an inspector of artillery in 1764, and lieutenant-general in 1765. In the same year he was created a Commander of the Order of St Louis, but shortly after he appears to have incurred royal disfavour for some reason; it was not until 1776 that he was named Chief Inspector of Artillery, and granted the Grand Cross of the Order of St Louis. Now at last he was able to put into full operation the reforms which he had been planning since 1764.

The core of Gribeauval's great reorganisation of the French artillery arm was the clear identification of the proper tactical employment of certain types of piece, and the production of the best guns and accessories for that purpose. He sponsored the introduction of lighter, more mobile field-pieces with thinner-walled barrels. Calibres were restricted to three for field use – four-pounders, eight-pounders, and twelve-pounders – and each gun was eighteen calibres long and weighed 150 times the weight of its ball. The charge was reduced to one-third the weight of the projectile by the application of more rigorous standards of shape and size for roundshot. He introduced a light six-inch howitzer for field use, and restricted the larger pieces to siege and coastal defence work; indeed, it was Gribeauval who established these three definite categories of artillery for the first time. He introduced horse teams shackled in pairs instead of singly, giving a great improvement in manœuvrability and speed, and designed a whole range of vehicles and ancillary equipment. He standardised many fittings and iron forgings on his guns and vehicles, so that they were interchangeable when damaged. He was unsuccessful, however, in his constant efforts to secure the introduction of true horse artillery, with the crew mounted on the limber. Although his guns were proving themselves on the battlefield by 1781, when Rochambeau (q.v.) used them in America, it was to be another ten years before true horse artillery was introduced in Lafayette's (q.v.) Army of the Centre.

Gribeauval gave considerable thought to the organisation of artillery units; having produced a rationally-designed kit of equipment, he gave his mind to a rational body of men to use it. As early as 1765 he introduced into the reserve or corps artillery (the units which operated the eight- and twelve-pounders) the concept of eight-gun batteries or divisions self-sufficient in men, guns, and ammunition supply. The logical conclusion, the provision of permanently attached horse-teams and drivers for each battery, was not to come for some years, however. Even so it was to Jean Baptiste Gribeauval that Napoleon (q.v.) and his marshals owed their greatest debt as they evolved ways of using artillery in the thick of battle ever more audacious and effective – and this was a debt which *le Tondu*, for one, gladly acknowledged. Gribeauval died in Paris on 9 May 1789, and was thus spared the sight of the collapse of the régime he had served all his life.

Guderian, Heinz (*1888–1954*), *General*

If any single officer should receive the credit for making possible Germany's lightning victory in the West in May–June 1940, the choice must fall on Guderian. He was not alone in his long advocacy of the new weapons, the new units, and the new tactics which made these victories possible in the face of Allied superiority; but he was the most visionary and single-minded of the German 'tank enthusiasts', and the father of the Panzer arm.

Guderian was born at Kulm (Polish Chelmno) on 17 June 1888, the son of an army officer. Originally commissioned in the infantry, he progressed through the junior ranks in a series of regimental and staff appointments, and served as a staff officer in the First World War. He stayed in the shrunken post-Versailles *Reichswehr*, and a posting to a telegraph battalion at Koblenz and subsequent work with radios started his technical education. From its earliest days as an identifiable group, Guderian was a leading member of the school of thought in Germany which pressed for the creation of a tank force supported by equally mobile formations of infantry and artillery and capable of fast independent action, being freed from the slow-moving bulk of the infantry army. These officers, among them Lutz, Reichenau, and Thoma, studied with great application the experiments and analyses in the field of armoured warfare which emanated from Britain in the late 1920s and early 1930s, notably the writings of Liddell-Hart (q.v.) and Fuller, and the exercises of the Experimental Mechanised Force. At first they met with the same conservative opposition among their superiors as did their 'co-religionists' in other countries, but the arrival on the scene of Adolf Hitler brought them the chance they needed. In 1933 Blomberg was appointed War Minister and Reichenau achieved a senior staff post; Hitler was receptive to new philosophies springing from a determination to give Germany a devastating edge over her numerically-stronger neighbours, and from a revulsion against the static slaughter of 1914–18. He was much impressed by a series of small demonstrations, and in 1934 the first tank battalion was formed at Ohrdruf.

Guderian was promoted rapidly from this stage onward. The technical problems of overcoming Germany's industrial handicaps and preparing a whole new technology, together with the still-vigorous opposition of the conservative element, slowed the growth of the new arm, and there were frustrating diversions down theoretical culs-de-sac. Nevertheless Guderian and his colleagues managed to avoid the false creed of 'different tanks for different battlefield tasks' which so crippled Britain's armoured forces in the early years of the Second World War. In 1938 Guderian was appointed General of Armoured Troops, and his embryo divisions gained much useful experience – and many badly-needed tanks – in the bloodless invasions of Austria and Czechoslovakia. By September 1939, when he led XIX Army Corps into Poland, there were seven Panzer divisions operational at various strengths. The crews were an *élite* group, and their vehicles, while numerically and technically still weak, had great potential for development in the light of technical advances and operational experience. Most important of all, the tactical doctrine, while still considered controversial, enshrined the main tenets of Guderian's creed: that armour, supported by mobile infantry and artillery and enjoying tactical independence from higher echelons, should not be split up in direct support of the mass of the infantry army, but should strike ahead relentlessly, making use of its local superiority of fire-power and mobility to achieve deep penetration and dislocation of the enemy's line.

The experience gained in Poland was invaluable; and Guderian, leading his corps from a tank in the first wave of 3rd Panzer Division, was well placed to react to the lessons learnt on the battlefield. It was found that many types of 'untankable' terrain could be penetrated or by-passed in practice, and that speed and unrelenting pressure were the keynotes of success. On the debit side, the fatal dangers of committing tanks to street fighting in urban areas were also clearly demonstrated. The possibilities of close co-operation between forward armoured elements and tactical airpower were shown to be exciting. The correct proportions of mechanised infantry and tank divisions within the armoured corps became clear. It was fortunate for the *Wehrmacht* that Poland possessed no tank forces worthy of the name, so that initial mistakes could be rectified without the simultaneous hazard of tank-versus-tank battles on a large scale.

There were still many German generals who were unconvinced, however, and it was the adoption of Manstein's (q.v.) brilliant 'left hook' plan for the invasion of France which gave Guderian and his colleagues the perfect opportunity to display to the full the potential of the Panzer divisions. Guderian's XIX Panzer Corps formed the spearhead of Panzer Group Kleist; it was Guderian's tanks which smashed through the fatally-dispersed French at Sedan on 14 May 1940, forcing the crossings of the Meuse in an incredibly short time and striking on across France supported by the *Luftwaffe*. Considerably outnumbered by the Allied forces, the German armies broke France and drove the British Expeditionary Force off the Continental mainland in a matter of weeks; and it was Guderian who gave them the weapon which made this possible, and the philosophy which paralysed the enemy's response. There is no space here for a detailed account of the gradual developments in armoured warfare during the rest of the War, in most of which Guderian played a major part; but it is true to say that after the victory in France his doctrine of the correct use of the armoured, all-arms mobile force was never challenged. Hitler himself became the greatest 'tank enthusiast' of all.

When Germany invaded Russia in June 1941 Guderian commanded the 2nd Panzer Group, in Army Group Centre. In the initial phase of staggering advances his divisions repeated their achievements of the previous year, and surpassed them. Pushing on audaciously far beyond their infantry support, knocking their tanks and trucks to pieces on the appalling roads, running the constant risk of encirclement and the exhaustion of fuel and ammunition – still the Panzers triumphed, protected by their mobility and by the paralysis of the Soviet command. In August and September occurred the diversion to the south which achieved another massive victory of encirclement in the Ukraine; but the cost in time had been too high, and now Moscow would never be taken before the onset of winter. The battered and worn-out Panzers became unserviceable with terrifying speed under the severe conditions of the Russian winter, and the impetus of the advance petered out. On 5 December the great Soviet counter-attack was launched, and the pursuers found themselves fighting desperately on the defensive. Hitler was incapable of appreciating the realities of the situation, and on Christmas Day 1941 Guderian was relieved of his command for making a militarily sound withdrawal against the *Führer*'s dogmatic prohibition. He was in good company – December 1941 saw a mass dismissal of army, corps, and divisional commanders unprecedented in history.

Guderian was brought out of retirement in March 1943 and made Inspector of Armoured Troops, a post in which he had authority over production priorities as well as purely military matters. He was dogged by persistent ill-health and by internal jealousies and vested interests in the upper echelons of the German command, as well as by the deteriorating situation on all fronts; the fairest summary of his performance is probably that no other man could have done as well, let alone better, under the same circumstances. The resilience of the Panzer arm in the face of repeated disasters was remarkable right up to the last days of the war, and much of the credit must rest with Guderian. On 21 July 1944 in the immediate aftermath of the bomb attempt on Hitler's life, Guderian was appointed Army Chief of Staff, and held the post until dismissed by the *Führer* in the course of an epic shouting-match on 28 March 1945. It says much for his character, talents, and obvious loyalty that he lasted so long, as he was by no stretch of the imagination a 'yes-man'. He was blunt and outspoken in his determination to bring home to the supreme command the military realities of the situation which faced Germany, and it seems that Hitler felt a genuine respect for him, however infrequently he listened to his views. Guderian was courted, cautiously, by the group of officers who led the attempt on Hitler's life, but while he agreed that the *Führer*'s conduct of military affairs was becoming disastrous he was not interested in any deeper involvement. Guderian was a straightforward soldier who concerned himself with military decisions and with the advancement of armoured warfare technology, and seems never to have questioned the political objectives of national policy. The fact that he had been frustrated so often, and over so many years, by the stubbornness of entrenched conservative opinion in the German military establishment may have tended to inhibit him from identifying very closely with that establishment when it found itself in open confrontation with Hitler.

Heinz Guderian died on 14 May 1954.

Gustavus II, Adolphus (*1594–1632*), *King of Sweden*

Gustavus's unchallenged title as one of the great military figures of history is due to a combination of factors: some genuine technical innovations appeared in his armies, but more significant was his overall philosophy of tactical co-operation between the arms. He was, in a very real sense, the first modern general.

Born in Stockholm on 9 December 1594, he was the eldest son of King Charles IX and his second wife, Christina of Holstein. He succeeded to the throne on 30 October 1611, a boy of sixteen beset by problems. There was a succession dispute involving his cousin John; a pact was in being which laid down that a regency should rule until the king was twenty-three years of age; his father had been unpopular with the nobility; and Sigismund of Poland, who had been deposed from the Swedish throne in 1600, was still actively engaged in trying to win it back. By the exercise of considerable personal abilities, however, Gustavus stabilised his position. He was a far more adroit handler of men than his father, and successfully negotiated the waiving of the regency agreement. He mollified the nobles by appointing Axel Oxenstierna as Chancellor, and achieved a close and cordial working relationship with that powerful man which lasted all his life. By the time of his death he had become the most popular and powerful Swedish king ever to reign. His reforms in many fields united the various classes and interests behind him; enthusiastic, gifted, endlessly curious, he laid the foundations for Sweden's prosperity over the next two centuries.

During the first years of his reign he was forced to follow a fundamentally defensive foreign policy based on blocking any possible Russian expansion in the Baltic area. In 1613–17 he fought campaigns in Novgorod and Muscovy which ended with the Peace of Stolbovo and the loss of the last Russian footholds on the Gulf of Finland. In 1617 Gustavus took the opportunity offered by Poland's simultaneous involvement in wars with Turkey and Muscovy to invade Polish Livonia; this was merely an episode in the incessant fighting caused by Sigismund III's refusal to abandon his claim to the throne of Sweden. Gustavus captured several Baltic ports and compelled Krzysztof Radziwill to sign a three-year armistice. He invaded Estonia with 12,000 men in 1621, gathering another 4,000 local troops and successfully besieging the capital, Riga. In 1625 the Swedes occupied the whole of Livonia and Kurland, and in 1626 Gustavus invaded the Polish territory of Prussia. He quickly occupied the whole northern part of the province, threatening Polish access to the sea, and returned to Sweden leaving forces investing the last stronghold, Danzig. A Polish attempt to relieve Danzig failed, and in May 1627 Gustavus returned with reinforcements. His 14,000 men repulsed Koniecpolski's 9,000 Poles at Dirschau, where the King was wounded. In 1628 the Swedes pushed the Poles far southwards, but retired into Prussia for the winter. Sigismund now received imperial troops, as the Emperor Ferdinand II (see under Wallenstein) wished to insure against the possibility of Gustavus's marching from Prussia into the area of the main campaigns of the Thirty Years' War. Gustavus was wounded again in 1629 at Sztum, an inconclusive cavalry clash; after this he could never again wear armour. The war dragged on, but Swedish control of the Baltic coast – Gustavus's major objective – was never shaken, and Poland needed peace. The Treaty of Altmark brought a six-year truce on favourable terms, and Gustavus was free to involve himself in Germany at last. He had long believed in the importance of Protestant solidarity, but felt that by pursuing his own vital interests against Ferdinand's Polish vassals he had contributed indirectly to the struggle. Now he was ready for decisive action. The quality of his army was well known, and his enemies awaited the Swedish intervention with some dismay.

Many treatises have been written on Gustavus's military reforms, and obviously only the barest summary can be given here. First, he instituted permanent units with a fixed hierarchy of command. The basic 'squadron' of about 400 men was divided into four companies, and three 'squadrons' were grouped permanently as a brigade, commanded by a colonel. Maintaining a standing army for periods longer than a single campaign required increased financial support from the Crown, and thus increased royal control and supervision. Discipline was firm, and there were frequent manœuvres and training sessions. The tactical formation of the squadron was a central block of some 200 pikemen with a wing of 100 musketeers on each flank. The men were drawn up in only six ranks. Gustavus so drilled his musketeers, and so improved their weapons, that a steady fire could be maintained; two ranks fired together, and then moved to the rear to reload while the next two ranks fired, and so on. The co-operation between pikemen and musketeers was brought to a peak of perfection; the entire formation advanced while this succession of volleys was being delivered, and the final impact of the pikes on shaken enemy ranks was thus enhanced. The pike was shortened and strengthened so that it was a truly offensive weapon. For his cavalry regiments Gustavus evolved a system using the best aspects of the German caracole – the successive wheeling attack by ranks of pistoleers – and the traditional sabre-charge. The first of three ranks of horsemen fired their pistols just before contact and then went in with the sword, while the other two ranks retained their pistols for emergencies. Direct support of the cavalry charge was given by detached units of musketeers interspersed along the line of mounted squadrons, and by field artillery. (Fuller details of Gustavus's improvement in field artillery and direct support will be found under Torstensson.) This comprehensive tactical use of all three arms was Gustavus's greatest achievement. His relatively shallow infantry formations, only six ranks deep, could cover a much longer front than the ponderous columns and squares still employed in other armies, and he is the father of the linear tactics of later generations of soldiers.

Gustavus's efforts to form a Protestant League failed, but the Holy Roman Empire's threat to Swedish interests on the Baltic coast was so clear that in July 1630 he crossed into Germany without an ally. His army of 13,000 was reinforced to 40,000 as it advanced. The Protestant princes were slow to ally themselves to him, but to the Protestant masses he was a conquering saviour, the 'Lion of the North' prophesied by Paracelsus. Magdeburg committed itself on the promise of Swedish support, and in November 1630 was besieged by Pappenheim and Tilly (q.v.). At length the Protestant princes, encouraged by the Swedish advance, presented Ferdinand with a manifesto demanding the remedy of many abuses and restrictions; when he ignored it they established an army under Hans Georg von Arnim. In April 1631 Gustavus carried out a brilliant surprise attack on Frankfurt-am-Oder and captured the city, hoping to distract Tilly from the siege of Magdeburg. He was unsuccessful in this, and prepared to march directly to Magdeburg, but on 20 May the city fell to the imperial forces and was subjected to such a hideous sack that it has been estimated that only some 5,000 of the 30,000 inhabitants survived. On 15 September Tilly took Leipzig, and two days later, with 36,000 men, he took up a position at Breitenfeld a few miles away. Gustavus had 26,000 Swedes and 16,000 Saxon allies under command when the battle began on 17 September.

Tilly placed his infantry in the centre and his cavalry on each wing; he commanded the centre and right, while Pappenheim commanded the left. Gustavus placed the Saxon horse and foot on his left wing, his infantry in the centre, and his cavalry on the right, with a small reserve of horse behind his centre. A galling cannonade by Torstensson's artillery lasted for two hours, and eventually Pappenheim lost his temper. He launched some 5,000 cavalry from the imperial left wing in an unauthorised charge on the Swedish right. Gustavus brought up his reserve and wheeled his manœuvrable units round to make a new front at right-angles to the centre, successfully repulsing the imperial cavalry until it was so weakened that a charge by his own right-wing cavalry drove it from the field. On the Swedish left, meanwhile, Tilly had advanced and driven the Saxons back, leaving the Swedes outnumbered and with their left flank exposed. Tilly now swung the mass of his infantry to his right, intending to envelop the exposed Swedish left flank, but was decisively foiled by the greater mobility of Gustavus's well-drilled professional soldiers. Gustavus drew men from the second line of his centre, constructed a new front at right-angles to his left flank to guard against the new direction of thrust, and brought his right-wing cavalry forward and to the left in a punishing charge at the imperial artillery – still in its original position – and at the left wing of Tilly's army. The superior tactical methods of the Swedish infantry held their threatened left flank steady while the cavalry captured the imperial guns; these were then turned on their erstwhile owners. Caught in a cross-fire, manœuvred off their line of communications with Leipzig, and outfought at every point, the imperial army collapsed. Tilly, badly wounded, lost some 13,000 dead and wounded against 2,100 Swedish casualties, and was forced to abandon Leipzig the next day. Breitenfeld was Gustavus's masterpiece, and it has been said that that single battle guaranteed the survival of Protestantism in Germany.

Within three months the Swedes and their allies controlled the whole of north-west Germany, and in the spring of 1632 Gustavus advanced southwards and then eastwards into Bavaria. At Lech on 15–16 April 1632 he attacked the rebuilt imperial army in their entrenched camp by crossing a bridge of boats; among the killed was Tilly himself, and Maximilian of Bavaria led the remainder of the army in a costly retreat. Gustavus occupied the whole of southern Bavaria. In the months which followed Wallenstein built a new mercenary army and led it northwards, threatening the Swedish line of communications and skilfully forcing Gustavus to follow. He drew Gustavus into a costly and unsuccessful attack on prepared positions at Alte Veste in early September, and by early November the armies were back in the Leipzig area. At Lützen on 16 November 1632 Gustavus attacked the imperial army across a sunken road which was captured and recaptured several times in the course of a confused see-saw engagement further obscured by fog and the smoke of a burning town. Half way through the battle, which ended in a Swedish victory, Gustavus was killed in a cavalry mêlée.

Haig, Douglas, 1st Earl of Bemersyde (*1861–1928*), *Field-Marshal*

Douglas Haig, British Commander-in-Chief on the Western Front during the First World War, was a resolute general of orthodox military views who came to regard as inevitable the use of massive armies in trench warfare.

Born in Edinburgh on 19 June 1861, Douglas Haig was educated at Clifton and Brasenose College, Oxford, before entering the Royal Military Academy, Sandhurst. Commissioned in the 7th Hussars, he served in the Sudan in 1898, and the following year shipped to South Africa as a major with the Inniskillings. He was a prisoner for a short time during French's march on Pretoria, and later commanded small cavalry columns on anti-guerrilla operations during the second invasion of Cape Colony; he ended the war as a colonel. In 1903 he was posted to India, where he remained for three years as Inspector-General of Cavalry; promoted lieutenant-general in 1910, he was appointed G.O.C., Aldershot, two years later.

On the outbreak of the First World War, promoted general, Haig commanded the British 1st Corps as part of the British Expeditionary Force sent to France. With the stabilising of the Western Front – after the initial breakthrough and containment of the German armies – the fighting in France degenerated into the semi-stalemate of slogging trench warfare. The failure at Loos, with 60,000 British casualties, was believed to be the fault of British Commander-in-Chief, Sir John French, who was replaced by Haig in December 1915. By this time the B.E.F. had grown to three armies, but after fighting the Battles of Loos, Festubert, and Neuve Chapelle Haig was faced with the abandonment of a war of movement and the adoption of an exhausting and costly war of attrition; at the same time he had to undertake the training and integration of new and territorial forces.

The terrible battles of the Somme in 1916, fought at the insistence of the French General Joseph Joffre against Haig's better judgement, and with insufficient artillery and ammunition, resulted in the exhaustion of the German forces and substantial relief for the French. But the very heavy casualties alarmed the Allied governments, infuriated the British Prime Minister, Lloyd George, and resulted in the inevitable enervating political controversy over the correctness of the military strategy. What the politicians had utterly failed to comprehend was that, having committed vast numbers of men to the cauldron of the Western Front, they had not the means to pursue a war of movement. The introduction of the tank was a belated and ill-handled attempt to remedy the tactical stalemate.

Never a man to suffer politicians gladly, Haig was constantly at odds with Lloyd George who, after the failure of the French General Nivelle's 1917 offensive, displayed open distrust of the Allied military command and sought to control strategy. It was largely the trust placed in Haig by King George V that saved him from following French. In 1918 he agreed to extend the British section of the Western Front southwards, but was denied the necessary reinforcements – with disastrous results when the expected German offensive opened. After Ferdinand Foch (q.v.) had been appointed supreme Allied generalissimo in France, Haig issued his famous message to British troops: 'With our backs to the wall and believing in the justice of our cause each one of us must fight to the end.' Finally the German advance was halted, its effort spent. The British counter-attack penetrated the Hindenburg Line. Quickly launching a series of heavy assaults, supported by the newly-arrived American Expeditionary Force, the Allies eventually forced the Germans to seek armistice terms in November 1918.

Haig was a humble man of great character and unshakeable loyalty to his King and country who, despite constant political efforts to achieve his downfall, never threatened to resign, steadfastly believing that his duty lay in the single-minded pursuit of victory unless he was actually relieved of his command.

After the war he devoted himself to improving the welfare of his men. He had been created G.C.B. in 1915, G.C.V.O. in 1916, and K.T. in 1917. In 1919 he was awarded the Order of Merit, created an earl, and given an award of £100,000. The post-war years were spent in initiating charities and travelling the British Empire in the interests of ex-servicemen. Exhausted by his efforts, Haig died on 29 January 1928.

Harald III, called Hardraade (*1015–66*), *King of Norway*

It would be pointless to try to assess in modern terms the generalship of this famous Viking warrior; the limited contemporary sources do not concern themselves with the necessary detail, and in eleventh-century Scandinavia generalship in the modern sense was unknown. Suffice it that in his day his fame as one of the great warriors of the age spread from end to end of the known world.

He lived in an environment of constant and merciless butchery, when the only road to fame and power lay through personal courage, strength in battle, and ruthless determination. In the eleventh century European crowns seldom passed peacefully from one head to another, or stayed on the same head for long. Warfare was the normal way of life, and to a large extent was seen as an end as well as a means. By modern standards mere survival for fifty years in such a world was a considerable achievement; a man's rise to a position of power indicates a degree of guile, physical prowess, and single-mindedness completely outside the experience of most twentieth-century men. Such a man had to dominate his surroundings – against fierce competition – to the point where he commanded a sufficient following to seize power. Having seized it, he had to embark on a programme of conquest or raiding to provide himself with the financial means to keep his restless followers loyal, while at the same time defending his possessions against other leaders similarly engaged. In this arena Harald Hardraade ('hard ruler') thrived to such an extent that his deeds, retailed by word of mouth, were the talk of Europe.

He was born in 1015, the son of Sigmund Syr, a petty king in eastern Norway, and Asta, mother of St Olaf. From the earliest recorded times Norway had been under the actual or nominal control of Denmark, which was the dominant Scandinavian power of the age. Sweden was sometimes allied with Denmark and sometimes at war with her, but generally tended to look on the Baltic as her sphere of influence. In the year 1000 King Sweyn Forkbeard of Denmark and King Olaf of Sweden defeated and killed the Norwegian King Olaf I Trygvesson at the sea battle of Svalde. In 1015 Olaf II Haraldson (St Olaf) of Norway drove the Swedes from their conquests in south-eastern Norway. Three years later Canute II came to the Danish throne on the death of his father Sweyn, and between 1026 and 1030 he prosecuted a bloody and eventually successful war against Olaf II Haraldson and Anund Jakub of Sweden, who allied themselves against him. In 1028 Canute won the sea battle of Helgeaa, and expelled Olaf from Norway. Two years later Olaf returned, but was defeated and killed at Stiklestad; it is known that the fifteen-year-old Harald Hardraade fought with his kinsman at this battle.

He survived the defeat, but was forced to flee for his life. He travelled south and east, and entered the service of Yaroslav the Wise, Prince of Kiev. (The Scandinavian presence in central Russia dated back to the 860s, when Rurik, a leader of the Rus or Varangians from Sweden, established himself as ruler of Novgorod. By the end of the ninth century Kiev was the capital of an extensive Varangian state stretching from the Gulf of Finland to the Carpathians and the northern Ukraine. The Varangians penetrated to the Black Sea, and made an attempt on Constantinople in 865.) Harald rose in the service of Yaroslav, and married his daughter Ellisif. At some point during the 1030s he travelled further south, and entered the service of the Byzantine Emperor Michael IV (who ruled 1034–41) – there was a long tradition of Scandinavian mercenary soldiering in Constantinople, and the axemen of the Varangian Guard were the *élite* household troops of the Emperor. Harald served the Empire for ten years, during which time his fame spread; he became the most noted Scandinavian warrior in Byzantine service. It is said that he fought in the Emperor's wars in Sicily and in Bulgaria, and that he made a pilgrimage to Jerusalem.

In 1045 Harald returned, by way of Russia and Sweden, to Norway. By this time his nephew Magnus I, son of Olaf Haraldson, was on the throne. It appears that Magnus agreed to share power with Harald as joint kings; considering the conditions of the day this probably represented an attempt to buy off Harald from making an outright bid for the throne. Magnus had succeeded to the throne of Norway in 1035, and to that of Denmark in 1042. In 1047 trouble broke out in Denmark when Sweyn Estrithson, a nephew of the late King Canute, led a rebellion against Magnus. Magnus and Harald took the field together against Sweyn, but Magnus died during the campaign, leaving Harald sole king of Norway. He was unable to crush the rebellion, and in 1050 Sweyn established himself on the throne of Denmark. Harald kept up sporadic offensives in an attempt to restore Norwegian rule over Denmark for many years, but finally came to terms with Sweyn after defeating him at Niz in 1062. In 1063 Harald turned his eyes towards Sweden, invading and seizing territory in the south-east of the country and defeating the ruler Steinkel. In addition, throughout his reign Harald was forced to give his attention to occasional revolts by ambitious local leaders, which he suppressed with a ruthless energy typical of his day. He also became embroiled in a serious dispute with Adalbert, Archbishop of Bremen, which brought him into disfavour with the Pope.

In 1066 Harald seized a promising opportunity to fish in the troubled waters of the English succession. The saintly King Edward had died in January, and the powerful Earl of Wessex, Harold Godwinson (q.v.) had been elected to the throne by the Witan. This earned him the immediate enmity of William (q.v.) of Normandy, who claimed that he had Harold's oath to support his own claim to the throne, and a Norman invasion army was prepared. Harald Hardraade now allied himself with Harold Godwinson's exiled brother Tostig, once the Earl of Northumbria, and anxious to regain his earldom by means of any alliance which offered. Tostig was promised Northumbria if he assisted in a Norwegian invasion of northern England. William of Normandy certainly knew of the Norwegian plan, but it is hard to see how actual collusion would have served his long-term interests unless he was confident that the Norwegian invasion would fail after distracting Harold's attention from the south coast. In September Harald Hardraade and Tostig sailed for England, and shipped a large army up the Humber river. On 20 September they met an English army, led by Morcar of Northumbria and Edwin of Mercia, at Fulford in Yorkshire. The English were defeated, and the city of York subsequently submitted to Harald Hardraade. On 16 September Harold Godwinson began to march north from London at the head of his army. On 25 September he came up with the Norwegian army at Stamford Bridge, and defeated it in a hard-fought battle which was finally decided by the mounted housecarls of Harold's army. Both the treacherous Tostig and Harald Hardraade were killed.

The battle cost Harold heavy casualties, and it was with a much-weakened array that he started the long march south again on 2 October, on receiving news of William of Normandy's landing at Pevensey six days earlier. The defeat of the English at Senlac Hill on 14 October, and the consequent establishment of the Norman kingdom of England, was Harald Hardraade's legacy to the world.

Harold II, Godwinson (*c. 1022–66*), *King of England*

England's last Saxon king was the second son of Earl Godwin of Wessex and his Danish wife Gytha, and the early part of his life seems to have been influenced by the Danish connection. In 1044 he became Earl of East Anglia; and two years later, when his elder brother Sweyn was banished, he received part of his lands. It is recorded that he opposed Sweyn's reinstatement in 1049. When Earl Godwin quarrelled with King Edward the Confessor in 1051 Harold shared his father's disgrace. He joined him in Gloucestershire and accompanied him to London; on hearing that they were to be exiled he reacted violently. While Godwin travelled to Flanders, Harold made his way to Ireland with his brother Leofwine, intending to raise an armed force to pursue the family's interests. He spent the winter with Dermot, King of Leinster and Dublin, and raised an army largely from the Norse settlers of the area. He sailed to England in the spring of 1052 with nine ships, and landed in Somerset; when landing parties sent out to gather provisions were resisted they harried the countryside. Harold sailed east up the coast, and accompanied his father in the advance upon London which brought about their reinstatement in 1053. At Easter 1053 Godwin died of a sudden seizure while dining at the king's table at Winchester – legend has it that he suffered a stroke immediately after challenging God to strike him dead if he had any responsibility for a death which was notoriously his doing. Harold succeeded his father as Earl of Wessex, as his brother Sweyn had died abroad in the meantime. At the age of about thirty-two he now became the second most important man in England, and the virtual ruler of the southern half of the country.

Harold's main rival for power was the house of Leofric, Earl of Mercia. In 1055 Leofric's son Aelfgar was outlawed on a charge of treason which Harold probably engineered. Later that year Aelfgar and his ally Gruffyd of Wales raided into Mercia and burnt Hereford. Harold led an army against them from Gloucester, but they refused battle and fell back into South Wales. A truce followed, but in 1056 there were more raids, concluded by yet another truce. In 1057 Leofric died and Aelfgar succeeded him; he allied himself even more closely with Gruffyd of Wales, however, and there was more trouble in 1058. It is believed that Aelfgar had died before 1062, when Harold settled the vexing problems of the Welsh border in ruthless fashion. Gruffyd's warriors began raiding again that year, and Harold made a surprise forced march through bitter winter weather which obliged Gruffyd to abandon his capital of Rhuddlan. Harold burnt Gruffyd's palace and beached ships, and returned to Gloucester to make more leisurely preparations for a campaign. In May he returned, sailing from Bristol and landing on the Welsh coast to link up with another force under his brother Tostig, Earl of Northumbria. Together they fought a skilled campaign which resulted in the ruin of Wales for a generation. Harold trained his men to march and fight light in a way they were not accustomed to, and to live off the land. They pursued the Welsh into the mountain refuges which had kept them safe for centuries, and wiped out all the able-bodied males they could find. It is thought that the devastation of Wales by Harold was responsible for the relative inability of the Welsh to interfere with the plans of the first three Norman kings. Seeing that their situation was hopeless, the Welsh dethroned Gruffyd, gave hostages for good conduct, and promised tribute; to underline their determination to co-operate, they sent Harold a present in August 1063 – Gruffyd's head.

It was probably in 1064 that Harold visited Normandy as the envoy of King Edward. The subsequent events are shrouded in uncertainty; the only detailed records which survive are Norman, and can hardly be considered disinterested, as it was partly on certain incidents which occurred during the visit that William (q.v.) of Normandy based his claim to the English throne. Harold seems to have been shipwrecked on the coast of Ponthieu, and only released from the dungeons of Count Guy on the intercession of William of Normandy. He stayed at Rouen as William's guest for some time, and marched with the Duke on his campaign against Conan, Count of Brittany. His strength, prowess, and good nature seem to have earned him the respect of many Norman knights during this campaign, and it appears that he risked his life to save some Norman soldiers who were trapped in a marsh or quicksand. On their return to Bayeux Harold seems to have sworn some sort of binding oath to William; the Norman version holds that he swore to uphold William's claim to the throne of England on Edward's death, but there is no independent evidence of this. Harold then returned to England.

In October 1065 the men of Northumbria rose in revolt against Tostig, and after a period of negotiation Harold was apparently obliged to accede to their request that the earldom be given to Morcar, son of Aelfgar of Mercia. Tostig, who was exiled, never forgave Harold for siding with his enemies;

he was an unstable man, and no doubt the Northumbrians chose wisely. Edward the Confessor died on 5 January 1066, after 'commending' the country to Harold. The following day the nobles of all England elected Harold king, although there was some initial reluctance by the Northumbrians. Early in May 1066 Tostig, who had been in Flanders, raided the south and east coasts, but was repulsed before Lindsey and retired to Scotland. Harold kept ships and men in the south coast ports until September to guard against the expected invasion by William of Normandy, but logistic problems forced him to disperse them that month. Almost immediately he learnt that Harald Hardraade (q.v.), King of Norway, had allied himself with Tostig and had brought some 300 ships and half the warriors of Norway across the North Sea to attack the north-east coast. Tostig joined him in the Tyne, and together they sailed south down the coast and up the Humber, ravaging the coastal areas as they went. The fleet anchored at Riccall on the Ouse; the Earls Edwin and Morcar brought an army against the invaders, but were defeated at Fulford on 20 September. York surrendered on terms about the 24 September by which time Harold had reached Tadcaster after advancing from London by forced marches. He surprised the Norwegians on the 25 September at Stamford Bridge. The *Heimskringla* gives a stirring account of the battle, unfortunately more poetic than factual. It is said that though Harold offered Tostig his earldom back if he would come to terms, he promised the Norwegian king only seven feet of ground, or as much more as his great size demanded. The battle began with a sudden attack by Harold on a part of the invading army which was separated by the Derwent from the main body. Harold pushed them back over the river, crossed after them, and, after a bloody fight lasting all day, defeated the rest of the Norwegians. Tostig, Harald Hardraade, and many hundreds of warriors on both sides were killed; it is said that the bones of the fallen still lay thick around the site of the battle many years later. Harold accepted the submission of Harold Hardraade's son Olaf, and allowed the remaining Norwegians to regain their ships and sail away. It was probably on 1 October, while recovering at York, that he heard the news of the Norman landings in Pevensey Bay on 28 September.

Harold's army was much weakened by the forced march and by the battle, but he at once marched south again. He reached London on the 5 October, and left it on the 12 October; the southern and eastern part of the country rallied to him, and he was able to array a force of some 6,000–7,000 men at Senlac Hill, where he was attacked throughout the 14 October by William's army. The Normans roughly equalled him in strength, but a much larger proportion of them were properly armed and equipped; they were also strong in cavalry, while Harold's Saxons fought on foot in close-packed ranks behind a wall of shields. Harold's dispositions favoured this defensive formation, and the Normans charged repeatedly and at great cost without breaking the Saxon line. The integrity of Harold's position was threatened more than once by the indiscipline of the *fyrd* (local levies), who broke ranks to follow apparent Norman withdrawals and were caught and butchered in the open. Harold's core of picked professional warriors, the axe-wielding housecarls, were not broken until the late afternoon. By that time their ranks were packed so tight that dead men were held up in the press. The shield-wall was apparently broken at last by a shower of arrows aimed to fall into the Saxon mass at a steep angle, followed and exploited by a last determined cavalry charge. Harold was cut down by the Norman knights in the mêlée; it is nowadays believed that the story of his being killed by an arrow in the eye is apocryphal, and based upon a misinterpretation of the Bayeux Tapestry. Harold's brothers Gyrth and Leofwine, and the majority of the housecarls, were also killed, and the survivors were pursued by the Normans as they scattered. Harold was buried, either at once or subsequently, at Waltham Abbey, which he had endowed. There were persistent rumours for many years that he had merely been wounded and had been spirited away, to return one day to save his people from the Normans. There is not the slightest evidence for these myths, which sprang from despair; there are obvious parallels with the Arthurian legends, and with traditions in other countries.

Harold was a man who commanded great love and loyalty, if the scanty contemporary accounts can be trusted. He was warm and human, a man of impulse more than of calculation, although by no means a mere berserker. He usually governed his temper, but could be provoked to rash action. He was physically impressive, very brave, stoic and uncomplaining in hardship, and energetic. The behaviour expected of an eleventh-century king was extremely lax by modern standards, but Harold seems to have been genuinely religious and to have enjoyed good relations with the Church. It is typical of him that he marched straight to the Norman beach-head instead of waiting for a more favourable opportunity for attack further inland; William's Normans were harrying Sussex, and Sussex had been part of his earldom. The efficiency of his administration during his short reign has been the subject of controversy, but many credit him with firm and imaginative government. In his campaigns in Wales and the north, particularly in the advance on Stamford Bridge, he showed himself to be a considerable leader of men. Finally, considering the type of personality which he seems to have projected, it may not be out of place to record that his mistress Edith Swanneck was widely reputed to be the most beautiful woman in England – and that she was inconsolable at his death.

Hart, Basil, *see* Liddell-Hart

Havelock, Sir Henry (*1795–1857*), *Major-General*

Havelock's career bears a certain superficial similarity to that of Robert Craufurd (q.v.), in that he did not have an opportunity to demonstrate his gifts to the full until a relatively late stage. His series of victories against the mutineers in India, achieved by awesome determination in the face of unusual difficulties, brought him a deserved fame which he did not live to enjoy.

Henry Havelock, the second son of William Havelock, a shipbuilder in Sunderland, was born at Ford Hall, Bishop Wearmouth, on 5 April 1795. After education at Charterhouse he decided to enter the law and became a student in the Middle Temple, but as a result of a misunderstanding with his father he was thrown on his own resources and forced to abandon his chosen career. Through the assistance of his brother, an Army officer who had distinguished himself in the Peninsula and at Waterloo, he obtained a commission in the 95th (Rifle) Regiment on 30 July 1815.

First posted to a company commanded by Captain (later Sir) Harry Smith, Havelock was encouraged to study military history and the art of war; this he did with great diligence, and went on to learn foreign languages, including Persian and Hindustani. Promoted to lieutenant on 24 October 1821, he spent his first eight years of military service on several postings in England and Ireland; anxious to see active service, he transferred to the 13th Regiment at the end of 1822.

He sailed for India with his new regiment, commanded by Major (later Sir) Robert Sale, in January 1823. At this period Havelock was influenced by a fellow officer, Lieutenant James Gardner, towards deep religious convictions – he had acquired strong principles in early youth but they had since lain dormant. Such was the strength of these convictions that they were to influence greatly much that occurred in his later life. When in May 1823 Havelock arrived in Calcutta he soon became known to Bishop Heber and other ecclesiastics through his visits to and interest in the missionaries at Serampore.

Shortly after his arrival in India war broke out against Burma, and Havelock was appointed deputy assistant adjutant general to the army under Sir Archibald Campbell. While campaigning in the Burmese jungle he became very ill after several stockade engagements, and was sent back to India, spending much of his time with his brother William, of the 4th Dragoons, at Poona. At the end of a year Henry had recovered fully, and rejoined his regiment at Prome in Burma on 22 June 1825. He was at the capture of Kemundine, Kumaroot, and Melloon, and fought in the engagements of Napadee, Patanago, and Pagahm Mew. When the King of Burma sued for peace, Havelock was selected to go to Ava to receive the ratification of the treaty. In February 1826 the army returned to India and Havelock rejoined his Regiment at Dinapore.

(His account of the First Burma War was published at Serampore in 1828.)

In March 1827 Havelock was appointed adjutant at the depot at Chinsurah, and was a constant visitor to the local Baptist mission; indeed, he married Hannah Marshman, daughter of one of the missionaries. In 1834 he was appointed interpreter to the 16th Regiment at Cawnpore, and the following year became adjutant to his own Regiment – a position he held for more than three years.

During the 1842 (First) Afghan War Havelock was aide-de-camp to Sir Willoughby Cotton, commanding the Bengal Division which made the eighteen-week march to Kandahar, and was also present at the blowing-up of the Kabul gate after the capture of the fortress. When the 13th and 35th Regiments were sent forward to open the passes, the force was heavily attacked on entering the Khoord Pass and Havelock was sent back to Kabul to fetch reinforcements. These and the force commanded by Brigadier-General Sir Robert Sale fought their way through and camped under the walls of Jellalabad; in the ensuing siege Havelock, as deputy adjutant general, was a leading spirit. In the meantime Kabul had fallen, and in August 1842 Havelock accompanied the Army of Retribution during its advance to retake the fortress, and was present at the Battles of Jagdallak and Tezin, taking a prominent part in the capture of Istaliff. In recognition of his services during these years Havelock was awarded the First Burma Medal (with bar Ava), the Jellalabad Medal, and the Kandahar-Kabul Medal; he was also made C.B. and promoted brevet major. His regular majority came on 30 June 1843, with appointment as interpreter to Sir Hugh Gough; and for his part in the Gwalior campaign he received the Star named to the Battle of Maharajpoor and was promoted brevet lieutenant-colonel.

In 1845 the First Sikh War broke out, and at the Battle of Moodkee Havelock had two horses shot under him; in the Battle of Ferozeshuhur his close friends Sir Robert Sale and Major Broadfoot were killed. Following the Sutlej campaign, for which he was awarded the Medal with bars Moodkee and Ferozeshuhur, he was appointed deputy adjutant general of the Queen's troops at Bombay. Colonel William Havelock, Henry's brother, was killed at the head of his troops at the Battle of Ramnuggur during the Second Sikh War, and this loss, together with ill health (after twenty-six years' continuous service in India), prompted Havelock to return to England with his family during the autumn of 1849.

Two years later, his health restored, Havelock once again went out to Bombay, being appointed quartermaster-general of the British Army in India by Lord Hardinge. In June 1854 he became adjutant-general and a brevet colonel. In the war with Persia, which broke out on 1 November 1856, Havelock was given command of a division in the force led by Sir James Outram (q.v.) and ordered to the Persian Gulf. The attack on Mohumra, on 26 March 1857, was the successful result of plans drawn up by Havelock, and throughout this campaign (for which he was awarded the 1854 India General Service Medal with bar Persia) his aide-de-camp was his son Henry, later Sir Henry Havelock Allan. Returning to Bombay on 29 May, Havelock was met with the news that native regiments had mutinied and that Delhi was in the hands of the rebels. Moving on to Madras, which he reached on 13 June, he found that following the death of General George Anson, command of forces in the Madras Presidency had passed to Sir Patrick Grant, and it was with a force under the leadership of Grant that Havelock travelled to Calcutta, arriving just two weeks later.

At Allahabad Havelock, again accompanied by his son as aide-de-camp, was placed in command of a column with orders to go to the support of Sir Hugh Wheeler at Cawnpore and Sir Henry Lawrence at Lucknow, and to destroy or disperse all mutineers in the course of his movements. Hearing of the massacre of the garrison at Cawnpore on 3 July, Havelock left a small garrison force at Allahabad under the command of Colonel Neill, and with a force of 1,000 bayonets comprising the 64th Regiment, the 78th Highlanders, the 1st Madras Fusiliers, the 84th Regiment, a few dozen Sikhs, six guns, and some volunteer cavalry, set out to recapture the town. Reaching Futtehpore on 12 July, this force defeated a group of rebels; it defeated another at Aong; at the bridge on the Pandoo-Nudee river there was a fierce engagement and once again Havelock defeated and put to rout the rebellious sepoys. News now reached Havelock that 200 European women and children still survived alive at Cawnpore, and he determined to push ahead in the hope of saving them. They were, however, murdered in cold blood, and Nana Sahib made a sortie with 5,000 men to meet Havelock; he was defeated after a brilliant flanking attack by the British force, which fought with great bravery – this after a forced march of 126 miles under a July sun, during which it had fought four successful battles.

Pushing on from Cawnpore, Havelock crossed the Ganges on 25 July in an effort to relieve Lucknow, fighting successful battles at Onao and Busseerutgunge. An attack from the rear by a reinforced enemy forced the British column to fall back on Mungulwar, where it received some replacements before pushing on towards Lucknow. However, Havelock had little ammunition remaining, and when cholera struck down some of his troops he had to make the difficult decision to return to Cawnpore. Still determined to succour the besieged garrison at Lucknow, Havelock left a token force of a hundred men under Neill at Cawnpore and marched out to attack 4,000 rebels, whom he defeated at Bithpoor and from whom he captured two guns.

It should be remembered that Havelock had been serving in an army in which senior staff command was usually bestowed upon titled officers whose own promotions had almost invariably been purchased by private means. Yet he had achieved great distinction without these advantages, and had risen to field commands as the result of his personal prowess. While at Cawnpore he learned of the appointment of Sir James Outram to the command of the district in which he, Havelock, had been operating with such success – as a result he was to be relieved of his command. Characteristically Havelock refused to allow his disappointment to affect his conduct of his duties, and he set about doing all in his power to assist Sir James when he arrived with substantial reinforcements. Notwithstanding this potentially embarrassing situation, there then occurred one of history's great acts of self-abnegation when Sir James, recognising the extraordinary achievements of Havelock, waived his own military rank in favour of his subordinate and offered to accompany the force in his civil capacity only, and to offer his military service as a volunteer until Havelock could effect the relief of Lucknow.

Building a bridge over the Ganges, which was completed on 19 September, Havelock marched out of Cawnpore at the head of 3,000 men of all arms and crossed the great river under heavy enemy fire. On arrival at Mungulwar he found the enemy army massed in his path, but immediately attacked, driving them out and beyond Onao on the 21 September. Resting for the night at Busseerutgunge, he set off the next day and laid siege to Bunnee before the enemy had time to destroy the important bridge. The following day he reached the Allumbagh, a dozen miles from Lucknow, where he again defeated the forces screening the approach to the besieged city.

On 25 September Havelock started his final advance on Lucknow, which was now visible on the horizon. Amid a storm of grape and musketry the small British force drove the enemy out of the Charbagh enclosure, and the neighbouring bridge was carried following a brilliant charge by the Madras Fusiliers, in which Havelock's son displayed great personal courage. Entering the outskirts of Lucknow under constant fire from all around which caused numerous casualties, the relief force crossed the last bridge and reached the Chattar Manzil, where Outram suggested it should remain for the night. Havelock however determined to push on, and in fact led his exhausted men up to the Residency before night fell. The following day, the 26 September, Havelock sent out a strong party to bring in his rearguard and to collect the wounded, and then handed over command to Sir James Outram. It now became clear that such had been the losses suffered during the march on Lucknow that there was insufficient transport to evacuate the civilians, sick, and wounded, and once more the city was besieged – a condition it endured for a further seven weeks before the second relief under Sir Colin Campbell (q.v.).

The much-strengthened garrison then set about clearing the enemy from the numerous buildings and enclosures surrounding the Presidency. The final relief of Lucknow had been marked by a meeting between Campbell, Outram, and Havelock – at which the latter had learnt that he had been appointed K.C.B. (His Indian Mutiny Medal, awarded posthumously, bore all five available bars for the campaign.)

Havelock's last active duty had been performed. On the day the withdrawal of the main force from Lucknow commenced, 20 November, much weakened by months of exertions and privation, he was struck down by a severe attack of dysentery and died four days later. He was buried at the Alumbagh, and Campbell, Outram, Inglis, and his son followed his body in the long procession to the graveside. Almost his last words were 'I die happy and contented'; turning to his son, he said, 'See how a Christian can die.'

News of Havelock's earlier victories had reached England, providing the first gleam of light in the darkness of massacre and revolt, an effect heightened by the earnest religious beliefs of this hitherto-unknown army officer who had fought for, not purchased, his promotion through the Army. His final promotion had been to major-general on 30 July, and his appointment as K.C.B. followed on 26 September. It was not until 7 January 1858 that news of his death reached England, by which time he had been created a baronet and awarded a pension of £1,000 a year by parliamentary vote. Thereupon the rank of baronet's widow was bestowed upon Lady Havelock (who died on 25 August 1882), and the baronetcy upon his eldest son and aide-de-camp, Henry Havelock; an annual pension of £1,000 was awarded to each. His statue was erected in Trafalgar Square, London, by public subscription.

Employed for the greater part of his career in subordinate positions, Havelock was gifted with military abilities seldom equalled in the British Army, but his lack of means, a certain sternness of disposition, and a somewhat narrow religious profession contributed to limit his progress. When the opportunity was eventually presented he was ready; he proved himself a resolute and brilliant military leader, and his singular achievements won the gratitude of his country.

Henry IV (*1553–1610*), *King of France; also Henry III, King of Navarre*

Widely referred to as Henry the Great, Henry of Navarre is one of the most popular figures in French history. Despite a poor grounding in military strategy, he succeeded in securing the kingdom of France by conquering the Holy League during the years 1589–98. His administrative achievements in the establishment of many French institutions which have survived four hundred years of republican upheaval must stand as his greatest memorial.

Born at Pau on 14 December 1553, the son of Antoine de Bourbon, Duc de Vendôme, and Jeanne d'Albret, Queen of Navarre (from 1555), Henry entered the French court at Amiens at the age of four and was educated as a Protestant at the Collège de Navarre in Paris from the age of eight. Joining the army at the age of sixteen, he gained distinction at the Battle of Arnay-le-Duc in Burgundy in 1570. Two years later, on the death of his mother, he became King Henry III of Navarre.

After the Massacre of St Bartholomew in August 1572 Henry renounced his Protestant faith to facilitate his escape from the court to Alençon – where he repudiated his renouncement, and put himself at the head of the Protestant rebels against Henry III of France. The death of the King's brother François, Duc d'Anjou, placed Henry of Navarre next in line of succession to the French crown; but, excluded from the succession by the Treaty of Nemours in 1585, he embarked on the 'War of the Three Henrys'. Eventually he was reconciled with Henry III of France who, dying by assassination in 1589, recognised Navarre as the heir to the French crown.

In the meantime the Holy League had gained effective control over almost all France, and from this time Henry of Navarre set about the reconquest of the country to secure his kingdom. His first success was at the Battle of Arques, fought on 21 September 1589; with a force of no more than 8,000 troops he lured Charles, Duc de Mayenne, with 24,000 Catholics into a defile of the Béthune river near Dieppe which he had prepared with extensive gun positions. By superb use of his slender resources Henry won this great victory and drove Mayenne from the field.

He then determined to secure Paris, and hurried towards the city with 20,000 men on 31 October. Here, however, he was repulsed by the garrison, which was largely Catholic in sympathy. Proclaiming himself King, he was obliged to establish a temporary capital at Tours while civil war raged throughout France. In February Henry laid siege to Dreux, but on the approach of Mayenne he decided to abandon the blockade and deploy to give battle near Evreux. In this, the Battle of Ivry, fought on 14 March 1590, the linear tactics adopted by Henry proved equal to the occasion. In this formation, first employed at the Battle of Coutras two years earlier, Henry employed two lines of arquebusiers (the front line kneeling, the rear standing), giving steadiness of fire and effectively doubling the fire-power of the line. This steadiness repulsed the Catholic cavalry charges, and Henry himself led a successful cavalry counter-charge which broke and put to flight the Catholic army. Only the Swiss contingent remained firm, threatening to fight to the death if they were not given honourable terms. Mayenne lost about 4,000 killed at the Battle of Ivry, compared to 500 killed in Henry's army; but the latter was slow to exploit his success, thereby enabling Mayenne to regroup his forces to assist in the defence of Paris, which was now invested by Henry.

The appearance of the Duke of Parma (see Farnese), who now joined Mayenne at Meaux, forced Henry to abandon the Siege of Paris (which had almost reduced the population to starvation), but no conclusive confrontation was forthcoming between the two armies, both of which numbered about 26,000 men. Two years of inconclusive manœuvring followed. At last in July 1593, Henry formally announced his return to the Catholic faith. At once the resistance of the Paris population to his rule – on the grounds of his Protestant allegiance – disappeared, and almost the entire French nation became united under the King against the threat of a full-scale Spanish invasion. On 21 March 1594 Henry made his long-awaited triumphal entry into Paris. (This was the origin of his widely-quoted remark, 'Paris is worth a Mass.')

On 9 June the next year Henry won a closely-contested battle with the Spanish at Fontaine-Française, but narrowly escaped being killed as the result of his own rash gallantry. He now turned his thoughts to bringing order and prosperity to his kingdom, and the Spanish took advantage of these internal preoccupations by seizing Calais (on 9 April 1596) and Amiens (on 17 September). The latter was not regained by Henry until a year later.

By granting religious freedom to French Protestants in the Edict of Nantes, signed on 13 April 1598, Henry brought about an end to the Wars of Religion and, in an apparent manœuvre around the interests of his English and Dutch allies, he went on to secure peace with Spain by the Treaty of Vervins, signed on 2 May 1598. Thereafter he abandoned all military enterprises to pursue a determined course of building national harmony. He brought about a total reformation of the nation's finances, and even transformed a situation of bankruptcy into solvency with reserves. He reorganised the army, founded a school for military cadets, established a regular corps of artillery, strengthened the frontier fortresses, and raised and regularised the scales of pay of his soldiers. Henry's services to the army in France were therefore roughly equivalent to those of Cromwell in England some fifty years later.

He met his death at the hand of a religious fanatic, François Ravaillac, in Paris on 14 May 1610 – an assassination which deprived France of one of her greatest kings.

Henry V (*1387–1422*), *King of England*

Though not destined for a long life, Henry V worked hard to achieve unity and peace in his realm. He redeemed a somewhat misconceived campaign in France by his able generalship in the famous Battle of Agincourt and in the conquest of Normandy; and his inspiring personal leadership was long remembered.

Eldest son of Henry IV, Henry of Monmouth (born at Monmouth in September 1387) was created Prince of Wales, Earl of Chester, and Duke of Cornwall on the accession of his father to the throne in 1399. At the age of fifteen he took command of the royalist forces in the war against the Welsh rebels, campaigning vigorously and successfully for the next five years. On succeeding his father in March

1413 he set about restoring order and unity in England and in the Church. In the long struggle to establish the English claim to the French crown, Henry at first tried peaceful negotiation to achieve what he had been educated to accept as his rightful inheritance; but the inevitable failure of this method determined his use of force. In April 1415 he declared war on France, and sailed with an army of 12,000 men on 10 August, laying siege to Harfleur three days later.

Having reduced Harfleur, Henry decided to march overland towards Calais, heading eastwards up the Somme and looking for an undefended crossing – in much the same manner as Edward III (q.v.) had done at the time of the Battle of Crécy. Eventually, at Athies, he found a crossing and started northwards. Meanwhile Charles d'Albret, Constable of France, took up a blocking position with 30,000 men near the castle of Agincourt, where they were found by the English on the afternoon of 24 October. Adopting much the same deployment as Edward III at Crécy, Henry positioned his army in three 'battles' in a narrow defile (between woodlands) about 1,000 yards wide; ahead of him the French also deployed in three 'battles', one behind the other owing to the narrowness of the defile. At daybreak on 25 October the two armies stood facing one another, each waiting for the other to attack – the French believing that the impetuous young English King would make the first move. Henry ordered a short, cautious advance to within about half a mile of the French, sending his archers into forward echelon between the battles of dismounted men-at-arms. Unable to restrain themselves, the undisciplined French men-at-arms moved ponderously forward, weighed down by their heavy armour, while mounted groups galloped forward on their flanks. But the English stood firm, and discharged showers of arrows into the disordered French ranks, inflicting enormous losses. Finding the soggy ploughed fields heavy going, the French quickly became exhausted, and despite repeated attacks, neither infantry nor cavalry could break the thin but solid English lines of archers. As the heavily armoured knights struggled the final hundred yards to the English line the increasing narrowness of the open field forced them into a packed mass, almost helpless when the archers burst from their palisades and, wielding hatchets, battle-axes, and swords, killed large numbers of the exhausted Frenchmen. Almost the whole of the first French battle was destroyed or captured. An attack by the second battle was not forced forward with the same vigour, and a third attack was easily thrown back. At this point Henry, hearing a report that the French were launching an attack on his rear, gave orders to kill all French prisoners – but the attack was found to be no more than a minor foray after plunder in the English baggage camp. Seeing that the enemy was now unable to mount any effective assault, the English King collected together a few hundred mounted troops and personally led them forward to disperse the remaining French. No accurate figures for total French losses have ever been established, but it is known that among the 5,000 men of noble birth killed was Charles d'Albret himself; while the surviving prisoners (who totalled about 1,000) included the Duke of Orléans and Marshal Jean Boucicaut (q.v.). It is said that the English losses were no more than thirteen men-at-arms and about a hundred footmen killed, but nationalist propaganda must be suspected. English losses were certainly negligible compared to those suffered by the enemy, however.

Henry took no strategic advantage of his devastating victory at Agincourt; he simply marched to Calais and then returned to England. The casualties he had inflicted in the battle had, however, assured the ascendancy of the Burgundians in France.

Returning to France in 1417, Henry systematically conquered Normandy in three distinct and well-planned campaigns. In May 1418 his ally John the Fearless, Duke of Burgundy, captured Paris and massacred most of the remaining Armagnac and Orléanist leaders, but was himself assassinated on 10 September. Taking advantage of the bitterness of the civil war which now raged, Henry marched on Paris and, supported by the Burgundians – now led by Philip the Good – negotiated the Treaty of Troyes, by which he gained formal recognition as heir to the insane King Charles VI, and as regent of France. He also married Catherine, daughter of the

French King, by whom he had one son, later to become Henry VI.

The victor of Agincourt and conqueror of Normandy dreamed of leading the combined forces of both realms to recover Jerusalem from the infidels; but the Treaty of Troyes was never accepted by the Dauphinists, against whom Henry was forced to continue campaigning. Falling victim to dysentery during the sieges of Meaux and Melun, he died at Bois de Vincennes on 31 August 1422, before his thirty-fifth birthday. The tragedy of his life and reign was that he squandered his great gifts and high ideals in committing them and his country's resources to a dubious foreign war – a course that could only be justified by the passionate sincerity displayed by him and by successive English monarchs in their belief in their right to the French throne. The sense of national pride and national identity which his victories engendered among the English people of all classes should not be underestimated, however. His personal courage was considerable, his tactical sense sound, and his personal manner and style of command seems to have appealed to the common soldiers.

Hideyoshi Toyotomi (*1536–98*)

Japan between the fourteenth and sixteenth centuries was a chaos of warring noble families and their adherents. The imperial family was recognised as pre-eminent among the clans, but had little actual power. Generation after generation of war-lords fought to establish their power over each other, but the first man to make solid progress towards unification was Oda Nobunaga (1534–82). He gradually extended his influence from his own estates to those of defeated neighbours, until he had brought half Japan under his sway and ruled as dictator (or *shogun*) from his capital of Kyoto. His principal general was Hideyoshi, who conquered most of western Japan for Nobunaga in the second half of the 1570s. Hideyoshi was a man of humble birth, one of the last to take advantage of the internal chaos of the country to raise himself to a position of influence before the reimposition of a rigidly-structured feudal society – a society which Hideyoshi himself did much to bring about.

Hideyoshi's victories were won partly because of his recognition that a time had arrived for change in Japanese military tactics. For centuries the form of war had been highly ritualised; battles were simply massed individual combats between *samurai* knights. Firearms had arrived in Japan in 1542 and their revolutionary value had been appreciated; large imports of guns and the beginnings of a local industry supervised by European missionaries and traders marked the second half of the sixteenth century. The career of Hideyoshi coincided with a more realistic attitude to warfare and the increased use of large armies of non-*samurai* soldiers, some with firearms. The art of fortification received belated attention, and from 1576 Nobunaga began to construct stone-walled castles with clear fields of fire at strategic points in his territory.

In 1582 civil war broke out within Nobunaga's territories after the rebellion of a general named Akechi Mitsuhide. Marching troops through Kyoto on the way to the Southern Honshu front, Mitsuhide turned aside and stormed Nobunaga's headquarters at Honnoji. Surrounded and without hope, the *shogun* committed suicide. His general Hideyoshi avenged him by defeating and killing Mitsuhide in the same year. He then leagued himself with another of Nobunaga's principal lieutenants, Tokugawa Ieyasu, and several other powerful elements, and by 1584 was unquestioned master of the *shogun's* territories, with Ieyasu as his main collaborator. Establishing a base at Osaka, Hideyoshi defeated the Oda family and extended his control over the whole of central Japan. In 1585 he fought a successful campaign in Shikoku, and in 1587 won a decisive victory on Kyushu over the Shimazu family, leaders of the Satsuma clan. Hideyoshi's successful consolidation of his position was crowned by victory in Odawara in 1590. Ieyasu based himself at Edo (Tokyo) and continued to support Hideyoshi as he began the work of ensuring peace by the deliberate structuring of society. Keeping only a proportion of his captured territories for his own use and income, Hideyoshi (who took the old title of *kampaku*, as his birth disqualified him for *shogun*) parcelled out the rest to vassals-in-chief, who in turn bestowed fiefs on lesser vassals. The system was similar to the classic European feudal society in several respects.

The vassals were expected to supply manpower in time of war, and labour for public service projects in peacetime. Members of important families were required to spend much of their time at Hideyoshi's court, more or less hostages for their families' obedience, and legal limits were set to the number and nature of the fortresses which vassals might build on their lands. The system was in fact more authoritarian than the feudalism of Europe, in that there was no agreed limit on the demands that the overlord might make upon his vassals, and the freedom of action enjoyed by the vassals was much more circumscribed. European missionaries and traders were initially encouraged by Hideyoshi, but he took steps to expel them in 1587 – possibly the native firearms industry was felt to be securely established by that time.

After successfully unifying Japan, Hideyoshi embarked on a gradiose scheme for the invasion and annexation of the whole of south-east Asia. He laid down a timetable which included the capture by the spring of 1592 of Korea, as a jumping-off point for an invasion of China. Peking was to fall by the end of 1592, and by the end of 1594 China was to be secure; at that time the imperial court was to move to Peking, while Hideyoshi established headquarters in southern China for the further extension of the empire. In actuality this great project never came near to success. The first expedition to Korea in 1592 met with impressive initial successes. In May Pusan was captured, and Seoul and Pyongyang fell in June and July; this advance included a stretch of 200 miles in three weeks. But in July the energetic Korean naval flotillas under Admiral Yi Sung Sin inflicted a devastating defeat on the Japanese fleet in the Battle of the Yellow Sea; fifty-nine Japanese vessles were sunk and a convoy of reinforcements for the ground forces was scattered. (It is worth noting that this action was the first known engagement by ironclad ships, of which Yi had two, of his own original and effective design.) In October 1592 the Ming Emperor of China sent troops to support his faithful Korean vassals, and Hideyoshi's forces, short of supplies and with their sea lines of communication cut, were pushed back down the peninsula in a dreary retreat harassed by local irregulars. They retained a small beach-head perimeter around Pusan when peace negotiations began in 1594; the Koreans observed this perimeter, and the Chinese army was recalled. There were repeated incidents in 1594–6, and in 1597 Hideyoshi sent another army to Pusan, which broke out of the perimeter and advanced once more. China once again sent troops, but this time the Japanese managed to maintain their advance and defeated several allied forces. Again the Korean navy struck; Admiral Yi died in the hour of victory at Chinhae Bay in November 1598 after sinking some 200 Japanese vessels. This disaster almost coincided with Hideyoshi's death. In December an armistice was declared in Korea, and the Japanese withdrew completely. Tokugawa Ieyasu was eventually successful in the power-struggle which followed Hideyoshi's death. A generation later a *shogun* of his line instituted the policy of total isolation which resulted in the complete stagnation of the military arts in Japan for more than two centuries.

Hill, Rowland, 1st Viscount Hill (*1772–1842*), *General*

Rowland Hill was known throughout the Peninsula Field Army as 'Daddy' Hill, a nickname of unusual affection which the British soldier has bestowed on few commanders. His claim to the distinction is hinted at in portraits, which show a homely, amiable, rather Pickwickian countenance and a mild and friendly gaze. Hill was quite simply one of the most decent English gentlemen of his day, a simple and honest man of endless kindness, patience, thoughtfulness, and charm. The anecdotes of his generosity and consideration to common soldiers are legion, and there is a legend that on only two occasions throughout the gruelling seven-year Peninsula campaign was he ever heard to swear. In a notably foul-mouthed army this alone would be enough to set him apart, but in fact he also has solid claims to military distinction.

Hill was the only one of Wellington's (q.v.) subordinates whom the Commander of the Forces trusted to command a detached corps manœuvring on his flank. He was no towering military genius, but an obedient, competent, common-sense soldier who neither panicked in the unexpected presence of the enemy, nor indulged in personal embroidery of Wellington's instructions in their absence. This calm reliability was worth diamonds to Wellington, who remarked that 'the best of Hill is that I always know where to find him'. Considering the inability of some of the generals (foisted upon Wellington by Horse Guards in accordance with a rigid seniority system) to obey the simplest orders, even when under the eye of the Commander-in-Chief, this praise is a lot warmer than it might at first appear.

Hill was born on 11 August 1772, the second son of Sir John Hill of Hawkstone in Shropshire. He was educated privately and at the Strasbourg Military School, and entered the Army in 1790 as an ensign in the 38th Foot. He distinguished himself at the Siege of Toulon in 1793; and when in 1794 Thomas Graham (q.v.), later Lord Lynedoch, was raising the new 90th Foot as its lieutenant-colonel, he secured young Hill as a major. In May 1794 he was promoted lieutenant-colonel, and was wounded during the first action of the Egyptian campaign in 1801. He was appointed brigadier in 1803 and, after a spell of apparently creditable service in Ireland, major-general in 1805. Hill went out to Portugal as a brigade commander in 1808, seeing action at Vimiero; he took part in the retreat to Corunna and the battle of that name in January 1809, and again led a brigade at the taking of Oporto in May 1809. In June Wellington gave him the 2nd Division, which he led without a break until January 1811. He distinguished himself at Talavera in July 1809; on the night before the battle proper the French made an audacious night attack on a vital sector of the British lines, and made considerable progress before Hill rallied the confused British and Hanoverian regiments and mounted a successful counter-attack. During the long months which separated this major action and Bussaco, in September 1810, Hill led a corps of all arms covering the southern flank of the Allied position in Portugal. His long march through difficult terrain to rejoin the main army in time for the battle of Bussaco was a masterpiece. In January 1811 he was forced to leave his division by illness; the detached southern command was taken over by Beresford (q.v.), with the result described in the entry for that general. Hill returned a week after Albuera in May 1811, and no doubt his familiar presence, a comforting reminder of the country squire to soldiers recruited in rural areas, was welcomed by his much-mauled command.

Hill was not merely an unimaginative plodder, though he was certainly a careful general – for which Britain should be grateful. On two occasions in particular he showed that he could exploit a situation with dash and aggressiveness when opportunity offered. In October 1811 he took a small picked force deep into Spain by secret forced marches through rough country, surprising and shattering the command of the French General Girard at Arroyo dos Molinos the following month. Five months later he pulled off a lightning blow at the vital forts and bridge of Almaraz on the Tagus, thus preventing Marmont and Soult (q.v.) from co-operating during the British Siege of Badajoz. Both these operations demanded quick and constant movement, a fierce and precisely calculated attack, and instant withdrawal before superior forces could be assembled for pursuit; and Rowland Hill proved himself entirely equal to the challenge. He commanded Wellington's right wing for the rest of the war, and particularly distinguished himself at St Pierre, near Bayonne, on 13 December 1813. With 14,000 British and Portuguese he was cut off by a rising river and washed-away bridges, facing Soult's six divisions of roughly 30,000 men. Hill was distinctly heard to swear on this occasion. He personally led his diminishing reserves into action, and spent the whole day moving from point to point in the forefront of battle directing and encouraging the desperate fighting. Somehow the Allies held out until late afternoon, when reinforcements at last began to come up, and Soult abandoned his attack.

During the Waterloo campaign Hill led the 2nd Corps, but this battle was fought so completely according to the Commander-in-Chief's orders that neither Hill nor any other general had an opportunity to shine. He was, however, in the thick of the fighting against the Imperial Guard on the Allied right wing, had a horse shot under him, and was for some time lost in the mêlée.

Lieutenant-General Hill was made a Knight

of the Bath in 1812, and was elected M.P. for Shrewsbury in the same year. On 17 May 1814 he was created Baron Hill of Almaraz and Hawkstone. When Wellington formed his first cabinet in 1828 he resigned his post as Commander-in-Chief of the British Army; Hill took over this appointment with the rank of general, and filled it until his resignation in 1842. He was then created Viscount Hill, and died on 10 December of the same year. On his deathbed he remarked that he had much to be thankful for, and that he did not believe he had an enemy in the world. The years have failed to produce evidence to contradict him.

Howe, George Augustus, 3rd Viscount Howe (*1724?–58*), *Brigadier*

The very able and popular Lord Howe died in action while still in his early thirties; the immediate result of his death was a disaster to British arms, and the long-term cost was incalculable. He was called 'the best soldier in the British Army'; he had both brains and influence, and might have risen to the heights of command at a time when Britain had sore need of imaginative generals – the War of American Independence.

Howe was born in either 1724 or 1725, the senior surviving son of Emmanuel Scrope Howe, second Viscount Howe, of Langar in Nottinghamshire. A rumour was current that his mother was an illegitimate daughter of George I, but this has since been discounted. Emmanuel Howe was Governor of Barbados from 1732 until his death in 1735, when his ten-year-old son succeeded to the title (in the Irish peerage). In March 1745, at the rather advanced age – for an ensign – of twenty, young Lord Howe entered the 1st Foot Guards (later the Grenadier Guards). In May 1746 he was promoted lieutenant, and captain in the Army – a dual rank system gave officers of Guards regiments an extra rank's seniority over their contemporaries in other units. In 1746–7 he served in the Austrian Netherlands as aide de camp to the Duke of Cumberland, commander-in-chief of a British expedition against the French, and fought at Laufeldt, near Maastricht, on 2 July 1747, when the great Saxe (q.v.) defeated the Allied army. The previous month, in his absence, Howe had followed his father into parliament by being elected M.P. for Nottingham. (He was re-elected in April 1754, and held the seat for the rest of his life.) On 1 May 1749 he was promoted captain, and lieutenant-colonel in the army. Promotion, like parliamentary seats, belonged in those days to the men who had the gold and the patronage to pay for them, and had nothing whatever to do with professional merit in the one case, or a democratic exercise of the franchise in the other. The remarkable thing was that while he was no doubt a most indifferent representative of his constituents, Lord Howe was a very gifted soldier.

On 25 February 1757 Lord Howe was appointed colonel of the 3rd Battalion, 60th Foot – the Royal Americans, an *élite* unit raised among immigrant communities in the eastern American colonies and largely officered by European professional soldiers. He arrived in Halifax in July, and joined his command at Fort Edward three days after the French had invested near-by Fort William Henry (for general background see under Montcalm). General Webb felt unable to send his relatively small force out in an attempt to relieve the heavily-outnumbered garrison of Fort William Henry, and considered the lines of communication back to Albany his first priority. In due course Fort William Henry fell; the grisly sequel is well known. In September 1757 Lord Howe was given command of the 55th Foot, then in upper New York; he commanded the belated relief column to German Flats in November. In December he was given the local rank of brigadier, and in February 1758 he led an abortive winter attack on the French forward base at Fort Carillon, Ticonderoga. A major offensive against this position was subsequently planned for the summer; its capture, and that of Crown Point, would open up an invasion road along Lake Champlain into the heart of French Canada. The rigid seniority system of the day gave command of the expedition to Abercromby, a lethargic officer of no great talent, but William Pitt saw to it that Howe was named as second-in-command and intended that he should be the moving spirit behind the campaign. Abercromby placed great reliance on Lord Howe's advice, and there was no reason why he should not be able to 'steer the campaign from the back seat'.

During the first half of 1758 Howe instituted sweeping changes in the army, starting with his own regiment and extending to all units under command. He was an open-minded and imaginative commander who appreciated that European training and tactics were useless in forest fighting. He trained first himself and then his men in fieldcraft, and studied the methods and techniques of the Ranger leader, Robert Rogers (q.v.), at first hand. He ordered that officers and men alike should carry the bare minimum of equipment and supplies with them on campaign. He had his own fashionably long hair cropped short, and ordered that all ranks should follow his example. The unwieldy cocked hats of the British soldiers were cropped down into small caps by the removal of most of the brim. The long tails of the red coats were cut off, and tomahawks and extra ammunition were carried in place of the cumbersome swords and half-pikes prescribed for full dress. Many officers were scandalised at having to carry their own gear, dine off a handful of meal and a strip of jerky, and roll up in a single blanket on the bare earth; but the gay young brigadier made a point of doing himself, publicly, everything which he ordered his subordinates to do. His energy and good humour were infectious, and he had considerable diplomatic skill; his charm and patience did much to break down the traditional barriers of ill-feeling and obstruction between regular and provincial volunteer officers. By the time the 15,000-man expedition was ready to leave the southern shore of Lake George in early July 1758 the whole army was devoted to Howe; officers and men adored him, and trusted him to lead them to victory. Considering that Montcalm had only some 3,000 men at Ticonderoga, their confidence was justifiably high. The army sailed up the lake, and disembarked on 6 July 1758. Howe led a column of troops into the forest to the north-west to scout a landward approach to Fort Carillon; they blundered into a small party of French scouts at Trout Brook, about two miles from the lake shore in thick woodland, and from a shot in the first confused volley Howe fell dead.

With his death, the heart went out of the army. Abercromby dithered, gave Montcalm time to improve his defences, and then inexplicably threw his infantry at the fort in an unsupported frontal attack against regular troops firing from cover. Losses in the first waves were very high, and the attackers were forced to withdraw. The British commander had only to invest the fort closely and bring up his artillery from the lake to ensure certain victory, for Montcalm could expect no reinforcements and was short of provisions. Instead Abercromby ordered a full retreat, for no reason which anyone has ever been able to explain, and thus prolonged the war in America by at least a year.

There are conflicting accounts of Howe's burial. One source states that his body was taken to Albany, and buried in St Peter's Church. Another states that he was buried where he fell, in thick forest, with only a simple stone marker; and that this and the remains were rediscovered in 1890. Wherever his body lies, it is certain that the colony of Massachusetts Bay paid him a most unusual compliment in subscribing for a monument to be raised to him in Westminster Abbey – this at a time when there was already great friction between colonists and King's officers. Howe's brothers William and Richard both became famous during the Revolutionary War, one on land and the other at sea.

Hunyadi, János, Count of Bestercze
(c. 1387–1456), Captain-General of Hungary

Hunyadi, the national hero of Hungary and one of the great figures of Christian resistance to Turkish expansion, was born about 1387, the son of a Magyarised Vlach named Vojk and of Elizabeth Morsina. In his youth he served Sigismund of Luxemburg, who was both the Holy Roman Emperor and the King of Bohemia and Hungary. The young Hunyadi saw action in some of Sigismund's numerous campaigns, including wars in Germany and Bohemia (see under Jan Žižka) and against the Turks in Serbia. In the latter campaign Hunyadi was prominent, expelling the Turks from the province of Smederovo (Semendria), and receiving honours and lands at the hand of the new king, Albert I, in 1437. In 1438 Hunyadi was named as the Ban of Szörény in western Wallachia, an appointment which carried with it the burden of constant frontier warfare against the Turks.

In 1439 Albert I died, and in the ensuing civil war over the succession Hunyadi was an important supporter of Wladislaw III of Poland against the partisans of the young László V of Bohemia. In 1440 Wladislaw established himself on the throne; Hunyadi was made captain of the great fortress of Belgrade, then in an exposed position, and Voivode of Transylvania. Hunyadi was the moving spirit behind Hungarian resistance to the Ottoman Turks, and in the four years of Wladislaw's brief reign he won a series of astounding victories against them. In 1441 he won the Battle of Semendria, and in 1442 the Battles of Hermannstadt and the Iron Gates. In 1443 he advanced across their Balkan provinces, capturing Nish and Sofia, and then linking up with another army under King Wladislaw to defeat the forces of Sultan Murad II at Snaim (Kustinitza). In February 1444 Hunyadi could take personal credit for smashing Turkish power in Bosnia, Hercegovina, Serbia, Bulgaria, and Albania. Hungarian preparations were under way for a further offensive to drive the Turks from Europe for ever, in conjunction with George Brankovic of Serbia and Skanderbeg (q.v.) of Albania, when Turkish envoys arrived to discuss a truce. A ten-year cessation of hostilities was agreed upon.

In July 1444 Wladislaw deliberately broke the truce by invading Bulgaria. A Venetian fleet was sent to prevent the Sultan from leading an army back into Europe from Asia Minor, and the Hungarians marched towards a projected rendezvous on the Black Sea coast. At this critical juncture Brankovic, for his own reasons, betrayed the Christian plan to the Turks and prevented Skanderbeg of Albania from marching to join the Hungarians. The Venetian fleet was defeated, and the Turks poured westwards once again. At Varna on 10 November 1444 the Hungarians were overwhelmed; Wladislaw fell, and Hunyadi only brought a remnant of the army to safety with great difficulty. Thus ended the last true Crusade – for this offensive had been planned and named as such.

Between 1444 and 1452 Hunyadi virtually ruled Hungary, as Regent for the young László V. The German Emperor Frederick III was the guardian of László, who was also King of Bohemia; Frederick prevented the young King from taking up his throne in Hungary, and Hunyadi went to war over the issue. He ravaged Frederick's border provinces of Styria and Carinthia, and even threatened Vienna before Frederick came to terms. In 1448 Hunyadi received the title of Prince from Pope Nicholas V, and resumed his campaigns against the Turks. On 17 October 1448 his army of 25,000 was badly beaten by a Turkish army four times the size in a defeat partly brought about by the treachery of Brankovic. The Turks lost over 30,000 men, and the Hungarians about 12,500; a feature of the action was a prolonged exchange of fire at quite short ranges between Turkish janissary bowmen and Hunyadi's German and Bohemian mercenary hand-gunners, both sides sheltering behind breastworks. This battle of Kossovo was most significant, in that it hastened the introduction of hand-guns in the Ottoman armies. In 1449 Hunyadi led a successful expedition into Serbia to punish Brankovic, and the following year ended a two-year truce with Frederick III by negotiating successfully for László's release. (There is no evidence that the young man was gratified, as he spent many subsequent years hampering Hunyadi's efforts against the Turks.) Now sixty-three years old but still extremely vigorous, Hunyadi was made Count of Bestercze and Captain-General of the Kingdom.

In 1453 Constantinople, defended by Constantine XI (q.v.), fell to the new sultan Mohammed II (q.v.), and a new wave of Turkish pressure washed against the rock of Hungarian resistance. At his own expense Hunyadi provisioned and garrisoned the strategic fortress of Belgrade, which was invested in 1456, while he assembled an army and a river fleet of some 200 vessels in Hungary. His forces were swelled by peasants inspired to volunteer by the eloquence of the Franciscan friar St John of Capistrano. On 14 July 1456 Hunyadi smashed the Turkish fleet on the Danube, and on 21–22 July he routed the army besieging Belgrade, forcing the Sultan to return to Constantinople and preserving Hungarian independence for another seventy years. The old general was planning to carry the war into Turkey itself when, on 11 August 1456, he died of plague in his camp. He was one of the handful of outstanding soldier-statesmen Europe produced under the stimulus of Turkish pressure, and as a general he was years ahead of his time in recognising the superiority of paid regular troops, whom he used in preference to feudal levies whenever possible.

Jackson, Thomas Jonathan (*1824–63*), *Lieutenant-General*

'Stonewall' Jackson ranks second only to Robert E. Lee (q.v.) among Confederate commanders of the American Civil War. He led half of the Army of Northern Virginia in 1862, and with brilliance; he enjoyed a rapport with Lee which the latter was never able to establish with any other lieutenant, and he was idolised by his men. After his needless death the Confederate army was never quite the same again.

Thomas Jackson was born on 21 January 1824 at Clarksburg, then in Virginia but now in West Virginia. His father Jonathan, a lawyer, and his mother Julia both died in poverty during his childhood, and he was brought up by his uncle Cummins E. Jackson. He added the 'Jonathan' to his name himself, when he was nearly grown up. He entered West Point in July 1842, and found himself much hampered by his poor preliminary education. He had to immerse himself in his studies in order to graduate seventeenth out of fifty-nine in the class of 1846. He was quickly shipped to Mexico, and so distinguished himself at Vera Cruz, Cerro Gordo, and Chapultepec that he was breveted major within eighteen months of graduating from the Academy, and was publicly congratulated by General Winfield Scott (q.v.). Returning to the United States in 1848, Jackson spent the next three years at Forts Columbus and Hamilton, New York, and in Florida. In February 1852 he resigned his commission to take up a professorship in military tactics and natural philosophy at the Virginia Military Institute, Lexington. He was apparently an uninspiring teacher, and the butt of much student humour. But at least the chair gave him a chance of travel, including a trip to Europe for five months in 1856. Apart from his presence as commander of the Cadet Corps at the hanging of John Brown on 2 December 1859, Jackson played no part in public life in the years before the outbreak of the Civil War; he deplored the idea of the war, but was a loyal Virginian. He was summoned to Richmond in April 1861, and was soon afterwards sent to Harper's Ferry as a colonel of infantry.

In June 1861 Jackson was promoted brigadier-general, and on 21 July, at the First Battle of Bull Run, his Virginian brigade distinguished itself by withstanding the pressure of MacDowell's Union army. His colleague General Bee, himself forced to fall back, coined the famous nickname when he remarked, 'There is Jackson, standing like a stone wall.' Promoted major-general in October 1861, he took command of the Confederate troops in the Shenandoah Valley the following month. At first dogged by bad weather, and frustrated in an abortive raid on Romney in January 1862, Jackson was harshly judged by some who considered that he was an unreliable and excitable commander. He silenced all critics, however, by his conduct of the Valley campaign of March–June 1862. It was a complex campaign of swift movement, forced marches, knife-edge retreats, rapidly shifting emphasis, feint and deception. Outnumbered by 70,000 to 17,000, Jackson seized and held the initiative by scientific use of his inferior resources, by a clear appreciation of the enemy's intentions and reactions, by logical reasoning, and by perfect timing. He prevented MacDowell and McClellan from linking up near Richmond to attack General J. E. Johnston by playing on Union sensitivity about a possible attack on Washington. Probably his greatest tactical triumph was at Port Republic on 9 June; in the course of his eventual escape from massive Union pursuit he turned on the advance guard of Shield's command and threw it back twenty miles.

Called to Richmond during the Peninsula campaign soon afterwards, Jackson – exhausted and moving through unfamiliar terrain – failed to show his usual energy and grasp during the Seven Days' Battle at the end of June. He soon restored his reputation, however, by a famous forced march before the Second Battle of Bull Run. He led his 'foot cavalry' for fifty-four miles in two days before falling upon General Pope's depot at Manassas. In the battle itself, on 29–30 August, Jackson again distinguished himself; he set his command up as a target for Pope, so that Longstreet could take him in the flank. After the tactical victory – but strategic defeat – of Antietam on 17 September, Jackson was promoted lieutenant-general and given command of one of the two corps into which the Army of Northern Virginia was divided. His men fought well at Fredericksburg on 13 December 1862.

Jackson's last and greatest battle came on 1–2 May 1863. While Lee faced Hooker's Union army at Chancellorsville, Jackson slipped around the right wing of the enemy line with more than half the Confederate army. At dusk on the 2 May Jackson fell on Hooker's right wing and destroyed it. Reconnoitring after dark to assess how best to exploit the advantage, Jackson was shot by his own men in one of the pointless accidents inevitable on battlefields. He died of pneumonia on 10 May 1863 at Guiney's Station, south of Fredericksburg. After lying in state in Richmond, he was buried at Lexington, Virginia.

Jackson was a slight man of medium height, with large hands and feet. He was quiet-spoken, inclined to be rather stiff in company, and a devout Presbyterian who got up to pray several times on the eve of battle. He was a good but not a graceful rider; he was often to be seen in the plainest of uniforms and a disreputable old hat, riding an ugly horse at the head of his quick-moving infantry. He disdained all outward show, and deliberately avoided cultivating a self-consciously close relationship with his subordinates. He was stern and uncommunicative, and even his most trusted subordinate, Ewell, thought him more than half mad; yet he had some gift which made his ragged soldiers cheer him whenever they saw him, and count it an honour to die under his command. What he would have achieved had he lived cannot be known; but it was certain that Lee could never replace him, and that the Union generals slept easier in their cots for knowing that 'Old Jack' was no longer roving out there somewhere beyond their patrols, waiting for the perfect moment to strike from some unexpected direction.

Jeanne d'Arc, called La Pucelle (*1412–31*)

Peasant girl visionary, patriot extraordinary, and leading influence in the revival of French national and military morale during the Hundred Years' War, Jeanne was a national heroine of the French after being burnt at the stake as a heretic by the English.

Born at Domremy in 1412, she led a normal peasant life until, at the age of thirteen, she claimed to have visions of saints and to hear voices encouraging her to assume the leadership of French forces opposing the English. With nothing but her naïve religious faith – and certainly without the slightest knowledge of military matters – she eventually gained an audience with the Dauphin and half persuaded him that she had a divinely-inspired mission to help him expel the English from France and have him crowned as the rightful king. In spite of fierce jealousy in the Court, the Dauphin placed the seventeen-year-old girl in command of an army sent to relieve the city of Orléans. Towards the end of April 1429 this army reached the city, and on 7 May Jeanne herself led an attack on the English-held fortified bridgehead on the south bank of the Loire. By nightfall the blockade was broken and the English, under Suffolk, withdrew. Jeanne herself was seriously wounded by an arrow.

Thereafter the English mistakenly divided their forces among numerous garrison towns in the Loire valley, and in the following weeks it was no difficult task for the French, their morale now soaring under Jeanne's inspiration, to recapture all these towns. This campaign was successfully concluded on 19 June when, in a surprise attack, Jeanne's forces defeated the English under John, Lord Talbot and drove them from the Loire valley, capturing Talbot himself. Fired with a passionate nationalism and a belief in Jeanne's invincibility, the French peasantry rose in revolt and conducted a harsh but effective guerrilla war against the English, and in July Jeanne's army captured Troyes, Châlons, and Rheims. It was in this period of euphoria that Jeanne engineered the triumphant coronation of the Dauphin as Charles VII in Rheims on 16 July 1429, an event which coincided with her forces' invasion of territory formerly dominated by the English.

Now thoroughly perturbed by the extraordinary influence enjoyed by this young girl, Charles and his Court did all they could to restrict her actions – at the same time keeping in mind the delicate relations which she enjoyed with the Church. Her military influence waned, but although she suffered from a lack of royal support she rashly led a small force in an attack on Paris – evidently unaware that the English had been strongly reinforced. She escaped successfully from the scene of action, though again she was wounded.

The following year, on 23 May, she led a small force to the relief of Compiègne, which was under siege by the English and Burgundians, but was captured by the latter the same day and handed over to the English. She was tried by a religious court on the orders of John, Duke of Bedford, Regent of France, who hoped thereby to discredit Charles's coronation. Now wholly uninterested in her future, Charles refused to intercede on her behalf, and she was convicted and burnt at the stake as a heretic.

Her memory nevertheless continued to inspire the French forces, while the morale of the English – concerned that they might have executed a saint – sagged. Within five years the French had achieved the ascendancy and recaptured Paris, while the Burgundians and Orléanists ended their civil war. Thereafter the French were successful in numerous minor actions, with the result that the English were thankful to conclude a five-year truce.

Jeanne's impact on her people is understandable, considering the religious and social climate of fifteenth-century France. Her military influence remains an enigma, for she was utterly unsophisticated in all matters of strategy and tactics. Nothing she did or professed (if in fact she professed anything beyond inspirational leadership) could counteract the fundamental technical superiority of disciplined forces supported by controlled weaponry. She was canonised in 1909.

Joffre, Joseph Jacques Césaire (*1852–1931*), *Marshal of France*

French Commander-in-Chief during the first two years of the First World War, a theoretician who was overshadowed by Foch (q.v.), Joffre was the victor of the critical Battle of the Marne in 1914 and the blind advocate of the disastrous and wholly ananchronistic Plan XVII, which cost France a generation of young soldiers.

Born at Rivesaltes in the Pyrenees on 12 January 1852, Joseph Joffre gained a commission in the French Army early in 1870 and served as a subaltern during the second Siege of Paris during the Franco-Prussian War. Later he participated in colonial campaigns in Indo-China, West Africa, and Madagascar. By 1905 he had risen to command a division, and in 1910 was a member of the High Council of War. The following year he was appointed Chief of the General Staff – a position which would in effect become Commander-in-Chief only in the event of war.

Joffre's position in the supreme seat of military power on the outbreak of the First World War was thus to some extent the result of precautionary administration rather than that of merit. Certainly he had gained little personal distinction on the field of battle, however inapplicable such experience might have proved later. Like his contemporary and equally inexperienced subordinate, General Foch, Joffre was a believer in the *offensive à l'outrance* – although, divorced from actual field command, he was not personally responsible for its tactical application. Utterly unprepared for – indeed, utterly unable to conceive the implications of – such an offensive as that outlined in the Schlieffen Plan, Joffre was responsible for launching the French strategic *offensive à l'outrance*, a calamitous frontal attack by waves of infantry over open ground which should have been discarded as suicidal seventy years previously with the invention of the machine-gun. Inflexible in the extreme in conditions of reverse, the French armies could do nothing but retreat in the north as the Schlieffen encircling movement through Belgium and northern France gained momentum. Joffre at first could not perceive the failure and cost of Plan XVII; nor could he at once understand the nature of the Germans' remorseless advance in the north.

Only when Plan XVII had ground to a halt and been abandoned did Joffre gather his resources. His armies fell back on the Marne and awaited the shock of attack by the German First, Second, and Third Armies; when it came the French Commander-in-Chief perceived a critical weakness on the German right flank and brilliantly created a new French Sixth Army to threaten this flank. Immediately Kluck's First Army faced to the right to meet this threat, thereby opening a massive gap through which Joffre threw the British Expeditionary Force. He did not, or was not able to follow this decisive advance with more than mounting pressure from his exhausted armies, and the Germans were able to disengage without suffering a major catastrophe. Nevertheless it was upon his success in averting the loss of Paris that Joffre's reputation depended.

In the face of the stagnant trench warfare which developed in 1915 and lasted throughout the war, all the elements of traditional French field tactics became outdated almost overnight. Yet in 1915 Joffre attempted – with a remarkable lack of understanding of the nature of the new battlefield warfare – to burst through the German trench system, whose depth he simply could not imagine, using waves of infantry. Each successive battle, launched with ever-increasing numbers, ended in costly failure.

It was Joffre, on the Allied side, who saw Verdun as the cornerstone upon which the French front depended for solidity. And so it was to Joffre that blame attached when the massive German attack of 1916 was made, and evidently found the French forces unprepared. If this was the first stage in the French general's downfall, his eclipse was effectively complete by the time the Battles of the Somme had inflicted near-catastrophic French and British casualties without significant benefit to the Allied cause. As Joffre's military eclipse was progressive, so was his removal from command. On 13 December 1916 he was appointed technical adviser to the Government and deprived of direct powers of command, his place being taken by Nivelle. Realising that he was now being tactfully retired, he resigned on the day after Christmas 1916, and was promptly made a marshal of France. This marked the end of his military career, and he died in Paris on 3 January 1931.

Unlike Foch, whose relatively subordinate command in 1914 to some extent shielded him from responsibility, Joffre shouldered the blame unflinchingly for the deteriorating situation in France that continued in one form or another throughout 1915 and 1916 – notwithstanding his survival on the Marne and his protection of Paris in 1914. He was inflexible to a degree; yet it must be said to his credit that he never shirked the blame, and thereby preserved intact the undoubted prestige of General Foch, whose outlook and policies he steadfastly endorsed in the early years, and who rose to become author of final victory on the Western Front two years after Joffre himself had been 'put out to grass'.

John III, Sobieski (*1629–96*), *King of Poland*

Eastern Europe was torn throughout the seveteenth century by civil wars, dynastic quarrels, rebellions, and invasions, while the shadow of the Turks lay heavy across the whole area. John Sobieski of Poland is remembered, properly, for his triumph over the Turkish army of Kara Mustafa outside the gates of Vienna in 1683; but throughout his career as an active military leader he displayed great energy and ability, whether he was fighting against Turks, Cossacks, Swedes, or fellow-Poles.

He was born on 17 August 1629 at Olesko, near Lwow; his father Jakob Sobieski was the castellan of Cracow. In 1646–8 he and his brother Marek made a 'grand tour' of France, England, and the Netherlands, returning to fight against the Cossack and Ukrainian uprising in 1648. He was present at the battle which brought this cruel war to an effective end, when John II Casimir and 34,000 Poles defeated 200,000 Cossacks and Tartars under Bogdan Chmielnicki at Beresteczko on 1 July 1651. In 1655 Poland suffered a brief period of Swedish domination, and Sobieski was one of many leaders who served the Swedes. He changed sides the following year, however, and with Stephan Czarniecki cleared the Swedes from the central provinces of Poland. Further distinguished services against the Cossacks and Tartars brought him the rank of grand marshal and the field command of the Polish armies in 1665, and the post of Commander-in-Chief three years later. In 1668 a vigorous power-struggle followed the abdication of John Casimir. Sobieski took bribes from Louis XIV of France to support a French candidate at the election diet of 1669, and when the indecisive Michael Wisniowiecki was elected instead, Sobieski was deeply involved in two plots against him.

In 1672 the Turks invaded Poland; some 200,000 of them were led into the south-eastern provinces by the Grand Vizier Ahmed, and they were aided by rebellious Cossacks and vassal Tartars. They took the strategic fortress of Kamieniec after an epic defence, and occupied Lublin with ease. Sobieski led every available man against the Turks, but the weak King Michael meanwhile concluded the craven Treaty of Buczacz, ceding large territories and tribute to the Turks and Cossacks. This Treaty was never ratified by the Polish diet, and Sobieski partially retrieved the situation by winning four victories in ten days. He rallied the Polish people, and with an army of 40,000 continued to defy the Turks throughout 1672 and 1673. In November 1673 he won the second battle of Chocim, wiping out a Turkish army of 30,000 and forcing other Turkish columns to withdraw from Polish territory. On 10 November, the eve of the battle, King Michael had died. On learning of this Sobieski left the frontier and hastened to Warsaw to advance his nomination as successor. His force of character, and the 6,000 veterans he brought with him, impressed the elective diet, and it named him King on 21 May 1674. His efforts to raise large forces to oppose the Turks were to be hampered throughout his reign by continuing internal disputes and intrigues among the Polish nobles, of whom a strong party were always hostile to him.

In 1675 some 60,000 Turks and 100,000 Tartars invaded Poland once again, taking back Chocim and threatening Lwow. Sobieski defeated them outside Lwow and in several subsequent engagements, gradually driving them back and liberating all Polish territory except Kamieniec. The next year another invading army crossed the border, 200,000 strong, under Ibrahim Pasha. At Zorawno in September–October 1676 Sobieski, with only 16,000 men, successfully repulsed Turkish attacks on his strongly-defended camp on the Dniester river for two weeks. The Treaty of Zorawno was concluded in October; by this the western Ukraine reverted to Poland, although the Turks kept Kamieniec and Chocim. For another seven years Sobieski was mainly concerned with continuing Cossack trouble in the Ukraine and with internal disputes.

In March 1683 he signed a defensive-offensive alliance with the Holy Roman Emperor, as Austria was now threatened by the Turks. The Emperor and the Austrian main army retreated in the face of an invasion by 150,000 Turks in the early summer, leaving Rudiger von Starhemberg with a mere 15,000 men to defend Vienna. Between July and September a savage campaign of storming parties, sorties, and raids by forces from outside the siege lines was waged around the city; the Turks took several breaches and established footholds in the streets. Half the garrison became casualties, and supplies were running low when, on 11 September, John Sobieski arrived with 30,000 Poles and advanced on the Turkish camp. He had covered 220 miles in fifteen days, a remarkable achievement for the period. He joined up with the Austrian and German forces lying west of the city, taking command of the combined army of 76,000 men. In the late afternoon of 12 September he attacked the Turkish position while the garrison of Vienna made a simultaneous sortie. The battle was decided by a Polish cavalry charge on Kara Mustafa's headquarters led by Sobieski in person, and Vienna was saved. During the last months of 1683 Sobieski led a vigorous pursuit of the Turks which freed most of north-western Hungary.

The relief of Vienna was the peak of Sobieski's career. The Emperor Leopold proved ungrateful; jealous of Sobieski's renown after his city had been saved for him, he failed to support the Polish King in Hungary or in his continuing campaigns in the Ukraine. Constant internal problems beset Sobieski, and he died a disillusioned man at Wilanow on 17 June 1696.

Joseph, Chief (*1840–1904*)

At the time of the Nez Percé War of 1877 Chief Joseph was credited by American sources with the whole conduct of his tribe's epic retreat, and this version of events has tended to persist. In fact Joseph was a civil rather than a war chief. The various engagements which took place during the campaign were directed at the tactical level by a series of war leaders – Looking Glass, Rainbow, Five Wounds, Lean Elk, and others. Nevertheless, Joseph has been included in this book: the tactical moves were of less significance than the skill and determination of the retreat as a whole, and Joseph is generally conceded to have been the moving spirit behind the doomed migration of his band.

The Nez Percé band, the leadership of which Joseph took over on the death of his father in 1871, inhabited the Wallowa Valley, a highly desirable tract of land in northern Oregon. They were a strong and intelligent tribe who had been on good terms with the white man since the days of the Lewis and Clark Expedition. It was their boast that no Nez Percé Indian had ever killed a white man, and Joseph's father had allowed himself to be converted by a local Presbyterian missionary. There is no place here for a detailed account of the sequence of deceptions, broken agreements, illegal intrusions, and other squalid moves which led up to the Government's order of spring 1877 that the Nez Percé Indians should leave their lands and move on to a reservation. Suffice it that they finally submitted, and were spending a few last days of freedom just outside the border of the reservation in early June when a few young braves broke away and killed three local whites notorious for their hostility to the tribe. Joseph, as always, counselled peaceful negotiation, but the tribe feared reprisals and hid in White Bird Canyon. Total available manpower in the camp was not more than fifty when Captain Perry and a hundred men of the 1st Cavalry and a civilian volunteer group arrived there to enforce the return to the reservation on 17 June 1877. The Nez Percé Indians sent out a truce party, which was immediately fired upon despite the white flag. The tribesmen then attacked the cavalry and routed them; thirty-four dead were left on the field, and Indian casualties were extremely light. The tribe retreated across the Salmon river, followed by General Oliver Howard with some 230 men. A detached force under Captain Whipple meanwhile attacked a peaceful Nez Percé village on the reservation, only to receive word that the main band had led Howard a frustrating chase in the wilds, lost him, doubled back over the river, and were now between Howard and Whipple. The Indians by-passed Whipple and, reinforced by the infuriated group from the reservation village, hid out on the Clearwater river. They now numbered some 200 warriors, hampered by about 550 women, children, and old people, and by their flocks and herds. It was already noted by the Americans that they had refrained from scalping or mutilating enemy dead, and had not molested civilians of either sex during their passage through inhabited country.

On 11 July Howard attacked the Clearwater encampment with 580 men supported by a howitzer and two Gatlings. Outnumbered at least three to one, the Indian warriors crossed the river towards the army force, and scaled bluffs under fire. They then encircled the expedition and kept it penned on a hilltop all one day and half the next, while the non-combatants and herds were led to safety. When the last warriors pulled back the army counted thirteen dead, two dying of wounds, and twenty-five wounded, against four dead and six wounded Indians. The Nez Percé now attempted, with pathetic optimism, to escape from the army's interest by crossing the Bitter-root Mountains into Montana. They crossed the mountains in nine days, to find a small post erected in their path and Howard following their trail. The civilian volunteer bulk of the post garrison departed rapidly, satisfied when the Nez Percé promised that they sought only passage through the territory and would harm no one. They by-passed the now-impotent fort and scrupulously respected their pledge while passing through the heavily-populated

Bitterroot Valley, paying cash for all food and supplies. But on the night of 8 August they were attacked while camped in Big Hole Basin by a new force under Colonel John Gibbon (see under Custer), numbering some 200 men. The warriors were driven from their camp, but rallied and retook it as the army were destroying it. Gibbon was forced back to a knoll outside the camp and encircled. The warriors maintained the siege while the non-combatants packed up and moved off southwards – indeed, they maintained it all the next day, and added insult to injury by capturing the force's howitzer and tipping it off a cliff. They pulled out the following night, after inflicting thirty-three fatalities and thirty-eight other casualties. They had lost about seventy-five Indians, including many of their best warriors, and many women and children butchered in the sleeping camp. Throughout the Battles of Clearwater and Big Hole Basin the Nez Percé had conducted themselves with a discipline and skill never before encountered among Indians by the troops. The Press was beginning to allow grudging admiration to creep into its reports.

The Nez Percé now continued their retreat south-east across Montana, aiming to cross the Continental Divide at Targhee Pass before swinging north and making for Canada and British sanctuary. Delaying Howard's furious pursuit by running off his pack horses in a night raid, they crossed into the Yellowstone National Park. Now they swung north up Clark's Fork, outwitting one of several columns of cavalry converging in the area; they took several white prisoners but harmed none of them, and eventually allowed them to escape. At Canyon Creek on 13 September they clashed heavily with a 7th Cavalry force under Colonel Sturgis, but managed to hold off the soldiers in the rocky terrain until the non-combatants were safe once more. Harried by soldiers and Crow Indians, the Nez Percé moved on across the Musselshell and Missouri rivers, always fighting their way clear or riding around the opposition. On 27 September, by now weak and exhausted, they looted an army depot at Cow Island, but did not harm the small garrison. In the Bear Paw Mountains, only thirty miles from the Canadian border, they halted to rest.

From Fort Keogh to the east Colonel Nelson Miles led out nearly 600 men, and on the morning of 30 September his scouts sighted the Nez Percé camp in a deep hollow. Miles charged the village, cutting the tribe into three groups. Two managed to join up in the smashed camp, although many of the horses had been run off and they were now encircled. Both sides dug in; snow fell, and the sufferings of both sides were intense. Miles shelled the camp with a howitzer; during a parley he attempted to seize Joseph by treachery, but had to exchange him later for a captured officer. On 4 October General Howard reached the battlefield, and the end was clearly in sight. On the following day Joseph surrendered to Miles and Howard to save his women and children from starvation; his speech of submission is one of the most moving ever made. He had led his people 2,000 miles, defeated much larger and better-equipped forces on several occasions, and outwitted the U.S. Army for four months. The campaign had been conducted with a humanity which conformed to the strictest codes of white warfare, and which earned the Nez Percé the respect of many Americans.

The tribe – those who survived – were sent back to reservations in Washington in 1885. Joseph died in 1904, after all attempts to persuade the Government to allow a return to the Wallowa Valley had failed.

Keitel, Wilhelm (*1882–1946*), *Field-Marshal*

Nazi minister of war and second-in-command of German armed forces under Hitler from 1938 until the end of the Second World War, when he was executed for war crimes. Keitel was neither an able strategist nor a competent administrator, yet managed to retain his post simply owing to his blind loyalty to Hitler while more imaginative subordinates were purged or dishonoured.

Born on 22 September 1882 at Helmscherode in central Germany, Wilhelm Keitel entered the artillery and served with this branch and on the staff during the First World War, rising to the rank of *Oberst* by the Armistice. Between the wars, under the patronage of Adolf Hitler, he became closely associated with the N.S.D.A.P., and during the early and mid-1930s held a number of influential posts in the *Reichswehrministerium*. Upon Hitler's assumption of personal command of all armed forces in 1938, Keitel was appointed to the new post of chief of the armed forces high command (O.K.W.) – in effect a military minister of war – a position he managed to retain until the defeat of Germany in 1945.

With no first-hand experience of strategic planning, nor any outstanding achievements in his past career, Keitel quickly earned for himself the contempt of his subordinates – a state of affairs that rarely encouraged him to attempt to restrain the extravagances of Hitler. Such attempts were invariably unsuccessful, and he never pursued any open criticism of his leader, for such questioning of the *Führer* invariably led to the questioner's removal from office – or worse. Keitel attended all significant conferences on the conduct of the Second World War; he dictated the terms of armistice to the French in June 1940; and he signed numerous operational orders – including those for the execution of hostages and prisoners of war. After the unsuccessful plot to assassinate Hitler on 20 July 1944, Keitel sat as a member of the 'Court of Honour' which sentenced so many high-ranking officers to death for their supposed part in the conspiracy.

His total and unquestioning subservience to Hitler cannot have entirely blinded him to the futility and self-destructiveness of many of his leader's orders but, having little imagination, he invariably acquiesced to Hitler's demands. He certainly lacked the brains of his subordiate Jodl, but together these two survived the frequent tantrums of the *Führer*, while almost every other senior commander suffered the consequences of disagreement – only a very few, such as Runstedt and Guderian (qq.v.), ever enjoyed restoration to favour.

After the war Keitel was brought to trial before the International Military Tribunal in Nürnburg, was convicted of planning a war of aggression and of committing war crimes and crimes against humanity, and was sentenced to death. He was hanged on 16 October 1946. There can have been few supreme commanders in history with less professional right to the control of great armies than had Wilhelm Keitel.

Kemal Atatürk, *see* Mustafa

Kesselring, Albrecht (*1885–1960*), *Field-Marshal*

Kesselring, the soldier-turned-airman-turned-soldier, was undoubtedly one of Germany's most able field commanders in the Second World War. His defensive campaign in Italy in 1943–5 was a classic demonstration of intelligent use of terrain and strictly limited military resources to delay the advance of a much stronger enemy.

He was born in Bavaria on 30 November 1885, the son of a middle-class family. Commissioned in the Bavarian artillery in 1906, he reached the rank of captain during the First World War. He stayed in the 'hundred thousand army' after the war, and although promotion was slow, he attracted attention by his ability in a number of staff posts. As a major he served in the department of the War Ministry concerned with training, and as a lieutenant-colonel he filled a staff appointment at army headquarters. Clearly a 'coming man', he was picked for transfer to a post of high responsibility in the new *Luftwaffe* in 1933, two years before the existence of that force was revealed to the world. After a period of preparatory work at the ostensibly civil Air Ministry, he officially transferred into the air force with the rank of major-general in 1935. In 1936–7 he was chief of the *Luftwaffe* general staff, and his influence helped to pave the way for Germany's impressive successes in the field of tactical air support in the early war years. He commanded *Luftflotte 1* in Poland in 1939, and in 1940 led *Luftflotte 2*, the formation which comprised roughly half of the entire strength of the German air forces, in the campaign in the West. During the Battle of Britain he was based in Brussels, and his units occupied Holland, Belgium, and north-eastern France. Although his record during the early phase of *blitzkrieg* warfare commands respect, his grasp of the situation in assessing the Royal Air Force's continued ability to resist daylight attacks in September 1940 was faulty. He is on record as having supported Göring in his decision to shift the emphasis of the attack from R.A.F. Fighter Command airfields to civilian and commercial targets. In this he was deceived by the notoriously inaccurate German intelligence assessment of British losses.

Kesselring continued to command *Luftflotte 2* with great ability in Russia between June and December 1941, and in the Mediterranean in 1942. Now a field-marshal, he was appointed Commander-in-Chief South in 1942, with responsibility for both land and air forces. He exercised complete control of the North African theatre during Rommel's (q.v.) occasional absences from the front. In 1943 he presided over such evacuation of German forces from Tunisia as was possible in view of Hitler's negative orders; and when the Allies landed in Sicily and Italy his personal leadership of the German defence brought into the open great strategic and tactical talents. For more than a year and a half he ensured that the British and American forces paid the maximum penalty in casualties and in time for every advance. He knew Italy well; he appears to have realised more clearly than the Allied commanders that it was a country where numerical superiority counted for little. He had the advantage of fighting on interior lines in mountainous terrain perfectly suited to stubborn defence, but this advantage should in theory have been cancelled out by the Allies' almost complete mastery of the air. German weakness in armour was less important; Kesselring could have made good use of any extra divisions which he had been able to secure in spite of urgent need in Russia, and later in the West, but Italy is poor tank country and the Allied superiority in this type of unit was not decisive. Kesselring's conduct of the campaign was characterised by holding on to each successive defensive line for far longer than his opponents had anticipated; by sealing off local Allied successes; by luring the enemy into accepting heavy casualties for the gain of positions which when finally taken proved to be of only local importance; and by well-timed withdrawals to secondary and tertiary defensive lines prepared well in advance. Even after Alexander (q.v.) had broken through the Cassino line and rendered the heroic resistance of the *Luftwaffe* paratroop regiments fruitless, Kesselring managed to withdraw important elements of his command to new positions in surprisingly good order.

In March 1945 Hitler seized on Kesselring as the new 'white hope' who would save the situation in North-West Europe, and appointed him Commander-in-Chief West. His task was impossible, however, and he could achieve no better results than von Runstedt (q.v.). When the Russians broke through south of Berlin Kesselring's command was extended to include all forces south of the breakthrough. He surrendered to American forces on 6 May 1945.

In 1947 Kesselring was convicted by a British Military Court at Venice of responsibility for the murder of 335 Italian civilians, and for ordering the shooting of hostages in reprisal for partisan activity; he was sentenced to death. This extreme interpretation of the doctrine of ultimate command responsibility, in a campaign characterised by intense civilian activity and by occasional drives by *Waffen-SS* commanders who were largely a law unto themselves, aroused deep misgivings among several senior Allied commanders. Alexander, among others, interested himself in the fate of his former opponent; and Kesselring's sentence was commuted to life imprisonment. He was released on grounds of ill-health in 1952. During his years of retirement he was active as the President of *Stahlhelm*, the ex-serviceman's organisation. He died at Bad Nauheim on 16 July 1960.

Kitchener, Horatio Herbert, 1st Earl Kitchener of Khartoum (*1850–1916*), *Field-Marshal*

Draconian British commander of the late Victorian era and victor at the Battle of Omdurman, Kitchener combined ruthlessness in battle with thoroughness in administration, thereby making his contribution to the achievement of British imperial ascendancy in Africa by the end of the nineteenth century, when he imposed his doctrine in bringing to an end the South African War.

Born on 24 June 1850 near Bally Longford, Co. Kerry, Ireland, Kitchener was educated privately and at the Royal Military Academy, Woolwich; but before graduating with a commission he served as a volunteer cadet with the French forces in the Franco-Prussian War of 1870–1. Returning to England, he was commissioned in the Royal Engineers in January 1871, and was employed as a subaltern on intelligence work between 1874 and 1882 in Palestine, Anatolia, and Cyprus. In the following year he was posted to Cairo as second-in-command of an Egyptian cavalry regiment. As such he served with distinction with the Nile Expeditionary Force (1884–5) which just failed to rescue General Gordon (q.v.) at Khartoum. He was then moved to Zanzibar as Governor of Suakin, and shortly afterwards returned to Cairo as Adjutant-General. Widely noted for his unmatched drive and thoroughness (while his undoubted harshness in the face of any opposition went unremarked), Kitchener was appointed to the command of the Egyptian Army in 1892 on the personal recommendation of the Prime Minister, Lord Salisbury.

Having no home ties – he was a bachelor – Kitchener was entirely motivated by an overwhelming sense of duty and mission; he lived for the day when he might avenge the death of General Gordon, and constantly advocated ruthless warfare against the Arabs. When in 1896, concerned by the increase in French and Italian colonial ambitions in the Nile Valley, Britain decided to reoccupy the Sudan, Kitchener embarked on a methodical campaign of conquest with a mixed army of well-trained British and Egyptian troops, moving south along the Nile, building a railway as he went, and supported by a flotilla of river gunboats. Facing fanatical resistance, but with his supply lines secure, he made good a steady advance, capturing Dongola on 21 September 1896 and Abu Hamed on 7 August 1897, and defeating the Mahdist forces at the Battle of Atbara River on 8 April 1898. At this point the Kalifa Abdullah and Osman Digna determined to halt and hurl back the advancing Anglo-Egyptian army, concentrating 40,000 men at Omdurman on the Nile, just north of Khartoum.

Kitchener's army of 26,000 men, of whom about half were British regulars and the remainder well-trained Egyptians, was encamped in a fortified position at Egeiga, four miles from Omdurman, when Abdullah's forces attacked on 2 September. The savage and fanatical courage of the attacking tribesmen, mostly armed with outdated muskets, swords, and lances, was of little avail against the steadfast discipline of Kitchener's men armed with light field-pieces, machine-guns, and modern small arms. The tribesmen were hurled back with heavy casualties, and the Anglo-Egyptian force promptly counter-attacked towards Omdurman. Under-estimating his enemy, however, Kitchener was suddenly faced with a complete reverse when the dervishes rallied, wheeled, and fell upon the British right and rear, while an unsuspected force of 2,000 enemy leapt from concealment and also attacked the right flank. Only the steadfastness of Sir Hector MacDonald's Sudanese brigade and the legendary cavalry charge by the 21st Lancers broke the enemy's strength and routed the dervishes; on Kitchener's orders the Anglo-Egyptian troops closely pursued the survivors off the field, cutting down any who offered the slightest resistance. Abdullah's losses at Omdurman amounted to over 10,000 killed, a similar number wounded, and 5,000 prisoners. Kitchener's losses were no more than 500 killed and wounded. Apart from the immediate importance of the Battle of Omdurman in re-establishing Anglo-Egyptian control over the Sudan, the engagement was the first in which the machine-gun demonstrated its power in a set battle.

At this time, the period of Queen Victoria's Silver Jubilee, Kitchener's series of victories thrilled the Empire, in spite of his acknowledged ruthlessness and brutality towards his enemies. In November 1898 he was raised to the peerage, and he received the thanks of Parliament the following year, together with a handsome grant of money.

Kitchener then returned to the Sudan, where he served for a year as Governor-General; but in the meantime war had broken out in South Africa. As a result of public clamour during the first months of disaster and defeat Kitchener was sent out to serve as Chief-of-Staff under Lord Roberts (q.v.), and in November 1900 he succeeded Roberts as Commander-in-Chief. Thereafter, by employing draconian methods characteristic of his earlier campaigns he turned the tide of events: during a period of nearly two years he gradually wore down the Boer guerrillas, always meeting force with greatly superior strength, and often burning settlements and directing civilians into ill-equipped 'concentration camps'. Although this phrase did not then have the frightful overtones of 1939–45, many inmates died of disease.

Returning home, Kitchener was acclaimed a national hero, created a viscount and, having declined service at the War Office, appointed Commander-in-Chief in India. There he set about a complete reorganisation of the army's administration, his particular achievement lying in reforms in hygiene, sanitation, and victualling. However, he quarrelled with the Viceroy, Lord Curzon, about the division of responsibilities between the Indian Civil Service and the military. The result was Curzon's resignation in 1905; but Asquith's Liberal Government refused to confirm the appointment of Kitchener to succeed him – a rebuff which was a lasting source of bitterness to Kitchener. Nevertheless he accepted appointment to the administration of Egypt, and over the following five years accomplished effective reform of the economies of both Egypt and the Sudan.

Created an earl in 1914, Kitchener was in London on leave when the First World War broke out, and reluctantly accepted the post of Secretary of State for War, serving as a field-marshal in the Cabinet. Quickly realising the inadequacies of Britain's regular Army, Kitchener warned his over-optimistic Cabinet colleagues that the war must last at least three years, and that victory over Germany would only be achieved by a very large, highly-trained army. Accordingly he set about a programme of widespread recruitment and training; his recruitment policies – and his poster-portrait bearing the slogan 'Your Country Needs You' – were the principal means by which Britain was enabled to field sixty-seven infantry divisions in 1916. But because of his constant outspokenness Kitchener at once became unpopular among his colleagues, and unfairly came to be considered untrustworthy. It was in this atmosphere of mistrust that, during an absence on a visit to Gallipoli, he was divested of all authority in matters of strategy in the war. His determination to continue serving his country and its people, to whom he was something of an idol, prevented him from resigning; but rejection by his colleagues deeply distressed him.

A mission to Russia, intended to strengthen the military co-operation of the Allies, offered a respite from the unhappy situation in London, and Kitchener set sail from Scapa Flow in the cruiser H.M.S. *Hampshire* on 5 June 1916. The same evening the ship struck a mine off the west coast of Orkney and sank almost immediately. The hero of Omdurman was not among the handful of survivors.

Always somewhat suspicious of politicians, and therefore inevitably a victim of their intrigues, Kitchener presented an austere image, severe of countenance and harsh of action. Yet he drove himself no less hard than he drove others, and thereby he gained the unqualified respect of his subordinate commanders and the loyalty of his troops. Much of his success on the battlefield had been achieved in the quartermaster's stores beforehand.

Komorowski, Tadeusz Bór (*1895–1966*), *General*

'General Bór' was the *nom-de-guerre* of the Polish underground army commander Tadeusz Komorowski, who led the tragic Warsaw uprising of August–September 1944. After his release from German captivity he adopted the style shown above.

Tadeusz Komorowski was born in 1895 near Lwow, the son of a landowner. He joined the army when the Polish Republic was resurrected in 1918, and in the 1919 Russo-Polish War fought as a captain in the 9th Cavalry (Malopolski Lancers). In 1920 he became the commanding officer of the 12th Cavalry (Podolski Lancers). During the years between the wars he climbed the ladder of promotion rather slowly and by a series of unremarkable postings, attracting more attention by his prowess in international horsemanship competitions than by his professional brilliance.

When the Second World War broke out in September 1939 Colonel Komorowski commanded the Grudziad cavalry training school, in an area quickly overrun by the Germans. Subsequently he managed to contact his old friend General Sikorski, now leading the Polish government-in-exile in France, and was instructed to organise an underground army for sabotage and intelligence work, and for the eventual liberation of the country. This he did, with such skill and effectiveness that at one time the Germans were pressing the respectable Komorowski to make some sort of contribution to the anti-Bolshevik crusade, while the mysterious 'General Bór' had a price of £400,000 on his head. He rose from Home Army Commander in the Cracow region to Deputy Commander, and finally in July 1943 to Commander-in-Chief, Home Army. On 1 August 1944, with the advancing Red Army only kilometres away across the Vistula, Komorowski responded to Moscow Radio appeals and ordered the Home Army to rise and strike at the Germans in Warsaw.

His decision has been much criticised, but at the time he was in a cleft stick. He wished above all to liberate the capital by force of Polish arms, and hoped to set up a patriotic government before the Russians could arrive and install a Communist régime. He was attacked later for his naïvety in believing that the Russians would lift a finger to save a potential non-Communist government; or alternatively, for sacrificing thousands in an obviously vain attempt to defeat the Germans without Russian help. The controversy is not clarified by the evident facts that the Red Army was at that time exhausted and, arguably, incapable of further advances; and that during the progress of the Warsaw fighting the Russians deliberately frustrated British and American attempts to air-drop vital supplies to the insurgents. As so often occurs in real life – as opposed to the neat logical progressions of text-books – it is probably fairest to say simply that the rising had to happen, and had to fail. At least it gave Warsaw one more brief hour of tragic but dignified freedom between the long nights of German and Soviet occupation.

The generals' bomb plot against Hitler had failed only days before, and the Army High Command predictably failed in their application to have the Warsaw rising treated as a front-line combat matter under Army control – an application amply justified by the facts. Instead Himmler and the Governor-General, Frank, were given the responsibility of putting down the revolt, for which purpose the *Reichsführer SS* dispatched his Chief of Partisan Warfare, SS Lieutenant-General von dem Bach-Zelewski. The men available to him – the word 'soldiers' is inappropriate – included the *SS Sturmbrigade Dirlewanger* and the Kaminski Brigade, two terrifying rabbles of cut-throats, renegades, sadistic morons, and cashiered rejects from other units, with a strong leavening of Russian traitors. These two units totalled roughly 10,000 men; during the two months of very bitter and bloody street-fighting which followed the uprising they brought to the streets of tortured Warsaw scenes not inflicted on Western European communities since the Thirty Years' War. Their excessive atrocities even offended the officers of other SS units present, and when the truth began to filter through to the Army High Command, Guderian (q.v.) made a strong plea for their withdrawal. (Many senior German officers have since tried to claim credit for this – and for the highly uncharacteristic decision by Hitler, when the last few Polish survivors staggered out of the smoking rubble of their capital, to treat them and their commander Komorowski as prisoners of war. It is likely that General von dem Bach-Zelewski had a wary eye on possible future war crimes charges; it was late enough in the day for the outcome of the war to be evident.)

After fighting to a standstill against superior numbers and vastly superior fire-power, the Polish Home Army in Warsaw was virtually annihilated. Their agony was immeasurably increased by the knowledge that by fighting in their home city they were bringing their families within reach of such barbarians as were commanded by Dirlewanger and Kaminski. The rising collapsed on 30 September, and 'General Bór' passed into German captivity. Surprisingly, he survived to be handed over to the Americans at Innsbruck in May 1945; it is said that Hitler's 'brother-in-law', the SS cavalry officer Hermann Fegelein, was instrumental in protecting Bór out of remembered friendship at pre-war horsemanship events.

After the war and the establishment of the Communist régime in Poland General Bór Komorowski lived quietly in retirement in Great Britain. He died of a heart attack on 24 August 1966 while out rabbit-shooting near Woughton-on-the-Green in Buckinghamshire.

Konev, Ivan Stepanovich (*1897–1973*), *Marshal of the Soviet Union*

Soviet commander of the Second Ukrainian Front during 1944–5, Konev won the race to reach German territory in the assault on the Eastern Front during the final year of the Second World War in Europe; his were the first Russian troops to link up with American troops just before the final German surrender.

Born the son of a peasant at the settlement of Lodeyno, near Archangelsk (in the far north), in 1897, Ivan Konev embarked on his military career at the age of fifteen as a private in the Tsarist army of the Grand Duke Nicholas, but joined the Bolshevik Party and the Red Army in 1918. Rising through the ranks and gaining a commission, he nevertheless survived the Stalinist purges of 1937, and the following year was given command of the Transbaikal Military District, a position he continued to hold until 1941.

As the Russian front in the West crumbled in the path of German advances in 1942, Konev was transferred to the southern front in command of an army corps, but it was not until January 1944 that he came to prominence. Given command of the Second Ukrainian Front – in effect a large group of armies – he launched his forces northwards to link up with the First Ukrainian Front under Zhukov (q.v.) in a giant envelopment of two German army corps around Korsun. Although this massive encirclement was not entirely effective, the Germans nevertheless suffered over 100,000 casualties.

Thereafter, in the face of desperate German resistance, the Russian armies hammered their way westwards, both sides suffering appalling losses in men and equipment. As summer approached, the three Ukrainian Fronts made assaults on a 350-mile front, supported by complete air superiority. Soviet armour tore a 250-mile-wide gap, encircling German strongpoints which were then eliminated by huge masses of infantry. During June and July German losses amounted to 381,000 killed, 158,000 captured, and more than 2,000 tanks, 10,000 guns, and 57,000 motor vehicles. Konev's Front captured Lwow on 27 July and reached the Vistula on 7 August, but over-extension of the Russian supply lines brought the offensive to a halt.

By the end of 1944 the Russian army had amassed even greater forces, and on 12 January 1945 the entire front again exploded into movement, with Konev's Second Ukrainian Front in the lead. Konev reached the Oder-Neisse line on 15 February, but once again overstretched lines of supply forced his troops to stop. At this point Stalin ordered a restructuring of the army command with the result that a vast new super-army-group, consisting of Konev's and Sokolovski's Fronts, was placed under the overall command of Marshal Zhukov; this great mass started forward once more on 16 April. Nine days later Konev's troops made contact with elements of Omar Bradley's (q.v.) American 12th Army Group. Konev shared with his great military rival, Marshal Zhukov, the honour of the capture of Berlin on 2 May.

In 1946 Ivan Konev replaced Zhukov as Commander-in-Chief land forces, a position he occupied until 1955. In 1953 he presided over the tribunal that sentenced Stalin's henchman Lavrenti Beria to death. In 1955 he became Soviet Deputy Minister of Defence, and so one of the architects of the Warsaw Pact and its supreme military commander. He retired from this position on account of ill-health, but was recalled to assume temporary command of the Russian forces in East Germany. Twice decorated as Hero of the Soviet Union and awarded the Order of Lenin five times, he died at the age of seventy-five in 1973.

Kosciuszko, Tadeusz Andrzej Bonaventura (*1746–1817*), *Major-General*

Kosciuszko, like Lafayette (q.v.), is remembered with great respect in the United States as a volunteer officer who fought with the Continental Army during the War of American Independence. He is more generally remembered as the tragic hero of the Polish civil war of 1792 and the uprising of 1794, and as a man of genuinely liberal convictions who practised what he preached.

He was born on 4 February 1746 at Mereczowszczyzna in Belorussia, the son of a small landowner. He entered the corps of cadets at Warsaw, and attracted attention in the highest quarters by his promise; in 1769 he was sent abroad to complete his education at the expense of the state. He studied painting and the art of fortification in France, returning to Poland in 1774. He became private tutor to an aristocratic family, but lost his position – and suffered professional injury – two years later as a result of a romantic entanglement with one of his charges. It may be that a broken heart, among less intimate motives, prompted him to volunteer for the American Continental Army. Before the end of 1776 he had been promoted colonel of engineers for his work in fortifying Philadelphia. In 1777 he distinguished himself in the withdrawal from Ticonderoga, contributing to the delays which eventually proved fatal to the plans of General Burgoyne (q.v.). In 1778–9 Kosciuszko planned and built the fortress of West Point, and in 1780 he was appointed Chief of Engineers to the southern army. In 1781–2 he played a considerable part in the Siege of Charleston, and in 1783 a grateful Congress rewarded him with financial grants, United States citizenship, and the rank of brigadier-general.

Kosciuszko returned to Poland in 1784 and pursued a promising military career, being appointed major-general in the Polish service in 1789. In May 1791 a liberal reformist constitution was forced through in Poland during a period when Russian domination had slackened, and when the Prussians had (for their own reasons) made demonstrations to discourage any direct Russian invasion of the country. The ultra-conservatives among the Polish ruling class were so opposed to this constitution that they actively enlisted the aid of Catherine the Great of Russia, and on this pretext Russian troops were sent into Poland. Kosciuszko fought with distinction against the invaders at Dubienka on 18 July 1792, in command of a division. When the reactionary party, the so-called Confederation of Targowica, succeeded in pressing their terms on the weak King Stanislaw II Poniatowski, Kosciuszko was among many Polish officers who resigned their commissions. In 1793 Prussia and Russia came to terms over Polish partition; Kosciuszko's mission to France in an attempt to secure French Revolutionary support for Polish aspirations failed. In 1794 Kosciuszko, living in retirement, was contacted by a group of conspirators and offered command, with dictatorial powers, of the forces for a national uprising against Russia and Prussia. He at first refused these offers, but after hearing that the insurrection had started and that Russian troops were concentrating to crush it, he finally accepted. He travelled to Cracow and there, on 24 May 1794, he consecrated his arms in the Capuchin church according to ancient custom. He then exerted himself to organise the large but motley patriot forces at his disposal, a high proportion of whom were peasants armed with scythes and pitchforks. On 3 April at Raclawice his army, comprising about 4,000 trained soldiers and 2,000 of these rustic volunteers, defeated a force of 5,000 Russian regulars. The Russian garrison of Warsaw was expelled after several days' street fighting on 17 April.

Russia and Prussia now operated in concert to crush the Polish uprising, and a series of desperate battles was fought by the outnumbered patriots. On 6 June at Szczekociny Kosciuszko was defeated, largely due to the unexpected arrival of a Prussian force. Another Polish division was beaten by the Russians at Chelm on 8 June, and on 15 June the Prussians occupied Cracow. Kosciuszko withdrew to Warsaw, which was invested by a joint Prusso-Russian army in mid-July. Kosciuszko was faced not only by the advance of some 90,000 enemy troops with 253 guns, but also by severe unrest in the city. Luckily he was able to restore order before the enemy tightened the ring. With about 20,000 troops and 18,000 armed civilians Kosciuszko conducted a brilliant defence of the city, beating off two major assaults and forcing the enemy to lift the siege on 6 September. It was a hollow victory: elsewhere Polish generals were suffering from lack of co-ordination, and were being defeated one by one by the Russian army of Suvarov (q.v.). On 10 October 1794 Kosciuszko, with 7,000 men, attacked 16,000 Russians of General Fersen's command at Maciejowice. One of his divisions disastrously failed to support him, and the Poles were routed. Kosciuszko was wounded and was captured as he lay unconscious on the field. The national uprising collapsed, and Kosciuszko disappeared into the grim fortress of St Peter and St Paul for two years. Poland disappeared as well, under a third partition plan.

Released in 1796, Kosciuszko travelled briefly to England, where he was treated with great respect. After a few weeks he took ship for the United States. He collected his accumulated pension, and instructed Thomas Jefferson to sell his American lands for him; the funds were used to buy Negro slaves, who were immediately set free. In 1798 Kosciuszko returned to France, where he lived in retirement; he distrusted Napoleon (q.v.) and resisted his attempts to secure his collaboration. In 1814 he pressed Poland's case in vain at the Congress of Vienna. He retired to live quietly at Solothurn in Switzerland, where he died on 15 October 1817. He was buried in Cracow cathedral; his mourners reverted to a custom from pre-Christian times, and raised a huge mound to his memory outside the city.

Kutuzov, Prince Mikhail Ilarionovich (*1745–1813*), *Field-Marshal*

Born at St Petersburg on 5 September 1745, the son of a lieutenant-general who had served under Peter the Great, Kutuzov entered the Russian engineering and artillery school at the age of twelve, and two years later was posted as a corporal to the artillery. Commissioned at the age of sixteen, he saw active service in Poland between 1764 and 1769, participating as a junior officer in the Russian invasion and anti-guerrilla operations of 1768–9. At his own request he was transferred to the Turkish front in 1770, where he gained much experience in the Russian army of Count Peter Rumiantsev which invaded the Balkans. It is believed that he was present at the battle of Karkal in which Rumiantsev severely defeated a Turkish and Tartar army attempting to drive the Russians from Moldavia in August of that year. In 1774 Kutuzov was severely wounded in the head and lost an eye, but he recovered and served for six years in Alexander Suvarov's (q.v.) Crimean army. He was promoted colonel in 1777, brigadier in 1782, and major-general in 1784.

In 1787 Russia precipitated a second war with Turkey by demands for the cession of Georgia and Bessarabia, while Turkey herself engaged in intrigues to incite the Crimean Tartars to rise against Russia. Suvarov's Crimean army repulsed a Turkish attempt to reconquer the province in 1788, and later the same year took the initiative by invading Moldavia. Kutuzov particularly distinguished himself during the capture of Ochakov at the mouth of the Danube, but again received severe wounds. (Characteristically, Suvarov ordered the massacre of all Turkish men, women, and children surviving in the captured territories.) Kutuzov was back on active service within a year and was present at the capture of Ismail by Suvarov in December 1790. After the treaty of Jassy ended the second Turkish War, Kutuzov was recalled from active service and spent the years between 1793 and 1798 in a succession of diplomatic appointments.

In 1802 Kutuzov fell into disgrace and retired to his country estate; but three years later Tsar Alexander I was obliged to participate in the third coalition with England and Austria against Napoleon (q.v.), and recalled Kutuzov, giving him command of the 55,000-strong army sent to block the French advance upon Vienna. Before the Russians could join with the Austrian army under General Mack, Napoleon won the great victory of Ulm (17 October 1805), thereby opening his most brilliant campaign. Kutuzov, faced by a greatly superior army, began a skilful withdrawal, fighting successful delaying actions at Dürrenstein (11 November) and Hollabrünn (15–16 November), and thereby preserving his army intact.

It was Kutuzov's plan to fall back to the Russian border and there await reinforcements before offering battle to the French, but Alexander overruled him, with the result that the Russo-Austrian army was heavily defeated at the great battle of Austerlitz on 2 December. Despite the apparently sound siting of the Allied forces and the undoubted bravery of the Russian troops, the sheer brilliance and daring of Soult's (q.v.) storming of the Pratzen Heights, the enveloping movements by Bernadotte and Davout, and the general fragmenting of the Allied forces decided the day. Only forty-eight hours later the Austrian Emperor Francis surrendered unconditionally, while Kutuzov shepherded his shattered forces back to Russia.

Not unnaturally Alexander blamed Kutuzov for the defeat at Austerlitz and summarily removed him from field command; nevertheless his previous achievement in preserving his army was considered as reason for mitigation, and in the absence of much talent among Russian commanders, Kutuzov was made Governor-General of Kiev in October 1806, and of Vilnius in June 1809.

Desultory fighting had continued on the smouldering borders between Russia and Turkey between 1806 and 1812, and in 1811 Alexander appointed Kutuzov to the command of an army sent to Moldavia. On 4 July he inflicted a heavy defeat on the Turkish forces at Rushchuk, and another later on the banks of the Danube, resulting in advantageous terms for Russia at the Treaty of Bucharest, concluded on 29 May 1812. For these achievements Kutuzov was created a count.

In June 1812 Napoleon's army embarked on its epic advance into Russia, while the Russian armies fell back before it. Reluctantly – but in the face of public clamour – Alexander appointed Kutuzov Commander-in-Chief of all Russian forces, creating him a prince the following day. Characteristically Napoleon sought to attract the Russian forces to a general engagement, but Kutuzov's strategy was to exhaust the invaders by incessant minor engagements as he fell back towards Moscow. The first major stand against the invaders occurred at Borodino on 7 September; 120,000 Russians attempted to stand firm against a similar number of French, and the threat of an envelopment of the Russian left flank led to a day-long battle of great severity during which Napoleon is reported to have suffered a stroke. Kutuzov's army was gradually forced back with almost 40,000 casualties, and the Russian commander abandoned his intention to stand firm before Moscow; but the French did not press forward energetically owing to the temporary loss of Napoleon's leadership.

The onset of the Russian winter and the constant attrition of Russian guerrilla tactics compelled Napoleon to withdraw after occupying Moscow for little more than a month. At first he attempted to move southwards via Kaluga, but at Maloyaroslavets he was repulsed by Kutuzov's 110,000-strong army and forced to turn towards Smolensk. Kutuzov, now a field-marshal, continued to harry the retreating French, engaging them at Vyazma and Krasnoi. In the latter engagement (on 16–17 November), finding Kutuzov's advance guard barring the westbound road, Napoleon brilliantly collected his surviving effective elements and drove off the Russians. In a gallant rearguard action Marshal Ney's (q.v.) corps sacrificed itself almost to annihilation to permit the main army to escape.

Ten days later at the Berezina River 144,000 Russians massed on both banks to prevent the French from crossing, and once again a single French corps (under General Victor) held a slender bridgehead in the face of huge odds to enable the main force to extricate itself. In January 1813 Kutuzov led his army into Poland and Prussia; but on 16 April he died at Boleslawice in Silesia.

Although widely credited, as the Commander-in-Chief of all Russian forces, with the Russian victory over Napoleon in 1812, Kutuzov never won a wholly decisive battle against that legendary conqueror; indeed, one is left to speculate whether he possessed any real qualities as a supreme commander. It is true that he managed his commands skilfully and conserved his armies intact in circumstances of great tribulation before

Moscow; but in view of the extreme exhaustion of Napoleon's army, one must doubt whether Kutuzov displayed any outstanding generalship when his enemy lay at his mercy. But he occupied high office in the Russian army for twenty years, and will be remembered, if for no other achievements, for his handling of the Turkish campaign.

Lafayette, Marie Joseph Paul Yves Roch Gilbert du Motier, Marquis de (*1757–1834*), *Lieutenant-General*

Lafayette was born into an ancient and aristocratic French family, and before he was twenty enjoyed great inherited wealth and privilege. It is a measure of his remarkable character that he is remembered as one of the great champions of social and political liberty.

He was born on 6 September 1757 at Chavaniac, Auvergne. He lost his parents at an early age, and in about 1773 he joined the wealthy and titled young men of his class at the court of Versailles. He married Adrienne de Noailles, daughter of the Duc d'Ayen, and purchased a captaincy in the Noailles Dragoons. Unlike many of his contemporaries, however, he took soldiering more seriously than the gilded circus of the court, and had great military ambitions. In December 1776, despite opposition from the highest circles, he obtained a commission as major-general in the Continental Army from Silas Deane, the American agent in Paris; and the following summer he sailed for America. He arrived at Philadelphia in July 1777, and soon struck up a relationship of great mutual respect with George Washington (q.v.). Although Congress confirmed his rank he served initially as an unpaid volunteer, without a field command. He distinguished himself on 11 September 1777 at the Battle of the Brandywine, where he was wounded. On his recovery he was given command of a division; he made a masterly withdrawal from Barren Hill on 28 May 1778, and fought at Monmouth Court House a month later.

Lafayette had acquired a considerable reputation both in America and in France by this time, and was responsible for the increased pro-American sentiment at home which led to the Franco-American treaty of February 1778. When Admiral d'Estaing arrived at Rhode Island with a French fleet in July 1778 Lafayette took on the duties of liaison officer, which he discharged with great zeal for American interests. He returned to France in 1779 with messages from Congress, and was promoted colonel in the French service. He rejoined Washington in 1780, at a time when Revolutionary fortunes were low, bringing the heartening news of Rochambeau's (q.v.) imminent expedition to aid the American patriots. He then again performed the role of liaison officer, presenting American pleas for men, material, and particularly naval support, with vigour. His own command was the *élite* Light Division, one of the most effective units of the Continental Army. Between April and September 1781 he commanded the Continental forces in Virginia; these only amounted to some 4,500 men, faced by twice that strength of British troops under Cornwallis. Lafayette manœuvred with great skill, leading Cornwallis on an exhausting chase and avoiding a confrontation. When Cornwallis fell back on Yorktown Lafayette followed, maintaining close contact and reporting to Washington. He was influential in bringing to bear the decisive French naval superiority which blockaded Yorktown and led to Cornwallis's surrender in October. In 1782 Lafayette returned to France, hailed as 'the hero of two worlds'. He was promoted major-general in the French service.

Lafayette's active part in the tortuous progress of the French Revolution has little place in this book, as his significance in that context was political rather than strictly military. Fired by his contact with the American Revolution he became a leading advocate of social and political reform, while remaining a moderate in comparison with many of his countrymen. He was a supporter of the concept of a constitutional monarchy, and attempted to save Louis XVI from the rapacity of the extremists and the consequences of royal folly. In the period 1784–92 Lafayette exerted great influence, which he used wisely and moderately; he was for a time the commander of the Paris National Guard and of regular troops stationed in the Paris area. He was, inevitably, attacked by monarchists as a radical and by radicals as a reactionary. In

October 1791 political events led him to the false assumption that the Revolution was over, and he resigned his command and retired to the country. Only months later Louis XVI summoned him to take command of the Army of the Centre in view of the apparent imminence of war with Austria; promoted lieutenant-general, Lafayette took up his command at Metz in December 1791. He found his troops shaken and demoralised by the recent political upheavals, and worked hard to bring them to combat readiness; apart from restoring their confidence, he instituted such tangible reforms as the provision of a horse artillery branch. He had hoped to bring his army into the field on 1 May 1792, but the unexpected declaration of war on Austria by the Assembly on 20 April caught him unprepared. He moved his troops from Metz to Givet by forced marches; but the defeat of elements of Rochambeau's Army of the North on 28 April near Lille greatly discouraged the armies of France, which went on to the defensive. Rochambeau resigned his command in disgust at the conduct of his men, and Lafayette succeeded him briefly. In August the Paris mob stormed the Tuileries and seized the King; Lafayette exerted what remained of his influence in an attempt to defend Louis, but was abandoned by his troops, relieved of his command, and proscribed by the Assembly. He fled across the lines and surrendered to the Austrians, hoping to find temporary asylum in America. In fact he was held in Austrian and Prussian captivity until 1797, when he was handed over to the U.S. consul in Hamburg.

Lafayette returned to France under the Consulate in 1799, but he strongly disapproved of Napoleon's (q.v.) ambitions and methods, and retired to private life. He took up an active political career once more in 1815, as he approved of Bonaparte's liberal constitution of that year. He was an influential figure in French (and international) politics from 1815 until his death in 1834. His credentials as a lifelong supporter of civil and republican liberties were spotless, and reformers from many countries sought his advice and sponsorship. He spoke out fearlessly for tolerance, democracy, civil rights, the abolition of slavery, and the abolition of the peerage: he had ceased to use his own title in 1791. He was received with tumultuous acclaim when he made a tour of America in 1824–5, being hailed as a living link with the founding fathers of the Republic. Since his death in Paris on 20 May 1834 his name has been honoured as a symbol both of Franco-American friendship and of the wider brotherhood of all who fight for freedom – notably in the title of the Lafayette Flying Corps of American volunteer airmen in France during the First World War.

Lake, Gerard, 1st Viscount Lake (*1744–1808*), *General*

Lake was Wellington's (q.v.) Commander-in-Chief in India at the beginning of the nineteenth century, and personally achieved a dazzling series of victories over much superior Mahratta forces in 1803–4. It must also be recorded that he crushed the Irish rising of 1798 with extreme ruthlessness.

He was born at Harrow on 27 July 1744, and entered the 1st Foot Guards (later Grenadier Guards) as an ensign in 1758. He served in Germany in 1760–2, and in January 1762 was promoted lieutenant, and captain in the Army, according to the Guards' dual rank system. He rose to captain and lieutenant-colonel in February 1776, and sailed for America in 1781. He fought at Yorktown that autumn, and distinguished himself on 11 October when he led the Guards' grenadier companies in a damaging sortie against the Franco-American lines. He spent the rest of the war as a prisoner on parole. He became a major in 1784, and was equerry to the Prince of Wales. In 1790 he was promoted major-general, and entered Parliament as member for Aylesbury, a seat he held until 1802. In 1792 he was appointed lieutenant-colonel of the 1st Foot Guards; generals drew separate pay for regimental appointments, and Lake was subsequently colonel of the 53rd and later the 73rd Foot. On the outbreak of war with France in February 1793 Lake led a brigade composed of the 1st Battalions of the three Foot Guard regiments to the Netherlands, joining the Allied army at Turnai in April; these were the first British troops actually to fight the French in the Revolutionary and Napoleonic Wars. During the siege of Valenciennes Lake and his guardsmen distinguished themselves on 18 August. Twelve French battalions drove the Prince of Orange's garrisons from a group of forts at Lincelles; the Dutch-Belgians could not be rallied for a counter-attack, so Lake and his three battalions recaptured the forts alone at the point of the bayonet, taking back six captured guns and six more French pieces which had been moved up.

In October 1793 Lake was invalided home; in December 1796 he was given the command in Ulster, and the following month was appointed lieutenant-general. When Abercromby (q.v.) resigned in disgust over the excesses of the militia in April 1798 Lake replaced him; he seems to have done nothing to curb the indiscipline of the troops. When the rising took place he dispersed the rebels at Vinegar Hill on 21 June, and authorised the most ruthless reprisals against any show of resistance in the weeks which followed. Lord Cornwallis took over the chief command shortly afterwards, but it was Lake who successfully encircled the French force of 1,200 men under General Humbert which landed at Killala Bay in late August, and obliged them to surrender on 8 September.

On 30 October 1800 Lake was appointed British Commander-in-Chief in India; he arrived in Calcutta in July 1801, and took immediate steps to improve the efficiency of the Honourable East India Company's troops. Areas in which his influence was quickly felt included the strengthening of the Bengal Native Light Cavalry, of which he made effective use in his later campaigns; the training of a strong skirmishing line within the foot regiments; and the provision of advanced support for the infantry and cavalry in the form of attached horse artillery elements. In 1802 Lake was promoted general, and the long-threatened war with the French-trained armies of the Mahrattas was sparked off by a civil war which cost the British-allied Peshwa his throne. While Wellesley moved against one Mahratta army, led by Sindhia of Gwalior, in the Deccan, Lake took an army of 10,500 against a second Mahratta army in the north. Lake's forces were largely composed of H.E.I.C. sepoys, with a small British contingent. The Mahratta army numbered some 43,000 men with more than 400 guns; it was commanded by the French General Pierre Perron and officered by other French adventurers. On 4 September Lake stormed and captured the walled city of Aligarh at a cost of 260 casualties. He then advanced on Delhi, and Perron rode into the British camp and surrendered. The Mahrattas, now led by Louis Bourquien, stood outside Dehli, where Lake defeated them on 16 September; their losses are unrecorded but were certainly heavy. British casualties numbered only 477, and Lake captured sixty-three guns and much treasure in this action. He occupied Delhi and moved south to take Agra, pressing hard on the retreating Mahrattas with his cavalry. On 1 November 1803 he won the decisive victory of Laswari, which, together with Wellesley's successes in the Deccan, led to the submission of Sindhia on 20 December. At Laswari Lake's cavalry overran the Mahratta forward line, which consisted of cannon chained side by side in the old Turkish manner. The British infantry, having advanced sixty-five miles in two days, then destroyed the Mahratta infantry. British casualties were just over 800.

In 1804 fighting broke out again when Jaswant Rao Holkar, another Mahratta leader, wiped out a small force under Colonel William Monson and moved on to lay siege to Delhi. Lake advanced to the relief of the city on 1 October, and Holkar fell back. On 17 November, after a 350-mile forced march, Lake defeated Holkar's cavalry at Farrukhabad and forced the Mahratta leader to flee into the Punjab. He then laid siege to Bhurtpore, whose rajah had allied himself with Holkar. He invested the city on 1 January 1805, and between then and mid-April made four assaults, all of which were repulsed with total casualties of 3,200 men. The rajah agreed terms on 17 April. In July Lake was superseded as Commander-in-Chief by Lord Cornwallis, but the latter died in October and Lake took command once again. Pursuing Holkar into the Punjab, he forced his surrender at Amritsar in December 1805.

Lake returned to England to receive the thanks of Parliament and a peerage. He died in London on 20 February 1808, and was apparently mourned by a wide circle of friends. He had the reputation for great calm and self-reliance, and the energy with which he pursued his missions speaks for itself. He was completely fearless in action, and seems to have been held in trust and affection by the rank and file.

Lattre de Tassigny, Jean de (*1889–1952*), *Marshal of France*

Posthumous marshal and senior French army commander during the latter part of the Second World War, Lattre de Tassigny later brought the Viet Minh offensive to a temporary halt in Indo-China in 1951.

Born at Mouilleron-en-Pareds (Vendée) on 2 February 1889, Jean de Lattre entered the military academy at St Cyr, graduating in 1911. After fighting as a junior officer during the First World War he served in French Morocco from 1921 to 1926, participating in the offensive under Marshal Pétain (q.v.) which led to the eventual surrender of Abd-el-Krim (q.v.). By 1938 he had risen to senior colonel in the metropolitan French army, and was appointed chief of staff in the Fifth Army, a post he held until the German offensive opened in May 1940. Promoted major-general, he was given command of the 14th Infantry Division, which he led into battle until the French armistice, after which he moved into the free zone of Vichy France.

On 11 November 1942 the Germans carried out their occupation of the free zone; Lattre resisted, but was taken prisoner and sentenced to ten years' captivity for his resistance. Within a year he contrived to escape, and reached North Africa in October 1943, where he assumed command of the French First Army. While the greater part of this force continued building up and training for forthcoming operations, some elements were used in the seizure of Corsica that month.

Lattre's first major operational command was his participation in the invasion of southern France (Operation 'Anvil') in August 1944, when large components of his command accompanied (as the French II Corps) the U.S. Seventh Army in its landing on the Côte d'Azur. While General Truscott fought his way up the Rhône Valley, Lattre's forces turned westwards to attack German forces in the Toulon and Marseille area. In a week's bitter fighting they captured the two ports, together with nearly 50,000 prisoners, for the loss of 1,500 dead and 5,000 wounded and missing. On 15 September the French First Army became part of the Allied Sixth Army Group under Lieutenant-General Jacob M. Devers.

There now arose a bitter personal feud between the two senior French commanders in Europe, Lattre and General Jean Leclerc (q.v.). Lattre, commanding the French First Army, believed for several reasons – mainly sentimental – that all French forces should be brought under his overall command for the ultimate assault on Germany. Leclerc, commanding the famous French Second Armoured Division which had been built up under his leadership in England, maintained independence from the greater bulk of French forces by remaining in the Seventh Army. Leclerc's liberation of Strasbourg on 23 November raised French morale to a new peak and appeared as something like a deliberate snub to Lattre – a snub apparently further aggravated when, fearful of the possible effect on French morale of a German counter-offensive in the area, de Gaulle insisted that Lattre's forces should be used to defend the city instead of accompanying the advance into Germany immediately. Nevertheless it was Lattre's French forces which, with some American assistance, were used to 'pinch out' the German Nineteenth Army in the Colmar pocket between 20 January and 9 February 1945. Thereafter, as Allied pressure mounted along the Rhine, Lattre's First Army moved forward, taking part in the river crossing at Gersheim on 31 March, forcing its way through the Black Forest during April, and reaching the Swiss border early in May. It was Lattre who, in Berlin on 8 May, received and signed on behalf of France Germany's final capitulation.

After the war Lattre de Tassigny, then a lieutenant-general, was appointed army inspector-general and, before the eventual ratification of N.A.T.O., Commander-in-Chief of the Western European Union ground forces.

French colonial troubles in the Far East, however, continued unabated after the defeat of Japan. While France had given a degree of independence to the state of Vietnam in

March 1946, this was only as a free state within the Indo-Chinese Federation and French Union. Continuing French limitations were unacceptable to the Communist-inspired factions in the area, and prolonged guerrilla war broke out between the Viet Minh and the French forces. When in 1950 the Viet Minh gained recognition from the Soviet Union and Communist China, the scale of guerrilla warfare was stepped up by the injection of regularly-trained troops and of equipment from southern China. Following a serious defeat at Fort Cao-Bang on 9 October 1950, the French were forced to abandon most of northern Vietnam to the Communist guerrillas and the situation in the area became desperate. It was at this moment, in December, that France's senior soldier, Lattre de Tassigny, was sent out to improve matters. He quickly restored French morale and, organising his available forces into effective combat columns, regained the initiative and reoccupied much of the territory lost to the guerrillas during the previous year.

A new treaty was then signed re-establishing Vietnam's status within the French Union – a singularly shortsighted instrument considering the danger posed by the internal factions which commanded such strong support from Communist neighbours. The inevitable result was renewed guerrilla activity, which now extended to outright terrorist attacks in the main cities. This in turn severely undermined the newly-strengthened morale of the French. Moreover, Lattre's military resources were utterly inadequate to cope with this type of guerrilla warfare; his forces, grouped as they were to meet conventional units in battle, were concentrated in garrisons dotted about the country. During his efforts to build and organise an effective Vietnamese regular army he became seriously ill and took the opportunity to return home to appeal for reinforcements. His illness forced his retirement, and he was succeeded by General Raoul Salan (q.v.). Lattre died in Paris on 11 January 1952.

Posthumously appointed a marshal of France, Lattre de Tassigny was widely respected in post-war France as a great army commander, and one of the true fighting patriots of the Second World War whose presence in France during much of the German occupation probably earned him more popularity than that enjoyed by Jean Leclerc who, for all his flamboyant leadership of the French Second Armoured Division, had become remote from the metropolitan public.

La Valette, Jean Parisot de (*1494–1568*), *Grand Master of the Order of the Hospital of St John of Jerusalem*

The most famous Grand Master in the history of the Order, la Valette conducted the successful defence of Malta by the Knights of St John against the Turkish invasion attempt of 1565. The harbour-city of Valetta is named after him. The importance of the siege of Malta was immense, and far greater than that of the earlier siege of Rhodes (see under Villiers de l'Isle Adam). It came at the high point of Turkish expansion westwards, and had the island fallen, nothing would have prevented Suleiman the Magnificent (q.v.) from launching an invasion of Sicily and Italy simultaneously with a major offensive on the Austrian border. With the powers of Christian Europe in disarray, it is hard to see how such a pincer movement could have failed to carry the banner of the crescent into the heart of Europe.

Jean Parisot de la Valette was born near Toulouse in 1494, and joined the Order at the age of twenty. From that day on this proud, handsome Gascon knight lived only for the Order; it is said that even when the Knights were in temporary residence at Nice in the late 1520s he did not once take leave of absence to visit the family estates. He fought in the Siege of Rhodes in 1522, and was among the survivors who accompanied Villiers de l'Isle Adam on the humiliating withdrawal of 1 January 1523. He rose steadily in the ranks of the Order during the thirty years of consolidation on Malta, becoming at one stage the Admiral of the Order's war fleet. He spent a year on the rowing benches of a Turkish galley after being captured in a sea fight, and survived that horrific experience with mind and body intact. He was a brilliant linguist, speaking Italian, Spanish, German, Turkish and Arabic; a totally dedicated knight of the old Crusader school; a merciless disciplinarian; and an energetic and resourceful commander.

He was elected Grand Master in 1557 at the age of sixty-three, and threw himself into preparing for the invasion which his spies – and his judgement – told him must come sooner or later. When the great test was laid upon him he was seventy years old, yet still capable of fighting in full armour throughout a blazing summer's day. History offers few examples of the hour producing so unquestionably the right man.

In the years between the Order's arrival on Malta and the Turkish attack of 1565, a great deal had been accomplished. The Grand Master, Juan de Homedes, with the technical assistance of the Italian engineer Antonio Ferramolino, had made great strides in the fortification of the harbour. This area may be described as a large open 'mouth' facing north-east. It is cut in two by the 'tongue' of Mount Sciberras protruding north-eastwards; this high ground dominates the area. The several inlets and peninsulas extending north-westwards into the southern stretch of the harbour like 'teeth' from the lower jaw were dominated by the fort of St Angelo, the main citadel of the Knights, on the tip of one 'tooth'. On the end of Senglea, the next 'tooth' to the west, was Fort St Michael. Facing them on the tip of Mount Sciberras was the new Fort St Elmo. Northing could be done about the way the ridge of Sciberras overlooked Senglea and St Angelo, but at least the three bastions could be rendered as strong as was physically possible. Massive new walls and outworks were constructed, and the peninsula of St Angelo was cut by a huge ditch separating the fort from the fortified township of Birgu at the base of the peninsula; an attacker by land would thus have to fight his way through Birgu and over the canal – which was large enough to harbour galleys – before even reaching the main positions. In their work of fortification the Knights were greatly aided by the skill of the Maltese masons. The tough, willing islanders also proved to be excellent material for manning the fleet; and their age-old hatred of the Moslem corsairs who had ravaged their islands for centuries made them far more enthusiastic collaborators than the hereditary aristocrats of the island, who shut themselves up in the old capital of Mdina and took no part in events. As the years passed the galleys and sailing ships of the Order began to take a fearful toll of Moslem shipping in the Middle Sea; this brought in useful revenue, as well as enraging the traditional enemy. The acquisition of the Great Carrack, a massive sailing ship with enough stowage space to stay at sea for six months and a battery of more than fifty heavy guns, helped the Knights to regain their position as the most efficient navy in the Mediterranean.

La Valette knew that the Order had to hold Malta or perish; once expelled from their refuge, they could not hope to be given another home. He also knew that he could expect little help, if any, from Europe. Although the Turkish threat was taken seriously, the resources of the Spanish in Sicily would be wholly committed to preparing a defence of that more important possession if Malta fell. Malta was the outwork, the isolated garrison which must hold out as long as possible to buy time. Although his available manpower numbered only about 540 members of the Order, 1,000 Spanish infantry, and 3,000–4,000 local irregulars, the Grand Master was aided by the very barrenness of the island. Food and water were scarce, the soil was thin, and there was virtually no timber. An invader would have difficulty in keeping his army from starving and perishing of thirst or disease, let alone in constructing siege-works. The Turks, despite marginal resources from the North African lands, would be to all serious intents isolated at the end of a thousand-mile line of communications. When the invasion was clearly imminent the Grand Master ordered the gathering into the citadels of all available foodstuffs and stock, and all willing local inhabitants. The few wells were poisoned, and the cisterns of the Order's forts were filled; stores of food, arms, and powder were assembled. A great boom blocked the inlet between Senglea and St Angelo.

The Turkish armada which was sighted on 18 May 1565 was one of the greatest the world had yet seen. Some 200 large ships and scores of smaller vessels conveyed an army of between 30,000 and 40,000 janissaries, sipahis, iayalars, and lesser troops, and a huge and efficient artillery. Command was – fatally, as it transpired – split between Mustapha Pasha, leading the army, and the Sultan's son-in-law

Piali Pasha, leading the navy. In their early quarrels over the initial objectives lay the seeds of Turkish failure. Mustapha wished to redeem earlier failures against the Knights by achieving a personal triumph. Piali, ignorant of local conditions, was needlessly apprehensive of storms wrecking the fleet with which he had been entrusted, and insisted on the early seizure of the northern arm of Grand Harbour – the arm above the 'tongue' of Sciberras, known as Marsamuscetto. This obviously entailed the silencing of Fort St Elmo on the tip of Sciberras; and the land forces were, in fact, committed to this attack. Inexplicably, Mustapha failed to occupy effectively the northern part of the island, thus allowing la Valette's messengers to maintain a perilous link with Sicily, and leaving the Order's cavalry, based in weakly-defended Mdina, free to harry Turkish foragers. Equally, he failed to grasp the futility of wasting men on St Elmo when the Knights' final defeat could only be achieved by the destruction of St Angelo.

Fort St Elmo, garrisoned by some 1,450 soldiers under the command of about fifty Knights, was brought under bombardment on 24 May. Against all reason it held out for thirty-one days, and cost the Turks about 8,000 men. They had great difficulty in establishing their batteries and trenches on the rocky spine of Sciberras, and the heat and bad water caused a steady drain of casualties from disease. The fort had to be battered down stone by stone; la Valette grimly consigned the garrison to certain death to buy valuable time for strengthening the main positions. Half way through the attack the great Turkish admiral and general Torghoud Rais ('Dragut' to the Europeans) arrived; although not formally given chief command, he was armed with the Sultan's order that his advice be heeded in all matters. The wiliest, most experienced lieutenant of the Grand Turk then living, Dragut was icy in his appreciation of the achievements of the younger commanders. His arrival electrified the Turkish effort – but he died of a stone splinter wound on the day St Elmo finally fell, 23 June. Even on that day it took the Turks an hour to overcome the last hundred or so defenders, nearly all wounded, and all weak with exhaustion and hunger. Two of the wounded Knights had themselves carried to the wall and placed in the breach in chairs, so that they could swing their swords till the last. Mustapha is reported to have wondered aloud what the father (St Angelo) would cost him, in view of the price he had been forced to pay for the son (St Elmo). He resolved that no quarter should be given, and the siege was marked by great cruelty – on both sides – from this point on.

Under the blazing sun of late June the Turks dragged their artillery off Sciberras and re-sited it to bring Senglea and Birgu under fire. Incredibly, some 1,000 reinforcements from Sicily managed to slip ashore and cross into Birgu on the night of 29 June. Uncertain of the strength of this force, whose arrival was triumphantly trumpeted from the walls, Mustapha offered la Valette the same terms Villiers de l'Isle Adam had accepted at Rhodes, but the Grand Master contemptuously refused. The Turks opened their bombardment of Senglea and Birgu in the first week of July, intending to attack the former as soon as it had been weakened, using a sea-borne force as well as land troops; some small galleys had been dragged overland and launched in the harbour, whose mouth was still covered from St Angelo. Maltese swimmers thwarted this plan by erecting – and later defending – underwater defences in the form of stake palisades. A massive attack was indeed launched on Senglea on 15 July, but it was repulsed with heavy loss. After intensive artillery preparation (which, it was said, could be heard in Catania, Sicily) another major assault was launched on the landward defences of Senglea and Birgu on 2 August, only to be thrown back with great casualties. Another attack on 7 August penetrated the outer lines of Birgu, but met disaster on the carefully-prepared secondary defences. Simultaneously, the most dangerous moment of the siege found la Valette's reserves fully stretched. The Turks managed to establish a lodgement into Senglea's defences, and despite the bridge of boats which linked the two peninsulas the Grand Master, under great pressure in Birgu, was unable to send reinforcements. With victory in their grasp, the Turks were suddenly recalled from Senglea and the lines before Birgu. A courageous sally by cavalry from Mdina had cut up the Turkish base camp, spreading reports that a relief force from Sicily had landed; Mustapha gave up his precious foothold in Senglea before learning the truth.

As August wore on, and the Turkish guns and sappers nibbled at the defences, some officers of the garrison argued for a withdrawal of all effectives within St Angelo. La Valette opposed this hotly. He was confident that the Turks would be unable to winter on the barren island so far from supply ports; and to concentrate the defenders would rob them of their advantage, allowing the enemy to concentrate in his turn. Every bastion was to be held until the last man fell; and to underline this philosophy, the Grand Master had the bridges linking Senglea, Birgu, and St Angelo destroyed. Although he placed no confidence in Spanish promises of a relief expedition by the end of the month, he remained adamantly committed to defence *à l'outrance*. On 18 August the Turks managed to bring down a vital bastion of Birgu – the Post of Castile – with a mine, and the Grand Master personally led the countercharge into the breach, remaining there, although wounded, until the line was secure again. Throughout the second half of August the suffering of both garrison and attackers was intense, as every weapon of siege warfare was brought into play along the walls. The numbers of dead and wounded mounted, the supplies dwindled, and – particularly in the Turkish camps – the dysentery which no army of that period could escape in summer laid men low in their hundreds and thousands.

The morale of the Turks had collapsed before they were actually forced to abandon the siege. On 6 September a fleet from Sicily was sighted; the relief amounted to no more than some 8,000 men, but the Turks were ready enough to heed the exaggerated reports which reached them. Mustapha attempted to recall his embarking regiments when he learnt the true figures, but failed. Less than a third of the original armada returned to Constantinople, slipping into harbour after dark to avoid undesirable public speculation. It was an unprecedented humiliation in the career of the Sultan. Some 30,000 men had been lost in the campaign, without counting the casualties suffered by auxiliaries such as the Barbary and Egyptian corsairs. Of the Knights, 250 had died and most of the remainder were disabled; only some 600 of the other troops were still fit to bear arms.

In the immediate aftermath of the siege, the Order found itself the darling of Europe – and not only of Roman Catholic Europe, but of the Protestant states as well. Taking swift advantage of a mood which was quite likely to evaporate before long, la Valette secured the backing of the Pope for an appeal for funds and technicians to restore the defences of the island. The Order moved its headquarters across to the commanding ridge of Mount Sciberras, where the impregnable new citadel of Valetta was soon to rise. The first stone was laid on 28 March 1566. La Valette, determined to maintain his independence from the Vatican, wisely declined the offer of a cardinal's hat. The ballad-mongers of Europe were still retailing his deeds when he suffered a fatal stroke in July 1568. He lies buried in the city which was named after him, and which could never have been built had it not been for his iron character and military expertise.

Lawrence, Thomas Edward (*1888–1935*), *Lieutenant-Colonel*

Welsh-born scholar-turned-soldier, military adviser to Arab guerrilla forces during the First World War whose undoubted influence on local affairs in Palestine was later greatly exaggerated; Lawrence's worldwide fame rested largely upon his subsequent renunciation of rank and reward.

T. E. Lawrence was born on 15 August 1888 at Tremadoc, Caernarvonshire, the second of five illegitimate sons of Sir Thomas Chapman, Bart. He was educated at Oxford, and in the five years prior to the First World War visited the Middle East to prepare material for his university thesis on the architecture of Crusader castles. However, an expedition which he accompanied to Sinai in 1914, ostensibly to explore the area, was in reality designed to gain information for the War Office about Turkish military dispositions on the Turco-Egyptian border.

On the outbreak of war he was sent to Egypt, where he was attached to the military intelligence staff concerned with Arab affairs. In

October 1916 he accompanied a mission to the Hejaz, where the Arabs had proclaimed a revolt against the Turks. The following month Lawrence, with the rank of captain, was ordered to join the amir Faisal's army as political and liaison officer. At once Lawrence's clear thinking and energy imbued the Arab revolt with new purpose, and he was instrumental in acquiring much material assistance from the British Army for the Arab cause. He also perceived that the key to the Turks' situation lay in the Damascus–Medina railway, by which the Turks could send reinforcements to crush the Arab revolt. Faisal's army therefore moved round to the north of Medina and started a series of guerrilla attacks on the railway itself. Lawrence set out with a force of Huwaitat tribesmen and captured the port of Aqaba on 6 July 1917; this became the new base for Faisal's army – a base at which it was well placed to co-operate with the British army advancing from Egypt in southern Palestine.

After the arrival in this theatre of war of General Allenby (q.v.) to take command of the British forces, Lawrence suggested to him that assistance could be rendered by the Arabs in cutting the railway from Dar'a; but in this task the Arabs failed, and in November Lawrence, now a major, was captured by the Turks and underwent a short period of humiliating torture (according to his own account). He soon escaped, and in January 1918 was present at the Battle of Tafila, for which Faisal had to move his forces up from Aqaba. For his own conduct in this battle Lawrence was awarded the D.S.O. and promoted to the rank of lieutenant-colonel. Subsequently he was present at the capture of Damascus on 1 October 1918, returning to England the following month.

During the last two years of the war Lawrence made a valuable contribution in the Middle East, in that his advice and influence effectively bound the Arab nations to the Allied cause, thereby tying down about 25,000 Turkish troops who would otherwise have opposed the British army. But Lawrence, for all his flamboyant poses and his adoption of Arab costume, was never a leader of Arab forces – command always remained firmly in the hands of the amir Faisal. Interest in his exploits was to a large extent self-generated through his own literary accounts and lecture tours, which were assisted by his post-war election to a research fellowship at Oxford University. In 1921–2 he was an adviser on Arab affairs to the Colonial Office, but left government service in protest at the conduct of the peace settlement. In August 1922 this enigmatic man enlisted (as John Hume Ross) in the ranks of the Royal Air Force, but was discharged six months later when his identity was disclosed by a London newspaper. He then enlisted as T. E. Shaw (a name he legally adopted) in the Royal Tank Corps, and transferred to the R.A.F. in 1925, remaining with that service until on 13 March 1935 he was seriously injured in a motorcycle accident; he died six days later.

Leclerc de Hautecloque, Philippe François Marie (*1902–47*), *Général d'Armée*

Leclerc was the dashing, modest, and much-loved young French commander of the Second World War whose famous march across the desert to join the British Eighth Army provided much-needed inspiration for the Free French after the apparent eclipse of General de Gaulle (q.v.) – temporary though that proved to be.

Born of an old Picardy family in 1902, Leclerc's true name was Philippe, Vicomte de Hautecloque. He graduated from St Cyr in 1924, and was promoted captain ten years later – the rank he held at the outbreak of the Second World War in 1939. During the Battle of France he was promoted major, wounded, and taken prisoner; but he escaped to England where he offered his services to de Gaulle, using the name Leclerc to avoid the likelihood of German reprisals against his wife and family in France. De Gaulle sent him to the Cameroons, where he played a prominent part in establishing a new régime under the Free French movement. His first outstanding military exploit was to capture the Kufra Oasis during General Wavell's 1940 offensive, but he had to retire when the British retreated soon afterwards.

Thereafter Leclerc was appointed general officer commanding in French Equatorial Africa, and in 1942 he set about organising a military column which he proposed to lead to join up with the British forces advancing after the Battle of El Alamein. This involved an extraordinary task of administration: providing huge reserves of labour to assemble stores, transporting heavy vehicles a thousand miles up the Congo, and training the soldiers to withstand the hardships of a long and hazardous journey across a hostile desert, defeating and destroying enemy outposts along the route. The desert march, which started out from Fort Lamy in the Chad, extended through 1,500 miles of almost trackless waste. Leclerc eventually joined up with the Eighth Army, which was held up before the Mareth Line. This achievement was of great morale value for the Free French forces following disagreements between General de Gaulle and other Allied commanders. Leclerc was promptly appointed a divisional commander in Tunisia.

After the end of the North African campaign Leclerc returned to England to raise and train the famous French Second Armoured Division, and to many it is as a brilliant divisional commander that he is best known. This division took part in the invasion of Normandy in June 1944, and led the spectacular dash to liberate Paris in August. Troops of Leclerc's division were also the first to enter Strasbourg on 23 November, and for a period during the following year Leclerc was military governor of the city.

In June 1945 (when he adopted the name Leclerc de Hautecloque) he was sent out to command the Far Eastern expeditionary force in Indo-China; but his tough measures (since seen to have been justified) against the Vietnamese insurgents excited much criticism among certain elements at home. He was recalled in 1946, promoted *général d'armée*, and appointed inspector-general of French troops in North Africa. He died in an aircraft accident on 28 November 1947.

Leclerc's leadership and achievements contributed much to the restoration of public faith in the French army, especially after the humiliating catastrophes of 1940. His extraordinary qualities of *élan*, simplicity, modesty, and prudence endeared him to Allied officers and men; but for his early death, he might well have brought to bear all the command qualities conspicuously absent during the French débâcles in Indo-China and Algeria during the 1950s which so racked the French political and military establishment.

Lee, Robert Edward (*1807–70*), *General in Chief of the Armies of the Confederacy*

Robert E. Lee, the embodiment of all that was fine in the culture of old Virginia, occupies a special place in the hearts of Americans. When he took command of the Army of Northern Virginia on 1 June 1862 he had never commanded in battle; until a year before he had never commanded any body of troops larger than four squadrons of horse. For three years he led the most important Confederate army in the field, and achieved results out of all proportion to the resources at his command. He was not always victorious, and in at least one important aspect of command he was deficient; yet his record includes several brilliant triumphs of generalship, and he had the gift for inspiring great loyalty and affection, and consequently super-human efforts, on the part of his subordinates. He went into retirement with the lasting respect of friend and enemy alike, which is not usually the fate of defeated revolutionary generals in civil wars.

He was born in 19 January 1807 at Stratford (his family's seat in Westmoreland County, Virginia), the third son of 'Light Horse Harry' Lee, a cavalry hero of the War of American Independence whose latter years were spent in reduced circumstances due to a complete lack of business sense. The family moved to Alexandria, Virginia, and there Robert was brought up. He entered West Point in 1825 and was apparently a model cadet, graduating in second place in 1829. Commissioned a brevet second lieutenant of engineers, Robert E. Lee led an exemplary but in no way extraordinary professional and social life over the next thirty years. He was promoted first lieutenant in 1836 and captain two years later. During the 1830s he was assigned to various military engineering projects such as forts and harbours, and served for a term in the chief engineer's office at Washington. In August 1846 he was posted to San Antonio, Texas, as assistant engineer to the army of General John E. Wool, with whom he served in the early part of the Mexican War. He distinguished himself before Buena Vista by making a notably courageous reconnaissance of the enemy positions. Transferred to the Vera Cruz expedition, he made an immediately favourable impression on Winfield Scott (q.v.), both by his professional diligence and skill, and by his manner. It was predictable that Scott, one of the old school, should approve highly of the young engineer officer who had made George Washington (q.v.) his ideal and who strove to emulate his hero. Lee made a name for himself during the Mexico City campaign; he located the heavy land batteries at Vera Cruz, provided the reconnaissance reports upon which the victory at Cerro Gordo was founded, and sited the batteries before Chapultepec, at which battle he was slightly wounded.

Returning to the United States in 1848, Lee was appointed brevet colonel for his gallantry. He spent four years at Fort Caroll in Baltimore, and in 1852 accepted – with some reluctance – the post of Superintendent of West Point. His term of office saw notable improvements in the curriculum and in instructional methods. In March 1855 Jefferson Davis, then Secretary of War, gave him the lieutenant-colonelcy of the 2nd Cavalry in Texas. He was much absent from his regiment during 1857–9, however, due to family probate difficulties and his wife's distressing illness. During his final period of active service with his regiment, between February 1860 and February 1861, he was given command of the Department of Texas. He had no sympathy with the secessionist sentiments then current in the southern states; but he admitted that if it came to a choice between Union and State, then his first loyalty must go to Virginia. He continued to await events in some anguish, hoping that he would never be called upon to make the choice, until in February 1861 he was called to Washington by Winfield Scott. In March he was made colonel of the 1st Cavalry, and it was obvious that he was being considered for senior command in the event of civil war. In April he was formally offered the field command of the United States army. He refused, explaining frankly to Scott that he could not contemplate taking up arms against the southern states. Scott advised him that resignation was the only honourable way out of this anomalous position, and the news of Virginia's secession decided Lee once and for all. He formally resigned on 25 April 1861, and was immediately appointed commander of Virginia State forces.

He organised the mobilisation of militia and the fortification of key positions with great energy and skill – as an engineer, his acute eye for terrain was always his strong point. In August 1861 he was made a general, and named as military adviser to President Davis. The next nine months were a period of frustration and some unpleasantness. He had a high title but little power, and was striving to bring some order out of the chaos that then attended Confederate efforts at co-ordinated action. Personalities were important, and he found himself caught between the mutually explosive temperaments of Jefferson Davis and the military commander, Joseph E. Johnston. In May General Johnston was wounded, and Lee was appointed as his replacement. He immediately renamed the army under his command the Army of Northern Virginia, and found himself faced with a testing situation.

General McClellan was within a few miles of Richmond with 100,000 men; three Union armies threatened Jackson (q.v.) in the Shenandoah Valley, and another lay on the Rappahannock river ready to support McClellan. Lee summoned Jackson to join him for an offensive while General Magruder held McClellan clear of Richmond. Jackson made a brilliant fighting withdrawal from the Valley, and in late June Lee launched the offensive known as the Seven Days' Battle. This was strategically sound but tactically faulty, being too complex in conception for the hastily-trained Confederate staffs, and too demanding logistically for the always weak Confederate commissary. Despite mistakes by many from Lee himself downwards, and despite the repulsing of almost every Confederate attack, the overall result was that McClellan was pushed back from Richmond and confined to a position on the James river at Harrison's Landing. Lee and the Army of Northern Virginia had gained valuable experience, considerable stocks of badly-needed weapons, and improved confidence and morale. Lee now made certain command changes as a result of recent experience, and watched with impeccable judgement the manœuvres of McClellan and a new Union army under Pope. In August he took his 55,000 men out against Pope, skilfully drew the two Union armies apart, and sent Jackson's command round to the rear of Pope to ravage his supply depot. Pope fell back to find and attack Jackson, and on 30 August was neatly caught between Jackson and Lee at the Second Battle of Bull Run. It must be mentioned, however, that Lee's great failing showed itself here for the first time. A mild-mannered Christian gentleman of impeccable breeding, Lee found it highly distasteful to force his opinions on anybody. He was unable to dominate and bend to his will any really stubborn senior subordinate, particularly General Longstreet. Lee also held the theory that the job of the Commander-in-Chief was to bring the army to the enemy in the most favourable manner, and that the tactical direction of the battle should be left to divisional commanders. This weakness led to fatal defeat at Gettysburg the following year, and was one reason for the less-than-complete exploitation of the victory at the Second Battle of Bull Run.

In early September Lee invaded the North, with the ultimate objective of encouraging European powers to recognise the Confederacy by a demonstration of military strength. He split his army in three parts over some twenty-five miles, and only the lethargy of McClellan's army – fully briefed by the discovery of a lost copy of Lee's orders – saved Lee from complete disaster. As it was he was given two vital days to reassemble, and established himself along Antietam Creek. The resultant battle on 17 September 1862 was the bloodiest single day of the war, with 12,400 Union and 13,700 Confederate casualties. Lee successfully defended his line, but although winning a tactical victory, had to admit strategic defeat, in that the invasion of the North had been halted.

Back in Virginia the army was reorganised in two corps under Jackson and Longstreet, and a period of waiting ensued. In November the Army of the Potomac advanced on Fredericksburg, and was repulsed there with great loss on 13 December. In April 1863 its new commander, Hooker, led it across the Rappahannock river in another attempt to destroy Lee; the main army crossed north of Fredericksburg, and a subsidiary attack through the town itself was led by Sedgwick. Lee was kept well informed of the enemy's movements by Stuart's (q.v.) cavalry. He left Jubal Early in the town with 10,000 men, and took the remaining 43,000 out against Hooker. He came up against the strong Union position at Chancellorsville on 1 May; facing Hooker's 73,000 men with only 17,000, he sent Jackson with 26,000 in a wide sweep around Hooker's right flank. Jackson fell upon the Union wing, destroyed the XI Corps, and rolled up the line in a classic attack, which he paid for with his life. Lee had to intervene to help Early when he was pushed back by Sedgwick's 40,000-strong force, and this prevented the fullest Confederate exploitation of the victory, but nevertheless Hooker fell back across the river on 5–6 May.

Anxious to retain the initiative, Lee once again invaded the North in order to strike at Unionist morale, to secure badly-needed supplies, and to draw pressure off Vicksburg (see under Grant). The army was now divided into three corps under Hill, Ewell, and Longstreet, and the new command structure had had no time to work itself in when it was faced with a severe test. As Lee advanced into Pennsylvania Stuart took advantage of Lee's habitual discretionary orders and left on a long raid, robbing Lee of his sensitive contact with the Army of the Potomac. Hill's corps ran into an enemy force of unknown strength near Gettysburg on 30 May, and beat it in stiff fighting with the support of Ewell. The Army of the Potomac came up to reinforce the rallying Union corps during the night of 1–2 July, and took up positions along Cemetery Ridge. On 2 July Lee's weakness for not hammering home his orders allowed Ewell scope for dilatoriness on the left wing, and he failed to take the anchor-point of Culp's Hill. All the time the Union positions along Cemetery Ridge were being strengthened, and only an attack by Longstreet on the morning of 2 July could have saved the Confederates. Longstreet, however, was apparently enraged by some imagined slight on his judgement by Lee, and delayed until the afternoon, by which time the Union line was strong enough to decimate his attack. This was renewed on 3

July with Pickett's famous charge, but was again repulsed with huge casualties. Union and Confederate losses were roughly 23,000 to 28,000, but Lee could ill afford such a butcher's bill. He fell back to Virginia in the second week of July, and offered his resignation, which was refused.

There were no major actions involving the Army of Northern Virginia between Gettysburg and May 1864; Lee's desire to press the offensive was frustrated by shortage of men and of every kind of supplies, and by the need to detach troops to Tennessee. In May 1864 Grant crossed the Rapidan river in the great attack towards Richmond described in the entry under his name in this book. Lee was outnumbered two to one in men, and by a much greater factor in equipment of every kind. His troops were ragged and hungry, his horses few and sickly. Nevertheless he achieved a magnificent series of defensive victories at the Wilderness, Spotsylvania, and Cold Harbor, always anticipating Grant's next move and guarding against it in time. The month-long campaign cost Grant some 50,000 men; Lee's losses are unknown, but were probably around 25,000. He never got the opportunity he was looking for – the chance of attacking Grant's army on the move or split up – and his burden was increased by the wounding of Longstreet and the illness of Ewell and Hill, so that he had to take on the tactical direction of the army as well as the strategic control. Once tied down on the Petersburg–Richmond line Lee's army could do little but endure. Their strength was sapped by hunger and illness, by constant Union probes, and by the need to detach troops to guard against new threats in West Virginia and the Shenandoah Valley. A probe towards Washington by Jubal Early failed to draw off Union forces in the quantities expected. In February 1865 Lee was named General-in-Chief of the Confederate armies, but his own situation only allowed of the most general direction of other commands.

Sherman's (q.v.) advance into North Carolina in March 1865 made Lee's defeat certain if he stayed at Petersburg, and on 2 April he was forced to abandon the line when an extension of the Union left compromised the entire position. Lee led some 35,000 men west in an attempt to link up with Johnston's small army in western North Carolina, but Grant paralleled his march to the south and harried him with cavalry. By the time the Army of Northern Virginia was trapped and forced to surrender at Appomattox on 9 April only some 7,800 men were left under arms.

Lee was paroled and allowed to return to Richmond; he was treated with great respect and courtesy by the Union army, and regarded with undiminished love and pride by his own defeated men. In September 1865 he accepted the presidency of the Washington College at Lexington, Virginia. His last years were spent in academic work; he was always a model of obedience to civil authority, and concerned himself with the economic and cultural rehabilitation of the South. Physically he was a man of great presence; nearly six feet tall, he had a fine head and carriage, dark brown eyes, a high complexion, and a grey beard and moustache. He had been fifty-five when he took command of the Army of Northern Virginia, and he had carried a heavier burden, for a longer time, than any man in America except Abraham Lincoln. He died on 12 October 1870 at Lexington, and his name is still revered in the South.

The special limitations under which he struggled make it difficult to assess his military skill against any absolute scale. He was always outnumbered at least three to two, and often three to one, by better-equipped, better-fed, better-trained, and usually better-disciplined troops. His achievements were quite remarkable, especially at Chancellorsville and in the 1864 campaign; but many of his most impressive victories were won only by an all-out effort, leaving no reserves to exploit success. Although a better strategist than a tactician, the death of his 'executive arm', Jackson, forced him to take more control at the tactical level, and he improved markedly as the war progressed. He was weakened as a general by just that quality which made him memorable as a man – his humility. He was a decent, considerate, pious, sensitive, and enduring man, forced by his sense of duty into a prominence he never sought, and making the very best of a doomed command for longer than anyone had a right to expect.

Lettow-Vorbeck, Paul Emil von (*1870–1964*), *General*

Commander of the German colonial forces in East Africa throughout the First World War, Lettow-Vorbeck pursued a brilliant campaign to the continued embarrassment of the Allies.

Eldest son of General Paul Karl von Lettow-Vorbeck, Emil was born at Saarlouis in the Saar in 1870. Completing his education at Berlin University, he was commissioned in the German army about 1891. He saw action in China during the Boxer rising, and by January 1914 had risen to the rank of lieutenant-colonel and had completed a field command of German colonial forces in Cameroon. In that month he was transferred to German East Africa to assume overall command of the colonial forces in that area.

Covering an area greater than Germany herself, German East Africa lay between British and Portuguese territories to the north and south respectively, and extended westward from the Indian Ocean to Uganda, Northern Rhodesia, and Nyasaland. To protect frontiers more than 2,200 miles in length, Lettow-Vorbeck commanded a force that initially numbered about 260 Europeans and 2,500 colonial troops. From the outbreak of the First World War this German force took the initiative. On 15 August 1914 it moved north into British East Africa near Kilimanjaro and seized the town of Taveta, the intention being to mount an attack on Mombassa with support from the German cruiser *Königsberg*. The vessel was, however, caught and trapped by British warships, and eventually ended her days in the Rufiji river.

The attack on Mombassa was abandoned, but Lettow-Vorbeck used Taveta as his base for raids upon the Uganda railway, and at one time even planned to attack Nairobi and capture the British Governor. These minor actions and the local threats they posed drew reaction from the British, who had utterly underestimated the efficiency and mobility of the local German command. Early in November 1914 an attack by 5,000 troops from India was launched to capture Tanga from the sea, but failed after suffering almost 800 casualties. A simultaneous attack on the German positions inside British territory to the north was mounted in an attempt to split Lettow-Vorbeck's forces, but his skilful deployment had rendered all key points sufficiently strong to withstand any likely threat, and the attack in the north also failed. In 1915 the British commander, Brigadier-General Stewart, who had shown little aptitude for colonial warfare, was replaced by Major-General Michael Tighe, veteran of numerous campaigns in Africa, and at once Allied pressure mounted against the German colony. In June a joint lake and land attack succeeded in capturing Bukoba, the principal German port on Lake Victoria.

By early 1916 the British forces in the area had grown to almost 100,000, including a South African expeditionary force commanded by General Jan Smuts. In March 1916 Portugal entered the war, and her troops now lightly manned the southern frontier of the German colony, thereby completely isolating Lettow-Vorbeck from his neighbours, while the Royal Navy mounted a blockade of his coastline. Against this the German forces, swelled by conscription, had grown to about 4,000 Europeans and 30,000 colonial troops and Arabs. Unlike the Allies, whose forces moved about in columns of several thousand men and were therefore slow and unwieldy, Lettow-Vorbeck organised his main field units in self-contained companies of 150–200 men, each with about a dozen European officers and two or three machine guns, for which ammunition was plentiful. By this manner of deployment Lettow-Vorbeck and his subordinate commanders were able to move around the countryside with great rapidity, constantly evading much larger forces and pursuing a highly effective semi-guerrilla campaign.

When in May 1916 British forces from Nyasaland invaded German territory the German commander avoided direct combat, merely leaving covering forces to delay the advance. Meanwhile Smuts attacked in the north near Taveta, and there started an extraordinary campaign which lasted more than two years, but never brought Lettow-Vorbeck to defeat or capture. Eventually he moved his main forces south into the lightly-defended territory of Portuguese East Africa (Mozambique), his men living off the land for much of the time. Despite frequent brushes with British colonial forces, and the effects of sickness, which forced his smaller columns to surrender in order to acquire medical assistance, the last months of the War found the German commander (by then promoted General), with forces reduced to about 8,000, still campaigning on the south-west borders of the German territory. One of his columns – a force of no more than 200 men under Leutnant Naumann – had marched more than 2,000 miles around the central area of German East Africa, evading every effort by Allied forces totalling 11,000 men to bring it to battle; eventually on 1 October 1918 Naumann was surrounded at Luita Hill, and when all his water holes had been captured or spoiled, he was forced to surrender.

In the south Lettow-Vorbeck had been pursuing a campaign of manœuvre in south-west Mozambique while the British forces, commanded by Lieutenant-General Sir J. L. Van Deventer, attempted to hem him in against the shores of Lake Nyasa. With great agility the Germans turned away and made northwards through Nyasaland but, after turning north-west into Rhodesia and reaching Kasama on 8 November, they were finally brought to action on 12 November. The breakdown of communications prevented Van Deventer from informing his opponent of the Armistice, and it was thus not until 14 November that the full terms could be handed to Lettow-Vorbeck, who formally surrendered on 25 November.

The extraordinary campaign conducted by Lettow-Vorbeck achieved all that anyone could have expected. Of course the war put an end to German territorial ambitions in Africa; yet the mere presence of an active enemy force in the area (to all intents and purposes self-sufficient) attracted to it Allied forces which were seldom less than four divisions in strength. While the great majority of these troops would never otherwise have been employed in the European theatre, their supply was nevertheless an added burden to the Allies out of all proportion to the actual threat. Lettow-Vorbeck himself was a much-respected commander who never appeared before his opponents where or when he was expected. He was a typical officer of colonial outlook: a disciplinarian, but nevertheless a considerate man. The intelligence with which he conducted his campaign may be judged by the fact that in his command there was one doctor to every 240 combat personnel; and that upon his surrender he still had half a million rounds of ammunition. He 'inflicted' total casualties of 62,220 on the British, of which 48,300 were caused by disease. His campaign cost the British Empire something in the region of £72 million: a greater cost in money, and three times as great a cost in men, as the Boer War of 1899–1902.

Lettow-Vorbeck lived to a great and honoured old age, dying in Hamburg on 9 March 1964 at the age of ninety-three.

Liddell-Hart, Sir Basil Henry (*1895–1970*), *Captain*

Liddell-Hart was the British military theorist who campaigned between the World Wars for the mechanisation of the British Army, and advocated the 'expanding torrent' form of attack and the armoured thrust. General Guderian (q.v.) described himself as his 'disciple and pupil' in tank tactics and strategy.

Born on 31 October 1895, the younger son of a parson, Basil Hart later joined the surnames of his father and his mother (*née* Liddell). After a university education he joined the King's Own Yorkshire Light Infantry, with which he served during the First World War. Wounds received during the war forced his retirement from the Active List in 1924, but already he had embarked on writing highly analytical studies of military history, as well as biographies of Scipio, Sherman (q.v.), Foch (q.v.), and T. E. Lawrence (q.v.). From his examination of the strategy and tactics employed during the First World War he evolved (and continued to advocate) the strategy of 'indirect approach' and the warfare technique later referred to as *blitzkrieg*, using a closely-integrated series of hammer blows by armour, mechanised infantry, and support aircraft. His 'expanding torrent' principle of forcing increasing masses of men and material towards and through a gap torn in an enemy's line was the cornerstone of German *blitzkrieg* planning in the early stages of the Second World War, and was demonstrated to near-perfection by the German General Guderian, perhaps the greatest tank commander in history.

In the prevailing atmosphere of pacifism that existed during the 1920s and early 1930s, it is not surprising that Liddell-Hart's ideas went unheeded in his own country, despite his appointment as military correspondent, first to the *Daily Telegraph* (1925–35) and then to *The Times* (1935–9). Only in the years immediately before the Second World War were his doctrines recognised as being in the nation's best interests, to meet the threat of total war posed by the Nazi menace. In 1937 he was appointed adviser to Leslie Hore-Belisha, with special responsibilities for the reorganisation and mechanisation of the British Army; but he found progress at the War Office deathly slow, and resigned the following year to pursue his campaign independently. Perhaps the best testimony to his perception lies in the successes of his German pupils, Guderian and Rommel (qq.v.). Their achievements alone would establish Liddell-Hart as the undisputed leader among British military thinkers of the twentieth century.

Ludendorff, Erich von (*1865–1937*), *General*

Ludendorff was a member of an impoverished land-owning family, born at Kruszewnia, near Poznan in Prussian Poland, on 9 April 1865. It has been said that his wholly humourless, austere nature was the outcome of this penurious environment. He entered the army through the cadet corps, being commissioned in about 1883 in the infantry, but from 1894 onwards he served almost continuously on the general staff, becoming head of the deployment section in 1908. As such he was engaged in contingency planning for the implementation of the Schlieffen Plan for the invasion of Belgium and Northern France.

In 1912 Ludendorff was given command of an infantry brigade, but on the mobilisation of German forces in July 1914, he moved to the Second Army as quartermaster-general under General Karl von Bülow. The Schlieffen Plan depended for initial success on the immediate capture of the Belgian fortresses, and in particular of that at Liège; as Ludendorff was intimately conversant with this objective, he was given personal command of the attack. After an unprecedented artillery bombardment and careful infiltration, the forty-nine-year-old general personally led the assault, and on 16 August the fortress town surrendered. Ludendorff remained with the Second Army only long enough to see it successfully embarked on its swift advance towards the Marne before being called to the Eastern Front as Chief-of-Staff in the Eighth Army under the elderly General Paul von Hindenburg (who had been recalled from retirement). As Chief-of-Staff Ludendorff was able to exercise almost total control of operations by his army, being responsible both to his direct superior and to the chief of the general staff at supreme headquarters. The knowledge of modern warfare and planning acquired by Ludendorff and the prestige and influence wielded by the aged Hindenburg combined to enhance the effectiveness of the famous partnership.

Studying reports from the Eastern Front even before he arrived on the spot, Ludendorff dispatched telegrams to the individual corps commanders ordering a concentration of forces against Samsonov's Second Army, while delaying Rennenkampf's First Army. Leaving one lone cavalry division to face the First Army, the bulk of the German forces moved south to the Tannenberg area, where on 25 August 1914 Samsonov's advancing Second Army ran full tilt into the entrenched German forces. On the following day the Russian right wing was rolled up from the north by the German XVII and I Reserve Corps, while the Russian left flank was turned by the German I Corps; by 29 August the Russian army was encircled. With no assistance forthcoming from the First Army, Samsonov's defeat was complete: Russian losses amounted to 125,000 men and 500 guns, compared with not more than 15,000 German casualties.

Turning his attention to the threat in the north, Ludendorff then moved the Eighth Army against Rennenkampf, and in the first battle of the Masurian Lakes (9–14 September), following another vigorous attack by I Corps, the German army gained its second resounding victory, although, under cover of a counter-attack by two divisions, Rennenkampf managed to extricate much of his army. Nevertheless the

Russians retreated, leaving behind a further 125,000 men, 150 guns, and half their transport. German losses were about 10,000.

Although the effects of these early defeats on the Russian army were never completely overcome, they were not in themselves decisive, and served only to stabilise the Eastern Front. On 1 November 1914 Hindenburg was appointed Commander-in-Chief of the Austro-German Eastern Front, and here both he and Ludendorff remained until August 1916. In the meantime the overall strategic policies of the chief of the general staff, Erich von Falkenhayn, had been discredited, and his place was now taken by Hindenburg, who brought Ludendorff with him as his first quartermaster-general.

Ludendorff's immediate task was to effect an improvement on the Western Front, where appalling losses at Verdun and on the Somme had created severe strategic problems and much unrest at home. This was alleviated by withdrawal to the 'Hindenburg Line' in March 1917. Revolution in Russia and the virtual ending of hostilities in the east enabled Germany to transfer substantial forces to the west towards the end of 1917, by which time America had at last entered the war against Germany. Ludendorff realised that he possessed no more than short-term superiority on the Western Front, and so determined to launch a massive offensive which must succeed before the flood of American forces started reaching France.

He selected for his point of attack the front of the British Fifth Army, and its immediate objective was Amiens, a vital centre of communications. The offensive was launched on 21 March 1918, but after initial successes and some advances, the impetus went out of the attack. Other supporting attacks were launched (Lys, 9 April; Chemin-des-Dames, 27 May; Metz, 9 June; and Rheims, 15 July), but by the end of July the offensive had ground to a halt. The Germans' last bid for victory had failed.

It should be said, however, that in planning and execution, Ludendorff's offensive was soundly conceived. He remained a fine campaign strategist throughout the First World War; but when he achieved the position of virtual supreme command, he was utterly lost when faced with all the wide implications of politics, finance, social problems, industrial supply, and foreign policy – all matters in which he felt obliged to interfere. The truth was that Germany's reversal (not merely her ignorance) of Clausewitz's (q.v.) theory of warfare as the natural extension of political means had resulted in war being waged by Germany as an end in itself from the moment when Ludendorff had returned to Berlin in 1916. He could not grasp the wider implications of strategic command, and left Germany helpless and exhausted after the collapse of his great offensive in 1918.

Dismissed from office by Prince Max of Baden on 26 October, Ludendorff fled to Sweden at the time of the Armistice, but later returned to Germany to lead a reactionary political group, concentrating his subversion against Freemasons, Jews, Jesuits, and Marxists. He supported Hitler in the 1923 *Putsch*, and the following year entered the Reichstag as a National Socialist. Later Ludendorff and Hitler quarrelled and became bitter enemies, and the ageing soldier never achieved any formal status in the N.S.D.A.P. He died at Tutzing on 20 December 1937.

Lyautey, Louis Hubert Gonzalve *(1854–1934), Marshal of France*

Born at Nancy on 17 November 1854, Lyautey entered St Cyr in 1873. While there he developed an enthusiasm for social reform, being to some extent influenced by the ex-soldier president MacMahon (q.v.), whose royalist leanings also inspired the young man. He was posted to a cavalry regiment during the late 1870s, went to Algeria in 1880, and returning two years later, was promoted captain. He expressed his royalist convictions widely, but was profoundly disturbed when, during an audience at Rome, he learned of Pope Leo XIII's sympathy for the French republican régime.

In command of a unit at St Germain in 1887, Lyautey paid particular attention to the welfare of his men and made something of a name for himself by his attempts at social reform in the army. In August 1894 he was sent to Indo-China; at Tongking he met Joseph Galliéni, whose ideal of conquest as a means of civilisation he adopted, and from whose experience he greatly profited. When Galliéni summoned him to Madagascar he went readily and, with the rank of lieutenant-colonel, was charged with pacifying about one-third of the island; promoted full colonel in 1900, he was given command of the whole of the southern part of Madagascar, and completely pacified the whole area in two years.

Lyautey returned to France in 1902 and assumed command of the 14th Hussars at Alençon. In 1904 Célestin Jonnart, Governor-General of Algeria, arranged for him the post of commandant of the sub-area of Aïn-Sefra. Now a brigadier-general, Lyautey never hesitated to encroach upon neighbouring Moroccan territory in order to 'flatten' the frontier; Jonnart gave him protection when Morocco protested to France, and Lyautey succeeded in reducing the frontier tribes to obedience. Appointed commandant at Oran in 1906, he continued to push the frontier westward.

In April 1912 Lyautey was appointed Resident-General in Morocco, over which a French protectorate had recently been proclaimed by the Treaty of Fèz. He was besieged in Fèz by dissident tribes, but managed to extricate himself, and then replaced the sultan Mulay Hafid by his more reliable brother Mulay Yusuf. He next embarked on the task of conquering and pacifying the whole country – a task in which he displayed the greatest respect for local institutions; he had an air of grandeur that appealed to the Arabs, and displayed great discernment in his choice of assistants.

In July 1914, on the eve of the First World War, the French Government ordered Lyautey to withdraw his forces from the interior of Morocco to the ports, thereby releasing troops for service in France. Although he at once sent home the number of troops demanded of him, Lyautey still managed to hold the whole country. He was recalled to France in December 1916 as Minister of War for several months; afterwards he returned to Morocco, and was promoted Marshal of France in 1921.

Lyautey considered leaving Morocco in 1924 when the leftist parties won the French elections. However, following the Riffian revolt under Abd-el-Krim (q.v.) in April 1925 he decided to stay on, and in the face of the border invasion by Abd-el-Krim, handled his slender resources with considerable skill in a holding campaign pending the arrival of Franco-Spanish reinforcements. However, he received little consideration from Marshal Pétain, who was sent out by Paul Painlevé's Government. Pétain displayed a degree of arrogance in his command of the counter-revolutionary expeditionary force, and Lyautey, in frustration and disgust, resigned on 24 September 1925, devoting the remaining years of his life to social work among the young. He died at Thorey (Meurthe-et-Moselle) on 21 July 1934.

Despite his high rank, Lyautey never commanded major forces in a pitched battle; but throughout his life he combined a rare gift of martial determination with consideration and respect for local institutions. He was, or at least tried to be, scrupulously fair, and helped to forward great material progress in Morocco. His greatest military achievement was his skilfully-judged series of holding operations in Morocco during the First World War, when German agitation among the tribes added to the problems of the wholly-inadequate garrison.

Lynedoch, 1st Baron, *see* Graham, Thomas

MacArthur, Douglas (*1880–1964*), *General of the Army, Field-Marshal of the Philippine Army*

MacArthur ranks with Marshall and Eisenhower (qq.v.) as one of the outstanding American personalities of the Second World War. His purely military career was longer than that of either of the other two, and his experience in the field was significantly broader. A figure of some controversy, he made a contribution to Allied victory in the Pacific in the Second World War, and American recovery at a perilous stage of the Korean War, which can never be questioned.

Douglas MacArthur was born in the barracks at Little Rock, Arkansas, on 26 January 1880. His father, Arthur MacArthur, was a remarkable officer who had greatly distinguished himself as the twenty-year-old colonel of the 24th Wisconsin Volunteer Infantry in the American Civil War; wounded and decorated several times, he was awarded the Medal of Honor for gallantry at Missionary Ridge. He joined the regular army after the Union victory and served extensively in the Indian and Spanish-American Wars. As major-general commanding the 8th Corps and military governor of the Philippines in 1900–1, he did much useful work in laying foundations for democratic government in the former Spanish colony. His son's career too was to be intimately bound up with the Philippines for many years.

Douglas MacArthur graduated from West Point in first place in the class of 1903. Between 1903 and 1906 he served with the military survey in the Philippines, and participated in studies of several Far Eastern countries, including Korea and Japan. He was aide-de-camp to President Theodore Roosevelt in 1906–7. He served on the general staff in 1913–15, and took part in the Vera Cruz expedition in 1914. After further general staff work he was appointed chief-of-staff of the U.S. 42nd (National Guard) Division ('Rainbow Division') in August 1917. On 6 August 1918 he was named as commanding general of the 84th Infantry Brigade, and took part in the major American offensives in France, being wounded twice. Appointed commanding general of the 42nd Division in November 1918, he commanded U.S. occupation forces in the Rhineland until April 1919. On 12 June 1919 he was appointed Superintendent of West Point, holding the post until 1922, and being promoted substantive brigadier-general in 1920. In 1922–5 he served in the Philippines once again, as commander of the Manila District, and in 1925 was promoted major-general. During 1925 he sat on the court martial on Brigadier-General William ('Billy') Mitchell, when that far-sighted officer's outspoken criticism of his country's neglect of air power brought his career to a premature end. Although details of the individual votes have not been released, MacArthur is widely believed to have voted for acquittal.

From 1928 to 1930 MacArthur was commander of the Philippine Department, and he was promoted general in 1930. From that year until 1935 he served as a notably energetic U.S. Army Chief of Staff, and in 1935 returned to the Philippines once again as director of national defence. In preparation for the independence of the Philippine Commonwealth a team under his leadership worked to organise national forces, and in 1936 he was appointed field marshal in the Philippine army. On 31 December 1937, at the age of fifty-seven and after a most distinguished career, MacArthur retired from the U.S. army, but continued with his duties in the Philippines. He could hardly have foreseen that his future held anything other than a few more years of congenial work in the land where he had spent so many years, followed by a well-earned retirement.

In view of threatening international developments, MacArthur was recalled to duty with the U.S. army on 26 July 1941, as commander of America's forces in the Far East with his former rank of general. Immediately after Pearl Harbor a Japanese air strike severely damaged U.S. air elements in the Philippines, and on 10 December the first of some 50,000 men of General Masaharu Homma's 14th Army landed on Luzon. MacArthur had about 10,400 American regulars and 12,000 crack Philippine Scouts, supported by more than 100,000 Filipino troops of varying degrees of readiness and reliability. Although they later recovered their spirit, many Philippine army elements crumpled before the first fury of the Japanese attack; and it was only with difficulty that MacArthur's forces succeeded in a planned withdrawal into the mountainous jungle of the Bataan Peninsula. Both American and Filipino troops greatly distinguished themselves in an epic defence of Bataan and Corregidor, hampered by lack of air and sea support and by serious logistic difficulties. Homma had reckoned on a fifty-day campaign; in fact the last Allied troops did not surrender until mid-May 1942, after five months' resistance. On 22 February 1942 MacArthur was appointed Supreme Allied Commander, South-West Pacific, and reluctantly obeyed President Roosevelt's repeated orders to make his way to Australia. He handed over command to Major-General Jonathan Wainright, and left the Philippines with the famous pledge,

'I shall return.' He arrived in Australia in mid-March after a perilous 3,000-mile journey by torpedo boat and bomber.

From the launching of the Allied offensive in New Guinea in the autumn of 1942 until he accepted the surrender of the Japanese on the deck of the U.S.S. *Missouri* in Tokyo Bay on 2 September 1945, MacArthur played a leading part in the formulation and execution of the successful Allied strategy in the Pacific island war. It is impossible in a book of this size to describe in detail the course of this long and complex campaign, so much of which was based on the co-ordination of ground, amphibious, naval, and air operations. Suffice it that MacArthur's judgement and handling of the forces involved has seldom been seriously questioned, and that Japan went down to final and devastating defeat despite the extraordinary determination and tenacity of her forces. MacArthur, naturally enough, was a constant advocate of all Allied plans designed to take the war closer to Japan, despite America's adoption of a policy of 'Germany first'. He was a commander who aroused strong feelings in military and political circles, being regarded either with intense admiration and loyalty or with bitter animosity. There was certainly a strong theatrical element in the man's make-up, and a complete confidence in his own judgement which he did not try to conceal; but many great generals have displayed the same character traits. They did not prevent him from exercising considerable diplomatic and political skill. Alanbrooke (q.v.) unequivocally pronounced him to be the greatest strategist the war produced, and Alanbrooke was by no means indulgent in his judgements on American commanders.

In December 1944 MacArthur was promoted General of the Army. After the Japanese surrender he spent five years as Supreme Commander of the Allied occupation forces in Japan, playing a major role in the restructuring of Japanese society and institutions and in the country's recovery from defeat. He was not a vindictive conqueror; although he pursued certain former Japanese commanders to the gallows on war crimes charges which probably would not have succeeded in more impartial courts, he won the friendship of Emperor Hirohito and the gratitude of many Japanese.

When North Korean Communist forces crossed the 38th Parallel and invaded South Korea late in June 1950, MacArthur was appointed commander of the United Nations forces in Korea. Although the initial North Korean advance proved too much for the American and Republic of Korea formations which faced it – they were by no means of high quality as regards combat readiness and logistics – a perimeter was set up around the port of Pusan and successfully held. It was MacArthur who conceived the plan for the American landings at Inchon on the south-west coast far behind the Communist lines, and pushed it through despite considerable opposition. The landings were made by U.S. Marine Corps and army formations on 15 September 1950, and were an unqualified success. Simultaneously the reinforced U.N. forces within the Pusan perimeter broke out and advanced to link up with the landing force, and within two months the North Korean forces had been routed and virtually destroyed. In late November 1950 massive Chinese Communist forces crossed the Korean border and forced the U.N. troops to retreat beyond the 38th Parallel once again. MacArthur succeeded in halting the retreat, and was ready to begin offensive operations by March 1951. He believed that China should not be allowed to hide behind the pretence that her forces in Korea were merely volunteers, and advocated a policy of blockading the Chinese coast, bombing Communist bases in Manchuria, and bringing Chinese Nationalist troops from Formosa to Korea to assist the U.N. forces. He expressed these views forcefully and publicly, despite the emphatic directives of his political superiors that America should seek no wider extension of the conflict. On 11 April 1951 he was relieved of all his commands by President Truman. After returning to the United States MacArthur made a memorable speech to both houses of Congress, in which he presented his own analysis of the Far Eastern situation and defended his proposals.

Douglas MacArthur died on 5 April 1964 in the Walter Reed Army Hospital in Washington, and is buried at Norfolk, Virginia.

MacMahon, Marie Edmé Patrice Maurice de, Duc de Magenta (*1808–93*), *Marshal of France*

Distinguished commander in Algeria, Italy, and the Crimea, MacMahon was the captor of Paris in the second siege of 1870, and second President of the Third Republic.

Descendant of an Irish Jacobite family, he was born at Sally (Sâone-et-Loire) on 13 July 1808, was educated at St Cyr, and entered the army in 1827. As a junior officer he served with distinction in the French army of invasion under Marshal Louis Bourmont in Algeria during the early 1830s and in the operations against the emir of Mascara, Abd-el-Kader (q.v.), displaying great personal courage and skill during the storming of Constantine in 1837 when the French leader, Marshal Damremont, was killed leading the final assault. Recognition of his services in Algeria brought steady promotion, and by the outbreak of the Crimean War he had risen to colonel. When General François Canrobert (q.v.) assumed command of the French siege corps investing Sebastopol during the autumn of 1854, MacMahon was given charge of an infantry division. Repeated attempts on the besieged town failed owing to the extraordinary strength and ingenuity of the Russian fortifications. After Canrobert had been replaced by General Aimable Pélissier (also a veteran of the Algerian wars) in April 1855, plans were drawn up to assault two key strongpoints vital in the defence of Sebastopol – the Malakoff and the Redan. An initial French attack on the former in June dwindled away owing to a lack of co-ordination with the British attack on the Redan, but in September plans were completed for a new French attack on the Malakoff.

The attack on 8 September was the one perfectly-planned and executed operation by the Allies in the Crimea. With meticulous attention to detail the French columns assaulted the Russian defences, the critical attack being carried out successfully by the division commanded by MacMahon. After eight hours' fighting the Malakoff was in French hands and, after the failure of the British to take the Redan, MacMahon turned his fire on the Russians there and drove them out also. The following day the Allies occupied Sebastopol.

In the war for Italian independence which broke out in 1859, King Victor Emmanuel II secured by secret treaty the collaboration of France in the expulsion of Austria from northern Italy. French forces under the personal command of Napoleon III invaded Lombardy, and at the Battle of Magenta a French corps commanded by MacMahon was primarily responsible for the defeat of the Austrian army under General Count Gyulai. The Austrian commander was dismissed, and a newly-reinforced army under the Emperor Franz Josef moved to meet the French along the Mincio River. Although in general the control of the French forces left much to be desired in the early stages of the Battle of Solferino (24 June), MacMahon's conduct during the great battle, which ended the first phase of the war, was recognised by bestowal on him of the title Duc de Magenta by King Victor Emmanuel, and by promotion to marshal in the French army.

In 1864 MacMahon was appointed Governor-General of Algeria, where he remained for six years while France under Napoleon III basked in the delusion of invincibility – fostered in no small part by the undoubted superiority of her army in the closing stages of the Crimean War and by the successes gained in the Italian War. Then, following Bismarck's diplomatic rally of the German states under Prussia in an anti-French coalition, the falsely-confident Napoleon decided to precipitate a war with Prussia (which he believed, in any case, to be inevitable), and war was accordingly declared by France on 15 July 1870.

Mobilisation by both countries followed: Prussia's according to a meticulous plan, France's haphazard and lethargic. Within six weeks three well-equipped German armies, totalling 475,000 men, faced across the Rhine less than half that number of French, distributed piecemeal in eight separate army corps. Of these the southernmost three corps were commanded by Marshal MacMahon as the Army of Alsace; facing his front was the Third German Army – 180,000 men under Crown Prince Friedrich Wilhelm. On 4 August this German force advanced and surprised the forward French positions. In a sharp action, the Battle of Weissenburg, MacMahon was forced to fall back, greatly outnumbered and lacking any strategic guidance from the French High Command. In contrast the German armies' aim was simple and straightforward – to drive westward and converge upon the French capital.

Two days after Weissenburg, while the German pressure built up on MacMahon's front, the French repulsed reconnaissance probes; but on their right wing, pulverised by the fire from 150 guns, they fell back on Fröschwiller, covered to some degree by gallant but costly cavalry charges and, to a lesser extent, by the desultory efforts of the reserve artillery. In the Battle of Fröschwiller, the first major battle of the war, MacMahon lost over a third of his 48,000-strong army in casualties, while German losses amounted to fewer than 10,000 out of 125,000 engaged. This defeat was not turned into rout, however, and the Army of Alsace retreated without interference over sixty miles to Châlons-sur-Marne during the following week. Nevertheless the Vosges barrier had been pierced and a path to Paris laid open.

The French troops, at this early stage, had fought well in the demoralising circumstances of a vastly superior enemy and quite inadequate artillery support. In contrast to the German reliance upon conventional medium-calibre cannon of modern design and fine quality, the French artillery was obsessed with the supposed potential of the *mitrailleuses* – somewhat recalcitrant, short-range, inaccurate machine-guns. On the other hand the French *Chassepot* rifle was found to be superior in accuracy and firepower to the German 'needle gun'.

While MacMahon remained at Châlons, French forces under Marshal Bazaine (q.v.) had been forced to retire across the Moselle, and it was hoped that the two armies could link up. The Prussian Second Army dispelled this hope, and the Battles of Mars-la-Tour, Vionville, and Rezonville decided Bazaine to retire to Metz. The Battle of Gravelotte-St Privat on 18 August was an example of French command inferiority for, at the moment when a French counter-attack could have carried the day, Bazaine remained passive and relinquished all control to his corps commanders. Moltke (q.v.), who had assumed command of the Prussian operations, waited for the counterstroke, but when it failed to materialise proceeded to seal the French forces into the Metz area. In answer to frantic appeals from Paris, MacMahon now moved out of Châlons with 120,000 men to the relief of Bazaine, but foolishly took a northerly route inviting a turning movement by the Prussians – an invitation willingly accepted by Moltke. Leaving the First and part of the Second German Army to invest Metz, the remainder of the Second Army thrust west to join with Friedrich Wilhelm's Third Army in barring MacMahon's approach. In a series of engagements the French army was forced into a bend of the Meuse at Sedan; MacMahon himself was wounded at Bazeilles on 31 August and was replaced by General Auguste Ducrot. On 1 September was fought the great Battle of Sedan in which the French army, now reduced to less than 90,000 men, faced almost 200,000 Prussians and a great arc of 426 German guns sited on the heights above the city. A day-long bombardment of the French positions and the destruction of the French cavalry by highly-disciplined German infantry fire decided Ducrot to retire his forces into the city, where they were blasted by the dominating artillery. The following day the French army's 83,000 survivors, now commanded by General Emmanuel de Wimpffen, surrendered. With it into captivity went the wounded Marshal MacMahon.

Freed by the peace, MacMahon was recalled by Adolphe Thiers to command the Versailles army which conducted the Second Siege of Paris, now in the hands of revolutionaries, and in May 1871 this army broke into the city to crush the Communes in savage retribution.

MacMahon was now widely esteemed as a veteran soldier of proven courage and distinction and, apart from Thiers himself, the only public figure in France capable of commanding general confidence and esteem. On 24 May 1873 Thiers resigned his post as head of state; and it was MacMahon that the people chose as his successor, although his acknowledged conservative, clericalist, and possibly royalist leanings made him particularly favourable to the anti-republican cause. Elected President the same day, MacMahon at the age of sixty-five embarked on a short-lived and highly controversial political career for which he was ill-suited, and in which he rendered poor service to the causes he had for so long favoured. He resigned as President on 28 January 1879, and died in Paris on 17 October 1893.

Mahmud of Ghazni (*971–1030*)

In the tenth and eleventh centuries southern and western Asia saw a ferment of rising and falling dynasties, of migrations and fragmentations of peoples. One of these separations, almost biological in its nature, was the process which detached the city-state of Ghazni, in central modern Afghanistan, from nominal allegiance to the Samanid amirs of Samarkand and Bokhara, and saw it become the focus of a new, impressive, but short-lived empire. The man who built that empire was Mahmud, the legendary Islamic hero of India, the archetype of the pious Muslim ruler and warrior. When he died his realm stretched from Ray and Isfahan in Persia to Lahore in India, and from the Oxus to the coasts of Makran; yet before he was ten years in his tomb all had been swept away. There could be few more convincing demonstrations of the dynamism and volatility of that age in Asia's history.

In 962 the ancient walled city of Ghazni was captured by one Alptigin, the Turkish former captain of the guard of the amirs of Samarkand and Bokhara. He was seeking to carve out for himself a safe refuge from the machinations of enemies at court, but once he was established at Ghazni he continued to acknowledge the suzerainty of the Samanid dynasty. He brought with him a favourite slave, one Subuktigin, who rose in his service and became his particular protégé. Subuktigin made himself useful and popular at Ghazni during the brief reigns of Alptigin's two successors, and in 977 the soldiery sponsored him as the next ruler. (It should be noted that in that culture one-time slavery was no bar to the highest position.)

At the time he took power Subuktigin had a six-year-old son by a local marriage – Mahmud. As a youth Mahmud fought in campaigns against the independent chieftains of the Ghor hills in western Afghanistan, and against the Hindu-Shahi Rajah Jaipal in the far south-east. There were also internal battles inside Khorasan – as the whole region was known – against rival Turkish officers in the service of the Samanids, and Mahmud served with his father in these affairs. In 994 the Samanid Amir Nuh appointed the young man commander of his troops in Khorasan; however, the nominal suzerainty of the Samanids had been a hollow sham for some years and Mahmud had far greater ambitions. His father died in 997, and his brother Ismail was named as ruler of Ghazni and Balkh; Mahmud immediately rode on Ghazni, defeated Ismail outside the walls, and imprisoned him. Two years later the last pretence of vassalage to the Samanids was removed when a Qarakhanid invasion took Bokhara and exiled the last of the dynasty. The Qarakhanids were centred in Turkestan, north of the Jaxartes river; and it was as independent ruler of the main cities of Khorasan that Mahmud successfully defended himself against them. As soon as the situation on his north-west borders was stabilised Mahmud began a career of conquest. He consolidated his position by annexing or establishing his suzerainty over neighbouring states such as Seistan, Qusdar (the north-east half of modern Baluchistan), Gharjistan, and Khwaresm, and fought two hard campaigns against the hillmen of Ghor in 1009 and 1020; but he is mainly remembered for his numerous forays into northern India. He sought no permanent territorial gains in India, as his motives were largely the acquisition of booty and the harrying of Hinduism for the sake of religious merit. He forced the Hindu princes of the Punjab to acknowledge his overlordship, however, and so weakened these states that later and more frankly acquisitive Islamic conquerors had little difficulty in overrunning northern India.

Beginning in the year 1000 Mahmud made some seventeen separate attacks into India, and his bold and skilful handling of his mobile armies – spearheaded by mounted archers – invariably defeated the far larger and more unwieldy Hindu armies assembled to resist him. In 1001 he defeated Jaipal, Rajah of Lahore, and in 1006 he took Multan. Later that year he was forced to turn west to defend himself against a Qarakhanid invasion by Ilak Khan. Muhmud defeated this and a subsequent invasion in 1008, and thereafter had no serious trouble from that quarter. In 1009 he plundered Nagarkot and smashed a coalition of Hindu princes at Peshawar – his success in this battle was partly due to his creating a stampede among the Indian elephants. In subsequent years he ravaged northern India, slaughtering and plundering; particularly notable were the sacks of Thaneswar and Kannauj in 1014 and 1018. This latter campaign was a devastating defeat for the Hindu princes; sacking Muttra, Mahmud swept down the Ganges and attacked the allied Rajahs of Kannauj and Kaniljar. The Pratihara dynasty of Kannauj was destroyed, and by 1022–3 Kaniljar had been reduced to a vassal state.

In 1025 Muhmud turned north again, briefly intervening in affairs in Transoxiana, the 'buffer' state between his borders and those of Turkestan. He crossed the Oxus on a bridge of boats, but no major fighting ensued. He met Qadir Khan of Kashgar, the most powerful chieftain of Turkestan, on 25 April, and with great ceremony an alliance was concluded between them which lasted until Mahmud's death. Later in that same year he mounted his most famous invasion of India, penetrating to the coast of Gujerat. He took an army across the desert from Multan to Somnath on the Kathiawar Peninsula, the main object being to sack the city and its temple, which contained a Hindu idol more than usually offensive to a True Believer. In the course of this campaign he is thought to have killed some 50,000 Hindus.

In 1029 Mahmud extended his empire westwards by conquering the principality of Ray; and in 1030 he died. He had succeeded in containing the growing threat of the Seljuk Turks (see under Alp-Arslan) during his later years, but by 1040 they had defeated his son Mahsud and had overrun Khorasan. Although his empire was short-lived, Mahmud was no mere butcher, as were so many Asian conquerors. In character his realm echoed the previous Samanid dynasty, with an *élite* of Turkish military officers and regional governors, many of them carefully trained and promoted slaves, backed by a Persian bureaucracy. Culturally Mahmud's court was impressive; he gathered around him many of the finest Islamic scholars and poets of his day, including Firdausi.

Mannerheim, Carl Gustav Emil, Baron von (*1867–1951*), *Marshal of Finland*

When Carl von Mannerheim was born on 4 June 1867 at Villnäs, near Turku, his native Finland was still part of Tsarist Russia, ceded by Sweden under the Treaty of Frederikshavn in September 1809. Accordingly he joined the Russian army in 1889, being commissioned as lieutenant in the cavalry, and as a major he served in the Russo-Japanese War of 1904–5. By 1917 he was a corps commander with the rank of lieutenant-general; but in December that year, following the Russian Revolution, he returned to his native Finland, which declared its independence on 6 December. The new Finnish Government appointed Mannerheim to command and organise an army in January 1918 – a task he quickly accomplished.

Bolshevik troops were active at that time in the former Russian state, and had already seized Helsinki, where a Red Guard was in formation. On 28 January, incited by the Bolsheviks, a revolt broke out throughout Finland – though opposed by the great majority of Finns. Quickly seizing the initiative, Mannerheim captured the Russian garrison at Vasa, thereby gaining valuable arms and ammunition for his levies. Thus equipped, his extemporised force moved south and captured Tampera, but he then found his way blocked by a large Red Guard force in mid-March.

It was at this point that in a surprise move a German force of 10,000 men under General von der Goltz landed at Hangö and on 18 April seized Helsinki from the Reds, thereby cutting the Bolshevik-held territory in two. Mannerheim, whose army was constantly growing, now moved east and cut off the Karelian Isthmus from Russia. Ten days later the Red forces, numbering about 12,000, surrendered, and although sporadic fighting continued against isolated groups of Bolsheviks for some months, the Treaty of Tartu (signed on 14 October 1918) finally secured Finnish independence. Mannerheim, a national hero, became Regent of Finland in December 1918, continuing to occupy this position for seven months until a republic was proclaimed.

At the age of fifty-two Mannerheim retired from the army, but in 1931 he was recalled from retirement to become chairman of the defence council. It was during his eight years in this post and under his overall direction that the fortified defence line of interconnected strongpoints, using all the features of the rugged country, and named the Mannerheim Line in his honour, was built across the Karelian Isthmus – this in spite of the Non-Aggression Pact between Russia and Finland signed on 25 July 1932.

With the engulfing of Poland, that useful 'buffer of convenience', by Russia and Germany in September 1939, the Soviets became apprehensive of German intentions, and sought mutual defence pacts with the Baltic States: Latvia, Estonia, and Lithuania acceded and were promptly occupied by Russian forces. Finland declined the invitation with the result that, perhaps recalling the earlier occasion when Germany had come to the aid of Mannerheim, Russia invaded Finland on 30 November 1939. Forces totalling almost a million men were hurled against Finland from the east, from the south-east across the Karelian Isthmus, and in a seaborne assault across the Gulf of Finland. Mannerheim was appointed Commander-in-Chief of forces which totalled no more than 300,000 – all but about 50,000 of them reservists. The attacking forces in the north reached and seized Petsamo, but were then halted. The amphibious assault in the south was repulsed. The heaviest attack came upon the Mannerheim Line – which held firm and threw back the invaders with appalling casualties.

It was inherent in Mannerheim's policy of command to equip and clothe his forces for a campaign fought in arctic winter over terrain familiar to them. Thus widespread and effective use was made of swift, mobile bodies of men, warmly clothed and equipped with skis. The Russians on the other hand – many of them from the Ukraine – were committed to battle with none but the most amateurishly-conceived strategy, with no thought given to logistics, and quite inadequately protected against the fearsome cold. Russian air raids were ineffective, and the fifth column in Finnish towns was useless.

An example of Mannerheim's tactics may be seen in the Battle of Suomussalmi of December 1939–January 1940. The Russian 163rd Division, fresh from the relatively mild Ukraine, advanced on this village in deep snow and in temperatures of minus forty degrees. On occupying the village the Russians were halted, bogged down by the snow. Finnish civil guards on skis then tied down the invaders with constant sniping attacks while the 9th Division moved up and, without waiting for artillery support, ambushed the Russian supply route and isolated the immobilised Russian force. Another Russian division, the motorised 44th, tried to fight its way up to the relief of the division in Suomussalmi, but was itself halted and isolated. Then the Finnish artillery arrived and proceeded to annihilate the 163rd Division, while civil guard units systematically chopped the 44th Division into numerous smaller groups, which the artillery equally systematically destroyed. By 8 January 1940 the battle was over; the Russians lost 27,500 men killed or frozen to death, and all the tanks, guns, and equipment of two divisions. The Finns suffered 900 men killed.

The Russians now paused to regroup for an all-out assault on the Mannerheim Line. On 1 February their Seventh and Thirteenth Armies – totalling no fewer than fifty-four divisions – were thrown into battle on the Karelian Isthmus. With brutal disregard for casualties the Russians came on again and again; the Finns were sickened by the numbers of men they were killing. The end was, however, inevitable: by sheer weight of numbers the Mannerheim Line was pierced near Summa and its right flank rolled back. On 12 March Finland capitulated to the invaders, but Stalin's terms were less harsh than expected – he had no wish to provoke an exhausting guerrilla war with the Finns. Mannerheim had lost about 25,000 men killed, but although no Russian figures have ever been published, it was estimated that the Finns and the Finnish winter killed over 200,000 men of the Soviet forces, and wounded 400,000 others.

At the time of the massive assault against

Russia by Germany (22 June 1941) Finland not unnaturally allied herself with the latter. A Finnish army corps under Mannerheim's command was ordered by the German High Command to re-occupy the Karelian Isthmus, thereby threatening Lennigrad from the north, and to attack eastwards to the north of Lake Ladoga as far as Lake Onega, thereby cutting the vital railway link between Leningrad and Murmansk. These objectives were attained by the end of 1941, but the Finnish command refused to press further than their original boundaries, and refused to participate in any open attack on Leningrad. Notwithstanding this limitation, when the Russians counter-attacked all along the Eastern Front from the Arctic to the Black Sea, only Mannerheim's army corps remained firm and threw back the Soviet attacks. In June 1942 Mannerheim was named Finland's only marshal.

In June 1944, after the tide had finally turned in Russia's favour, a heavy Soviet attack mounted by General Govorov penetrated the old Mannerheim Line and captured Viipuri on 20 June. Gradually the Finnish front was overrun, and a truce between Russia and Finland was signed on 4 September 1944. Marshal Mannerheim became President of the Finnish Republic late in 1944, and after peace came to Europe the following year embarked on the delicate policy of maintaining a relationship finely balanced between East and West. In 1946 he was obliged by ill-health to retire, and he died at Lausanne in Switzerland on 27 January 1951 at the age of eighty-three.

Mannerheim was a remarkable man who, unlike so many of his contemporaries who had served much of their active lives in the combat environment of horsed cavalry, constantly adapted his military outlook to take account of modern developments. The architect of a highly-effective fortified defence line, his grasp of military technicalities was all-embracing, while his tactics in mobile warfare were at least a match for the Soviet army. As an administrator he was also a master in the always-difficult circumstances of warfare in the far north. Little wonder that he was regarded by the Finns as a national hero – and as a much-respected opponent by the Soviet Union.

Manstein, Fritz Erich von (*1887–1973*), *Field-Marshal*

Manstein has been called the foremost military intellect of Nazi Germany. He was the architect of the Germans' lightning victory in the West in May–June 1940, and a considerable stabilising influence on the Russian front during several periods of great peril.

The outbreak of the Second World War found him, newly promoted to major-general after a period as a colonel on the O.K.H. planning staff, chief of staff to Runstedt (q.v.) at Army Group South. It was Manstein who suggested the brilliant 'left hook' plan for the invasion of France: a surprise armoured penetration of the Ardennes area in strength and at speed, while the Allies were lured into committing their forces to the Low Countries, followed by a north-westward drive to trap them against the coast. This plan was supported enthusiastically by Guderian (q.v.) and the other tank generals, although it was regarded with reservation by the more conservative elements of the *Wehrmacht* command. In the event it succeeded brilliantly, and Manstein himself commanded an infantry corps with great distinction during the campaign. He was given command of the 56th Panzer Corps in East Prussia in the spring of 1941, and again distinguished himself in June on the northern flank of the German invasion of Russia. On the first day of 'Barbarossa' his corps – comprising the 8th Panzer Division, the 3rd Motorised Division, and the 290th Infantry Division – advanced faster than any other German formation, penetrating fifty miles by sunset. By 26 June he was across the river Dvina, and the advance of Hoepner's 4th Panzer Army, of which his corps formed a forward element, was threatening the Leningrad front. Later in the campaign this momentum was lost due to ill-judged decisions by Hoepner and his superior Leeb, commander-in-chief of Army Group North; in fairness it should be mentioned that this command had been assigned very ambitious objectives and insufficient means to achieve them.

In September 1941 Manstein was given command of the 11th Army on the south-east front. In ten months he cleared Soviet forces from the Crimea, despite inferior strength, and took 430,000 prisoners. The climax of this campaign was the successful storming of the great fortress complex of Sebastopol. He maintained his isolated forces throughout the severe winter, and in July 1942 was promoted field-marshal. In the autumn of 1942 he was given command of Army Group Don at a time when the encirclement of the 6th Army in Stalingrad threatened a disaster. His cool assessment of the situation was in sharp contrast to the unsound emphasis placed on the importance of the Stalingrad position by Hitler, but his mission was vaguely defined and the forces at his disposal were inadequate and widely dispersed. His command included – nominally – the trapped 6th Army itself, the 4th Panzer Army, and the almost useless 3rd Rumanian Army. He quickly determined on an attempt to link up with the Stalingrad garrison, co-ordinated with a break-out to the south-west, as the only hope of saving Paulus's army. His energetic attempts to concentrate a realistic relief force and to link up with the 6th Army were frustrated, however. Paulus hesitated fatally over the desirability of committing his forces to an all-or-nothing break-out attack, and Hitler's personal refusal to countenance this abandonment of Stalingrad – with which he had become obsessed – was the final nail in the coffin of the 6th Army. Had Paulus shown greater determination Manstein might have presented Hitler with a *fait accompli*, and there is reason to believe that a considerable part of the garrison might have been saved.

As Commander-in-Chief of Army Group South during 1943, Manstein achieved some impressive results. The shock which spread through the higher command after the fall of Stalingrad gave him some room to manœuvre, and he used it to advantage. He brought the Russian advance to a halt at Kharkhov in March, and adopted a policy of 'mobile defence'. He appreciated that the inexorable increase in Russia's material superiority could only be countered by a flexible defence which traded ground for time, sucking Russian armies into deep penetrations which could be cut off with great loss by intelligent manœuvre. This was anathema to Hitler, who justified his insistence on holding every mile of captured territory by pointing to the correctness of this attitude of his in the winter of 1941. Despite this difference of attitudes Hitler had a great respect for Manstein, who was on several occasions surprisingly successful in persuading the *Führer* to sanction strategic withdrawals, such as the retreat to the Mius after Stalingrad and the retreat to the Dnieper after Kursk. Manstein was opposed to the '*Zitadelle*' offensive against the Kurst salient in July 1943, pressing an alternative proposal which envisaged a withdrawal in the south to lure Soviet forces into a penetration which could be 'lopped off' and destroyed against the Sea of Azov. In the event Kursk was a disaster. Manstein's Army Group South and Model's (q.v.) Army Group Centre suffered terrible tank losses in the deep belt of prepared Soviet defences, but never came near to linking up in their simultaneous drives from north and south. After carrying out a masterly withdrawal and partial recovery, Manstein demanded once too often that Hitler allow him flexibility of manœuvre. On 25 March 1944 the *Führer* relieved him of command, albeit with courtesy and with an award of the Swords to his Knight's Cross to lessen the sting. He lived quietly on his estates for the rest of the war, and finally died in retirement in 1973.

Liddell-Hart (q.v.) described Manstein as 'the Allies' most formidable military opponent; a man who combined modern ideas of manœuvre, a mastery of technical detail, and great driving power'.

Marlborough, John Churchill, 1st Duke of (*1650–1722*)

The British general and master tactician John Churchill, later Duke of Marlborough, was a military leader second only in the eighteenth century to Frederick the Great of Prussia, and the most successful of all the Allied commanders in the War of the Spanish Succession.

Born the son of an impoverished Royalist squire at Ashe in Devon, John Churchill seems always to have been embroiled in the schemes of the aristocracy. As a young man he became page to the Duke of York (his sister Arabella had become the Duke's mistress) and the lover of Lady Castlemaine, later Duchess of Cleveland. John Churchill received the sum of £5,000 for services rendered to Lady Castlemaine, this sum being used to purchase an annuity; and in 1667 he was given a commission in the Foot Guards. He served in Tangier between 1668 and 1670 during the Consolidation by Rashid II. Promoted captain in 1672, Churchill served with the Allied fleet which was badly mauled by the Dutch (in the Third Anglo-Dutch War of 1672–4) at the Battle of Sole Bay on 28 May 1672. In December that year he accompanied English troops sent to assist Louis XIV against the Dutch, and particularly distinguished himself during the capture of Maastricht on 29 June 1673, finding personal favour with the Duke of Monmouth. Louis appointed Churchill commander of the English regiment which was present at the Battle of Enzheim, an inconclusive battle fought between an Anglo-French army under Turenne (q.v.) and Bournonville's army of 38,000 men on 4 October 1674. Tiring of foreign service under Louis XIV, Churchill returned home and served the Court on various minor diplomatic missions. It was during this service that he was attracted to Sarah Jennings, a lady-in-waiting and childhood friend of Princess Anne (who later became queen).

John Churchill and Sarah Jennings were married during the winter of 1677/8, and thereafter he rose rapidly at Court, being created Baron Churchill in the Scottish peerage in 1682, and promoted colonel of the 1st Royal Dragoons the following year.

In 1685 James II ascended the throne, and Churchill was sent to France to seek subsidies for the British crown from Louis XIV. He was made lord of the bedchamber, and Baron Churchill in the English peerage. Promoted major-general in the same year, he was appointed second-in-command of the force sent to oppose the Duke of Monmouth, who had landed in Dorset at the head of a Protestant revolt as pretender to the English throne; at the Battle of Sedgemoor (6 July 1685) the revolt was crushed, and Monmouth was captured and afterwards beheaded. Despite this service under James II, Churchill was against Catholicism in principle and also in the context of his wife's close association with Princess Anne. King James's policies made him so many enemies that a powerful faction was on hand to support the Protestant Prince William of Orange when he landed in England, by invitation, on 5 November 1688. The ever-realistic Churchill at once deserted the English army and joined William at Axminster in November 1688. James fled to France in the following month, and William and Mary were proclaimed joint sovereigns; Churchill was sworn a privy councillor and created Earl of Marlborough in 1689.

In the war of 1688–97 between France on the one hand and Spain, England, Holland, and the German States on the other, Marlborough first commanded an English contingent of 8,000 men in the army of Prince George Frederick of Waldeck, contributing to the defeat of the French army in the Battle of Walcourt in Flanders on 25 August 1689. At the same time support for James's Catholic cause and his desire to regain the English throne resulted in the outbreak of war in Ireland, compelling King William to lead a continental army against James and Tyrconnel in Ulster. During the absence of William from London, Marlborough served on the Committee of Nine, managing the government of the country in the King's place. It was during the absence of William from London that Marlborough fell into disfavour with Queen Mary, as a result of his and his wife's conspiring to assist the queen's sister Princess Anne in obtaining a settlement of £50,000 a year in defiance of the Queen's wishes. A breach occurred between Mary and Anne when the latter refused to dismiss Sarah Churchill in 1692; in spite, Mary promptly had Marlborough stripped of his offices and imprisoned in the Tower.

Queen Mary died in 1694 and the Churchills were reconciled with William. In 1698 Marlborough was made governor to Anne's son, the Duke of Gloucester, and also annually appointed one of the Lords of Justices until 1700. When in 1702 Anne ascended the throne Marlborough was created Knight of the Garter, and promoted Captain-General of the Forces and Master General of the Ordnance.

In 1701 the War of the Spanish Succession had broken out following the death of Charles II, the last member of the Spanish Hapsburg line. The contenders for the Spanish throne were Philip, grandson of Louis XIV of France, and Leopold I, the Hapsburg Emperor. Unwilling to see France united with Spain, England declared war, and John Churchill, Earl of Marlborough, was sent to Holland as Captain-General of the combined English and Dutch forces on 15 May 1702. The following month he led this army in an invasion of the Spanish Netherlands, and with an army of 50,000 men (of whom 12,000 were British) attempted to force a French army under the Duke of Boufflers into battle. However, suspicious of his reckless tactics, the Dutch governmental deputies repeatedly refused to give sanction for the Dutch troops to join battle – to the anger and frustration of Marlborough.

By September Marlborough had moved southwards and opened the lower Rhine and Meuse rivers. As such movement was in the interest of Dutch trade routes and there was no risk of open warfare, the Dutch deputies then agreed to the seizure of the Meuse fortresses. Accordingly Churchill captured Venloo, Ruremonde, and Liège – for which successes he was created Duke of Marlborough.

While Marlborough was thus engaged, Bavaria entered the war on the side of France, thus endangering communications with Austria. He therefore made plans to secure these lines through the Rhine, and then to penetrate and shatter the line of French forts in the Spanish Netherlands and take Antwerp. His first objective was achieved when in May 1703 he captured Bonn, thereby opening the Rhine valley; but when he turned to his second, he was again frustrated by the Dutch refusing their sanction for their forces to face the greater French army under Villeroi.

Realising that the situation in central Europe would hinge on the survival of Austria, whose armies were fighting on the Rhine and in Italy, Marlborough decided to mount feint attacks in several directions. As Captain-General of the Anglo-Dutch forces he sent an expeditionary force under the Archduke Charles to Lisbon for an invasion of Spain; he sent another diversionary force to attack Toulon in support of the Huguenot rebels in southern France; and he moved a small force against Alsace. While the Austrian forces, under Prince Eugène (q.v.), were recalled from Italy, Marlborough decided to effect a major concentration of Allied forces in the Danube valley to save Vienna, to eject the French from Germany, and knock Bavaria out of the war. (There is much controversy as to whether this bold strategy was that of Marlborough himself; some opinion leans to the belief that its architect was Prince Eugène of Savoy. Whoever was the originator, it was Marlborough who was to carry the greatest burden of logistics and whose lines of communication were most tenuous.)

Cleverly deceiving both friend and foe during May–June 1704, Marlborough left 60,000 men to protect Holland and marched towards the Rhine with an Anglo-German army of 35,000 men (of whom about 11,000 were British, and the remainder Germans in English pay). Reinforced by a further 35,000 men under Prince Louis of Baden, Marlborough then pushed on towards the Danube and towards Eugène's 30,000 Austrians. On 2 July, in a surprise attack, Marlborough seized a hill overlooking Donauwörth and captured the town without a siege. By the time Marlborough and Eugène joined forces on 12 August the Allied army had been reduced to a total of about 56,000 men, as further feints had been made and garrisons provided *en route*. Encamped beside a small stream near Blenheim, just north of the Danube, was Marshal Tallard's Franco-Bavarian army of about 60,000 men.

The lines of battle lay roughly south-east to north-west, the Franco-Bavarian army facing north. Tallard's main strength occupied the southernmost section of the French line, based upon the village of Blenheim, and facing Marlborough's main strength. Eugène's forces occupied the right flank of the Allied line, facing Marsin and Maximilian. Marlborough's plan was to launch his major attack frontally against Tallard, forcing the French to commit their reserves in their centre and right; at the same time Eugène was to press forward in an enveloping movement folding back the French left on to the Danube. Marlborough and Eugène attacked simultaneously at 12.30 p.m. on 13 August; the main attack suffered heavily but achieved its purpose: Tallard poured in his reserves. For four hours savage fighting continued between the main forces, but at 4.30 p.m., seeing that Eugène was making steady progress round the French left flank,

Marlborough mounted a heavy attack with his cavalry squadrons, and inside an hour these had broken through the French centre. Although he was unable to turn the French flank, thereby closing the jaws of the huge pincer, Eugène did manage to swing his flank round, with the result that many of the fleeing French and Bavarians were drowned in the waters of the Danube. Tallard himself was captured, and it fell to Maximilian and Marsin to extract the remnants of the Franco-Bavarian army from the field. Allied losses amounted to 12,000 (of whom 4,500 were killed), compared with 38,500 casualties in the Franco-Bavarian army.

The Battle of Blenheim provides one of history's great examples of co-operation and co-ordination between armies of allied nations. The prestige of the French military was shattered and Maximilian was forced to flee his country, Bavaria being annexed by Austria. Notwithstanding his brilliant victory, Marlborough was again frustrated by the faint-hearted Dutch: when he returned to the Netherlands, he was once more deprived of their sanction to proceed against Villeroi. However, a grateful Emperor created Marlborough Prince of Mindelheim. At home, by Act of Parliament, a grant of £100,000 was awarded to him, accompanied by an order from Queen Anne for the building of a palace for Marlborough at Woodstock, north of Oxford. (In the event a considerable sum in excess of this amount was spent from public funds in the completion of what came to be called Blenheim Palace.)

After a year of stalemate and desultory campaigning, Marlborough remained convinced that the French could be decisively beaten in Flanders, and moved with an army of about 60,000 men to seize the city of Namur during April 1706. Rightly interpreting this aim, Villeroi assembled an army of equal strength to protect the city, but was intercepted by Marlborough at Ramillies. The French formed their line of battle on high ground in a defensive position, partially entrenched. In a manœuvre typical of Marlborough, the Allies feinted against the French left, thereby drawing reserves from the centre, whereupon the British commander threw his main attack against the weakened right. After severe fighting, the entire French line wavered and, in the face of renewed attack, fled the field. Marlborough ordered a vigorous pursuit and heavy casualties were inflicted, the French losing about 15,000 men. Allied losses were 1,066 killed and 3,633 wounded. Quickly following up this heavy enemy defeat, Marlborough set about consolidating the Spanish Netherlands, capturing Antwerp (6 June), Dunkirk (6 July), Menin (22 August), and Ath (4 October). Once again there followed a year of recrimination and debate by the Dutch, effectively causing stalemate in the north and allowing the French to reorganise their commands. The incompetent Villeroi was replaced by the French marshal, Louis Joseph, Duke of Vendôme (q.v.), victor of a brilliant campaign around Turin in the autumn of 1706.

By early 1708 Marlborough was at Brussels with an army of almost 70,000, while Eugène commanded an army of 35,000 at Coblenz. It was planned that as soon as a new army had been raised in central Germany it would take Eugène's place on the Rhine, and he would then move north to reinforce Marlborough. Realising that the junction of these two Allied armies would create an overwhelming threat, Vendôme seized the initiative in May 1708 and started to move against the Allies near Louvain, suddenly swinging westward, seizing Ghent and Bruges, and threatening the garrison at Oudenarde. Marlborough, realising that he had been outmanœuvred and facing falling Dutch morale, decided to face Vendôme and seek battle without further delay.

At this point, learning that Eugène's army had started its march north from the Rhine, the French command became nervous since its army lay scattered in the field north of Oudenarde. While his army was still some distance away Eugène himself joined Marlborough and was given command of the Allied right wing, and the two commanders hastily prepared a plan of battle reminiscent of Blenheim.

As the Allied army swarmed across the river Scheldt, the Duke of Burgundy at first ordered a general retreat but, being persuaded by Vendôme that he would be cut off from France if he did not stand and fight, ordered instead a defensive battle. At 2 p.m. on 11 July both armies attacked simultaneously and a vicious battle ensued, some advantage being gained by the Allies from the determination of their commanders; the French, on the other hand, seemed dispirited by the continuing vacillation in their command. The Battle of Oudenarde was to some extent a repetition of the envelopment at Blenheim; by dusk Marlborough had enveloped the French right while Eugène continued his swing round from the Allied right. As dusk fell the French were driven from the field, losing 4,000 killed, 2,000 wounded, 9,000 prisoners, and 3,000 deserters. Allied losses were 2,000 killed and 5,000 wounded. In spite of this humiliating defeat, Vendôme displayed his brilliance in the field by rallying his defeated troops; he counterattacked and repulsed the pursuing Allies at Ghent on the following day, thereby retaining control of western Flanders. Three days later Eugène's army joined Marlborough, who promptly determined to invade France with his army, which now numbered about 120,000 men. This plan was equally promptly forbidden by the Dutch. Instead, Marlborough laid siege to Lille, which surrendered on 11 December.

Following further setbacks during the winter of 1708–9, the French armies fell back to their own borders while King Louis sought peace terms from the Allies. But such was the severity of these terms that Louis, supported by the French populace, decided to continue the war. This was a period of intrigue among the Allied Governments – intrigue which, in view of Marlborough's inherent opportunism, led to a deterioration in confidence in the military control and aims of the Allies. In the early stages Marlborough engaged in active diplomacy to retain cohesion among the Allies, being himself thoroughly opposed to any talk of peace with France. In this he was supported by the Dutch (for all their faintheartedness) who still hoped to make good their claims on the Barrier Forts. Further discord was aroused when the Emperor offered the governorship of the Barrier Forts to Marlborough – an appointment for which he longed; however, in view of Dutch opposition, he was forced to decline it in the interests of Allied cohesion – though not without hopes that he might yet achieve it.

Preparing for the continuation of the war with France, Marlborough and Eugène, commanding an army of about 110,000, occupied most of the Spanish Netherlands and faced a French army in a fortified line, lying roughly along the line of the border. The French commander, Villars, with about 90,000 men, was under orders from Paris to avoid general battle if possible. During the summer of 1709 the Allied army indulged in some inconclusive manœuvring, but as their plans took shape some measure of success materialised. Tournai fell after a month-long siege on 29 July, and the Allied army moved towards Mons, thereby threatening the French garrison. Villars now received orders from Paris to protect and hold Mons, to which the Allies laid siege.

Villars concentrated his 90,000 men at Malplaquet, knowing that by so doing he would attract the Allied army to him and away from Mons. Marlborough, leaving 20,000 men to continue the siege, eagerly accepted the challenge and, with Eugène and 90,000 men, advanced to meet the French on 9 September. As usual Marlborough advanced on the left, with Eugène on the right. The Allied plan, effective at Blenheim and Ramillies, was for holding attacks to be launched on the French flanks to draw reserves from the centre. The main blow was to fall in the centre. But this was the first time the Allied commanders had been opposed by Villars, a highly competent general whose ability almost proved equal to their own. When the battle opened on 11 September Villars, fearing an envelopment on his left flank by Eugène, weakened his centre to oppose the Austrian commander, and bitter fighting followed in which Eugène was twice wounded (but refused to leave the field), and Villars himself was so badly wounded (while leading an attack himself) that he had to turn over his command to Boufflers. Marlborough now launched a tremendous attack by both cavalry and infantry against the French centre and burst through. Villars had anticipated this and had retained a measure of reserve, which Boufflers now threw into the battle and with which he managed to re-form his line. At this point both Marlborough and Eugène committed the last of the Allied reserves, and again the French centre was pierced. Without fresh forces to withstand this new threat, Boufflers ordered a general withdrawal; the Allies, however, having suffered very heavy casualties, were unable to give chase.

In this Battle of Malplaquet the French suffered about 4,500 killed and 8,000 wounded, compared with 6,500 killed and 14,000 wounded among the Allies. The battle had little result except to enable the Siege of Mons to continue: the town was eventually captured on 26 October. It was the last major battle fought by Marlborough, for though the two Allied commanders captured the border fortresses of Bethune and Douai during the summer of the following year, the English commander was recalled to London on the last day of 1711.

Three years earlier, amidst the controversy that raged over continuation of the war with France, the Whigs had emerged as the 'War party' and so had attracted Marlborough's support. When the Tory party was returned at the next election, the cabinet recalled the Duke and relieved him of all commands. Following the turn of the tide of public feeling he was reviled in the Press for extravagance in the building of Blenheim Palace – for which he certainly was not personally responsible. Accordingly he travelled abroad and made contact with the Court at Hanover, rightly locating the likely line of succession to the English throne. When in 1714 George I acceded to the throne of England Marlborough returned and was permitted to resume his military offices, but two years later he suffered a stroke and his health broke down. He continued to perform his public duties, but suffered a second and later a third stroke, and died at Cranbourne Lodge, Windsor, on 16 June 1722.

No one will deny that Marlborough possessed tactical genius, and he was undoubtedly one of history's great commanders. His battlefield successes, almost invariably in partnership with Prince Eugène of Savoy, bore the stamp of absolute determination – often in the face of less resolute opponents and against a continuing background of Dutch pusillanimity. He was the first really competent English commander to lead an army based upon Cromwell's 'New Model' on the continent of Europe, and his astonishing march to the Danube was nothing short of brilliant – its due reward was reaped at the decisive Battle of Blenheim. Yet one cannot escape the conclusion that throughout his career at command level he was constantly a victim of his own insatiable ambitions.

Marshall, George Catlett (*1880–1959*), *General of the Army*

Although he did not make his name a household word through the well-publicised exercise of a field command, as U.S. Army Chief of Staff throughout the Second World War Marshall made a greater contribution to the Allied victory than most of the field commanders. His considerable diplomatic activities and post-war political career have no place in this book, but mark him nevertheless as a considerable figure in American foreign policy in the decade 1940–50.

Born in Uniontown, Pennsylvania, on 31 December 1880, Marshall graduated from the Virginia Military Institute at Lexington in 1901 and was commissioned second-lieutenant in February 1902. He saw service in the Philippines in 1902–3. During the First World War he rose to be Chief of Operations of the U.S. 1st Army, and Chief of Staff of the 8th Army Corps. He was intimately connected with the planning of the major American operations at St Mihiel, and with the movement of 500,000 men and nearly 3,000 guns to a new sector of the front in readiness for the Meuse-Argonne operations. From 1919 to 1924 Marshall served as aide to General John J. Pershing (q.v.). He spent the years 1924 to 1927 with the 'Forgotten Fifteenth' – the 15th Infantry, stationed for many years at Tientsin, China. An appointment of major significance followed – assistant commandant in charge of instruction at the U.S. Army's infantry school at Fort Benning. Many future American commanders of the Second World War, including Bradley, Stilwell (qq.v.), Hedges, and Bedell Smith, were influenced by the doctrines put forward by Marshall in this period.

On 1 September 1939 Marshall was sworn in as U.S. Army Chief of Staff, a post he retained until his resignation on 21 November

1945. He began to push through a vigorous reorganisation plan as soon as he took up the appointment, and it was largely due to his vision and energy that when America was brought into the Second World War by Pearl Harbor she already possessed a sound framework upon which the largest army in her history could be quickly and effectively built. For six years Marshall bore the ultimate responsibility for the selection of officers for top field commands, for the raising and training of new units, for the development of equipment and weaponry, and for the entire administration of the U.S. Army all over the world. Roosevelt relied upon Marshall's military advice to a great degree, and the general accompanied his President to the top-level conferences at Casablanca, Quebec, Tehran, Yalta, and Potsdam. He was a firm believer in a unified Allied command, and in the importance of maintaining the major effort against Nazi Germany, as the most dangerous enemy the Allies faced. In December 1944 Marshall was promoted General of the Army, a newly-created five-star rank equivalent to British Field-Marshal, and also carrying with it retention on the active list for life.

On 21 November 1945 Marshall resigned as Chief of Staff, and the post was taken over by his protégé Eisenhower (q.v.). Marshall was immediately appointed special representative of President Truman and sent to China with the rank of ambassador in a vain attempt to mediate between the two sides in the civil war. On 21 January 1947 Marshall was sworn in as U.S. Secretary of State, a post he held for exactly two years. During this period he was instrumental in founding the doctrine of systematic American aid to rehabilitate the economies of the war-shattered European nations; for this 'Marshall Plan', as it became known, he received the Nobel Peace Prize of 1953. During his period in office, and during his tenure of the post of Secretary for Defense (21 September 1950 to 12 September 1951), Marshall was a firm advocate of the North Atlantic Treaty Organisation and of increased United States military presence in Korea.

Marshall died in Washington on 16 October 1959; he is buried in Arlington National Cemetery.

Masséna, André, Duc de Rivoli, Prince d'Essling (*1758–1817*), *Marshal of France*

In modern history it is rare to find a military leader whose record commands respect despite his enthusiastic indulgence in both financial greed and lechery. André Masséna somehow managed to pursue a lifelong and monumental enthusiasm for gold and women, while at the same time achieving an enviable professional reputation. He regarded men and events from the standpoint of a worldly cynicism which, it has been said, was second only to that of Talleyrand among his contemporaries. This merciless realism stood him in good stead on many battlefields. He was capable of the bold stroke, and was tenacious and cool in defence, but the rash grabs at glory of a Murat or a Ney (qq.v.) were foreign to him. In the end he was ruined – professionally – by Wellington (q.v.), but even this was not unforeseen. The Duke paid him a pretty compliment when they met in Restoration Paris, to the effect that Masséna had made him lie awake at nights – a practice which the Duke made a point of avoiding.

Physically small, with an animated Mediterranean face and flashing dark eyes, Masséna was born in Nice on 6 May 1758, and spent part of his boyhood as cabin boy on a merchant ship. Enlisting voluntarily in 1775, he rose to the rank of *adjudant sous-officier* by the time of his discharge in 1789. He married, and spent two years in business in Antibes, allegedly sailing rather close to the law. He resumed his military career via the National Guard, and was elected lieutenant-colonel in 1792. He served with the Army of Italy until 1798, rising to the rank of general of division in 1793. He saw much action, but his most dazzling successes did not come until Bonaparte took over the army in 1796. Masséna appears to have recognised the extraordinary qualities of his young General-in-Chief immediately, and in that and the following year suddenly stepped – or charged – into the first rank of Revolutionary generals. His victories included Cherasco, Lodi, Milan, Verona, Lonato, Castiglione, Arcola, San Michele, Rivoli – after La Favorita Bonaparte called him 'the darling child of victory', and he went on to win at Neumarkt and Unzmarkt in April 1797. In 1799 he commanded the French troops in Switzerland, and won the vital battle of Zürich in September. A Russo-Austrian invasion of Switzerland from Italy was foiled when Masséna smashed General Korsakov, who lost 8,000 of his 20,000 men. Suvarov (q.v.), who was advancing through the St Gotthard with another army, found himself blocked, and only cut his way through to the upper Rhine at great cost; as a direct result Russia quitted the Allied coalition.

Early in 1800 Masséna was given the 40,000-strong Army of Italy, but soon found his command scattered and hard-pressed by 100,000 Austrians under Mélas. He himself was shut up in Genoa with 12,000 men by the Austrian General Ott, with twice that number. From 5 April to 4 June Masséna was the backbone of an epic defence, in the face of disease, famine, and bombardment by British warships. On his eventual capitulation he demanded, and was granted, the full honours of war. His resistance had bought time for Bonaparte to raise and lead a new army through the St Bernard Pass, and contributed to the classic victory of Marengo on 14 June. Masséna was briefly given command of the Armies of Italy and the Reserve, but lost them through shameless plundering and peculation. He was one of the first creation of marshals on 19 May 1804.

In 1805–6 Masséna commanded the Armies of Italy and, later, Naples, taking Capua and Gaeta. In 1807, after a further corruption scandal, he was given a corps of the *Grande Armée*, but soon went on leave. In 1808 his Emperor shot out one of Masséna's eyes in a hunting accident, for which Berthier (q.v.) was promptly blamed. Masséna led the IV Corps in 1809, distinguishing himself at Landshut and Eckmuhl, and particularly at Aspern-Essling on 21–22 May. Bonaparte pushed part of his army across the Danube against fierce Austrian resistance, but was unable to reinforce them adequately and eventually withdrew, losing 20,000 men including Marshal Lannes. Masséna conducted a stout defence of the village of Essling, and eventually covered the French retreat with cool brilliance; it is said that he did not leave a single one of his wounded behind. At Wagram on 5–6 July he commanded the French left, which successfully held off the main Austrian pressure on the first day; injured in a fall, Masséna rode around the field in a light carriage, fearlessly exposing himself to great danger. Early in 1810 he received his princedom, and a command he would much rather have forgone – the Army of Portugal. Pressed by Napoleon, he took up a mission which he regarded with some foreboding.

The Army of Portugal had been unceremoniously bundled out of that country into Spain by Wellington. Now it advanced again, by the classic northern route, 65,000 strong. Masséna took the key fortresses of Ciudad Rodrigo and Almeida in July and August 1810. Wellington, with 18,000 British and 14,000 Portuguese, fell back steadily before him along prepared routes. He stood and fought on the ridge of Bussaco on 27 September, and by brilliant use of previously-scouted terrain bloodily repulsed Masséna's attacks. Eventually Masséna turned the position, and Wellington resumed his retreat – into the impregnable Lines of Torres Vedras, a massive fortified zone around Lisbon, about which Masséna knew nothing. Masséna sat down before the Lines, and managed to keep his army in being for a surprisingly long period considering the bareness of the country and the harassment of the *guerilleros*; but eventually he had to retreat into Spain, nagged by the mistress he had been foolish enough to bring with him on campaign. He attempted to gather reinforcements from the other marshals in Spain, but few were forthcoming. In May 1811 he made a determined attempt to relieve the French garrison which still held out in Almeida, but was checked in the bloody but indecisive battle of Fuentes de Onoro.

Bonaparte never forgave failure, and Masséna was replaced the same month and given a sinecure home command. His Emperor reportedly greeted him with undeserved contempt, and during the Hundred Days Masséna gave a masterly demonstration of deliberate dilatoriness which saved him from making any serious public stand. He devoted his few remaining years to his main hobbies, and died, old before his time, on 4 April 1817. It is a profitless but fascinating exercise to imagine the outcome of the Waterloo campaign if Masséna had commanded one of the French wings.

Maurice, Prince of Orange and Count of Nassau (*1567–1625*)

The father of the independent Netherlands, the moving spirit behind the weary struggle for freedom, and the greatest martyr for the cause was William the Silent, Prince of Orange and Count of Nassau. The greatest military figure of the Netherlands' turbulent childhood was, however, his son Maurice of Nassau. William pointed the way, but it was Maurice who achieved the victories which made success possible. He was one of the ablest generals of his century, and had considerable importance on a level far beyond the Dutch struggle for independence.

Maurice was born at Dillenburg on 13 November 1567, the son of William and his second wife Anne. After studying at Heidelburg, he travelled to the Netherlands and lived with his father at Breda and at Antwerp. In 1583 he read mathematics and the classics at Leiden university, partly at public expense. In July 1584 his much-loved father was assassinated, and the seventeen-year-old Maurice was appointed President of the Council of State of the United Provinces. In November 1585 he succeeded his father as Stadtholder of the provinces of Holland and Zealand; by 1591 he had also been named as Stadtholder of Utrecht, Overijssel, and Gelderland, and was captain-general and admiral-general of the forces of all these provinces. (He could not properly use the title Prince of Orange until the death in 1618 of his elder half-brother Philip William, who had been held in Spain from 1566 to 1596.)

Maurice's claim to an importance beyond the local context of the Dutch revolt rests on the nature of the army he created. During his career the Netherlands enjoyed a dramatic expansion of commerce and wealth, and this was used to create a fairly large standing army of paid, long-term professional soldiers. These were not mercenaries, but locally-recruited troops who were trained and disciplined to a standard not found in any contemporary army except the best of the Spanish royal units. Maurice employed the principles of the Romans, still the soundest basis for the movement and disposition of formed infantry, together with the latest tactical developments in the handling of weapons. The infantry of the day was the conventional mixture of pikemen and arquebusiers; each had a role in attack and in defence, and the introduction of heavier, longer-range muskets during Maurice's career added to the offensive shock-power of his formations. The lighter arquebus was gradually limited to use by his skirmishing line. Maurice was a master of siege warfare and had an excellent grasp of the potential of artillery in the field; and he sent his engineers and other senior officers to the university of Leiden to study the science of warfare. By the time his reforms were complete his new army was responsive, flexible, fast-moving, reliable, and homogeneous: the equal of the best Spanish royal infantry, and superior to the Spanish mercenaries used in large numbers in the Netherlands. Nor did he neglect the development of a strong cavalry arm – despite the marked excellence of the Spanish horse, it was the Dutch cavalry who won the day at Turnhout in 1597 and at Nieuwpoort in 1600.

Maurice began to play an important role in military affairs after the withdrawal of Leicester's English force in late 1587. The Duke of Parma (see Farnese) was distracted by the planned expedition to England in 1587 and 1588; and in 1589 Philip II, confident that the reconquest of the Netherlands was near enough to completion to allow it, withdrew Parma from his primary task and ordered him to France to take part in the civil war between Henry IV (q.v.) and the Catholic League. At this time the Dutch held only Zealand, Holland north of the Waal, Utrecht, and a few isolated cities in the east. But when Parma and the majority of the Spanish troops were marched off to France, Maurice struck. Within ten years, aided by his cousin and close associate William Louis, Stadtholder of Friesland, he was to free the north and east for ever; an important factor in his success was the English control of the Channel from 1588 onwards and the Dutch control of the coastal waters.

In 1590 Maurice opened his offensive with

the surprise capture of Breda; hidden assault parties were brought right up to the wharves in turf-barges. In June 1791 he took Zutphen in seven days, and Deventer in eleven days the following month; both these successes were marked by swift deployment and the skilled application of superior artillery. Often the rapid opening of a practical breach, coupled with offers of honourable capitulation and protection from pillage, won him possession of towns without storming. In August and September 1791 Maurice manœuvred cautiously up and down the Waal river opposite a Spanish army led by Parma, who had hurried back from France. But to the latter's intense frustration, no doubt, his King ordered him to return to France to relieve Rouen before he was able to bring matters to a conclusion with Maurice. The Stadtholder immediately mounted a lightning advance on Hulst, west of Antwerp, using both land and water transport; Hulst fell in five days. The army was then rushed back to Nijmegen, at the far end of the line of Spanish fortresses; Nijmegen fell on 21 October after six days' artillery preparation.

From 1592 to 1596 Maurice carried out a series of campaigns, winning and consolidating the northern provinces. Parma died in France in December 1592, and the new Viceroy, Mansfeldt, was no match for the Stadtholder. Maurice now made intensive training of his new regular units a priority, in preparation for more challenging campaigns in the southern provinces. There they would be far from the comforting presence of naval support; the open, drier terrain favoured the Spanish regular formations, and the population was by no means unanimously friendly to the Protestant cause. On 24 January 1597 Maurice won a remarkable victory at Turnhout. He brought up his 7,000 men under cover of bad weather to attack 6,000 Spaniards led by Varas. The Dutch cavalry drove the Spanish cavalry from the field, and then co-operated with the Dutch infantry to break the Spanish foot. Spanish losses were about 2,500 killed and captured, against only some 100 Dutch dead.

In 1600 the States General ordered Maurice to invade Flanders and complete the conquest of the coastal zone as far as Dunkirk. He had many objections to this plan, but obeyed orders. On 2 July 1600 he fought the Battle of Nieuwpoort, one of his most famous victories over the Spaniards. It was a hard-fought battle, which took place on the beach and continued in the sand-dunes and on the coast-road when the rising tide forced the armies inland. Early Spanish success was checked by accurate Dutch artillery fire, and fine charges by the Dutch reserve cavalry finally decided the day; Spanish losses were about 3,500 men, and Dutch about 2,000. But Maurice's lines of communication were still vulnerable, and he raised the siege of Nieuwpoort and retired into Dutch territory.

Maurice showed little inclination to carry the war into the south after Nieuwpoort. The Spanish general Spinola achieved several successes in the early years of the new century, notably the costly three-year Siege of Ostend in 1601–4. Dutch naval superiority was decisive during desultory compaigns in 1604–7. Eventually negotiations were followed by a twelve-year truce (1609–21) which Maurice strongly opposed. His last years were embittered by his political struggle with the Advocate of Holland, Johan van Oldenbarneveldt, whom he finally had executed in 1619. He refused to renew the truce in 1621, but failed to defeat Spinola. The Spanish siege of Breda was still in operation when, on 23 April 1625, the Stadtholder died at The Hague of a liver ailment. Maurice was unmarried, but left seven bastards.

Model, Walther (*1891–1945*), *Field-Marshal*

Model was one of the most energetic of Hitler's generals, and was employed in the later stages of the Second World War as 'the *Führer*'s fireman' – a commander who could be trusted to retrieve ugly situations. He was one of the 'new men' who had risen to senior command under the Nazis, not a member of the traditional conservative establishment of the German officer corps.

Born in 1891, Model served as Chief of Staff of the 4th Corps in Poland in 1939, and Chief of Staff of the 16th Army during the French campaign of 1940. In 1941 he led the 3rd Panzer Division – an *élite* unit known as the 'Bear' Division from its badge, the

Berlin bear – in the invasion of Russia, operating under the 24th Panzer Corps in Guderian's (q.v.) 2nd Panzer Army, the spearhead of Army Group Centre. In September 1941 Model, who had shown an aptitude for the thrusting style of leadership appropriate to the Panzer units, led the southward drive which encircled vast Soviet forces around Kiev. In 1942 he was given command of the 9th Army. During the spring of 1943, when the advisability of the '*Zitadelle*' offensive against the Kursk salient was being considered, Model advised against the plan: it was clear from reconnaissance that the Soviet command were well aware of the planned attack, and Model felt that available German forces were insufficient to break through. He was opposed by Kluge and Zeitzler, but in the event Hitler decided to delay the offensive until units equipped with the new PzKw V Panther tank were available. Ironically, a case can be made out that this delay, while strengthening the attack, also allowed a parallel strengthening of the defence and thus led to heavier German casualties when the offensive was finally launched in July. Model's 9th Army, spearheaded by no fewer than three Panzer Corps, was responsible for the northern pincer of the offensive south of Orel; his forces quickly lost momentum, freeing Russian formations to reinforce the southern sector where Hoth's 4th Panzer Army was slaughtered.

During the Soviet advances of 1943–4 Model was switched rapidly from threatened sector to threatened sector. After checking the Russian drive towards the Baltic states as commander of Army Group North, he was transferred to Army Group South, fighting Zhukov (q.v.) around Lwow in Poland. Finally he was moved to Army Group Centre, and halted the Soviet winter offensive near Warsaw. Known to his soldiers as 'The Lion of Defence', Model had a talent for inspiring troops and subordinate commanders to stubborn resistance.

On 17 August 1944 Model was appointed Commander-in-Chief West, taking over from Kluge after the latter's very brief tenure of this command. On 4 September Runstedt (q.v.) returned to this post, leaving Model free to command Army Group 'B'; his surviving documents show that on the day of his appointment to this post he foresaw the eventual outcome of the war clearly. Nevertheless he conducted the stubborn withdrawal from Belgium into Holland with marked ability. On 17 September, when the British 1st Airborne Division was dropped near Arnhem, Model was at his headquarters in the Tafelberg Hotel at Oosterbeek and was able to watch the drop with his own eyes. He reacted instantly, deploying the unexpectedly formidable *Waffen-SS* formations in the area to good effect before removing his headquarters to a less perilous position. His energy and calmness in this moment of danger were a decisive factor in the defeat of the British attack.

It was Model's Army Group 'B', comprising the 6th (S S) Panzer Army, the 7th Army, and the 5th Panzer Army, which launched the Ardennes offensive of 16 December 1944. While Manteuffel's 5th Panzer Army enjoyed considerable success in the southern sector, Dietrich's (q.v.) 6th Panzer Army was quickly bogged down in the north; by 18 December Model accepted that the offensive had failed, but Hitler characteristically refused to allow the salvaging of what remained of Germany's painfully rebuilt Panzer forces. Model's command was forced inexorably back and finally surrounded in the Ruhr pocket. When Paulus had surrendered at Stalingrad two years previously Model had said that surrender should be unthinkable for a field-marshal. Now he faced the same choice; and despite the American General Ridgway's (q.v.) urging that he accept honourable terms to save the lives of his men, he refused to capitulate. On 15 April 1945 Army Group 'B' was formally dissolved. On 21 April Walther Model shot himself.

Mohammed II, called Fatih (*1432–81*), *Ottoman Sultan of Turkey*

Mohammed II, known as 'the Conqueror', was born at Edirne in Turkey on 30 March 1432. When he was only twelve years old his father Murad II retired to Manisa following the grave Turkish setbacks in the Balkans and the conclusion of a ten-year truce with the Hungarians (see under János Hunyadi), and Mohammed came to the throne. It was to be a brief taste of power – the high officials of the

Porte prevailed on Murad to return to the throne in the same year when the Hungarians and Venetians broke the truce and launched the Last Crusade. Mohammed finally succeeded to the throne on his father's death in 1451, and from the earliest years of his reign the emphasis was on consolidation of Turkish power. He earned his title many times over, although in Europe it is natural that those few heroic leaders of Christendom who held out against him should be remembered with more admiration than the infinitely more powerful, sophisticated, and visionary Sultan.

In 1453 the Sultan's armies battered in the walls of Constantinople and wiped out the small garrison (see under Constantine XI), removing for ever the traditional bastion of Europe's eastern frontier, and the last link with vanished power. Mohammed had the walls rebuilt and erected impressive public buildings; communities of Muslims, Jews, and Christians from various provinces of the Ottoman Empire were brought in to resettle the city and to begin the slow process of changing its character. With Constantinople started on its path of metamorphosis into 'Istamboul', Hungary was the next line of defence for Christian Europe.

In 1456 Mohammed was sharply defeated before the Hungarian fortress of Belgrade by the mighty Hunyadi, and forced to retire to Constantinople. In 1459 he brought Serbia's brief and tenuous independence to an end, turning what was left of it into the frontier province of Semendria. In 1458–61 he extended Ottoman rule to Greece; his fleet drove the Genoese from the Aegean, while combined land and naval forces conquered Athens and the Morea. In 1463–4 he overran Bosnia; subsequently many Bosnian nobles were accepted into the Muslim faith, and became ferocious harriers of their Christian neighbour states in the Sultan's name. Mohammed broke the power of Genoa and of Venice in the eastern Mediterranean and Black Sea, destroying or capturing their trading colonies and defeating their fleets. His war against Venice lasted from 1463 to 1479; opening campaigns included attacks on Dalmatia and Croatia, and the invasion of Euboea followed in 1470. Venice managed to bring Persia into the struggle to distract the Turks, and Uzun Hasan invaded Anatolia in 1473, only to be defeated and repulsed at Erzinjan. The death of Skanderbeg (q.v.) led to a swift Turkish reoccupation of most of Albania, including Venetian coastal enclaves; Turkish troops even crossed the Alps from Croatia into north-east Italy, causing great alarm. Although the Venetian garrison at Scutari held out against many attacks in an epic defence, Venice conceded defeat in 1479 and concluded a peace treaty recognising Mohammed's conquests. The milestones in Mohammed's parallel campaigns against Genoa were the capture of the trading centres of Amastris and Sinope – together with the Greek state of Trebizond – in 1471, and the capture of the Crimean enclave of Feodosiya in 1475. (In that year the Khan of the Krim Tartars became an Ottoman vassal.) In 1480 a Turkish expedition crossed the Adriatic and seized Otranto, and in the same year Mohammed attempted to crush the greatest irritation in the eastern Mediterranean – the fortress and naval base of the Knights of St John of Jerusalem on the island of Rhodes. Despite a lengthy siege and very heavy losses this attack failed, due to the remarkable valour, skill, and endurance of the Knights. On 3 May 1481, on the eve of a new campaign in the East, Mohammed the Conqueror died, having greatly assisted the advance of the Ottoman Empire towards the summit of power which it was to achieve a hundred years later.

Moltke, Helmuth Karl Bernhard, Graf von (*1800–91*), *Field-Marshal*

Chief of the Prussian and German General Staff, organiser of victory in the Seven Weeks' War and in the Franco-Prussian War, Moltke was a member of the great military triumvirate under Bismarck which was to ensure French humiliation and the construction of the German empire under Prussian leadership during the latter half of the nineteenth century.

Moltke was born (like so many of his military contemporaries) of an aristocratic but impoverished family on 26 October 1800 at Parchim in Mecklenburg. His great mental powers and organising ability were undoubtedly inherited from his mother, who came of an old Lübeck family. He was educated with the royal cadet corps at Copenhagen, after which he joined a Danish infantry regiment; but a visit to Berlin decided him to seek a commission in the Prussian army, and he became a *leutnant* in a Leibgrenadier regiment. During the late 1820s he spent a period in the topographical office, and in 1833 was transferred to the Prussian general staff as a captain. In 1835 he was sent to Turkey to advise Sultan Mahmud II's government on the modernisation of the Turkish army; contrary to the original intention, Moltke soon entered Turkish service, and in 1839 accompanied Hafiz Pasha to Syria. The Turks' failure to dislodge the Egyptians from Syria was to some extent the result of ignoring Moltke's sound military advice.

Returning to Prussian service at the end of 1839, Moltke started to interest himself in the potentialities of a railway network in the conduct of a war – the usefulness of this study was not to be fully realised for almost twenty years. Staff appointments continued throughout the 1840s, during which time he gave much thought to the military unification of the German states – under Prussian domination. His delight at the victory of reactionary ideas after the revolution of 1848–9 was unconcealed, for the one obstacle to Prussia's dominance was democracy.

Promoted full colonel in 1851, Moltke was appointed personal aide-de-camp to Prince Friedrich Wilhelm, the future German Emperor, a position which demanded much travelling and at the same time much useful surveillance of current military tactical and technical thought. Thus emerged the man selected in 1857 to become chief of the Prussian general staff – an appointment confirmed the following year. In this post Moltke became an essential member of the triumvirate (with Otto von Bismarck as Chancellor and Albrecht von Roon as Minister of War) – the military hierarchy which within fourteen years was to redraw the map of Europe, bring France to total humiliation, and establish Prussia as the undisputed leader of the German empire for forty years.

The rapid growth of the Prussian railway system during the late 1850s had been closely studied by Moltke, whose searching reorganisation of the army included the fundamental use of rail for the rapid deployment of forces – thereby largely eliminating the traditional time-consuming and exhausting movement of armies by marches on foot and on horseback. The first trial of this new strategic flexibility came about in the Austro-Prussian War (the 'Seven-Week War') of June–July 1866. Moltke's plan was to move forces on the widest possible front to converge upon a chosen battlefield; and the speed with which he was able to concentrate his armies utterly confused his opponents.

As Falkenstein entered Hanover with 50,000 men in the west, Prince Friedrich Karl and General von Bittenfeld's armies moved south through Silesia and Saxony. Converging southwards, they surrounded General von Arentschildt's Hanoverian army and compelled it to surrender at the Battle of Langensalza on 27–29 June. Meanwhile another manœuvre of concentration was taking place to the east under the personal control (by telegraph) of Moltke. Converging Prussian armies totalling about 220,000 men faced a similar number of Austrians under General Ludwig von Benedek at Königgrätz on 3 July. This great battle proved that while the plan of concentration may have been sound, management of large armies attacking in concert from different directions demanded a degree of control not yet achieved. The battle almost went in the Austrians' favour owing to the early effect of their massed artillery fire; only the last-minute counter-barrage by the Prussian reserve artillery crushed the Austrian centre, though the newly-introduced breech-loading 'needle' rifle gave the Prussian infantry a crucial fire-power advantage over the Austrians with their muzzle-loaders. The Austrians made an orderly retreat covered by their artillery, leaving 45,000 men dead or prisoner. The Prussians lost 10,000 men. These two battles constituted virtually the sole operations of the Austro-Prussian War (apart from a limited campaign by Italy in the south). They had confirmed the value of a strategy of battlefield concentration, but they also demonstrated to Moltke the vital necessity of improving command co-ordination and communication – lessons he quickly applied in the next four years.

Meanwhile relations between Prussia and

France deteriorated over a move, supported by Bismarck, to place a Hohenzollern prince on the Spanish throne, thereby apparently posing a threat to France on two fronts. In his false confidence in the French Army's invincibility, Napoleon III determined to precipitate hostilities with Bismarck in 1870, and on 15 July declared war – even before his own mobilisation had been completed.

Instantly adopting a plan of concentration, Moltke quickly assembled three well-equipped armies, totalling almost half a million men, beyond the Rhine. By contrast the French armies, of about 220,000 men, were widely scattered in eight army corps. In contrast to almost non-existent French intelligence and detailed war aims, the Prussian army's intelligence was complete and its aims uncompromising – the destruction of the French army in the field and the capture of Paris. As the Prussian armies invaded, French defeats mounted and culminated in the disaster at Sedan on 1 September. Thereafter, with half of France's regular army in captivity and most of the remainder locked in siege at Metz, the Prussians advanced to lay siege to Paris. On 27 October 173,000 Frenchmen laid down their arms at Metz, and Paris capitulated on 28 January the next year.

Behind the Prussian field victories lay the organising genius of Moltke, whose professionalism in such matters as transport administration, supply, reinforcement, and intelligence utterly crushed the outdated military prestige of France. That is not to say that the conduct of the campaign in France had been without its difficulties for the Prussians; and the emergence of *francs tireurs* behind the advancing armies sorely tried the ever-lengthening lines of communication and introduced a new element of military conduct into modern warfare.

After the Franco-Prussian War the German states continued their process of unification as a stabilising influence between the major powers of France and Austro-Hungary. The retention of the leading position seized by Prussia in this move towards unity was ensured by the so-called diplomacy of Bismarck and the proven military genius of Moltke. Having decisively beaten their giant neighbours, the Prussians' strategy was thus wholly Clausewitzian – that war was the extension of diplomacy.

Montcalm, Louis-Joseph, Marquis de Montcalm-Gozon de Saint-Véran
(1712–59), Major-General

Montcalm's fame rests on a short career during which he never commanded more than some 4,000 regular troops, and which ended with his death and defeat, and the loss of France's considerable American colonies for all time. It might be thought strange that he should be remembered as one of France's most respected soldiers, yet few dispute his right to the name. This respect is partly due to his extremely attractive personality, and the unusual affection he inspired in his subordinates; and partly to the impressive successes he achieved with indifferent resources in the face of a superior enemy.

Born at Candiac, near Nîmes, in the south of France, Montcalm was the son of an obscure country nobleman with land but little wealth. He entered the Army as an ensign at the age of fifteen, and when he was seventeen his father bought him a captaincy. Six years later he inherited the title, and a mass of debts. An arranged marriage eased his financial position to some degree and, it is pleasant to record, brought him enormous happiness; surviving letters show that husband and wife enjoyed a tender and loving relationship. Montcalm fathered ten children, of whom five survived him, and was never happier than when living quietly on his rather run-down estates in the south, surrounded by his large family. He was a Mediterranean personality, warm and open, with the gift of commanding affection and great loyalty from his officers and men. Humorous, impulsive, and excitable, he was never unjust in his dealings. Physically he was short and inclined to stoutness, with thick black brows over a pair of flashing dark eyes, and a great predatory beak of a Roman nose. As he was without great influence, and never touted for position among the gilded moths circling the flame of Versailles, he seemed destined for a decent but obscure career as an infantry officer.

Montcalm served with the army of French 'volunteers' which Marshal de Broglie led into Bohemia in 1741 to repulse a Bavarian invasion. In 1743 he became colonel of the *Régiment d'Auxerrois*, and in the following three years saw much action in the confused campaigns of the War of the Austrian Succession. He distinguished himself in Italy in 1746, serving under Marshal de Maillebois in the Franco-Spanish army which was fighting an Austro-Sardinian coalition. An indecisive battle was fought at Piacenza on 16 June, and Montcalm was captured after receiving five sabre wounds while rallying his men under the walls of the town. He was paroled and exchanged in time to see more action – and receive another wound – before the Peace of Aix-la-Chapelle in 1748.

Early in 1756 Montcalm was promoted major-general and sent to command French troops in Canada. The appointment was not greatly sought after by officers of more influential connections, who regarded it as a professional dead-end in view of the European situation. Both sides faced special difficulties in the bitter war which was being fought along the frontiers of Britain's thirteen American colonies. The trackless forests could only be penetrated by small craft moving along the lakes and rivers. French and British struggled to establish, and to capture, strategic forts commanding these waterways; the terrain was so difficult, and the distances were so great, that front lines in the European sense were meaningless. In summer the forests and lakes provided some food, but staple provisions had to be freighted up long and vulnerable lines of communication. The Indians and European forest rangers employed by both sides could infiltrate and ambush almost at will. Isolated forts were hard-pressed to survive the harsh winters in the face of hunger, illness, and raiding parties. Both sides used provincial troops of uneven quality, around a core of European regulars who sometimes had difficulty in adapting their tactics to wilderness conditions. The French, numerically much the weaker, were aided by the feudal discipline of their colony; but Montcalm was hindered by squalid local politics, administrative corruption, and the obstruction of Governor-General Vaudreuil, who bitterly resented the wide powers granted to this commander sent out from France. British military strength far outstripped that of the French, and the Thirteen Colonies had greater resources of manpower; but the provincial administrations were inexperienced at mounting operations on a large scale, and tended to be unco-operative – even twenty years before the American Revolution there was considerable restlessness among the colonists. The huge areas of threatened frontier country also led to a fatal dispersal of British strength.

Montcalm, who never had more than about 4,000 regular infantry available, mounted a series of expeditions using mixed forces of regulars and provincials supported, very unpredictably, by Indians. But he achieved surprise and local superiority of numbers in well-conceived attacks against strategic targets such as Fort Oswego on Lake Ontario (July 1756) and Fort William Henry on Lake George (August 1757). The latter was forced to capitulate after a prolonged bombardment; Montcalm transported some thirty cannon down the waterways on platforms lashed across whale-boats. Despite his emphatic orders to the contrary, the garrison and their families were attacked by the Indians who accompanied the expedition, and several hundred were massacred. Montcalm showed considerable courage in intervening physically to save the rest of the wretched prisoners from the drunken savages. In July 1758, 15,000 British and provincial troops under Abercromby advanced against Montcalm's base at Ticonderoga, commanding the junction of Lakes Champlain and George. With only 3,600 effectives, Montcalm organised a spirited defence and inflicted severe casualties on the first waves of attackers. Abercromby was apparently so shaken that he abandoned the offensive and withdrew – a militarily inexplicable decision.

Montcalm's skilful deployment of his very limited resources to achieve local victories enabled him to defuse the successive threats of major British offensives in three campaigning seasons; but the refusal of France to send adequate reinforcements eventually doomed the colony. William Pitt authorised, and intelligently directed, an inexorable increase of British strength in America. The 'wings' of the French colony's defences were neutralised with the capture of Louisbourg in the north and of Fort Duquesne in the Ohio valley – the original flash-point of the whole conflict. Montcalm had no option but to abandon his forward bases and withdraw to defend the heart of the colony against the Quebec expedition of 1759. He directed this defence with skill and energy, and it must be remembered that Wolfe's (q.v.) attack on the Heights of Abraham was in the nature of a desperate last gamble to achieve something before the onset of winter and the deterioration of his command. Montcalm's advance from the city against Wolfe's line was certainly premature; had he waited, strong reinforcements could have been gathered both from beyond the city and from forces some miles behind the British position. Montcalm's southern impetuosity must have played a part, although his staff backed the decision unanimously. It is thought that he was anxious to crush what he feared was merely the advanced party of a much larger British force before it became too firmly established on the Heights. In any event, two British volleys and a charge broke the French line, and Moncalm was shot through the body in the confusion. He died in the early hours of 14 September 1759, with courage and great dignity, and was buried that night in a shell-hole under the floor of the damaged Ursuline convent. Soldiers and citizens wept at his passing, and the affection and respect which he had earned from friend and enemy alike has survived both the slanders of the wretched Vaudreuil, and the passage of more than two centuries.

Montfort-l'Amauri, Simon de, Comte de Toulouse (*c. 1160–1218*)

This leader of the so-called Albigensian Crusade was born in about 1160 at the family castle of Montfort-l'Amauri between Mantes and Rambouillet in the Île-de-France. His father, of the same name, was Baron de Montfort-l'Amauri; his mother was Amicia, sister of the English Earl of Leicester, Robert de Beaumont. Little is known of Simon's early life, but it may be assumed that he acquired a considerable reputation as a soldier. In 1202 he set out on the Fourth Crusade, a disgraceful exercise which damaged Christendom more than Islam. The original intention had been to sail direct from Europe to Egypt in Venetian ships; but on the death of the original leader of the enterprise it was decided instead to travel to Constantinople by way of Zara. This former Venetian enclave was held by the Hungarians, and the price of sea transport was lowered on the understanding that the crusaders would first help the Venetians recover it. The Pope denounced this plan and threatened to place the ban of the church on any who shed the blood of fellow-Christians rather than fight the infidels, but the joint expedition to Zara took place nonetheless. Simon played a part in the capture and sack of the city. The majority of the crusaders then entered into another agreement with the son of a deposed Emperor of Byzantium to the effect that they would capture Constantinople and restore his family to power in return for support on the eventual expedition to the Holy Land, and large financial inducements. This was apparently more than Montfort could stomach, and he and a party of crusaders took ship direct for Palestine; they fought for some three years in Syria. The remainder of the army attacked and sacked Constantinople, causing great damage and loss of life.

Simon returned to France in 1207. The following year Pope Innocent III proclaimed an internal 'crusade', and though King Philip Augustus played no direct part, he encouraged his northern barons to participate. Certain heretical sects known as the Catharists and Waldensians had become firmly established in southern France, particularly in Languedoc, and the clergy had failed to uproot them by any less drastic means. The 'crusade' offered the northern barons the chance of rich plunder – a consideration which probably weighed with them far more heavily than a tenderness for religious orthodoxy – and King Philip saw it as an opportunity to increase the power of the central government at the expense of the local barons. Between 1208 and 1213 Simon de Montfort-l'Amauri led the 'crusaders' in a series of able campaigns which destroyed the strength of the southern barons in western Languedoc. He gradually captured most of the towns and strongholds of the heretics, including the Catharist centre of Albi; vast numbers of the inhabitants of these places were slaughtered in an orgy of cruelty.

He also took the opportunity to add to his personal holdings. In 1209 the Viscomte de Beziérs et de Carcassone, Raymond Roger Trencavel, was stripped of his lands, and Simon appropriated them despite the clear claim of Raymond's infant son. Raymond VI of Toulouse eventually rose against the invaders, but was defeated at Castelnaudary in the summer of 1212. By this time only he, the Comte de Foix and the Comte de Comminges still resisted among the southern aristocracy, and Simon had taken all the main centres except Toulouse and Montauban. Most of the other northern nobility had gradually left the crusade and returned home, leaving Simon with fairly weak forces with which to conclude the campaign.

In 1213 matters took an unexpected turn with the intervention of Pedro II of Aragon, who made common cause with the heretics despite his staunch Catholicism. His religion was outweighed by his alarm at the increased power of the King of France over the southern provinces, and anxiety for his own feudal holdings north of the Pyrenees. Pedro led a Spanish army into Languedoc and linked up with Raymond of Toulouse in 1213, storming and capturing several of Simon's strongholds. In September, with some 4,000 knights and 30,000 infantry, they invested the key fortress of Muret which was garrisoned by only 700 of Simon's men. Just as the joint army arrived to take up siege positions Simon arrived in person with another 900 horsemen, and managed to join the Muret garrison. He found that the fortress was low on supplies and could not possibly hold out until reinforcements arrived from the far north. He therefore settled on an audacious plan, which won him the victory on which his military reputation largely rests. He lured part of the attacking army to assault an apparently weakly-defended gate, ambushing them inside the walls with his cavalry and causing heavy casualties. While the enemy's attention was firmly focused on this south-eastern sector of the defences, he led his 900 horsemen out of the south-west gate and rode westwards, apparently abandoning Muret to its fate. In fact he circled to the north, surprising and smashing a stronger force under the Comte de Foix which he encountered. The sounds of battle warned the main Aragonese army, but they only just had time to form line when 600 of Simon's knights debouched from cover and hurled themselves straight into the centre of the enemy mass. The Aragonese, outnumbering this force by thirty to one, were closing around this little group of horsemen when Simon himself led his last 300 men in a desperate attack on the enemy rear. Incredibly, the Spanish broke and fled; in the course of the short pursuit they lost many dead, including King Pedro himself. Simon then turned upon to so-far-unengaged force led by Raymond of Toulouse, and smashed it.

Although isolated strongholds held out for another ten years or so, Muret was the last major battle of the Albigensian Crusade. Simon de Montfort-l'Amauri occupied Toulouse, declared himself the Count, and concentrated on consolidating his personal grip over his newly-conquered territories. At the Fourth Lateran Council of 1215, after heated exchanges, Pope Innocent reluctantly confirmed the new Count in his position, and Simon did homage to Philip Augustus for his lands. He styled himself Viscomte de Beziérs et de Carcassone, Comte de Toulouse, Duc de Narbonne; but his victims were not yet cowed, and he did not live long to enjoy his success. Although failure could only bring them another horrifying massacre the people of Toulouse rose against him, and it was while besieging the rebellious city on 25 June 1218 that Simon was killed. His third son succeeded to the Earldom of Leicester through the female line, and was prominent in leading the revolt of the English barons against King Henry III.

Montgomery, Bernard Law, 1st Viscount Montgomery of Alamein (*1887–1976*), *Field-Marshal*

One of the outstanding military leaders of the Second World War, architect of British victory at El Alamein, and land commander of the Allied forces in the victorious invasion of northern Europe in the last year of the war, Bernard Montgomery was a controversial figure for the last thirty-five years of his life.

Son of Bishop H. H. Montgomery and of Irish stock, Bernard Montgomery was born in London on 17 November 1887. He was educated at St Paul's school and the Royal Military Academy, Sandhurst, and served with distinction during the First World War, being promoted brevet major and winning the D.S.O. Staff and command appointments followed between the wars; he commanded the 1st Battalion, Royal Warwicks, from 1931 to 1934, and the 8th Division in 1938–9. On the outbreak of the Second World War he commanded the 3rd Division in France until the fall of that country, returning to Britain to take command of the 5th Corps. Already Montgomery had gained recognition for his thoroughgoing methods of troop training, with the emphasis upon physical fitness; perhaps reflecting the rather puritanical background of his own youth, he presented a somewhat austere image and – unusual in an army officer – was a strict abstainer and a non-smoker.

General Rommel's (q.v.) swift advance in Libya early in 1942 was quickly followed by changes in the command of British forces in the area, both Auchinleck and Ritchie being relieved by Churchill, and replaced by General Harold Alexander (q.v.) in overall command, and General Gott in command of the British Eighth Army. Gott was killed *en route* to take up his post when his aircraft was shot down, and the appointment then fell to Montgomery (who had been scheduled to become Eisenhower's (q.v.) deputy for the forthcoming 'Torch' landings in North Africa). Fortuitous though his appointment may thus appear, Montgomery immediately brought to the threatened battlefield exactly the qualities demanded by the situation. His meticulous attention to detail, his impeccable staff work, and his cautious, painstaking, and inflexible generalship promised to stem the tide of misfortune that was running in the Western Desert.

Auchinleck had succeeded in halting Rommel, who had outreached his lines of communication, at El Alamein on 7 July 1942, and when Montgomery took command on 13 August he found the British forces exhausted, but steadfastly facing the combined Italian and German armies, which were clearly reinforcing preparatory to launching a new attack before the Eighth Army could recover its strength. Rommel, however, attacked too soon – at Alam-el-Halfa on 31 August; and by using the British 7th Armoured Division to meet and block the frontal attack, and a dug-in tank brigade to repulse a flanking attack, Montgomery held his line while the German armoured forces, often stranded by lack of fuel, were punished from the air. The opportunity was taken to strengthen the line at El Alamein on interior supply routes, while Rommel desperately tried to rebuild his forces, isolated at the end of long and vulnerable lines of supply. Characteristically, Montgomery set about obtaining massive reinforcements, supplementing his armoured strength with the latest American Shermans. He assembled no fewer than 150,000 front-line troops, superbly trained and physically fit, organised in three army corps; he amassed 1,114 tanks in three armoured divisions and seven additional armoured brigades; and he brought up no fewer than 1,000 guns. Against him, Rommel assembled 96,000 men (half of them Italian) and 600 tanks. But while Montgomery's fuel and ammunition supplies were plentiful, the Axis reserves were critically low. At the same time the British Desert Air Force had gained air supremacy over the area. Montgomery resisted Churchill's pressure for an early offensive.

On 23 October, preceded by a devastating barrage from 1,000 guns, the British attack was launched under a full moon. The enemy minefields were opened by the infantry, who were followed through by an armoured corps. The following day German counter-attacks halted the British, and although a penetration by the Australian 9th Division threatened to pin the German 164th Division against the sea, no significant progress was made on either side, and a ferocious infantry and artillery battle raged for a week. Then, as the final German reserves were committed, Montgomery opened a heavy barrage and sent in the New Zealand 2nd Division against the fortified Kidney Ridge on 2 November; a final desperate armoured counter-attack used up the last of the German tanks so that, when Montgomery hurled a further attack against Kidney Ridge, the final breakthrough followed. Rommel's entire line crumbled, and after a pause of twenty-four hours the cautious

Montgomery gave his army orders for headlong advance. The Axis losses were 59,000 men killed, wounded, and captured, 500 tanks, and 400 guns; Montgomery's losses were 13,000 killed, wounded, and missing, and 432 tanks knocked out.

The Battle of El Alamein marked a turning-point in the Second World War. Rommel's arrival at this position had been the high tide of the Axis operations in the West. The destruction of his army in this great battle was quickly followed by an exhausting retreat over 1,500 miles of desert, with Montgomery in close pursuit. El Alamein placed Montgomery on a pedestal of military prestige (despite some criticism of his failure to wipe out Rommel's army at the end of the battle). He followed up with a masterly pursuit through the desert, being occasionally brought to pitched battle – as at Medenine and Mareth in February/March 1943. But when the Eighth Army joined forces with Eisenhower's Allied army which had landed in Algeria and Morocco (Operation 'Torch'), the final destruction of Axis forces in North Africa was assured. His caution has been criticised, but Montgomery was always aware of the risks involved in lengthening lines of supply, and the conservation of his troops was his constant preoccupation.

With the final defeat of the German forces in Africa, Eisenhower assumed supreme command of Allied forces in the Mediterranean theatre, and Montgomery retained command of the now-famous Eighth Army in its seaborne invasion, with General Patton's (q.v.) Seventh Army, of Sicily. Given the task of overrunning the eastern part of the island, Montgomery met with the stiffest resistance, while the Americans rolled up the Axis forces in the western half of the island. Eventually the two armies crushed the remaining forces in the north-eastern corner and ended the invasion with the capture of Messina on 17 August 1943. Then followed the long-drawn-out Italian campaign, which opened in September with a seaborne assault on Calabria by Montgomery's Eighth Army, simultaneous with General Mark Clark's amphibious landing at Salerno. This two-pronged campaign only narrowly succeeded, for it depended upon the Americans' establishing and holding their beach-head long enough to allow Montgomery to fight his way two hundred miles from the south.

In December 1943 both Eisenhower and Montgomery left the Mediterranean theatre to command the great offensive against Northern Europe; command of the British Eighth Army was assumed by General Sir Oliver Leese. With their headquarters in England, Eisenhower was appointed to the command of the Allied Expeditionary Forces, and Montgomery to the overall command of all ground forces, British and American, amounting to forty-five divisions – about one million fighting men and almost as many again in logistical and support roles. Unquestionably the greatest military project ever undertaken, Eisenhower's assault on northern Europe involved the airborne and seaborne landing of these forces in Normandy with an element of surprise (in itself a formidable task, having regard to enemy reconnaissance and the assembly of vast quantities of material); the establishment of a beach-head before German reinforcements could be brought to bear; and the destruction of enemy lines of communication throughout north-west Europe from the air.

The landing in Normandy involved putting ashore about 80,000 British and British Commonwealth troops in the area of Caen and about 100,000 American troops some twenty miles to the west. In the event the British landings went largely according to plan, but those of the Americans met much stiffer resistance and lost heavily on the beaches. But it was the Americans who eventually burst out of their bridge-head, while the British were held around the French town of Caen. While the American VII Corps smashed its way across the Cotentin Peninsula, capturing the large port of Cherbourg, and swung west and south, the British held large German forces in what was to become the pivot of a gigantic enveloping movement. By the first week in August 1944 Montgomery's right flank (spearheaded by General Patton's American Third Army) had swept round and threatened the rear of the German Seventh and Fifth Panzer Armies, forcing them into a pocket which was mercilessly assaulted from the air. Despite a vigorous counter-attack (ordered by Hitler in person) the German forces were unable to withstand the tremendous strength of Allied forces, and were powerless to oppose the crushing Allied air superiority over and behind their lines; by 20 August the German armies were in full flight eastwards across northern France. The battle around the Falaise 'pocket' cost the German armies 10,000 dead and 50,000 prisoners; Allied casualties amounted to 29,000. On 25 August Paris was liberated.

The speed of the pursuit through France brought its problems of supply. The British forces, being on the inside of the wheeling advance, used less fuel and possessed shorter supply lines, and therefore moved in a compact, well-supported mass; the Americans, with much greater areas to cover, and determined to reach the great traditional barriers to Germany before the enemy forces had time to concentrate their strength, crossed the Meuse but, with fuel tanks empty, ground to a halt. There now occurred serious differences of opinion in the Allied Command, and these almost certainly had the ultimate effect of lengthening the war in Europe by several months; indeed, they possibly influenced the whole political structure of post-war Europe.

The American forces were stationary on the Meuse, but the British forces in the north were still under way, with little but piecemeal resistance on their front. Montgomery and the American field commander, General Omar Bradley (q.v.), both agreed on a single thrust, supported by invincible air power, into the depths of Germany, but neither commander would agree to subordinate command. The delay was critical, and was only overcome by the tact of Eisenhower himself, who assumed overall command on the ground – thereby leaving Montgomery free to handle his 21st Army Group in the north. By-passing the port of Antwerp, his plan was to speed north-eastwards through Holland and then break south round German forces massing in front of the almost-stationary American forces. In pursuance of this plan he launched a series of air drops ahead of his forces in Holland, aimed at capturing the vital bridges in preparation for the British armoured thrusts; but the drop at Arnhem proved too ambitious, and led to the virtual annihilation of the British 1st Airborne Division. The whole front now came to a standstill, with the Allied armies dependent upon stretched supply lines, and the Germans feverishly consolidating as winter approached. Montgomery now set about eliminating the numerous pockets left isolated by his forward rush, and Antwerp became fully operational as a main supply port in November.

It was this great port of Antwerp that was the target for Hitler's final offensive in the West. In the Ardennes the American armies had failed to consolidate their positions after severe fighting at Schmidt, and were in the process of replacing their battle-weary 28th Division when the blow fell. Driving between the American V Corps and the 4th Division (which held firm in the north and south respectively), twenty-four German divisions struck deep into the American front. Led by eight Panzer divisions, the attack penetrated fifty miles before being halted. Eisenhower immediately transferred command of all forces (the American First Army under Lieutenant-General Courtney Hodges, and part of the British Second Army) north of the 'bulge' to Montgomery, leaving only the American Third Army to General Patton. Together Montgomery and Patton threw their combined weight against the shoulders of the 'bulge' and, assisted immeasurably by the heroic defence of the American forces which had survived at Bastogne and St Vith, succeeded in repulsing the German forces. The last great attack in the West had cost Germany 120,000 dead, wounded, and missing, 600 tanks and assault guns, and 6,000 other vehicles. Allied losses amounted to 60,000 dead, wounded, and missing, and more than 700 tanks and tank-destroyers (most of them American). There is little doubt that Montgomery's overwhelming blow in the north was the key to this defeat; nevertheless, the tone he adopted in his public analysis of these events damaged Anglo-American relations quite severely.

Pausing only to reconsolidate the front and make good battle losses, Montgomery maintained his pressure in the north while the two American Army Groups in the south smashed their way forward to the Rhine. Thereafter the final months of the war witnessed a series of crushing pincer movements beyond the Rhine by the Americans, while Montgomery drove his First, Second, Ninth, and Fifteenth Armies round the north of the Rühr and then south-eastwards to link up with the Americans.

Montgomery's final assault on the German homeland was a series of brilliant set-piece battles, forging onwards with the main thrust and enveloping enemy forces in pockets in the process. With almost undisputed air superiority he was able, like the Americans in the south, to take city after city with relatively light casualties – despite ferocious resistance by the Germans who were defending their homeland. But to oppose eighty-five well-equipped Allied divisions the German commander, Field-Marshal Albrecht Kesselring, could muster fewer than sixty half-strength divisions, short of fuel, ammunition, and equipment. During the period 5–9 May 1945 German resistance collapsed in the West, Montgomery accepting the unconditional surrender from General Johannes von Blaskowitz of all forces in the north. The Second World War in Europe officially ended at midnight on 8–9 May 1945.

Like many other commanders of the Second World War, Montgomery was haunted by the recollection of the appalling casualties suffered in the stagnant warfare of 1914–18, and his battlefield tactics developed accordingly. Never committing himself to offensive risks until possessed of superior strength, and being an ardent believer in 'set-piece' warfare, he attracted fairly widespread criticism from other commanders; but one cannot deny that his caution eliminated costly setbacks such as that which occurred in the Ardennes. Only the fiasco at Arnhem, which may be attributed to the enervating squabble beforehand, marred a brilliant campaign in northern Europe, while his achievement in North Africa assures him of a place in any list of outstanding British commanders. Many observers attribute his greatness to his ability to assess changing conditions and make his plans accordingly. He was a rigid disciplinarian, never reluctant to take harsh action against an inefficient subordinate commander, but was much respected by his troops. He had a clear understanding of the importance of morale; prior to his arrival in Egypt the British forces had been 'messed about' continually, and he immediately improved their outlook by a number of reforms, including the inculcation of a strong *esprit de corps* within individual divisions. He carefully projected a personal 'image' to the troops, which laid him open to charges of theatricality, but paid off in terms of morale. His well-founded self-confidence and rather Olympian manner made him enemies, but his achievements far outweigh all the petty criticisms he attracted. He died on 25 March 1976.

Montrose, James Graham, 1st Marquis of (*1612–50*), *Captain-General*

Although his campaigns were marked by as much ferocity as any others in an age of religious intolerance and general cruelty, Montrose is remembered with admiration in Scotland. He was a brave, generous, and cultivated man, consistently loyal to his King and to his religious convictions, and a guerrilla general of great skill. His steadfastness in adversity and his courage in betrayal are qualities calculated to appeal to his countrymen.

He was born in 1612, succeeding his father to become fifth Earl of Montrose in 1626. He was educated at St Andrews University, and at the age of seventeen married Magdelene Carnegie, daughter of the future Earl of Southesk. In 1637 Charles I's insensitive attempts to enforce uniformity of religious observance led to massive unrest in Scotland. Montrose was an early adherent of the party of resistance, and signed the document protesting against the ecclesiastical innovations – the Covenant – in 1638. He was loyal to his King, but believed that the King should be above religious faction, and that the clergy should concern themselves purely with spiritual duties. When opposition to the Covenant arose in Aberdeen and the Gordon country in the north, Montrose suppressed it with great firmness. Although he remained a staunch supporter of the Covenant, however, and fought bravely at Newburn on the Tyne in 1640, Montrose became increasingly alarmed at the line taken by many who had placed themselves at the head of the Covenanters, notably the party of the Earl of Argyll. Personal ambition and anti-monarchism were the plain motives of the Argyll clique, and Montrose wanted no part of them.

When the English Civil War broke out in

1642 the Scottish reaction was delayed. The Covenanters generally sided with the English Parliament, but did not immediately aid that party. Montrose remained loyal to Charles I, but was prevented by the Duke of Hamilton from opening a new front in support of the King in Scotland. In 1644, however, a Covenanter army marched south to aid Cromwell (q.v.), and Montrose struck. His first attempt was abortive, but in August he went into the Highlands and raised an army of some 3,500 men, built around a core of Irish warriors led by Alaster Macdonald ('Colkitto'). The Covenanters made great efforts to locate and destroy Montrose, but he evaded them with bewildering skill, allowing contact only when it was to his advantage. Created first Marquis of Montrose and a lieutenant-general by his King, he fought a masterly campaign of hit-and-run. He consistently avoided pursuers and ambushers, crossed supposedly impossible terrain, fell upon unsuspecting enemies from unexpected directions, and melted away into the Highland wilderness before he could be pinned down.

At Tippermuir on 1 September 1644 he beat Lord Elcho and captured his guns and ammunition, subsequently sacking Perth. His Highlanders tended to drift away if they were sated with loot, but his Irish had nowhere to go – Argyll had destroyed their ships – and stayed reasonably close to Montrose's standard. He now fell back northward, and eluded Lord Balfour's 2,700 Covenanters at the Bridge of Dee. Getting behind them, he defeated them in a running fight which took both sides into the streets of Aberdeen, which was sacked with great ferocity. Montrose evaded an advancing army under Argyll, and eventually took refuge in the hills of Badenoch, giving it out that he was going to winter there. In mid-December, however, his men made a nightmare forced march through the snow-covered mountains and debouched into the plain of Argyllshire without warning; Argyll himself (now a Marquis) only escaped by taking to an open boat, and until the end of January 1645 Montrose laid the county waste with fire and the sword. Argyll mustered his Campbells and many supporting clans, and a series of battles ensued. Montrose, whose victory at Inverlochy on 2 February brought him the belated support of the Gordons (who remembered his deeds from 1638), was soon in virtual control of the north. He advanced on Dundee by way of Aberdeen and Dunkeld, evading a strong force under Baillie and escaping into the hills with much booty, although his losses were quite severe. On 9 May 1645 he won another victory, leaving some 2,000 of the Covenanters stretched on the field of Auldearn. He won again at Alford on 2 July, and at Kilsyth on 15 August he is reputed to have killed nearly 5,000 of the enemy. All Scotland lay open to the Royalists – but all England lay open to Parliament, for in the south King Charles had failed conspicuously to match the achievements of his fine Scots general. On 14 June, at Naseby, Cromwell's Ironsides smashed the Royalist army. Montrose tried to stay in the field, but his Highlanders deserted in large numbers, and the people of the Lowlands had been too greatly antagonised by the plundering of both the Highlanders and the Irish to supply recruits now. The able Covenanter General David Leslie, with a much stronger force, routed Montrose at Philiphaugh on 13 September.

Montrose escaped deep into the Highlands once more, but failed to rally the clans. In 1646 he travelled to Norway, and in 1647 he was in Paris. In 1650, burning to avenge his beheaded King, he persuaded the exiled Charles II to make him nominal Lieutenant-Governor of Scotland. (Charles later disowned his family's most faithful servant in order to meet the terms demanded by the Argyll party for their recognition of him as King.) In March 1650 Montrose landed in the Orkneys, and then on the mainland, with a mere 1,200 men – mostly mercenaries hired in France. He again failed to rally the clans. The influence of his enemies was too strong, and he was held to be a Royalist but no true Covenanter. At Carbiesdale on 27 April 1650 his little force was routed. He himself wandered on the moors for some time before taking refuge with the Macleods of Assynt. Neil Macleod of Assynt betrayed him for a reward of £25,000. On 20 May 1650 he was condemned to death for treason at Edinburgh, and he was hanged the following day. He died as he had lived, with courage, dignity, grace, and humour, unswervingly loyal to his code.

Moore, Sir John (*1761–1809*), *Lieutenant-General*

It is inevitable, but unjust, that the name of Sir John Moore should be popularly associated with the dismal retreat to Corunna, the disintegration of a British army, and a lonely death and burial. This tragic ending came after a career of enormous distinction; Moore was greatly admired and respected by his contemporaries, and his finest legacy was the great contribution to final victory over Napoleon (q.v.) made by the British light infantry arm which he created.

Moore was born in Glasgow on 13 November 1761, and educated locally until the age of ten. His father was the personal physician to the eighth Duke of Hamilton, whose feeble health made extensive travelling in Europe necessary. The young Moore accompanied his father and his father's patient all over the Continent for five years; he benefited greatly from the experience, becoming a most accomplished young man. At the age of fifteen, in March 1776, he entered the 51st Foot as an ensign, joining his regiment on Minorca. In January 1778 he transferred to the newly-raised 82nd Foot and served with the regimental headquarters as a junior officer in Nova Scotia throughout the War of American Independence. He saw some action, in particular a fierce combat on the Penobscot river in August 1779. As a company commander he went on half pay when the Regiment was disbanded in 1783, and entered Parliament as member for the Linlithgow Boroughs in 1784. Unlike many members he took a keen interest in his parliamentary duties, generally but not invariably supporting Pitt. Between 1785 and 1787 he served with the 100th Foot; and in January 1788 he served briefly with the new 4th Battalion of the 60th (Royal American) at Chatham. Soon afterwards he moved on to his old unit, the 51st, then in Ireland. They were apparently in a poor state, but he succeeded to the lieutenant-colonelcy in November 1790 and applied himself to improving the regiment. He had achieved considerable results by the time the unit was shipped to Gibraltar, where they served in 1792–3.

Early in 1794 Moore served with the 51st in Corsica. He fought at Martello Bay, and distinguished himself in fierce hand-to-hand fighting in the attack on the Convention Redoubt. In May he was given command of the grenadier reserves; he was prominent in the Siege of Calvi, leading his grenadiers in the storming of the Mozzello fort, at which he was slightly wounded. But the time Calvi fell in August the British commander, Stuart, had appointed Moore adjutant-general. Personality clashes led to his recall to England some months later, but he was well received by Pitt and the Duke of York. In August 1795 he was appointed to the brevet rank of colonel, and in September was given a brigade then preparing to sail for the West Indies, with the local rank of brigadier-general. His command was still preparing when in February 1796 he was switched to another brigade, which he led to the West Indies in the expedition of Sir Ralph Abercromby (q.v.). He distinguished himself at St Lucia, being given charge of the island after its capture, and restoring order; this involved complex operations against bands of brigands in difficult country, and Moore's popularity was enhanced by his practice of living and eating under the same harsh conditions as his soldiers. Yellow fever sent him home in mid-1797, but he recovered in time to answer Abercromby's summons to Ireland that winter. In 1798 he became a major-general, with a command of 3,000 men based on Bandon. He fought at New Ross on 5 June 1798, and marched on Wexford. He beat off an attack by 7,000 rebels at Taghmone, reaching Wexford on 21 June; he seems to have been a moderating influence in an ugly campaign. In June 1799 he returned to England, and his mentor Abercromby gave him a brigade for the Helder expedition. On 9 September 1799 he was hotly engaged with the advance guard in the action in which Abercromby defeated a Franco-Dutch army, and was wounded in the right hand, being saved from death only by the ball's striking his telescope. On 2 October at Egmont-op-Zee his brigade was engaged in heavy fighting in the sand dunes for several hours, and lost some 650 dead and wounded; Moore himself was wounded in the thigh and the face, being carried from the field unconscious by two 92nd Highlanders, whom he later made energetic but vain attempts to find. He was shipped home, but recovered in time to resume command of his brigade at Chelmsford in December.

When Abercromby went to the Mediterranean in 1800 Moore went with him, as a divisional commander. He led the reserves in the Egyptian expedition, and his troops were the first to land at Aboukir in March 1801. They also bore the brunt of the fighting at Alexandria on 21 March, and Moore was once again seriously wounded. He resumed command of the reserves before Cairo in June 1801, and after the French garrison surrendered he escorted them to the coast. He stayed in Egypt until the fall of Alexandria in September 1801. In July 1803 he was given a brigade in Dundas's division guarding the south coast of England against French invasion. This brigade originally comprised the 4th, 59th, 70th, and 95th Foot, and also the 52nd, of which Moore had been colonel since May 1801; it was designated as a light corps. There was some subsequent reorganisation, and he later got the 43rd Foot as well. The brigade was based at Shorncliffe, and it was here that Moore put into practice the training of light infantry which he had been planning since he had first been impressed by the French *voltigeurs* on Minorca. Briefly, he felt that while the range and accuracy of the musket of the day dictated that the mass of infantry should manœuvre and fight in close order, there was an important place in the army for specially-trained light units. These men, armed either with rifles or with lightened muskets, should be trained to operate in extended order, using their initiative, and making maximum use of cover; they should be flexible in manœuvre, self-sufficient in the field, and expert marksmen. Since infantry training was necessarily based on a rigid code of discipline designed to keep men in their ranks without thinking while facing the hideous dangers of battle, Moore could only produce his new type of soldier by evolving a completely new training syllabus. This he did; his riflemen and light infantry were later to vindicate the theory triumphantly in the Peninsula (see under Craufurd), and it may truly be said that John Moore gave the British Army its first modern soldiers. Central to his method was a humanitarian approach; the type of soldier he wanted had to be capable of individual judgement and initiative, and that could only be achieved by giving him a high sense of his own worth. The lash was banished from Shorncliffe, and the result was a magnificent *esprit de corps*.

Moore was made a Knight of the Bath in November 1804, and became a lieutenant-general a year later. He now commanded the Kent area, but kept his Shorncliffe headquarters. His reputation and popularity were very high; Pitt often came over from Walmer to seek his advice on military matters. In June 1806 it was felt that the invasion threat had receded and that Moore could safely accept an overseas posting – a step which had been publicly resisted a year or two earlier on the ground that his services were too valuable at home in a time of emergency. He returned to the Mediterranean as second-in-command to General Fox, with headquarters in Sicily. In due course Fox was invalided home, and Moore took over the command. In September 1807 he received orders to hand over to John Sherbrooke, and lead 7,000 men to Gibraltar in readiness to assist Portugal against the French invasion under Junot. In the event the Portuguese royal family preferred to flee to Brazil, abandoning their country to the invader, and Moore brought the expedition back to England.

Another rather harebrained scheme sent him off to Sweden in May 1808, to explore the possibility of landing troops to help King Gustavus against French, Russian, and Danish pressure. Moore anchored and went ashore to present himself to the King; that personage turned out to be more than a little mad, and demanded that the British army should land to help him in grandiose schemes of conquest. When Moore declined he was detained, only escaping back to his squadron by disguising himself as a peasant.

He returned to England once again, angry at being assigned to such time-wasting duties, and in July was told that he must sail for Portugal once again to serve under Sir Hugh Dalrymple and Sir Harry Burrard. The initial British landings under Sir Arthur Wellesley (see Wellington) were to be heavily reinforced, and these two officers of great seniority but little battlefield experience had been grafted on to the top of the local chain of command, as it were. Moore protested at this subordinate command as a retrograde step in his career, but followed orders, landing with his troops at Vigo, but not coming up with the main army until after the Convention of Cintra (see under Wellington) had been signed. That ill-considered piece of diplomacy very soon removed the obstacles to Moore's occupation of the chief command. Wellesley, Dalrymple, and Burrard were all swept away by a furious country (the former, providentially, only for a year); and Moore was promised heavy reinforcements to bring his command up to a strength of about 35,000. In September Lord Castlereagh ordered him to arrange to advance into northern Spain by whatever route he thought best, there to co-operate with Spanish armies under Blake or Romana in operations against the French. He was assured that Spain would provide for all his logistic needs, and that he would be acting in concert with Spanish forces totalling at least 60,000 men. From the outset Moore appears to have had serious misgivings about this vague order, the successful execution of which must rely so heavily upon Spanish promises. His friendly relationship with Wellesley is confirmed by the fact that the latter was at this time corresponding with him and acting as his advocate with the Government at home.

By 14 November Moore had reached Salamanca with the bulk of his army, and was forced to await the arrival of his cavalry and artillery – who had taken another route because of the state of the roads – and of reinforcements under Sir David Baird, marching down from Corunna. In the meantime Napoleon had taken personal command of a reconquest of Spain, and his armies had inflicted a series of defeats on the badly-equipped, badly-led and badly-handled Spanish armies opposing him. Moore had great difficulty in getting reliable intelligence, as Spanish co-operation at every level had fallen dismally short of what had been promised. Gradually it dawned on him that he was isolated in an advanced position, threatened by massively superior forces, and that the Spanish armies with whom he had been ordered to collaborate were no longer in existence. They had been smashed during October and November at Zornoza and Espinosa, Gamonal, and Tudela. Baird reached Astorga by 28 November, and Moore ordered him to retrace his march through the punishing Galician mountains to the coast. It was no part of his brief to throw away most of Britain's field army in a hopeless stand unsupported by the Spanish, and he prepared to fall back into Portugal. His political masters had other plans, however.

It was decided in London that some move had to be made to help the Spanish patriots who had risen against the Ogre, and Moore made a brave choice. His strength lay in the fact that the French armies were pouring across his front under the impression that he was in full retreat for Portugal. If he pressed on to the north-east he might inflict a serious defeat on them, and would at least distract French attention from Madrid for a time. Once again the sceptical general was assured that Spanish forces were being assembled after the disasters of the past two months, and on 11 December Moore struck out over the bleak winter hills for Valladolid. A captured dispatch soon enlightened him as to the situation. Madrid had already fallen, but Soult (q.v.) lay at Saldana, about 100 miles away, with 18,000 men. Moore, his presence unsuspected, had some 27,500 men; if he could come upon Soult before he was reinforced there was the chance of a useful victory. The British army surprised and beat a French cavalry force at Sahagun on 21 December (see under Paget), but two days later Moore learnt that the element of surprise was lost. Soult was moving towards him, and the Emperor himself was leading another 20,000 men up from Madrid to trap him. There was nothing for it but to cut and run for the coast.

The sufferings of Moore's army on this retreat have become legend. The terrain was bare and harsh, the roads appalling, and the organisation inadequate. Stocks of vital equipment were destroyed at the outset of the retreat, for want of which hundreds of men perished later. Officers and men alike were dissatisfied at having to retreat without a fight, and discipline soon began to break down. There was drunkenness, looting and murder of the civil population, and every kind of dereliction of duty. Moore seems to have been obsessed with the need to preserve the army by pushing on as quickly as possible; but there is strong evidence that had he halted more frequently to fight rearguard actions at strong defensive points along the road, the army would have held together much better. When the pursuing French were engaged, as at Benavente, the British army performed very well. It cannot be denied, however, that despite Moore's efforts, and those of a few dedicated officers, the army deteriorated with horrifying speed. It should be added that some units, particularly those of the Reserve Division or rearguard, performed prodigies and maintained their discipline to the last.

After this nightmare march through the snow-covered hills the head of the army reached Betanzos on 10 January 1809 and began the last lap of the retreat to Corunna. In this area the climate was milder, and depots were available to restore the combat readiness of most of the units. But the transport ships were not ready, and Moore was forced to fight a defensive battle against Soult's advancing army outside Corunna town on 16 January in order to gain time for embarkation. Some 5,000 men had been lost on the retreat, and the Light Brigades had marched by another route to Vigo, leaving Moore with some 15,000 men. Soult had about the same number of infantry, with some 4,500 cavalry. The battle lasted all day, and several positions were lost and retaken on both sides; nightfall found the armies in much the same positions as they had occupied in the morning, and Soult, who lost some 1,500 men against British casualties of around 800, was unable to interfere with the embarkation of the army. During the battle a round-shot nearly tore off Moore's left arm and opened his chest. He was carried from the field in a blanket, lingered in agony until late evening, and died during the night. His last words indicate that he knew all too well how England would react to his handling of the retreat. At dawn he was wrapped in his cloak and buried in a grave hastily dug by men of the 9th Foot.

Moore was indeed much criticised by a country shocked at the condition of the army which staggered off the transports in the ports of the south coast. There are aspects of his conduct of the campaign which invite censure, but a proportion of the blame heaped on his head when he could no longer defend himself stemmed from the wish of other officers to distract attention from their own less than admirable behaviour during the campaign. Moore's tragedy was that he was given a mission which made no military sense, and which was doomed from the start. His past record had been so brilliant that his failure was all the more humiliating to the country and the Army. Although more realistic and tolerant opinions soon replaced the first storm of criticism, it is worth noting that his enemies paid more generous tribute to his courage and skill than did his countrymen. Both Soult and Napoleon himself had admiring words to say of the handsome, brilliant, doomed general; and Wellington, who was too good a soldier and too much of a gentleman to join in the chorus of ill-informed abuse, gave Moore full credit for his contribution to the subsequent Allied victory. The most demanding critics any general can have are his men, and it should be noted that many years after his death the common soldiers in the ranks still regarded it as a matter for boasting that they had been among Moore's 'Shorncliffe Boys'.

Murat, Joachim, Prince Murat, Grand-Duc de Berg et de Cleves, Roi de Naples (*1767–1815*), *Marshal of France*

It is unthinkable to exclude Joachim Murat from any list of the great names of the Napoleonic Wars, yet his claims for inclusion are curiously unsatisfactory to the student of military history. In the final analysis Murat stands as the ultimate symbol of a particular type of soldier: the outrageously-uniformed officer of light cavalry, swaggering and duelling and wenching, laughing at danger, tailor's bills, and furious husbands, and quite prepared to start his own wars if his masters are slow to provide him with entertainment. Murat was by no means a great commander; there were several among his subordinates who displayed a far greater mastery of the art of manœuvring large bodies of horse in the field. In keeping with his Gascon heritage he was extravagantly boastful and vain, and even in an age of conspicuous military glamour his dazzling personally-designed uniforms were considered a trifle ridiculous. Hot-headed to a fault, he often crossed the line which divides a fiery spirit from sheer unthinking stupidity. He was as deficient in loyalty as in common sense, finally betraying, out of simple self-interest, the Emperor who had loaded him with honours and wealth, given him his own sister for a wife, and eventually raised him – the son of an inn-keeper – to a throne. Yet despite all these defects, Murat possessed, to a degree unequalled among his contemporaries, one great talent which occasionally proved invaluable to his Emperor: the gift of inspiring cavalry troopers – cold, saddle-weary, hungry, exhausted – to draw sabres and follow him hell-for-leather straight down the cannon-barrels of any enemy on earth.

Murat was born on 25 March 1767 at La Bastide-Fortunière, and in his adolescence studied theology – a uniquely unsuitable vocation, one must conclude. He was coldly received by his family after being expelled from the Toulouse seminary, and in February 1787 he enlisted as a cavalryman. He achieved commissioned rank in October 1792; he was lucky in enjoying the patronage of an aristocratic general and, subsequently, the confidence of the Committee of Public Safety. During the Terror he expressed violent political convictions. He had achieved the rank of *chef d'escadron* in the 16th Chasseurs by 4 October 1793 – the day when he brought forty cannon from Sablons to the Tuileries, thus enabling General Bonaparte to fire his legendary 'whiff of grapeshot'. His reward, though delayed, was handsome; as soon as Bonaparte received the command of the Army of Italy he appointed Murat as his chief aide de camp, with the rank of colonel.

During 1796 Murat served at Dego, Ceva, and Mondovi; he was promoted general of brigade on 10 May, and led the cavalry at Borghetto on 30 May. In September he was wounded, and he was given the cavalry of Rey's division in December. In Rey's absence he led the division in the Tyrol, and fought at Tagliamento in March 1797. He held several different commands in the months which followed, and led a dragoon brigade to Egypt in the spring of 1798. Murat distinguished himself at the storming of Alexandria, fought at the Battle of the Pyramids, and commanded the cavalry of the expedition to Syria in February 1799. He was promoted general of division on the field of Aboukir on 25 July, after leading an impressive charge and suffering a painful face wound. He accompanied Bonaparte back to France late that summer, and it was he who sent troops to eject the deputies from the Orangery on 2 December. Bonaparte made him commander of the Consular Guard, and in January 1800 he married Caroline Bonaparte.

During 1800 he commanded the cavalry of the Army of Reserve, and displayed great courage at Marengo on 14 June. After holding a home command, he was given a corps in Italy in November, and operated successfully against the Neapolitans. In 1804 he was made Governor of Paris, and promoted marshal. In February 1805 the newly-crowned Emperor made him a prince. He led the Reserve Cavalry of the *Grande Armée* in the 1805 campaign; in Napoleonic terms this was not an 'insurance' formation kept in the rear, but a strong force which was often thrown into battle to tip the balance at the decisive moment. In October Murat was given temporary command of the corps of Ney (q.v.) and Lannes as well, during the closing of the trap around General Mack's Austrian army at Ulm. Murat mishandled this command, senselessly antagonising Ney when his mistake was pointed out, and drawing on himself a stinging imperial rebuke. Murat regained favour by leading the energetic pursuit of the Austrians and taking many prisoners. In November he again misread the situation, and by a precipitate advance on Vienna, allowed the Russian General Kutuzov (q.v.) to inflict quite heavy casualties before withdrawing safely. On this occasion his brother-in-law and Emperor accused him of proceeding 'like a bewildered idiot, taking not the least notice of my orders', but Murat again made amends by a display of energy. He served at Austerlitz, and at the end of the campaign was rewarded with the Grand Duchy of Berg and Cleves.

Murat played a major part in the successful pursuit of the Prussians after Jena, taking many thousands of prisoners. His graetest moment came at Eylau on 8 February 1807, when he took the pressure off the mauled French centre at a moment of crisis by leading a charge of nearly 11,000 men which covered 2,500 yards to smash the Russian centre. In 1808 he entered Madrid to the cheers of the crowd as the Emperor's Lieutenant in Spain. On 2 May the same crowd came into the street to hunt down French soldiers with knives and clubs, and Murat suppressed the rising ruthlessly. He then resigned his post on health grounds, and thus escaped further entanglement in the deadly Peninsular campaigns. When Joseph Bonaparte was given the crown of Spain, Murat was offered and accepted the vacant throne of Naples; he occupied himself harmlessly with minor campaigns in southern Europe until 1812. In Russia he again led the Reserve Cavalry, and was responsible for costly mistakes at Krasnoi (on 15 August) and Borodino. Nor did he shine during the retreat from Moscow. On 5 December Napoleon left the *Grande Armée* and handed over command to Murat; the King of Naples held the appointment – most indifferently – for only six weeks before deserting in his turn, handing over to Eugène de Beauharnais and making off for the more hospitable climate of his kingdom.

Murat then busied himself with making secret contact with the Allies, in order to ensure that he could retain his throne in the event of French defeat. He reluctantly rejoined the *Grande Armée* in August 1813, and fought at Dresden and Leipzig with some honour. In November 1813 he absented himself again, and the following January concluded treaties with the Allies whereby he guaranteed troops to fight against France as the price of keeping his kingdom. Napoleon's reaction to this shabby deal may be imagined. During the Restoration period, and after Napoleon's return from Elba, Murat indulged in increasingly desperate and contemptible scufflings to secure the continuation of his own prosperity, and in May 1815 he had the impertinence to offer his sword to Napoleon once again. The offer went unanswered. The Italians, in whose affairs he had meddled too often, put a price on his head after Waterloo; he made a hopeless attempt to regain his lost kingdom in October 1815, and was captured, court-martialled, and condemned to death. On 13 October, within half an hour of sentence being passed, he was executed by a firing-squad. He died with courage.

Mustafa Kemal Atatürk (*1881–1938*), *General, 1st President of the Turkish Republic*

'The Father of Turkey', one of the most remarkable national leaders of the twentieth century, was also a military commander of great ability. While there is no space in this book for even a summary of his extraordinary achievements in bringing his country into the modern world, his part in the Allied defeat at Gallipoli in 1915 deserves a mention in any compilation of notable military men.

Born in Salonika, Greece, of humble parents, Mustafa decided on a military career early in life. He attended the Harbiye Staff College at Istanbul, and displayed such a natural ability for mathematics that he was nicknamed 'Kemal' – 'perfection'. In 1906 he was posted as a captain to the Damascus cavalry regiment, and in September 1907 he joined the staff of the 3rd Army in Salonika. During the turbulent years before the outbreak of the First World War, when many powers were attracted by the potential plunder of the ramshackle Ottoman Empire, he served with ability in a number of staff posts and combat commands. In 1909 he took part in Mahmud Shevket Pasha's march on Istanbul to depose the sultan, Abdul-Hamid II, and in 1911 he served in a staff appointment in the War Office. He saw action in Libya during the 1911–12 Italo-Turkish War, and a year later was chief of staff of a division based at Gallipoli during the Balkan War. He was with the force which recaptured Edirne (Adrianople), and by 1913 had risen to the rank of lieutenant-colonel. He was still quite unknown in Turkey, and his latent promise was overshadowed by the flamboyant rise of his contemporary Enver Pasha – a dashing politically-minded soldier, leader of the 'Young Turks', and a remarkably inept general. Mustafa and Enver were men of strongly contrasting character who disagreed on every subject; and on none more violently than the encouragement of German influence in the Turkish government and armed forces. Mustafa believed that Turkey should remain neutral in the First World War, doubted the chances of the Central Powers, and resented Enver's invitation to Germany to send a strong military mission not only to advise but actually to command Turkish forces. After a period of dreary exile as military attaché in Sofia, Mustafa was recalled and appointed to command the 19th Division, based at Rodosto on the Gallipoli Peninsula, with the rank of colonel. His unit was to form the mobile reserve in this sector under the direct authority of Otto Liman von Sanders, the able German commander-in-chief. Mustafa's rise can be traced from the morning of 25 April 1915, when Allied troops began to land in 'Anzac Cove' near Gaba Tepe.

Mustafa personally reconnoitred the hills above the beachhead, and without referring to higher authority, launched the regiments of his division piecemeal in a series of very fierce counter-attacks. He immediately grasped the importance of the Chunuk Bair and Sari Bair hills to the outcome of the campaign. He greatly distinguished himself under fire while personally leading and directing a number of attacks in a successful attempt to pin down the Australian and New Zealand troops within a narrow perimeter – attacks in which he deliberately sacrificed great numbers of his men in order to gain time for reinforcements to come up. The first day's fighting cost each side about 2,000 casualties, and although some 20,000 Anzacs were landed by the end of the second day, they never succeeded in destroying the cordon of Turkish positions which Mustafa had established. Mustafa never left the front until the campaign was virtually over, months later. He continued to display an impressive grasp of terrain, and to expose himself to enemy fire with complete indifference. He was constantly at odds with his superiors, and with his outspoken opinions and contempt of misguided authority he must have been an uncomfortable subordinate indeed; but he was virtually always right. He foresaw the Suvla Bay landings weeks before they occurred, but his warnings were ignored; when fighting began in that sector he was given command there as well. He ruined his health by his exertions, but his abilities were so obvious that he was promoted general despite his many political and professional enemies. One of his narrowest escapes occurred when his staff-car was straffed by the famous British naval pilot Charles Samson; the chauffeur was killed, but Mustafa survived the bombs and machine-gun fire untouched. At last, in December, his health broke down completely, and he was evacuated to Istanbul. He had been living on injections – and his nerves – for months.

His later appointments included command of the 16th Army Corps in eastern Anatolia from March 1916; he fought a successful series of actions against the Russians near Lake Van. Trouble in the eastern Arab provinces took him to Diyarbakir to command the 2nd Army, and led to his famous report of December 1917 in which he declared that it was not worth the life of a single soldier to try to hold on to the non-Turkish areas of the Empire, and that all forces should be withdrawn and concentrated to defend the Anatolian homeland. His feud with Enver Pasha, the pro-German War Minister, continued, and his report was rejected and he was sent on indefinite sick leave. As the military situation deteriorated, however, he was recalled and posted to command the 7th Army in Palestine. He managed to extricate the bulk of this command and withdrew first to Aleppo, and then to the Anatolian frontier, late in 1918. Politically active after the armistice, he used his appointment as inspector-general of the armies in eastern and north-eastern Anatolia to strengthen those elements who were working for a free and independent Turkish nation. He became President of the National Assembly at Ankara in 1920, and successfully directed the Turkish defence in the war with Greece in 1921–2. The Lausanne peace conference granted almost all the concessions which Turkey demanded, and on 30 October 1923 Mustafa was elected President of the new Turkish Republic. His rule was dictatorial, but under the circumstances this was quite inevitable, and he never turned his back on the idea of democratic institutions. His achievements in every field of national life were extraordinary, and virtually single-handed he inspired Turkey to take her place among the modern nations of the world. He took the title Atatürk in 1934; on 10 November 1938 he died at the Dolmabahche Palace in Istanbul, and was mourned by the entire nation. It is worth pointing out that he enjoyed almost universal respect and support among the peasant classes. Mustafa Kemal was a solitary, introverted figure whose early frustrations and later exertions took a great toll of his physical and nervous strength. For the last years of his life he was without family or friends, uncomforted by religious belief, and handicapped by the after-effects of hard campaigning and hard drinking.

Napoleon I, surnamed Bonaparte (*1769–1821*), *Emperor of the French*

It would require far more space than this book can afford to present anything but the briefest and most simplified summary of the career of the man who dominated European military and political life for twenty years – dominated in a very personal sense, not seen before since the days of the Caesars and not seen again until the triumphs of Hitler. His place in political history has been discussed by countless historians, but is not relevant to this book. His military reputation stands uniquely high, and certainly few generals have fought so many campaigns and won so many devastating victories in the course of twenty years.

Napoleon Buonaparte was born on 15 August 1769 at Ajaccio, Corsica, the fourth son of the lawyer Carlo Buonaparte and Letizia Ramolino. His father was of the stock of the Tuscan nobility; it is helpful to remember that Napoleon's origins were Italian, while his education was French. In 1778 he briefly attended the College of Autun, and between 1779 and 1784 the military college at Brienne; the year 1784–5 he spent at the military academy in Paris. He passed out in forty-second place in a class of fifty-one, and was commissioned a *sous-lieutenant* of artillery in the *Régiment de La Fère*, based at Valence; the Regiment transferred to Auxerre in 1788. His activities during the Revolution have no place here; suffice it to say that he was associated with the Jacobin movement, was politically active, and rose steadily in responsibility and reputation among men of affairs. In 1791 he was a lieutenant in the Fourth Artillery at Valence, and was elected lieutenant-colonel in the National Guard. The following year his regular rank rose to captain, and in 1793 he was based at Nice to supervise the assembly of supplies for the Army of Italy. In September 1793 the forces of the Convention – the Revolutionary Government – were engaged in a campaign against royalists supported by British troops who held the port of Toulon. Through a personal contact Buonaparte obtained command of the artillery in the army of the Convention. On 29 September he was promoted *chef de bataillon*, and on 7 October adjutant-general and head of brigade. He took an active part in the bombardment of the British positions, and on 16 December suffered a bayonet wound in close fighting. On 22 December, after the capture of Toulon, he was promoted *général de brigade* and was much praised in important circles. In February 1794 General Buonaparte commanded the artillery of the Army of Italy, and took part in operations around Nice. In 1794 and 1795 he languished without a command as a result of political difficulties following the *coup* of 9 Thermidor. In October 1795 a royalist faction rose against the Convention, rejecting the constitution which it was trying to introduce. The Convention was dubious about the reliability of the commander of the army of the interior, Menou; young General Buonaparte was available in Paris, and was appointed deputy commander as a precaution. On 5 October his 'whiff of grapeshot' ensured his immediate promotion to high command by a grateful Government. After a few months as commander of the troops of the interior, close to the seats of influence and power, he was given chief command of the Army of Italy in March 1796.

Bonaparte's command (he changed the spelling of his name in 1796) was supposed to comprise some 43,000 men; in fact he found, on his arrival at the Savona headquarters, that it consisted of about 30,000 poorly-clothed, hungry, diseased, ill-supplied, and pessimistic soldiers. They were spread along the Riviera from Nice almost to Genoa; the coast was blockaded by British ships, and two enemy armies – 25,000 Piedmontese under Colli and 35,000 Austrians under Beaulieu – lay to the north. These were widely separated, and the energetic young general quickly launched an offensive designed to defeat them in detail before they could link up – one of his classic tactics throughout his career. Taking the offensive on 12 April, he widened the gap between the two enemy armies by driving in Beaulieu's right flank at Montenotte. Three days later he chased Beaulieu out of Dego, inspiring his tattered soldiers to great exertions; when the Austrians retreated north-east he did not follow but turned west against Colli. On 21 April he beat Colli at Mondovi and marched on Turin; the Piedmontese (Sardinians) sued for peace, and the armistice of Cherasco took them out of the war on 28 April and led to permanent French annexation of Nice and Savoy. Bonaparte now turned on the Austrians again; by this time Beaulieu had spread a defensive system for sixty miles along the north bank of the Po. On 7–8 May Bonaparte crossed at Piacenza, endangering the Austrian left and communications with Mantua, after confusing Beaulieu with feints at other points. On 10 May Bonaparte personally led a bayonet charge which cleared the stubborn Austrian rearguard from the bridge at Lodi, earning the soldiers' nickname 'Little Corporal': the common soldiers were unaccustomed to inspiring personal examples of this sort from their generals. On 15 May he entered Milan, and six weeks later the Austrian garrison of the citadel surrendered. Beaulieu was hurriedly bundled over the Adige, and all northern Italy was in French hands except Mantua. The French army invested this city on 4 June 1796, spreading defences along the Adige to screen the siege. While Bonaparte reorganised the political framework of the area to suit his taste, four separate Austrian attempts to relieve Mantua were defeated at Lonato, Castiglione, Caliano, Bassano, Arcola, and finally at Rivoli on 14 January 1797. In all these operations Bonaparte displayed great energy and speed of movement; his vigorous attacks on several enemy forces simultaneously threw them into confusion and prevented mutual support, and his forced pace enabled him to surprise the enemy and to cut them off from their communications. He was not infallible, however, and was saved from an ugly situation at Rivoli only by the arrival of Masséna (q.v.). On 2 February Mantua finally capitulated. In March Bonaparte invaded Austria and penetrated to within less than a hundred miles of Vienna; the Austrians sued for peace in mid-April, and the resulting Peace of Campo Formio echoed the terms Bonaparte had demanded – without reference to his Government – at Leoben in April. Europe was at peace after five years of fighting; the map of both northern and southern Europe was altered in France's favour. Bonaparte had shown himself capable of applying the classic principles of warfare with such energy, such audacity, and such offensive spirit on the part of his citizen-soldiers, that the more hide-bound – if better trained – Austrians were simply bullied out of his path.

In February 1798 the young lion inspected his new command: the army assembled on the Channel coast for an invasion of Britain. Bonaparte, convinced that such a plan had no chance of success without the establishment of naval superiority – which was not seriously in prospect – suggested an alternative plan to the Directory. He maintained that the sources of British wealth – her colonies in India and her sea trade – could be threatened by a French occupation of Egypt. The Directors were not sorry to see the back of this dynamic and popular young general, and he was given command of the new Army of the Orient which formed around Toulon, while the Channel demonstration continued to keep British naval strength near home. In May Bonaparte sailed for Egypt with 40,000 men, captured Malta, and landed near Alexandria on 1 July 1798. That port was immediately stormed and taken, and the Nile delta was overrun. At the Battle of the Pyramids on 21 July the wild Mameluke horsemen of the 60,000-strong Egyptian army were repulsed by steady fire from French infantry squares, and the Egyptians were dispersed with great loss. Cairo fell the following day; but on 1 August the situation was dramatically altered when Nelson destroyed Admiral Brueys's fleet in Aboukir Bay. The French army was now cut off in the midst of a hostile population, while a Turkish army assembled in Syria. In February 1799, after attending to the reform of the Egyptian administration,

Bonaparte marched into Syria with 8,000 men. In February and early March he captured El Arish and Jaffa, and on 17 March invested Acre, which was defended by a garrison under the royalist *émigré* Phélippeaux and the British sailor Sir Sydney Smith. A Turkish attempt to relieve Acre was defeated at the Battle of Mount Tabor in mid-April; but the siege made no progress and cost Napoleon heavy casualties from disease, and on 20 May he raised it. His retreat to Cairo was gruelling and costly, and in all the expedition suffered more than twenty-five per cent. casualties. In July a second Turkish army, some 18,000 strong, was transported from Rhodes in British ships, and entrenched itself at Aboukir. Bonaparte attacked on 25 July with some 6,000 men; Lannes and Murat (q.v.) distinguished themselves, and the Turks were routed.

Meanwhile affairs in Europe commanded Bonaparte's attention. There was internal unrest in France and danger abroad, and Allied victories in Italy threatened to undo all Bonaparte had achieved there. A power vacuum seemed to exist. Bonaparte could see no future in Egypt, and abandoned his army to its fate. He slipped home in a frigate with a few colleagues, and was in Paris by October 1799. Masséna (q.v.) had stabilised the situation in Switzerland, as had Brune in the Netherlands; but although the immediate pretext no longer existed, Bonaparte carried out the *coup* of 18 Brumaire (9 November 1799) and installed himself as virtual ruler of France under the title First Consul. Just thirty years old, he was the victor of more than a dozen battles in the period of less than four years since he had taken the Italian command.

In the winter of 1799–1800 Bonaparte raised a new Army of Reserve at Dijon and planned a campaign against the Austrians. He decided to strike at them in Italy, where Masséna was operating, and held to this decision in the spring when Austrian victories scattered Masséna's army and left him besieged in Genoa. In mid-May 1800, before the snow had melted, Bonaparte led his army through the St Bernard Pass and pushed on across the plain of Lombardy, taking Milan and Pavia. On 4 June Masséna was forced to capitulate at Genoa, but Bonaparte's advance cut off from his communications the Austrian general, Baron Mélas, at Turin. On 14 June Bonaparte, with a weakened and dispersed army, ran unexpectedly against the Austrians at Marengo; the odds were almost two to one in the Austrians' favour, and they attacked vigorously. The French right was enveloped and Bonaparte was driven back two miles, but he rallied his shaken troops and gathered the other dispersed elements of his army. The Austrians, assuming that the day was theirs, were advancing without adequate precautions when they were struck by a heavy counter-attack, in which Kellermann's cavalry were prominent. Mélas lost some 9,000 dead, and his army was routed; he capitulated at Allessandria the next day. Marengo gave France control of the Po valley as far as the Adige; while Moreau's victory at Hohenlinden in Germany the following December over another Austrian army led to the Peace of Lunéville of February 1801.

Britain was now standing alone against France, and in March 1802 she too ended hostilities at the Treaty of Amiens. In May 1802 Napoleon was voted First Consul for life by referendum. But a year later his unchecked ambitions to expand French influence, on the Continent and elsewhere, led inevitably to the resumption of hostilities with Britain. The following year, as a measure ostensibly designed to preserve the security of the state from royalist conspiracies, the consulate for life was transformed into a hereditary empire under Napoleon I, Emperor of the French. In due course, with the Pope hovering in the background, Napoleon crowned himself – a gesture of such blazing conceit that there could hardly be any point in trying to preserve an appearance of revolutionary ideals. From this point on Napoleon ruled a military dictatorship.

Between 1803 and 1805 Britain fought without allies; the *Grande Armée* was concentrated around Boulogne and some hundreds of vessels were prepared in Channel ports for a possible invasion. Napoleon's plans to reduce British naval superiority to an acceptable level by the use of the combined French and Spanish fleets were shattered at Trafalgar in 1805, after which battle Britain enjoyed complete freedom of movement at sea. She could still only hurt Bonaparte on land by means of coalitions of continental allies, however; her army was far too small to play more than a supporting role at this stage. In 1805 Austria, Russia, Sweden, and Naples entered one such coalition; the *Grande Armée* was ordered to the Danube, and advanced in secrecy and at speed. Backed up by a superb logistic organisation, the French army encircled the Austrian General Mack at Ulm in October, and he was forced to surrender 30,000 men and sixty-five guns. In early November Bonaparte invaded Austria, guarding his flanks and communications with detached corps. Vienna was occupied on 13 November, and leaving 20,000 men to hold the capital, Napoleon concentrated his remaining 65,000 or so near Brünn. Some 18,000 Austrians lay to the north-west at Prague; some 90,000 Russians and Austrians were commanded by Kutuzov (q.v.) at Olmutz to the north-east; and another 80,000 Austrians were south of the Alps, blocked by Ney (q.v.) and Marmont who guarded the passes. Napoleon was concentrated in the midst of these superior forces like a ripe bait, and he anticipated with complete accuracy the next Allied move. Kutuzov moved south, intending to swing around the French right flank and cut Napoleon off from his lines of communication with Vienna. Near Austerlitz on 2 December 1805 Napoleon won his greatest victory. He deliberately dangled under the noses of the enemy a right wing so foolishly extended that they could not resist an attempt to crush it. It was forced back, and eventually about a third of the Allied army was concentrated against it, while other elements were crossing the front of the French army to join in this movement. At the perfect moment Napoleon smashed forward in the centre, cutting the Allied army in two and defeating it in detail. Soult (q.v.) delivered the key attack on the Pratzen Heights, and then rolled up the Allied left. Lannes and Bernadotte enveloped the Allied right, and by nightfall the Austro-Russian army had lost some 26,000 men to 9,000 French casualties. Before the month was out Austria had withdrawn from the war, making valuable concessions in Italy and Germany. During 1806 Napoleon formed his client Confederation of the Rhine, and in September of that year he turned to face Prussia.

Prussia, appalled at French domination of so much of Germany, was preparing for war when in October 1806 Napoleon led the *Grande Armée* northwards from Bavaria into Prussia. The army moved in three columns over a front of some thirty miles, screened by a strong cavalry advanced guard, and moving at about fifteen miles a day. It was prepared to concentrate quickly in any direction according to circumstances, and moved with such speed and secrecy that by the time the Prussians recognised their danger the French were past their left flank and closer to Berlin than they were themselves. Napoleon now swung towards Jena, ordering Davout (q.v.) and Bernadotte to turn west behind the Prussian army; Bernadotte moved in the wrong direction and only arrived on the battlefield of 14 October in the nick of time. By noon on that day 100,000 French under the Emperor had smashed 51,000 Prussians at Jena. At Auerstädt, to the north, the mass of 63,000 Prussians attacked Davout's 27,000 for six hours without breaking them. When the Prussians showed signs of wavering – at the news of Jena, and of the fall of their commander – Davout counter-attacked, and Bernadotte eventually arrived in the Prussian rear to seal the victory. The twin battles cost the Prussians 25,000 dead and wounded and nearly the same number captured; total French casualties were some 8,000 men.

During the winter of 1806–7 Napoleon overran Prussia and occupied Poland. In January 1807 the Russian general Bennigsen advanced into Prussia, and Bonaparte attacked him at Eylau on 7 February. The Russians had some 67,000 men on the field, and were expecting 10,000 Prussian reinforcements; Bonaparte had only 50,000, and awaited the support of Ney (q.v.) and Davout (q.v.). Both sides fed these progressive reinforcements into the battle during the day, but neither gained a decisive advantage. Bennigsen withdrew after suffering 25,000 casualties and inflicting 19,000. Both sides went into winter quarters and prepared for renewed hostilities in the spring. On 14 June they met again at Friedland. Lannes's 17,000 men pinned down a Russian advance by 46,000 men, while Napoleon concentrated 80,000 men for the decisive blow. Bennigsen had another 42,000 men in reserve. Late in the afternoon the Emperor launched his main attack, smashed the Russians' left flank, and drove them from the field leaving 25,000 dead and wounded and eighty guns behind them. French casualties were about 8,000. This defeat took the heart out of Alexander I, and at Tilsit in early July he agreed peace terms with Napoleon in the famous meeting on the raft moored on the Niemen. Frederick William III of Prussia also took part in the negotiations, which ended in a Franco-Russian peace treaty and massive reparations by Prussia; wide Prussian territories passed under French control, including the Grand Duchy of Warsaw, which was to provide Bonaparte with some of his finest troops. Napoleon was now the acknowledged ruler of most of Continental Europe.

Unable to invade Britain, Napoleon sought to starve her into making terms by a process of commercial quarantine; he issued his Berlin and Milan Decrees in 1806 and 1807, forbidding any contact whatever between the other European nations and Britain. Portugal was unable to comply with these instructions without committing fiscal suicide, and in 1807 a French army passed through Spain and occupied Portugal. The weak Spanish monarchy was manœuvred into abetting French ambitions, and was subsequently removed; Spain, too, was occupied, and in 1808 Napoleon set his brother Joseph on the throne. A general insurrection followed, and early attempts to crush it failed spectacularly. The Peninsular War of 1808–14 presented Britain with an opportunity to intervene directly on the Continent with substantial land forces, and the vigorous resistance of the Spanish civilian population lit a beacon of hope for all the foreign vassals of the French Empire. Napoleon played a personal role in Spain only during the reinvasion of the winter of 1808, subsequently turning the pursuit of Sir John Moore's (q.v.) British army over to Soult and returning to France. (For a more detailed summary of the Peninsular War see under Wellington and Moore.)

The French reverses in Spain encouraged Austria to make another attempt against Napoleon, and it was the report of his efficient spy service on this possibility which brought the Emperor back from Spain in January 1809. In April the Archduke Charles led an Austrian army into Bavaria, and another under the Archduke John crossed the Julian Alps and invaded Italy. A popular rising in the Tyrol added to French difficulties. Arriving at Stuttgart on 16 April 1809, the Emperor took over command of the *Grande Armée* from Berthier (q.v.), and a series of brisk actions followed. Napoleon crossed the Danube, and in the course of operations marked by his usual energy and initially – before exhaustion slowed down the French troops – by his usual speed, he forced the Austrians back at Abensberg, Landshut, and Eckmühl. He was slightly wounded at Ratisbon on 23 April. In seven days he had inflicted some 30,000 casualties and suffered 15,000; although the bulk of the Austrian forces were still in being, the French had seized the initiative. Capturing Vienna on 13 May, Napoleon attampted to force his way across the Danube in the face of concentrated Austrian opposition on 21–22 May. The crossing was attempted at the island of Lobau; although part of the French army crossed successfully, Napoleon was unable to reinforce the bridgehead sufficiently and was eventually forced to withdraw with heavy casualties – some 20,000, compared to 23,000 Austrian casualties. This Battle of Aspern-Essling was his first major defeat, and cost him the life of his trusted subordinate Lannes. In June, while his stepson Eugène de Beauharnais forced the Austrian armies out of Italy into Hungary, Napoleon concentrated nearly 200,000 men and adequate bridging materials near Vienna. Before the Archduke John and his 50,000 men could link up with the Archduke Charles and his 140,000, Napoleon made a surprise night crossing of the Danube on 4–5 July. With 140,000 French troops north of the river, Napoleon attacked Charles at once. He planned his attack to prevent any last-minute junction of the two Austrian armies, and after pounding the Austrian centre with the heaviest concentration of artillery yet seen, launched infantry attacks which proved decisive. This campaign of Wagram, which cost the Austrians and the French some 45,000 and 34,000 casualties respectively, forced Austria to sue for peace. The Treaty of Schönbrunn was onerous, and left Britain once more alone under arms against Napoleon. 1810 marked the climax of Napoleon's power and influence.

After a period of worsening relations between France and Russia, Britain persuaded both Russia and Sweden to renounce Napoleon's Continental System and to sign a treaty with her in June 1812. Napoleon gathered troops totalling more than 450,000 in Poland during the spring, and on 24 June he crossed the Niemen to crush Russia once and for all; it should be noted that less than 200,000 of his men were French, the remainder being Austrian, German, Polish, and Italian. Russian armies in the immediate area of operations totalled some 218,000 men under Barclay de Tolly, Bagration, and Tormassov. The French advance was seriously hampered from the outset by the torrid weather and by the logistic problems of supplying such vast forces, and the Russians fell back, destroying all natural resources as they went. Napoleon was unable to advance with the speed necessary for one of his classic left-and-right victories of penetration. There was some indecisive fighting, but the Russian forces were still strong, and still retreating, when Kutuzov took overall command in late August. Napoleon had intended to winter at Smolensk, but the logistic problems and the continued existence of strong Russian forces now forced him to attempt a decisive encounter during 1812. The Russians stood at Borodino on 7 September; Napoleon's handling of his army in this engagement was noticeably less impressive than on previous occasions, and he was distracted by painful illness. In a bloodbath which served little strategic purpose the Russians lost some 40,000 men and Napoleon some 28,000. Moscow was now naked to the invader, and was completely abandoned. The stores necessary to bring the French army up to full efficiency were not found in the city, which was largely destroyed by fire as soon as the French entered it on 14 September. The forward elements of Napoleon's forces numbered about 95,000 exhausted men, and the allied contingents were displaying a serious breakdown of morale. The rest of the army was spread along the line of advance all the way back to the Niemen. Kutuzov had 110,000 fit and aggressive troops near Kaluga, and the whole line of communication was harassed by Russian irregulars and by detached and by-passed Russian regular corps. On 19 October the French began a withdrawal to Smolensk, having failed to force the Tzar to the conference table. The winter snows came early in 1812, and the army was already hungry when the march began. Conditions soon deteriorated still further, and the retreat became an epic military disaster.

Starving, freezing, and rapidly disintegrating into a disorganised mob, the *Grande Armée* was repeatedly attacked by irregular formations and by elements of Kutuzov's command. Individual French units fought off many attacks, and the exploits of Ney's rearguard were particularly heroic; but although Napoleon showed a brief flash of his old brilliance at Krasnoye on 17 November, the initiative had been lost forever. French strength dwindled at terrifying speed, and only some 37,000 effectives were still under command when the Berezina was reached in late November. Despite Russian attacks on the bridgehead areas on both banks, the majority of the French got across, but this last effort cost such heavy casualties that by 8 December only some 10,000 men were left capable of serious military activity. On that day Napoleon left the survivors and travelled to Paris. The Russians, themselves exhausted by the pursuit, halted at the Niemen. The fighting throughout the campaign had been particularly merciless; total French casualties of 300,000 were almost balanced by Russian losses of about 250,000. During the final stages the survivors of the *Grande Armée* were commanded by Eugène de Beauharnais, since Murat, whom Napoleon had entrusted with the task, had deserted his command.

The disaster in Russia led to a general rising among Napoleon's vassals, and massive defections among his allied troops. In February and March 1813 a new coalition of Russia, Prussia, Sweden, and Britain was formed; some 100,000 Allied troops were gathered in the Elbe Valley, while Beauharnais commanded some 68,000 *Grande Armée* survivors and reinforcements at Magdeburg. Napoleon's system of conscription and mobilisation showed its astounding efficiency in providing him with a new army of nearly 200,000 by April, but its quality fell far below that seen in previous years. Joining up with Beauharnais, the Emperor advanced towards Leipzig at the end of April 1813. He intended to defeat the Allied armies in detail in the old style, but was taken by surprise on the road on 2 May. Wittgenstein attacked Ney's corps with some 75,000 men at Lützen; Napoleon arrived in the nick of time, and again displayed what he was still capable of achieving. Gathering his shaken troops in a brilliant demonstration of intelligent improvisation, he opened up a massed artillery bombardment on Wittgenstein's centre, and led the subsequent counter-attack in person. Each side lost about 18,000 men; the Allies would have lost many more if Napoleon had commanded veterans instead of tyros, especially in the cavalry arm. Capturing Dresden, the Emperor followed the retreating Wittgenstein to Bautzen. The Allied army was drawn up in a strong position on the far bank of the Spree; Napoleon had sent Ney in a classic sweep far to the north with nearly half his army, and when he launched a frontal attack on 20 May Ney was in the perfect position to fall upon the Allies from the north and smash into their flank and rear. Amazingly, Ney missed his opportunity; both sides suffered about 20,000 casualties, but Wittgenstein withdrew in good order.

On 4 June 1813 Napoleon secured an armistice which lasted until August; meanwhile both sides frantically built up their forces. When Austria declared war once again on 12 August three Allied armies were in being, all heavily subsidised by British gold: Schwarzenberg had 230,000 men in Bohemia, Blücher (q.v.) had 195,000 in Silesia, and the renegade Bernadotte led 110,000 Prussians and Swedes in the north. With some 300,000 men Napoleon had established himself between the Elbe and the Oder, protected by the strongholds of Hamburg and Dresden and ready to react to Allied manœuvres in any direction. The Allies now pursued a policy of engaging Napoleon's outlying corps at every opportunity, while avoiding a confrontation with the Emperor and the bulk of his field army. Bernadotte and Blücher achieved victories at Grossbeeren and Katzbach during August; and on the 26 and 27 of the month Napoleon won a brilliant tactical victory over Schwarzenberg when the latter attacked Gouvion St Cyr at Dresden. Vandamme attempted to follow up the French success, but after good initial results was annihilated at Kuln. In September Ney was beaten at Dennewitz by Bernadotte, and the following month the Allies closed in on the remainder of the French main force. At Leipzig ('The Battle of the Nations') on 16–19 October 1813 Napoleon fought off the combined armies of Schwarzenberg, Blücher, and Bernadotte for three days before being forced back; the bridge over the Elster river was blown prematurely, and many French troops were trapped in the city. Although decisively beaten at Leipzig, Napoleon kept his army in being, and his escape is not greatly to the Allies' credit. Casualties were about 60,000 on each side. Napoleon withdrew to the Rhine, brushing 40,000 Bavarians out of his path at Hanau (30–31 October) with great style. The *Grande Armée* crossed the Rhine early in November.

Napoleon could have made peace on advantageous terms at this point, but he foolishly rejected the Allies' approaches. In January 1814 the invasion of France began. The Emperor had some 118,000 troops west of the Rhine, and some 100,000 more guarded his back in the Pyrenees; some 50,000 still held out in German garrisons, and a similar number faced an Austrian army in Italy. The Allies invaded along three routes, all aimed ultimately at Paris: Bernadotte led 60,000 men westwards through the Netherlands, Blücher led 75,000 up the Moselle into Lorraine, and Schwarzenberg led 210,000 from Switzerland through the Belfort Gap. In his defensive manœuvres Napoleon proved once again that at his best he could still dominate any battlefield in the world. Although forced steadily back throughout February and March, he won eight separate victories against Blücher and Schwarzenberg (Brienne, La Rothière, Champaubert, Montmirail, Château-Thierry, Vauchamps, Montereau, and Craonne), all at bad odds and all with a largely untrained army of boys and national guardsmen. Blücher's steady resistance at Laon on 9–10 March cost the French dear, however, although honours were evened up by Napoleon's audacious recapture of Rheims three days later. In the last ten days of March the quicksilver defensive campaign suddenly collapsed.

Schwarzenberg and Blücher penetrated beyond Napoleon's main field forces, and refused to be deterred from their advance on Paris by movements threatening their lines of communication. On 30 March, united at last, they forced an army of 22,000 under Mortier and Marmont back to within artillery range of the capital, which was surrendered to them the following day.

On 11 April 1814 Napoleon, ill and exhausted, abdicated the throne of France unconditionally. His attempt to persuade the Allies to accept his infant son as his successor had failed. Fours years after exercising the undisputed mastery of most of the European landmass, his domains were reduced to the island of Elba, his armies to a token battalion of guardsmen. Louis XVIII was placed on the throne of France, and the Allies settled down at the Congress of Vienna to redraw the map of Europe. Their tortuous diplomacy was disturbed in March 1815 by the news that Napoleon had escaped from Elba, and landed at Cannes on 1 March. The Bourbon régime sent troops to arrest him; they rallied to his eagle banners and their generals, for the most part, offered their swords to Bonaparte once more. He made a triumphal progress to Paris, and prepared to face the armies which the Allied nations now assembled to crush him for ever. His administrative genius was unimpaired, it seemed. Within a couple of months he had built up a field army of 188,000, and had another 100,000 men in depots and garrisons; some 300,000 raw peasant levies were under training. The main field command was the Army of the North, 124,000 strong, positioned around the capital. It was a formidable force, but too many of the old faces were missing – from marshals' conference and infantry campfire alike. Several key commands had been filled unwisely, and though the Emperor still enjoyed enthusiastic loyalty, the weapon under his hand was not as perfectly balanced as in the old days. He himself was handicapped by intermittent attacks of agonising illness, and at critical moments was subject to periods of torpor during which urgent requests for decisions were rejected, sometimes with a petulance unworthy of the issues at stake.

In early June Napoleon moved to crush his enemies before they could concentrate. Allied armies lay along the Rhine, in Germany, in Italy, and in the Low Countries. The Prussian and Anglo-Allied armies in this last theatre presented the most important target. Blücher had 124,000 Prussians under command, while Wellington led a polyglot army of British, Dutch-Belgian, Hanoverian, and Brunswick troops, some of them of dubious quality. When Napoleon seized Charleroi on 15 June Blücher reacted with characteristic energy, concentrating his army near Sombreffe. Wellington, some fifteen miles to the west, kept his options open until he could identify Napoleon's exact strategic objectives, and concentrated more cautiously – and mistakenly. The key crossroads of Quatre Bras lay between the Prussians and Anglo-Allied armies. On 16 June Ney was sent with 25,000 men of the French left wing to seize this pivotal point, while Napoleon hurled 77,000 men of the centre and right wing against Blücher's 83,000 Prussians at Ligny. By the late afternoon of 16 June the Prussians were in full retreat, and the stage seemed set for a classic Napoleonic victory of penetration. Ney could finish the Prussians off, and finally dash any hope of a junction between Blücher and Wellington, by driving north from Quatre Bras against the Prussian right flank; and Wellington could be caught and destroyed before his army was fully assembled. The key had eluded the Emperor's grasp, however; Quartre Bras had not fallen.

Ney had hesitated fatally, possibly persuaded by the bitter resistance of the single Allied brigade at Quartre Bras that far stronger forces lay concealed behind cover – Wellington's favourite tactic, which had cost Ney's corps so dear at Bussaco. A hastily-gathered reinforcement brought Allied strength up to 36,000 men, and a counter-attack forced Ney back. This seemed to confirm his earlier appreciation, and even when Wellington withdrew from Quatre Bras towards Brussels on 17 June, it was after noon before Ney could be persuaded to pursue him closely. Quatre Bras cost each side about 4,500 casualties. Both Ney and his Emperor were denied the support of the 20,000-strong corps of Drouet d'Erlon, which marched and countermarched between Ligny and Quatre Bras on the afternoon of 16 June, prevented by a confusion of orders from contributing to either engagement. The failure of Ney to break through, and Napoleon's delay in committing his right wing of 33,000 men under Grouchy to the pursuit of the Prussians as they fell back from Ligny, enabled Blücher to withdraw to Wavre in reasonable order.

In heavy rain, which turned the ground to a mire and prevented a serious pursuit of Wellington on a wide front on each side of the single main road, the French army followed the Anglo-Allied forces throughout 17 June. That night Wellington halted and began to dispose his troops along a line long since selected as the best potential defensive feature in the area – the low ridge-line running through Mont St Jean near Waterloo. Both armies passed a miserable night in the open. On the morning of 18 June Napoleon might still have smashed through Wellington's front before Blücher had a chance to intervene – as intervene he must, for Grouchy's pursuit of the Prussians was misrouted and dilatory, and presented no obstacle to a junction between the two Allied armies. In the event Napoleon waited until noon before launching his attack. There is no place here for a detailed account of this most famous and most studied battle in Western history. Napoleon had 72,000 men, Wellington 68,000, spread along the reverse slope to avoid the massive barrage launched by Napoleon's artillery. Infantry and cavalry of both sides attacked and counter-attacked all along the line; by late afternoon the Allied army was badly weakened and all its reserves were committed. Ney misjudged the degree to which Wellington's front had been weakened, however, and committed the greater part of the cavalry reserve to attacks on unbroken British infantry squares, in which the French suffered great loss for no appreciable advantage. There will always be controversy over Napoleon's apparently inexplicable failure to take a firm personal grasp of developments at this stage of the battle. By early evening the Prussians were beginning to make their presence felt on the extreme right of the French position, and remaining reserves were insufficient to hold them off while simultaneously delivering a decisive attack on Wellington's stubborn but mutilated line. The Imperial Guard infantry, robbed by the afternoon's bloodbath of the cavalry support they needed, made a last attempt to pierce the British line, and were thrown back in a matter of moments. Wellington ordered a general counter-attack, and this coincided with a breakthrough on the French right by the bulk of Blücher's forces. The Army of the North disintegrated into a fleeing rabble; while the exhausted Anglo-Allied regiments camped among the dead, the Prussians mounted a merciless pursuit which lasted throughout the night of 18 June. Wellington lost some 15,000 men at Waterloo; Blücher, 7,000; and the French 30,000, including 7,000 prisoners.

On 21 June 1815 Napoleon Bonaparte abdicated for the second time, and subsequently surrendered himself to the British. He was exiled on the island of St Helena, with a very small household, and lived in a discomfort unnecessarily aggravated by his vindictive jailer until his death on 5 May 1821.

To attempt to summarise the extraordinary qualities of this monumental figure in a few words is a daunting task. He was, first, a man of great intelligence with a phenomenal capacity for hard work and sustained application. He was driven by a burning ambition, and was confident of his destiny; his energies were never sapped by self-doubt, and he showed enormous resilience in the face of apparent disaster. He was impatient of external limitations, and ruthless in sweeping them aside. He was a complete autocrat, caring nothing for the popular will or for the principle of democratic debate. He believed that a man of vision, backed by sufficient bayonets, could achieve anything. He seems to have enjoyed extraordinary personal magnetism, which he wielded quite cynically. He had the gift of the grand gesture which would restore the morale of hard-pressed troops; at the same time he was unimpressed by casualties, and sacrificed the lives of hundreds of thousands of men with cool calculation. His administrative abilities were considerable, and the reforms which he made in the legal, educational, fiscal, and governmental fields formed the basis for the modern French state.

He always went to war with a clear strategic

objective – the destruction of the enemy's forces in the field, and thus of their ability to continue hostilities, rather than any limited territorial gains. Territorial concessions could be wrung from the defeated nation at the conference table once its ability and will to resist had been destroyed on the field. By fast movement and intelligent manœuvre he usually gained the strategic initiative before the tactical battle was ever joined. He excelled, tactically, at both the battle of penetration – placing himself in a position from which he could destroy separate enemy forces before they could concentrate – and the battle of envelopment.

The armies which he led to victory on a score of battlefields were not remarkable for any technical innovation or unorthodox tactical doctrine; they simply represented a conventional combination of arms brought to a high peak of co-ordinated efficiency. He assembled his armies by the old Revolutionary method of forced conscription; recruits and veterans served side by side within each regiment, and the long tradition of French victory, together with the deliberately glamorous trappings which characterised Napoleon's armies, produced a remarkable morale. The infantry attack was launched in massive columns, weak in firepower but strong in physical and psychological impact. These were screened by strong forces of skirmishers, and accompanied and supported by horse artillery batteries of great efficiency. In defence the square formation was used, or, less frequently, the linear, depending upon the nature of the threat. The cavalry was greatly strengthened under Napoleon; apart from providing, most effectively, the conventional scouting, screening, penetration, and flank forces, it gave him a massive mobile reserve for use in the main battle. These enormous divisions of aggressive and well-trained horsemen often turned the tide; sometimes they were committed to vigorous attacks on the enemy flank to coincide with infantry pressure on the centre, and sometimes they were sent against the enemy centre in massive frontal charges. Frequently they sealed the doom of an enemy already shaken and dislocated by infantry and artillery attacks, either by exploiting a gap torn in the centre, or by rolling up a flank driven loose from its anchor and left in the air. Napoleon's cavalry also excelled at the pursuit to destruction of a beaten enemy. As an ex-artilleryman, Napoleon greatly increased the strength of this arm, which excelled both at forward support of the attacking infantry, and at breaking up enemy concentrations, which could then be overrun by the cavalry. In several of his battles Napoleon used an enormous concentrated battery, made up of most of his available guns, to bring an intense fire to bear on key points of the enemy line.

Every schoolboy knows that the battles of the Peninsular War and Waterloo proved the fundamental superiority of the line formation over the column. What the schoolboy finds hard to understand, if he is thoughtful, is how Napoleon smashed Prussian, Austrian, Russian, and Spanish armies employing the linear formation, if this is so. The answer is of course that the greater firepower of the line did not *of itself* provide security against the column, which was a great deal more flexible as a mobile formation than is sometimes supposed; Napoleon proved this on a score of battlefields, and his marshals on a hundred. The Peninsular victories were won under the special circumstances created by Wellington: an infantry which was generally conceded to be the steadiest in the world, commanded by a general who carefully neutralised every element of the traditional Napoleonic attack by the intelligent use of terrain and specialist troops. But this formula, which removed all other factors and reduced the engagement to a straight contest of infantry firepower, was the exception, not the rule, in Napoleonic battles. The linear formations of Russia, Prussia, and the rest were most effectively dismembered by Napoleon's own formula, and it was there that his genius lay. Rushed into imperfect dispositions by his lightning advances, blinded by his *tirailleurs*, thinned and demoralised by his cannon, the blue-, green-, or white-clad lines were torn apart again and again by his infantry columns, and butchered by his wild-riding cavalry. Where his enemies were rigid, he was imaginative; where they were unwieldy, he was swift and decisive; where they were predictable, he was dynamic and audacious. Now they are buried in obscurity, and he is immortal.

Ney, Michel, Duc d'Elchingen, Prince de la Moskova (*1769–1815*), *Marshal of France*

In an age when the humblest company officer was expected to behave at all times with complete indifference to his own safety, Michel Ney earned the sobriquet 'the bravest of the brave'. The most famous of all Napoleon's marshals, he displayed the qualities of a true leader of fighting men. He was an impulsive, passionate, headstrong commander whose brilliant leadership at corps level was periodically marred by the type of blunder to which men of his temperament are often prone. That he died the victim of judicial murder has, if anything, enhanced his status as the symbol of a romantic military tradition.

Born on 10 January 1769 at Sarrelouis (Moselle), Ney was originally destined for the cooper's trade. This lacked appeal, however, and he enlisted voluntarily at the age of eighteen in the cavalry regiment which shortly thereafter became the 5th Hussars. He achieved commissioned rank in October 1792; he served as a *sous-lieutenant* with the Army of the North until June 1794, part of the time as an aide-de-camp to General François-Joseph Drouot. In 1794, as *adjudant général chef de brigade*, he served with the Army of Sambre-et-Meuse, and was wounded at Maastricht in December of that year. He fought at Opladen in 1795, and at Altenkirchen in 1796. In July 1796 he captured Würzburg, and after distinguishing himself at Forcheim, was promoted general of brigade in August. In 1797 he saw considerable action as commander of the hussars of the Army of Sambre-et-Meuse, but was captured at Giessen in April; he was soon exchanged. During 1799 he rose to general of division (28 March) and was again heavily engaged, particularly at Mannheim and at Winterthur, where he was wounded on 27 May. During the year he held several senior commands for short periods, including provisional command of the Army of the Rhine, In 1802 he led the French army in Switzerland, capturing Zürich. On 19 May 1804, at the age of thirty-five, he was one of the first eighteen marshals created by Napoleon.

Ney commanded the VI Corps in the 1805 campaign against the Austrians. His command was placed to cut off the potential escape route along the north bank of the Danube of the Austrian General Mack, encircled at Ulm with 30,000 men. Ney achieved a brilliant victory over a break-out attempt at Elchingen on 14 October – but only after a violent altercation over dispositions with his hot-headed superior Prince Murat (q.v.). Ney advanced into the Tyrol, capturing Innsbruck on 7 November. In the 1806 Prussian campaign he displayed one of his weaknesses: at Jena, again on 14 October, he imperilled his command by advancing prematurely. He distinguished himself in the subsequent pursuit, however, taking some 36,000 prisoners and 700 guns at Erfurt and Magdeburg. In 1807 the unauthorised activity of his corps played a part in provoking a winter attack by the Russian General Bennigsen, which achieved some success until defeated at Eylau on 8 February. Ney's command arrived on the field late in the day, but sealed the Russian defeat. He led the French right at Friedland on 14 June 1807, and earned great credit for his considerable contribution to the victory.

In 1808–10 Ney commanded the VI Corps in Spain. He captured Bilbao in September 1808, and his corps led the pursuit of Sir John Moore's (q.v.) British army that winter. He achieved some successes against Spanish armies in 1809, largely in the north-west; and his corps was assigned to Masséna's (q.v.) Army of Portugal for the invasion of that country in 1810. He took Ciudad Rodrigo and Almeida, and pushed Craufurd's (q.v.) British Light Division back over the Coa, but his command was badly cut up at Bussaco on 27 September. Ney commanded the rearguard during the eventual French retreat into Spain, and with some brilliance; but the bitter personal antagonism between himself and Masséna led him into open insubordination, and he was removed from his command and sent home in March 1811.

Ney commanded the III Corps of the *Grande Armée* in the invasion of Russia in 1812; his troops were in the forefront at Smolensk, and he distinguished himself at the bloody and mismanaged Battle of Borodino, where Napoleon ignored his and Davout's (q.v.) urgent advice and in consequence failed to exploit the costly victory. The Russian armies remained in being, and on 19 October the epic retreat from Moscow began. On 3 November, as the first snows fell, Ney replaced Davout in command of the rearguard; and it was in this onerous and dangerous post that he displayed his greatest qualities. With a command which dwindled around him at alarming speed, dangerously isolated and more than once given up for lost, hampered by the wretched mass of stragglers, Ney conducted a brilliant and aggressive withdrawal through the freezing desolation of the Russian winter. At Kowno Bridge on 13 December 1812, with a rearguard which had shrunk from some 6,000 to about one hundred men, Ney held off the pursuit for several hours, musket in hand among his soldiers. He was the last Frenchman to leave Russian soil.

'*Le rougeaud*' – the redhead – had covered himself with glory, and was rewarded with honours, titles, and a huge pension; but he had little opportunity to enjoy them. Heartened by the Russian disaster, Napoleon's enemies formed a new coalition. On 2 May 1813, at Lützen, Ney's corps was surprised on the road by the Allied army of Wittgenstein, and he himself was wounded in the leg. His wound may have contributed to his blunder at Bautzen on 20–21 May, where Napoleon sent him on a wide sweep to the north of Wittgenstein's position, with about half the French force. He failed to spring the trap when Napoleon attacked frontally and drove Wittgenstein back, and the Allied army escaped in reasonable order. Ney fought with some success at Dresden on 27 August, but his attempt to capture Berlin was foiled by Bernadotte at Dennewitz on 6 September. Ney commanded the northern sector at Leipzig (16–19 October 1813), but was wounded on the 18 October. He saw further action during many of the battles in France in 1814. After the fall of Paris on 31 March, he was the spokesman for the group of marshals who urged Napoleon to abdicate.

Under the restored Bourbon monarchy Ney received further titles and continued employment, and when his former Emperor landed in the south of France it was Ney who was sent to arrest him. He accepted the mission with the much-quoted assurance that he would bring Bonaparte back in an iron cage; but on 12 March 1815 he changed sides, and offered his sword to Napoleon once more. The 'iron cage' remark was not forgiven, however, and it was not until 11 June 1815 that Bonaparte summoned him to join the Army of the North. He was given command of the left wing, and the mission of securing the vital crossroads of Quatre Bras, the potential link between Wellington's (q.v.) Allied and Blücher's (q.v.) Prussian armies. Unaccountably, Ney hesitated. The position was held only by a 'scratch' Allied brigade, which his 25,000 men could have overrun with ease. It is thought that he was inhibited by his memories of Bussaco, where he had suffered from Wellington's favourite tactic of exposing only a small part of his strength until the enemy was committed. At all events, his hesitation at Quatre Bras allowed Wellington vital hours in which to assemble his forces for a stand at Mont St Jean.

At Waterloo (18 June 1815) Napoleon displayed an inexplicable and fatal indecision and lack of energy. He was painfully ill; his orders were unclear, based on faulty appreciation, and petulant. Ney conducted the greater part of the action in the afternoon and, through a combination of misunderstandings, sacrificed the French cavalry in a long series of gallant but hopeless and costly charges at the unbroken squares of the British infantry. The final infantry attack by the Imperial Guard was thus robbed of cavalry support. As the French army collapsed and fled the field Ney was seen attempting to rally a few men to 'see how a Marshal of France dies'. He had led the cavalry in person, and had five horses killed under him. An eyewitness account leaves us with the unforgettable picture of 'the bravest of the brave' in the thick of the carnage, alone and on foot, bareheaded and filthy with the grime of battle, beating a broken sword on a cannon barrel in an agony of frustration.

Although the Bourbon party may be forgiven for hating the man who had defected from their cause so spectacularly, it was a vindictive revenge to put on trial for his life a man who had served his conception of his country's glory so long and so courageously. Michel Ney was tried for treason by the Peers of

France on 4 December 1815; a verdict of guilty was returned on the 6 December, and the following morning he was executed by firing-squad in the Luxembourg Gardens.

Nivelle, Robert Georges (*1856–1924*), *General*

Nivelle, a French general of the First World War, was successor as Commander-in-Chief to Joffre, and architect of the abortive 1917 offensive, the failure of which resulted in his eclipse.

Born in 1856 at Tulle, Nivelle studied at the École Polytechnique. A member of the Conseil Supérieur de la Guerre in the pre-war years, he first came to the fore at Verdun in 1916. In May that year, against heavy odds, he resisted German efforts to overwhelm Fort Vaux, but eventually both this strongpoint and that at Douaumont fell. Nevertheless he launched a successful series of counter-attacks in October and November on the east bank of the Meuse, and recovered both positions.

After Joffre's eclipse following the costly stalemate at Verdun, Nivelle was appointed Commander-in-Chief on the strength of his own localised success. It was during the general German withdrawal to the Hindenburg line in March 1917 that he conceived the bold plan of launching a massive offensive – 'the decisive blow, to be performed with violence, brutality and rapidity' – and, abjuring the method of attrition, prepared a combined assault by all his armies from the Aisne heights; this was to be the most ambitious plan of attack since the Battle of the Marne in 1914. However, unity of command was non-existent, and Nivelle and Haig quarrelled violently over their command relationship. Predictably Lloyd-George, naïve as ever in matters of military strategy, was captivated by the charming, English-speaking Nivelle, and was at the same time suspicious of Haig: he acquiesced in French supreme command, to the horror of Haig and the newly-appointed British Chief of the Imperial General Staff, Sir William Robertson. Not only did this disunity become public knowledge, but Nivelle himself made imprudent statements which pierced the veil of secrecy, with the result that all element of surprise was lost before Nivelle's grand assault opened in April 1917.

Although the British, in their opening assaults, gained tactical victories at Arras and Vimy Ridge, there was no breakthrough. Immediately afterwards five French armies totalling 1,200,000 men attacked on a forty-mile front in the Second Battle of the Aisne and the Third Battle of Champagne during 16–20 April. Well prepared for the offensive, the German artillery destroyed much of the French armour before it arrived on the battlefield, while the French rolling artillery barrage moved forward too fast for Nivelle's infantry, who met counter-artillery and machine-gun fire at point blank range. Repeated attacks gained little ground, and the whole so-called offensive was an appalling failure. French losses in five days were about 120,000 men, compared with a quarter of this number lost by German forces. At other times such losses would have been considered commonplace by the French, but on this occasion Nivelle had promised victory ('We have the formula!'). Widespread mutiny rocked the French army; and the nation itself was shaken by the political repercussions, Nivelle's methods being contrasted unfavourably with the more cautious approaches of Pétain and Foch (qq.v.).

Nivelle was utterly disgraced but, realising that to sack him on the spot would destroy the last vestige of morale in the field, the French revived the post of Chief of the General Staff and appointed Pétain to fill it. It was no great problem thereafter to promote Pétain over the head of Nivelle, and the latter was obliged to step down and disappear into obscurity.

Outram, Sir James (*1803–63*), *General*

British soldier and gifted political agent in India, opponent of Sir Charles Napier (who nevertheless respected him as 'the Bayard of India'), Outram is perhaps best known as the defender of Lucknow during the Indian Mutiny.

Born at Butterley Hall, Derbyshire, on 29 January 1803, James Outram was educated at Aberdeen and won an Indian cadetship at the age of sixteen, entering the Bombay army. Displaying outstanding qualities as an officer (and also as a big game hunter), he was chosen as a subaltern for the task of pacifying the restless Bhils of Khandesh and raising a Bhil light infantry corps. He served with distinction early in the First Afghan War (which he considered to be a folly from the start), on one occasion making a hazardous journey in disguise through hostile country. Outram's understanding of native affairs brought an appointment as political agent in the Sind, and he managed to maintain peace in that region during the troubles in Afghanistan. He won the confidence of the Mirs (amirs) of Sind, whose case he frequently pleaded until he was replaced on the appointment of Sir Charles Napier. Much to Outram's disgust and opposition, Napier imposed a harsh new treaty upon the Mirs, and only as a result of tactful intervention and persuasion by Outram were they reconciled to the terms.

In 1844 Outram gained distinction in a minor campaign in the south of Bombay province, and was afterwards appointed Resident at Satara and Baroda – but in 1847 he was removed by the Bombay government for exposing official corruption. Thereafter he was appointed Resident in Oudh, whose annexation he advocated and carried out in 1856. The following year he commanded the successful Persian expedition, but was recalled on the outbreak of the Indian Mutiny to the commissionership in Oudh and command of two divisions. With these he joined Sir Henry Havelock (q.v.) at Cawnpore in the march to relieve Lucknow – quixotically handing over command of the operation to Havelock. He succeeded in holding the city until the second relief by Sir Colin Campbell (q.v.) and thenceforth defended the Alambagh (a sort of walled park) against great odds until the eventual capture of Lucknow.

Outram received a baronetcy in 1858 in recognition of his tremendous services in India, and was appointed military member of the Viceroy's council; but the years of campaigning and state administration in the steamy heat of India had taken their toll. His health broke down and he returned to England in 1860; he died at Pau in the Pyrenees on 1 March 1863, at the age of sixty.

Paget, Henry William, 1st Marquis of Anglesey (*1768–1854*), *Field-Marshal*

Henry Paget was the most able British cavalry commander of the Napoleonic Wars; indeed, one is tempted to say that he was the only British cavalry commander of this period of real ability, except possibly Le Marchant, whose career was so tragically cut short at Salamanca before he could show his true worth. Paget was also the *beau idéal* of the English cavalry officer of popular imagination – tall, handsome, laconic, fearless, and a great lover of women.

He was born on 17 May 1768 in London, the eldest son of the Earl of Uxbridge, and educated at Westminster and Christ Church, Oxford. In 1790 he became M.P. for the Caernarvon Boroughs, holding the seat until 1796; he was also Member for Milborne Port in 1796, 1802–4, 1806, and 1807–10. He served in the Staffordshire Militia in his youth; and in September 1793 he raised a regiment of Staffordshire Volunteers, largely from among his father's tenantry, himself having the temporary rank of lieutenant-colonel. This was one of the twelve regiments added to the regular Establishment on the outbreak of war with France, becoming the 80th Foot. In December 1793 Paget led the regiment to Guernsey, and in June 1794 to Flanders to join the Duke of York's army. During the British withdrawal to Bremen which followed the French victories at Fleurus and Charleroi in that month Paget commanded a brigade. His position – that of a lieutenant-colonel without permanent rank and with but a year's experience of soldiering – was obviously anomalous, and in the spring of 1795 he began a rapid rise through the commissioned ranks; this presented no obstacle to a man of his wealth in those days of purchased commissions and promotions. On 11 March 1795 he was commissioned lieutenant in the 7th Royal Fusiliers; on 25 March he became a captain in the 23rd Fusiliers; on 20 May he became a major in the 65th Foot; and on 15 June lieutenant-colonel of the 16th Light Dragoons. May 1796 saw him a colonel in the Army, and in April 1797 he moved to the lieutenant-colonelcy of the 7th Light Dragoons. He led a cavalry brigade of four regiments in Holland during the Anglo-Russian expedition of 1799; it was poor cavalry country, but he did well at Bergen on 2 October, recapturing some British guns and putting to flight a French cavalry force under Vandamme. Four days later the cavalry took 500 prisoners at Kastricum, but the expedition as a whole was doomed to failure, and embarked once more on 18 October.

Paget saw no more active service for several years, but devoted himself to his regiment, making it one of the best in the Army. He became its colonel in May 1801, and in April 1802 was appointed major-general. He had a genuine talent for handling horse in the field, and his character seems to have been a happy combination of that dash and audacity proper in a light cavalryman with the ability to assess coolly the consequences of his actions. Promoted lieutenant-general in April 1808, he went out to Portugal in that year. At first he was without a specific command, but at the end of the year was given the cavalry division which was sent out to join Sir John Moore's (q.v.) army. After a difficult march from Corunna he joined Moore at Salamanca on 24 November 1808, and from then until the re-embarkation of the army at Corunna in mid-January 1809 he handled the cavalry with consummate skill and economy. On 21 December he led the 10th and 15th Light Dragoons (Hussars) against 600 French dragoons at Sahagun, routing them and taking 167 prisoners. Three days later the retreat began, and Paget's regiments covered the tail of the army brilliantly. The much-superior French forces pursued closely and there were several clashes with their advanced cavalry. Notable actions were fought at Mayorga on 26 December, and at Benavente on 29 December. On the latter occasion the French General Lefébvre-Desnouettes pressed across the Esla river with a strong force of the *Chasseurs à cheval de la Garde Impériale*. The British rear piquets held the enemy until Paget arrived with the 10th Hussars and led them in a most effective charge which drove the enemy back across the river and captured seventy prisoners, including Lefébvre-Desnouettes himself. Moore praised his cavalry highly, and they suffered at least as much as the infantry on the dreary retreat; they had to look after their mounts as well as themselves in a bare countryside, and fully half the horses died *en route*. There was no room in the ships for the rest, and they had to be slaughtered at Corunna – a sad task for cavalry soldiers. Later in 1809 Paget commanded an infantry division in the Walcheren expedition, staying on the pestilential island until 2 September.

For the next five years Paget had no military employment; Wellington (q.v.) could have made great use of a cavalryman of his talents in the Peninsula, where the quality of British cavalry commanders was unimpressive, but the rigid seniority system, together with a personal scandal, made such employment impossible. Paget had married in 1795 Lady Caroline Villiers, and gave her no less than three sons and five daughters. He then became involved, to the scandal of society, with Charlotte, wife of Wellington's brother Henry Wellesley. There was a highly-publicised duel, and the couple eloped; eventually each obtained a divorce and they married, and by Charlotte the tireless cavalryman had another three sons and three daughters.

Paget, who had become Earl of Uxbridge on the death of his father in 1812, was ordered to Flanders in the spring of 1815 and was given command of the entire cavalry and horse artillery of Wellington's army – although the Prince of Orange retained the Dutch-Belgian mounted troops until the very morning of Waterloo. At Quatre Bras on 17 June Paget conducted a magnificent rearguard action, covering the withdrawal of the Allied infantry to Mont St Jean throughout the afternoon. It was, in his words, 'the prettiest field-day of cavalry and horse artillery that I ever witnessed.' Given complete discretion by Wellington to conduct his own operations, he leap-frogged horse and guns back along the single passable road through the muddy fields, pausing every now and again to discourage the pursuing French cavalry with a few shells and a brief charge. He was particularly successful in discouraging the French lancers in a stiff action outside Genappe, although he was angered by the conduct of his light horse in that combat. On 18 June the most important British cavalry operation was the charge of the Union and Household Brigades, commanded by

Ponsonby and Somerset, against the great infantry attack by d'Erlon's corps on the British left. Paget sent Ponsonby against the infantry, who were already checked and shaken by the fire of Picton's (q.v.) infantry, while he led Somerset's regiments against the supporting brigade of Milhaud's cuirassiers. This famous attack was completely successful; some 3,000 prisoners were taken, together with two eagles, and thirty guns were disabled. However, instead of re-forming, the British cavalry displayed their old fault – wild with excitement they charged straight on into the enemy lines, and were terribly cut up by fresh enemy cavalry while they were scattered, on blown horses, far from help. Paget tried hard to recall the heavy brigades, and brought up the too-weak reserve line to cover the withdrawal, but they lost about half their strength. He later admitted that he should never have charged in person with the Household Brigade, but should have retained his control of the overall situation with strong reserves under his hand. Nevertheless, the aftermath could not take away the triumph of the charge itself.

As is well known, in the final stage of the battle a shot struck Paget in the knee as he sat on his horse next to the Duke; this is supposed to be the occasion of the probably apocryphal exchange between the two which has been widely quoted as an example of British phlegm – Paget's 'By God, sir, I've lost my leg!' and Wellington's reply, 'By God, sir, so you have!' In fact the leg was not blown off, but the knee was shattered, and amputation was inevitable. This was performed in a house in Waterloo village, and Paget displayed a cold-blooded courage which was impressive even to those used to that murderous age. He chatted with friends and wrote to his wife while waiting for the operation; nobody so much as held his hand during the amputation, which was of course performed without anaesthetic, but he lay still and silent, apart from remarking at one point that the scalpel seemed rather blunt. It is recorded that his pulse rate was unchanged at the end of the operation. The leg was buried in the garden, and for many years the householder made a modest income from tourists. In later years Paget visited the house himself, and insisted on dining at the table on which he had been mutilated.

Paget was greatly acclaimed for his service and his heroism, and in July 1815 was created Marquis of Anglesey. He became a Knight of the Garter in 1818, and a full general the following year. He was appointed Master-General of Ordnance, with a seat in the cabinet, in April 1827, filling the post until January 1828; he held it again between July 1846 and February 1852. In 1828 he was appointed Lord Lieutenant of Ireland at a time of great unrest; his calm, tolerant, and even conciliatory attitude to Catholic activism enraged hard-liners in England and Dublin, and he was recalled in March 1829, but held the same post again in 1830–3. In December 1842 he became colonel of the Royal Horse Guards, and in November 1846 he was made a field-marshal. He died on 29 April 1854 at the ripe old age of eighty-six, and was buried in the family vault in Lichfield Cathedral.

Parma, Duke of, *see* Farnese

Patton, George Smith (*1885–1945*), *General*

'Old Blood and Guts' Patton was probably America's most able front-line commander of the Second World War. His unusual, not to say eccentric personal style was the cause of much controversy, but his enormous energy and boldness as a commander of armoured forces was of inestimable value during the American advance across France in mid-1944. He was flamboyant, opinionated, inclined to be reckless both of authority and of established procedures, and one of the greatest military 'grand-standers' of all time; but his successes outweighed all these legitimate criticisms. In simple terms, he won – and in situations which defeated more orthodox commanders. No general needs any greater justification than that.

The descendant of a Virginian family with military traditions, Patton was born in San Gabriel, California, on 11 November 1885. He was educated at the Classical School for Boys, Pasadena, and later at the Virginia Military Institute. He graduated from West

Point in 1909 and was commissioned a second lieutenant in the 15th Cavalry. The cavalry leaders of the Civil War were a special study of his, and the mounted branch was clearly his true *métier*. He served as weapons instructor at the Cavalry School in 1914–16, and in 1916–17 saw active service in Mexico as aide to General Pershing (q.v.); during this campaign, according to anecdote, he once fought his way out of an ambush by dropping from their saddles three *Villistas* with three pistol-shots. Still as aide to Pershing, he sailed for England in May 1917. This fine horseman and shot was the first man detailed to the new U.S. Tank Corps in November 1917. He organised and commanded the Tank School, and later led the 1st Tank Brigade in 1918 during the St Mihiel and Meuse-Argonne offensives; he was wounded in action in September 1918. Patton, unlike some dyed-in-the-wool cavalrymen, took to armour with enthusiasm. He grasped the possibilities of fast, highly-mobile tank forces, and between the world wars was America's foremost exponent of armoured warfare.

He commanded the 1st (later 304th) Armored Brigade at Camp Meade, Maryland, in 1919–21, and the 1st Squadron of the 3rd Cavalry at Fort Meyer, Virginia, in 1921–2. He graduated with honours from the Command and General Staff School in 1923, and served with the General Staff from 1923 to 1927. He filled an appointment in the office of the Chief of Cavalry between 1928 and 1931, returning to the 3rd Cavalry as executive officer for the period 1932–5. After another two years with the staff, he commanded first the 9th and later the 5th Cavalry in 1938; the 3rd Cavalry in 1938–40; and then the 2nd Armored Brigade. In November 1940 he was appointed commander of the 2nd Armored Division at Fort Benning, and was soon promoted to command I Armored Corps. In 1942 he commanded the Desert Training Centre, instilling combat techniques for all arms. Late in 1942 Major-General Patton was selected by General Marshall (q.v.) to command the ground forces in the American landings in French North Africa – Operation 'Torch'.

After the American defeat at Kasserine, Patton was given personal command of the U.S. II Corps, replacing Fredendall. Patton rapidly stabilised the situation, and his highly personal style of leadership proved most valuable at a time when the morale of the shaken American troops was at a low point. He only held this post for a short period in March-April, before being promoted to command the U.S. 7th Army, but he led most ably during the later stages of the Tunisian fighting. He commanded the 7th Army with great success in the invasion of Sicily, leading a wide sweeping advance through the western part of the island and capturing Palermo, before driving on to Messina. During the early part of 1944 he commanded the 'shadow' American army group in south-east England which threatened an invasion of the Pas-de-Calais area to distract German attention from Normandy. He did not have a combat command during the opening stages of the Normandy fighting; there is a suggestion that this period in the 'wilderness' was a punishment for some characteristically outspoken excess which angered his superiors.

In July 1944 Patton took over the U.S. 3rd Army, at the head of which he achieved his most spectacular successes. Pressing ahead with reckless boldness, his tank columns raced across France from Avranches all the way to Metz; disregarding many tenets of orthodox generalship – and many orders from above – Patton prevented the establishment of any important German defensive system. He threw all his reserves into the fighting line; diverted logistic support specifically earmarked for other commands in order to fuel and rearm his tanks; ignored frantic orders that he show more caution; and very nearly came to disaster when his forces were stalled with dry petrol tanks on the Meuse. Results justified his gamble. During the Ardennes counter-offensive it was Patton who made a forced march under severe conditions to relieve the beleaguered garrison of Bastogne; and by the end of the war in Europe the 3rd Army had reached the frontier of Czechoslovakia. In October 1945 Patton was given the command of the 15th Army in France; but on 21 December 1945 'Old Blood and Guts' died after a car-crash in Germany.

Patton was an extraordinary man who stirred up extraordinary passions. He was a skilled professional soldier; one of the early 'tank enthusiasts' at a time when such views were highly unpopular; and a gifted commander in the field. Nevertheless, he sometimes displayed an extraordinary disregard for the details of his profession, and he was an uncomfortable subordinate. He was outspoken and colourful in his public opinions, to the delight of newspapermen but to the horror of diplomats and 'political' generals. He openly enjoyed and worshipped the profession of arms, and felt himself, in a very real sense, to be mystically linked with great soldiers of the past. He did not disguise his appetite for the trappings of command and of personal success. His theatrical style, his pithy remarks, and his hell-for-leather victories made him the darling of the more unsophisticated sections of the American Press and public. Such traits did not endear him to British observers, brought up in a tradition of self-effacing discipline; and it was inevitable that his 'press-on-regardless' philosophy should have brought him into a conflict of personality with the careful, deliberate Montgomery (q.v.). Nevertheless, it is unjust that his superficial faults should be remembered so vividly. There was a great deal more to George Patton than a nickel-plated six-shooter with ivory grips.

Pershing, John Joseph (*1860–1948*), *General of the Armies of the United States*

'Black Jack' Pershing was a dedicated and determined professional soldier who commanded universal respect, if not affection; and his career includes two achievements unique in the annals of the United States Army. His active service spanned an important transitional period in military developments. He served as a young officer of mounted cavalry in the last stages of the Indian Wars; he reached his professional peak in command of an army supported by machine-guns, aircraft, and tanks; and he lived to see the atomic bomb used against his country's enemies.

Pershing was born near Laclede in Linn County, Missouri, on 13 September 1860. Graduating from the U.S. Military Academy, West Point, in 1886, he spent the next four years as a junior officer of cavalry, taking part in operations against the Indians in South Dakota and in the south-west. From 1891 to 1895 he was an instructor in military science at the University of Nebraska, where he was awarded an LL.B. in 1893. He returned to West Point in 1897–8 as an instructor in military tactics. He served with distinction in the Spanish-American War, first in Cuba and later in the Philippines, where he saw action against the Moro tribes of Mindanao. In 1903 he returned to the United States, and for two years served on the general staff. As a military observer attached to the Japanese forces in Manchuria he studied the Russo-Japanese War at first hand. In 1906 he gained the first of the two unique distinctions which set him apart from all other officers of his service; as a mark of special confidence President Theodore Roosevelt authorised his promotion from the rank of captain direct to the rank of brigadier-general, over the heads of 862 more senior officers. The resentment which this caused may be imagined, and must have played a part in forming the rather rigid and forbidding face which Pershing presented to the world in later years. Between 1906 and 1913 Pershing served in the Philippines once again, and in 1913 he returned home to command the 8th Cavalry Brigade, with headquarters in San Francisco.

In 1910 the Mexican President Diaz abdicated, and that unhappy country was plunged into bloody factional war once more. One powerful anti-American faction was led by the famous Pancho Villa, and anxiety in the United States over the safety of American property and nationals south of the border increased. There were several incidents, and these came to a head in 1916 when Villa crossed into New Mexico, raided the town of Columbus, and indulged in a gun-battle in the streets with men of the U.S. 13th Cavalry. A punitive column commanded by Brigadier-General Pershing was sent into the state of Chihuahua, with the grudging acquiescence of the ineffective Mexican Government, to hunt down Villa dead or alive. This mission failed, although the cavalry of Pershing's command inflicted quite heavy casualties on the *Villistas* in a number of skirmishes. (Serving with the column was one Lieutenant George S. Patton (q.v.) of the 8th Cavalry.) At the conclusion of the Mexican operation Pershing was promoted major-general.

In May 1917 Pershing was appointed Commander of the American Expeditionary Forces, and led them to France to participate in the First World War. The Allies, exhausted by three years of unprecedentedly costly warfare, had expected to be able to reinforce shaken sectors of their line with fresh American divisions. Pershing, however, was determined that when American troops went into action it should be as a unified national army under their own commanders. He was under constant pressure from the British and French to release American units for amalgamation in Allied formations, and also to release contingents from France to serve on other fronts. Pershing resisted all these pressures with characteristic tenacity, although at a time of crisis during the German offensive of March–June 1918 he did make American troops temporarily available to the Allied Commander-in-Chief, Foch (q.v.). Although the A.E.F. was never truly self-sufficient and relied upon the other Allies for certain categories of equipment and support, it was as a separate national command that it took part in the St Mihiel offensive in September 1918. After this had been brought to a successful conclusion Pershing, at Foch's request, abandoned his original intention of an advance on Metz, and regrouped the A.E.F. for a new offensive in the Meuse-Argonne sector. The progress of this latter offensive was delayed somewhat by incomplete preparation and by a measure of inexperience, and Pershing must bear some of the responsibility for logistic and operational errors. There has never been a general who did not make occasional mistakes, however, and Pershing must be accorded much of the credit for the successful assembly and employment of the American forces in France. (It may perhaps be of interest that at one stage Pershing's chauffeur in France was Eddie Rickenbacker, the American racing driver who was at that time still trying to secure a transfer to the aviation branch, of which he became the leading fighter ace before the end of hostilities.) In October 1917 Pershing was promoted to the rank of general.

Pershing finally returned to the United States and an enthusiastic reception, and in September 1919 he received the second unique distinction of his career: he was appointed to the rank of 'General of the Armies', which had been put forward for, but never actually held by, George Washington (q.v.). Pershing served as U.S. Army Chief of Staff from 1 July 1921 until his retirement from the service in 1924. He died in Washington on 15 July 1948, and is buried in Arlington National Cemetery.

Pétain, Henri Philippe Omer (*1856–1951*), *Marshal of France*

Pétain was the leading 'second generation' French commander of the First World War and 'the Saviour of Verdun' – despite the loss of 350,000 casualties. He turned defeatist early in the Second World War, made himself chief of state throughout the period of German occupation, and remained as such despite senility. Pétain was one of the few men in modern European history whose lifetime encompassed national adulation as a patriot hero, and national contempt as a defeatist traitor.

Born at Cauchy-à-la-Tour in the Pas de Calais on 24 April 1856, Philippe Pétain represented almost exactly the same enigma as his near-contemporaries Joffre and Foch (qq.v.) in that by the time the greatest war mankind had ever known broke out in 1914, the most powerful army opposing Germany was commanded by generals who had acquired virtually no active experience in the field, and whose claim to the leadership of millions of men rested largely upon an academic appraisal of colonial campaigns. So it was that after training at St Cyr, Pétain rose steadily – by his obvious academic acumen – through a succession of staff appointments, scarcely hearing a shot fired in anger before reaching the age of fifty-eight.

Nevertheless, while colonial commanders had been testing and apparently confirming the doctrine of *l'offensive à l'outrance* against somewhat secondary opposition, Pétain had evolved a more realistic tactical stance and one that would lend itself more aptly to the straits in which France found herself after the successive failures of Joffre in 1915–16, and Nivelle in 1917. This was, in effect, that much could be achieved by maintaining a defensive strategy on the ground, provided the forces were faithfully supported both in supply, and tactically with adequate artillery. This stance was put to the test at Verdun in 1916.

Soon after the outbreak of war Pétain had been given command of the XXXIII Army Corps in Artois, and particularly distinguished himself near Arras in May 1915. The following month, as a full general, he was appointed to command the French Second Army. When in 1916 both Joffre and Falkenhayn determined to break the deadlock on the Western Front with an attack at Verdun, the German Fifth Army struck first, and seems to have caught the French unprepared. Early German objectives were successfully gained, and the great French fortified area seemed doomed. A colossal artillery bombardment inflicted heavy casualties upon the French, who seemed on the point of collapse when Joffre, recognising in Verdun a symbol of French resistance, forbade further retreat ('Ils ne passeront pas') and ordered Pétain to assume command and retrieve the situation. Pétain's answer was to rush forward large reinforcements and virtually all France's reserves of artillery. Attack after attack was launched by the Germans; most were initially successful, but all were blasted to a halt by the French batteries. By skilfully 'rotating' his exhausted troops Pétain held out against repeated attacks over the mounting heaps of dead which littered the cratered landscape. Left with only a forty-mile secondary road to his rear supply bases, Pétain brilliantly and methodically organised a supply system which involved a constant stream of lorries moving at fourteen-second intervals under constant artillery fire. Renewed German attacks in June and July almost broke the French lines; phosgene gas was used for the first time; Pétain even suggested abandoning the Verdun salient, but Joffre refused. The Germans, faced with the problem of supplying reinforcements for the Brusilov Offensive, decided that the drain of fifteen divisions at Verdun was too extravagant and, by withdrawing some of them, cut their losses and abandoned the campaign. French casualties at Verdun amounted to 542,000 men (compared with 434,000 Germans), and while Pètain was lauded as 'the Saviour of Verdun', Joffre shouldered the responsibility for the shocking cost in dead and maimed.

Verdun represented the high point of Pétain's military achievements, although in the final year of the war his abilities were recognised by his appointment to command the French armies under Foch during the successful counter-offensive which ultimately ended the carnage on the Western Front. A fortnight after the Armistice he was promoted a marshal of France.

Between the wars his advancing years did not deter Pétain from continuing interest in the military affairs of France. Adopting the universal pacifist attitude of the age, successive French administrations seized upon any defence strategy that would be thrifty of French lives and money. As if refusing to acknowledge that a Schlieffen Plan had ever been drafted – let alone that it had almost succeeded – the French constructed the Maginot Line, which extended no further than the Belgian frontier. Pétain, uneasy about a defensive plan dependent upon vast systems of immovable concrete, did manage, as minister of war in a short-lived government under Gaston Doumergue during 1934, to secure increased military expenditure to acquire modern weapons, however.

When the Second World War started and the collapse of the French armies revealed the defeatist attitudes adopted by civil and military leaders alike, it was perhaps inevitable that France should turn to the man whose name had for so long been associated with victory – despite the fact that recently even this man's voice had been heard all too often voicing defeatist opinions. At eighty-three Pétain returned to France – from Spain, where he had been ambassador since 1936 – charged with arranging an honourable, but speedy armistice with Germany. In the event France was partitioned, and Pétain created for himself the post of head of state with a government at Vichy, while the Germans occupied the north and west of France. Political intrigues continued behind his back, until on 20 August 1944 he was arrested by the Germans and imprisoned. In April 1945 he voluntarily returned to France, where he was put on trial by the provisional government under General de Gaulle. He was sentenced to military degradation and death, but the penalty was commuted to life imprisonment. He was moved to a villa in Port Joinville a month before his death on 23 July 1951.

One of the most controversial figures of French twentieth-century history, Pétain was simultaneously regarded by some as a national traitor and by others as a patriot who had stood by a stricken France. Militarily he epitomised the transition between the anachronistic French *élan* of 1914–15 and the carefully-planned strategy of 1918.

Picton, Sir Thomas (*1758–1815*), *Lieutenant-General*

Picton's personality and style of command provide a neat contrast with those of his Peninsula contemporary Rowland Hill (q.v.). Each was a valued subordinate of Wellington (q.v.), and each made a great contribution to Allied victory; yet, despite the fairly small and sharply delineated social class from which British generals of the day were drawn, no two men could have been less alike. 'Daddy' Hill was mild in manner and speech; Thomas Picton was violent and intemperate, and known for his ability to out-blaspheme the foulest-mouthed sergeants of his division. Hill was loved for his character, while Picton was respected despite his. It is dramatically satisfying that the one should have died in the peaceful bed of honoured old age, while the other fell in battle, probably with a curse on his lips.

Born at Poyston, Pembrokeshire, in August 1758, Picton was educated privately. After a period spent at a private military academy in France he was gazetted ensign in the 12th Foot late in 1771, at the age of thirteen. He went overseas two years later, and served at Gibraltar. He was promoted captain in 1778, transferring to the 75th Foot. The announcement of the disbanding of that regiment in 1783, at the time of the Treaty of Versailles, provoked a riot which Picton helped to put down. He resumed his military career and attained the rank of lieutenant-colonel in 1794, joining the staff of the Commander-in-Chief, West Indies. He served at the capture of St Lucia and St Vincent in 1796, and at Trinidad in 1797. With the rank of colonel he was appointed Governor of the latter in 1797; he was promoted brigadier in 1799, and saw further action on St Lucia and Tobago while holding the governorship. The former Spanish colony of Trinidad was a lawless chaos, and Picton did much to improve its stability and trade; but while never corrupt, his administration was certainly brutal. His brief specifically ordered him to apply the old Spanish legal code, and it was in accordance with this that he allowed a woman accused (and guilty) of a serious cash theft to be questioned by 'picketing' – i.e. being forced to stand with the weight of the heels on a wooden peg. His many enemies seized upon this incident, and he was obliged to resign in October 1803. He returned to England and beggared himself fighting a long and eventually inconclusive law suit; the inhabitants of Trinidad put up £4,000 for his defence (which sum he later donated to a relief fund after a disaster at Port of Spain) but the slur clung to his reputation for many years.

Picton was promoted major-general in April 1808, and Wellington subsequently requested that he be sent out to the Peninsula to command a division. He took up command of the 3rd Division in February 1810 and held this post until June 1812, and again between May and September 1813, and between December 1813 and April 1814. He soon acquired a reputation as a rough, foul-mouthed commander of great energy and complete fearlessness. His style was to lead his Division from the front, often wearing the slightly eccentric civilian costume normal among many Peninsula officers, roaring terrible oaths, and heading for the thickest part of the battle. He was no mere berserker, however; his career offers many examples of cool and skilful handling of his Division in moments of danger, and he was unquestionably a most able soldier. He ruled his command with an iron fist, and handed out many floggings; nevertheless his style seems to have appealed to the rough soldiers in the ranks, and he had a reputation for approachability. Picton distinguished himself at Bussaco, Fuentes de Onoro, and Ciudad Rodrigo, and one of his most famous exploits was the capture of the castle at Badajoz. He suggested to Wellington that his 3rd Division should assault this position, as he was not convinced that the main breaches on the other side of the town could be carried by the troops assigned to them. In fact the major assault on the breaches failed bloodily, and the eventual fall of Badajoz was only accomplished by Picton's secondary attack storming through the castle and into the streets of the town in the rear of the main body of defenders. This sort of bloody work was the kind of soldiering at which the stormy Welshman excelled.

Invalided home in August 1812, Picton was knighted the following February; he was promoted lieutenant-general in June 1813 and returned to Spain, once again distinguishing himself at Vittoria. His impetuosity caused unnecessary casualties on more than one occasion, but paid dividends at that battle. The division assigned to advance over a vital bridge was badly delayed; Picton, ordered – to his fury – to 'support' its advance, took matters into his own hands and led his own command straight at his senior colleague's objective, carrying it by fearless and determined fighting spurred on by his usual uninhibited vocabulary. Nevertheless, as his record during the Pyrenees, Orthez, and Toulouse campaigns makes clear, Picton had little talent for independent command or for handling any larger formation than a division. Like so many of Wellington's generals, he was a valuable tool in his Commander-in-Chief's hand, but erratic if left to himself.

Picton rejoined the Army for the Waterloo campaign, and was given command of the 5th Division. At Quatre Bras on 16 June 1815 his command made a spirited defence which successfully held off the superior forces of Marshal Ney (q.v.), and Picton himself was struck in the side by a musket ball which broke two ribs. Unwilling to be sent to the rear, he kept his injury a secret and had his soldier-servant bandage him sufficiently to enable him to sit his horse. He was constantly in the saddle during the next two days, and must have been in great pain. At Waterloo on 18 June it was Picton who led the decisive charge of the British infantry on the left, which smashed Erlon's great infantry attack on the Allied lines. He was wearing his well-known broad-brimmed top hat and a civilian coat as he led the Gordons, the Black Watch, and the 44th Foot down the gentle slope on to the bayonets of 8,000 Frenchmen. There a musket ball struck him in the temple and tumbled him out of his saddle, dead before he hit the ground. The massive blackened wound in his side was only discovered when he was stripped after the battle. He was buried on 3 July 1815 at St George's, Hanover Square, London.

Pontiac (*c. 1720–69*)

Throughout the history of the red–white confrontation in North America the Indian was handicapped as much by his temperament as by his technical inferiority. His loyalties were essentially local: few Indians ever saw the struggle as a fundamental battle for the whole continent between red and white, and while they clung to a tribal view of events they could be defeated (or bought off) in detail. Pontiac was one of a tiny handful of Indian leaders who saw matters in other than purely local terms. He was one of the very few who managed to build an effective confederation of tribes for co-ordinated action against the whites, and almost unique among the Indians in being a genuine strategist – one who planned a long-term campaign for long-term prizes. (The greatest pan-Indian leader of all was Tecumseh of the Shawnee; but though his vision and abilities were astonishing, his true significance was political rather than military.)

Pontiac was thrust into power by the situation west of the Allegheny Mountains in 1760–3. In 1760 the French, with whom the tribes of the region had always enjoyed good relations and for whom they had fought the British colonists and redcoats, surrendered to the British and abandoned the Indians to their new masters. This was a severe psychological shock, which was followed by more tangible hardships. Despite the advice of his experienced Indian agents, General Amherst refused to follow a low-key policy during the first years of British rule. Men like Sir William Johnson and George Croghan tried to minimise the impact of the new order on the Indian way of life by perpetuating practices started long ago by the French – the distribution of arms and ammunition for hunting, gifts of food in winter, the use of the services of fort smithies for the repair of guns, and so forth – and now firmly established in tribal custom. Amherst would have none of it, however, and seemed to be deliberately seeking a confrontation. He gave his agents no leeway for negotiation, and ignored their warnings of unrest. Other provocations were offered by the traders and settlers from the coastal colonies as they expanded westwards. The traders, who had treated the Indians well during the years when they had been trying to undercut their French competitors, now began to make ruthless use of their monopoly position. The French had not been land-hungry, requiring merely a few strategically-sited forts to administer their vital fur trade; the British colonists wished to take over the land itself, and now began to press through the mountains in increasing numbers. A prophet rose among the Delawares, preaching a return to the old ways and an end to dependence on the white man's goods; and the French peasants who still farmed along the Detroit river whispered that a rising would be supported by French troops from Louisiana, sent by the Great Father Over the Sea, who had been sleeping, but who was now awakening once again. Pontiac of the Ottawas was among the Indians who listened to these promises, and believed them.

Little is known of his early life. He was born among the Ottawas (but one of his parents was probably a Miami or an Ojibway) about 1720 or a little earlier, and was brought up on the north bank of the Detroit river not far from the French post at Fort Pontchartrain (which later became Detroit). His rise to influence was gradual, but he was a noted war-chief during the French-Indian War. He is reputed to have led an Ottawa contingent in the massacre of Braddock and his two redcoat regiments in 1755, and to have been presented with gifts of unusual value by Montcalm (q.v.). That civilised Frenchman's distaste for Indians was well known, so any special marks of favour showered on Pontiac from that source would seem to indicate that he was a genuinely important figure. War leadership did not by any means guarantee a high place within the tribes in peacetime, however; the Indian social structure was sophisticated in matters of power. Pontiac had to spread his influence between 1760 and 1763 by patient diplomacy, secret persuasion, endless debate, and rousing eloquence. But it seems that he managed to communicate his vision with extraordinary effect, and to harness the tribes, each with its own discontents and fears, into a loose federation ready for synchronised action. He recognised the impossibility of expelling the white man completely, and limited his aims to driving the British back behind the Alleghenies, and re-establishing generous and unintrusive French authority around the Great Lakes.

In early May 1763 Pontiac struck at Detroit with his Ottawas and their close associates the Hurons and Potawatomis. Simultaneously a number of other tribes – some eighteen in all – struck at forts, settlements, convoys, and isolated groups of British in their own regions; prominent were the Delawares, Shawnees, Senecas, Ojibways (Chippewas), Miamis, Mingoes, Kickapoos, Sauks, Weas, and Mascoutens. Pontiac tried to take Detroit by a ruse, but was twice foiled by the commandant, Major Gladwin, a very able young veteran of frontier warfare. Finally the Ottawa leader settled down to besiege the fort and its garrison of some 200 soldiers and civilians. Elsewhere the forts fell one by one; by the end of June eight of the twelve British posts in Indian territory had fallen, one had been abandoned as untenable, and two – Detroit and Pitt – were under siege. Civilian settlements all along the frontier suffered greatly in the general uprising. Amherst was at first unable to appreciate the scope of the disaster, but then swung on to the counter-offensive with great ruthlessness. Relief columns were sent out to Detroit and Pitt; some 260 men under Dalyell rowed up the river to Detroit under cover of fog one night in late July, and gained the fort. Dalyell was killed leading an unwise sortie two days later, and the retreat of the party was covered with some distinction by Major Robert Rogers (q.v.). Further reinforcements reached Detroit by water in August, and it began to dawn on Pontiac that French support was unlikely. Some of the tribesmen lost heart, and made individual peace with the British before slipping back into the forest. Nevertheless Pontiac somehow kept the bulk of his men in position around Detroit well into October; it had been a considerable feat to persuade the impatient warriors to attempt a static siege at all, and its duration was remarkable. Meanwhile, despite isolated successes all over the area, the Indian revolt was generally running out of impetus. In August the famous Swiss officer Colonel Henri Bouquet raised the siege of Fort Pitt. The colonists were clamouring for revenge, and the fighting along the frontier during August and September was notably savage, even by local standards. A price of £200 was put on Pontiac's head, and Amherst made the suggestion that blankets infected with smallpox be planted in Indian camps. His orders were expressed in terms which described the Indians as vermin to be exterminated on sight – an unprecedented attitude. Atrocities were committed by both sides.

As it became obvious that the early gains could not be held, Pontiac lost support and prestige. His hopes of French troops from Louisiana were finally dashed when he received a letter from that colony urging him to make peace with the British. His first overtures to Gladwin were passed to higher authority for a decision, but he did not wait for an answer. With a small number of faithful followers he went south and west, preaching revolt to the tribes of the Illinois and Mississippi country. He continued to foment trouble throughout 1764, with some local success; but the British were pacifying faster than he could inflame. Most of the tribes made separate peace treaties, anxious for the revival of the British trade which they had missed badly. Amherst had been replaced by General Gage, a much more subtle man, who smothered the last centres of rebellion by a mixture of fair words and overwhelming military strength. By the spring of 1765 Pontiac realised that the cause was doomed. He came in to make peace, and on 18 April announced his final capitulation. He was completely sincere; when in the years which followed his fellow-tribesmen complained about just those injustices over which he had led them to war, he counselled peace. The British treated him with honour and favour, and it was rumoured that he drew a cash pension. It is believed that this elaborate deference and courtesy was a deliberate plot to make him unpopular with his fellow Indians. If so, it worked; he lost prestige and authority, and in 1768 was forced to leave his village. On 20 April 1769, for no known reason, he was struck down and killed by a Peoria Indian as he came out of a French store in Cahokia, a hamlet on the east bank of the Mississippi opposite St Louis.

Pulawski, Kazimierz (*1747–79*), *Brigadier-General*

Count Pulawski was the hero of the 1768 Polish rising against Russia, and later gained fame as the commander of the Continental cavalry in the American War of Independence.

He was born on 4 March 1747 at Winiary in Mazovia, the son of Jozef Pulawski, a Catholic nobleman. In 1768 a league of minor Polish Catholic gentry, the Confederation of Bar, was formed as a reaction against abuses by the pro-Russian party in the country, and against Russian domination. Its subsequent armed rising was militarily hopeless, but by dint of much individual heroism the insurgents managed to keep up the fight for four years, thus drawing international attention to Russia's cynical tyranny. All three of Jozef Pulawski's sons took an active part in the fighting; Franciszek was killed in Lithuania in 1769, and Antoni was captured. Kazimierz greatly distinguished himself in the defence of Berdichev in the spring of 1768, and later led guerrilla operations near the Turkish frontier. In 1769 he again drew attention to himself by his able defence of Zwaniec and Okopy Swietej Trojcy. In September 1770 the insurgents occupied the fortified monastery of Czestochowa, and this became their base for wide-ranging operations over the whole area between the Carpathians and Poznan. Pulawski's defence of Czestochowa against Russian troops won him fame which spread beyond the borders of Poland. In October 1771 he led a party which attempted to kidnap the weak pro-Russian King Stanislaw II Poniatowski and bring him to the rebel base. The attempt failed, and in the spring of 1772 the invasion of Poland by Prussian and Austrian armies – prompted by considerations of European power politics, not by any interest in Poland's aspirations – finally doomed the insurrection. In May 1772 Pulawski fled, first to Saxony and later to France. Many Polish soldiers escaped over the Turkish border, and in 1774 Pulawski travelled to Turkey to help organise them to fight against the Russians alongside the Turks; but this, too, came to nothing when Turkey and Russia made peace. Pulawski's estates were confiscated, and for two years he lived in poverty in France.

In December 1776 Pulawski was introduced to Benjamin Franklin, one of the American commissioners in Paris. He was attracted by the idea of helping the American colonists in their revolt against Britain, and in June 1777 arrived in America with a letter of recommendation from Franklin to George Washington (q.v.). In due course he was made Chief of Cavalry by Congress. The patriot cavalry was extremely weak, short of weapons, and poorly trained, and used only for courier duties and the most undemanding scouting. The dashing Polish officer worked hard to train the men in classic cavalry tactics, and with considerable local success, although Continental resources were never sufficient to underwrite the organisation of a serious battlefield cavalry on the European model.

At the Brandywine on 11 September 1777 the British general Howe skilfully turned Washington's right and forced the patriot army back towards Philadelphia with heavy losses; these would have been heavier but for Pulawski's leadership of his small cavalry force, which he split up into small groups and led against the advancing enemy in repeated charges, buying time for the endangered American infantry. He fought at Germantown, and stayed with the army throughout the bitter winter of 1777–8. In 1778 he was replaced as Chief of Cavalry by General Stephen Moylan, but was given the rank of brigadier-general and was authorised to raise and command his own corps. This mixed force was known as Pulawski's Legion, and consisted of some 300 men of various nationalities and backgrounds; there is evidence that, with his hard experience of guerrilla warfare in terrain threatened by superior regular forces, Pulawski led this little command with some success. They distinguished themselves in an action against a British landing party at Egg Harbor, New Jersey, in October 1778. In May 1779 Pulawski took part in the defence of Charleston against General Prevost's expedition. In October of that year he was severely wounded by a shell splinter during bitter fighting outside Savannah, which was under attack by a Franco-American army, and he died on 11 October aboard the vessel *Wasp* while being taken back to Charleston. He was thirty-two years old, had been exiled from his native land for seven years, and had been engaged in fighting against his conception of tyranny since his twenty-first birthday.

Raglan, Fitzroy James Henry Somerset, 1st Baron (*1788–1855*), *Field-Marshal*

Raglan, the old one-armed general who commanded the British forces in the Crimea, has suffered at the hands of history. In his youth he was a most promising young officer of unshakeable integrity; it was his tragedy that forty years of peaceful stagnation followed the glory of Waterloo (as it was that of the Army in which he rose to senior command).

Born at Badminton on 30 September 1788, Lord Fitzroy James Henry Somerset was the youngest son of the Duke of Beaufort. He was educated at Westminster, and was commissioned as cornet in the 4th Light Dragoons in June 1804. After four years, during which he reached the rank of captain, he went to Portugal as aide-de-camp to Wellington (q.v.), being present at Rolica and Vimiero; he was wounded at Bussaco. He was appointed military secretary to Wellington in 1811, thus gaining direct contact with the battalion commanders and an intimate knowledge of each regiment. His brevet-majority was promulgated on 9 June that year after Fuentes de Onoro. At Badajoz he obtained the surrender of the French after the breaches had been stormed and before the defenders could organise further resistance. He gained his brevet lieutenant-colonelcy at Wellington's request, after being awarded the Army Gold Cross with five clasps, and taking part in all the battles conducted by his Commander-in-Chief. His appointment as K.C.B. was announced on 2 January 1815, and he married Emily Harriet, second daughter of the third Earl of Mornington and a niece of Wellington.

Somerset accompanied Wellington to Paris after Napoleon's first abdication, Wellington taking up the post of ambassador and Somerset being appointed his secretary. He remained in Paris until Napoleon's return on 20 March 1815, and then went to the Netherlands as Wellington's military secretary. Towards the close of the Battle of Waterloo, whilst standing beside his Commander-in-Chief, he was struck by a bullet and his arm had to be amputated. Shortly afterwards, on Wellington's recommendation, he was appointed aide-de-camp to the Prince Regent, and returned to the British Embassy in Paris, where he remained until the end of 1818. 22 January 1827 saw the death of the Duke of York, and Wellington succeeded as Commander-in-Chief. Somerset was made military secretary at the Horse Guards and retained this post for no less than twenty-five years; during this period he became known for his efficiency, tact, and utter truthfulness. He became a colonel in the 53rd Foot in November 1830 and was promoted lieutenant-general in 1838. On 14 September 1852 Wellington died and was succeeded as Commander-in-Chief by Hardinge. Somerset received the G.C.B., taking the title Baron Raglan of Raglan, Monmouthshire, and was appointed to succeed Hardinge as Master-General of the Ordnance.

When England and France declared war against Russia in 1854, Raglan was sent out east in command of the British forces. He had the strength and vigour of a much younger man although by now he was sixty-five. Yet it is true that this was to be the first time he had actually led troops in the field. By the end of June that year most of the British and French troops were camped at Varna; Raglan was instructed to lay siege to Sebastopol if his information suggested that he would thereby achieve success. On 14 September 50,000 men were landed at Kalamita Bay on the west coast of the Crimea, and the subsequent victory at Alma raised hopes for the capture of Sebastopol. Under Raglan's supervision trenches and batteries were prepared and the Allies opened fire with 126 guns on 17 October. The Russian batteries had in the meantime been strengthened, and replied with counterfire from 341 guns which soon silenced the French. Thereafter all thoughts of early success had to be abandoned.

Disaster followed disaster. On 25 October there occurred the eclipse of the Light Brigade which accompanied a Russian attempt upon Balaclava, and for which Raglan laid the blame upon Lord Lucan. 55,000 Russians were thrown into an attack upon the Allies' right flank on 5 November, and the British only barely escaped destruction at the Battle of Inkerman; indeed, the Allies were so heavily outnumbered that they were unable to reach Sebastopol and were forced to dig in where they stood, exposed to the rigours of winter. To this misfortune was added the loss of twenty-two store ships, destroyed by a savage storm while on passage through the Black Sea on 22 December.

Up to this time the commissariat was a department of the Treasury, not of the War Office, and growing criticism was levelled at this department for the paucity of supplies sent to the Crimea and for the dreadful suffering of men and horses in the field. As information about the army's condition filtered home to England a storm of indignation blew up, prompting the Government to lay the blame upon the quartermaster-general, James Bucknall Estcourt, and the adjutant-general, Richard Airey (Lord Airey). But Raglan refused to allow these two officers to bear the blame. In Parliament Roebuck's motion for an inquiry into the army's conditions brought defeat to the Government. The formation of a new ministry by Palmerston followed, and in this Lord Panmure was appointed Secretary for War. Panmure wrote to Raglan to inform him that commissioners would shortly visit the Crimea to report on conditions, continuing: 'It would seem that your visits to the camp were few and far between, and your staff seems to have known as little as yourself of the conditions of your gallant men.' In a private letter to Raglan he told him that only a radical change of staff members would appease the public.

Raglan's reply was entirely characteristic: 'I have served under the greatest man of the age more than half my life, have enjoyed his confidence, and have, I am proud to say, been ever regarded by him as a man of truth and some judgement as to the qualifications of officers, and yet, having been placed in the most difficult position in which an officer was ever called upon to serve, and having successfully carried out different operations, with the entire approbation of the Queen, which is now my only solace, I am charged with every species of neglect; and the opinion which it was my solemn duty to give of the merits of the officers, and the assertions which I made in support of it, are set at naught, and your Lordship is satisfied that your irresponsible informants are more worthy of credit than I am.'

There was some foundation for the charges brought in that Raglan did not frequently visit the camps – this was due in the main to his previous service's demanding long periods at a desk. As he was reticent by nature, however, the visits he did make went almost unnoticed.

Meanwhile the end of February 1855 brought a resumption of pressure against Sebastopol, with attacks by the much-reinforced French forces under Canrobert; but these attacks were relatively ineffectual as the French commander insisted upon recalling his forces too soon. Pélissier replaced Canrobert, and at once achieved success in an expedition to cut the Russian lines of communication at Kertch. The main attack upon Sebastopol was planned to take place after a two-hour artillery barrage, but at the last moment Pélissier decided, much against Raglan's wishes, to attack at break of day. Rather than weaken the alliance, Raglan acquiesced. The French forces attempted to storm the Malakhoff, but were driven back with appalling casualties. In an attempt to retrieve the situation, Raglan ordered the British troops to assault the Redan, but they were immediately pinned down by the murderous Russian musketry and grape. It has usually been averred that Raglan gave these orders much against his own professional judgement and inclination, but in the strong belief that otherwise the French would have blamed the British for a defeat in which British forces were withheld.

This failure deeply affected Raglan, by now a Field-Marshal and afflicted by severe bouts of dysentery. He had disciplined himself constantly to act with composure and, under the difficult field conditions, with dignity – a code of behaviour in which he had schooled himself from the example from his great chief, Wellington. But the repeated assaults upon his prestige quickly undermined this composure, while the raging dysentery destroyed his physique. On the evening of 28 June 1855 he died.

Richard I, called 'Cœur de Lion' (*1157–99*), *King of England*

Although a cocoon of romantic nonsense has been woven about the figure of Richard I, his reputation as one of the foremost soldiers of his age is unassailable. He was a disastrous king: careless of England and the future of the English monarchy, he damaged English institutions and squandered English resources to finance his foreign adventures. Nevertheless his exploits brought great military prestige to the country in which he spent only six months of his ten-year reign.

Richard was the third son of Henry II of England and Eleanor of Aquitaine, and was born on 8 September 1157. Through his mother he received the duchy of Aquitaine in 1168, being enthroned as Duke at Poitiers in 1172. Like all Henry's sons he was turbulent, restlessly energetic, irresponsible, and without foresight. Although he engaged in constant quarrels with his father and brothers during his youth, however, he quickly gained a solid reputation as a war leader of courage, skill, and resource. He handled his hot-headed barons of Gascony and Poitou with some skill, and their frequent petty rebellions and civil wars provided him with much practice in the arts of war. One of his most famous early exploits was the capture of the supposedly impregnable castle of Taillebourg in Saintonge. After the death of his brother Henry in 1183 he became the heir to England, Normandy, and Anjou; when his father tried to persuade him to give up Aquitaine to his brother John he resisted furiously, playing off the very able Philip Augustus of France against Henry, and eventually allying himself with Philip to harry his father to his death in 1189. His talent for political intrigue was impressive, but he came to the throne with a single obsession which blinded him to the needs of his kingdom – the Crusade against the Saracens whose recent recapture of the Frankish states of the Holy Land had culminated in the taking of

Jerusalem by Salah al-din in 1187 (see Saladin).

After nearly bankrupting England – and prostituting her institutions – in order to pay for the expedition, he sailed for Palestine via Sicily in 1190. Finding the Sicilians hostile he stormed Messina, subsequently concluding a treaty with them which caused bitter enmity with the German contingent in the crusader army. He then sailed on to Cyprus, which he wisely conquered as an advanced base before landing in the Holy Land in June 1191. During his stay on the island he concluded a marriage of convenience with Berengaria, daughter of Sancho VI of Navarre; the match was more than usually loveless, since Richard was a lifelong homosexual.

When he landed near Acre on 8 June the siege of Salah al-din's garrison by the crusader army and the remainder of the local Frankish nobles under Guy of Lusignan was nearly two years old. Richard was tacitly accepted as the leader of the Crusade upon his arrival, and his vigour and determination quickly transformed operations. The Muslim relieving army was driven out of the area, and on 12 July 1191 Acre at last capitulated. Richard showed the callous side of his nature by allowing the massacre of some 2,600 Muslim prisoners. Soon after this the incessant quarrels between the leaders of the Crusade led to King Philip's return to France. For the rest of the course of the Third Crusade the military effectiveness of the crusaders was seriously impaired by continuing disputes between Richard, the German Conrad of Montferrat, Leopold of Austria, and the local Frankish leaders. Richard made enemies of all of them, and there was general ill-feeling between Europeans and local Franks due to the aggressiveness and bloodthirstiness of the former and the more relaxed, pragmatic attitudes of the latter. Nevertheless there was no denying Richard's military brilliance.

In August–September 1191 Richard, determined to take Jerusalem, led a multi-national army of some 50,000 towards Ascalon. He displayed great ability in enforcing discipline and order in his unruly army, and approached the practical problems of a long march in semi-desert terrain with an intelligence unusual for his age. His logistic preparations, his precautions against fatigue and disease, and his conduct of the advance in the face of constant hit-and-run attacks on the column were all admirable. He advanced slowly down the coast with his fleet keeping pace offshore, and so skilfully was the column ordered and led that the Saracens were quite unable to break its ranks; nor did individual barons or contingents respond when the Saracens tried to lure them into hasty counter-attacks – a remarkable tribute to Richard's powers of command. On 7 September Salah al-din set an ambush at Arsuf, and made a series of attacks on the rearguard, hoping to provoke the knights into charging back at their persecutors and thus breaking the column. Richard kept the army under tight discipline, marching stolidly forward, until the Turks grew bolder and committed themselves too heavily to the attack on the rearguard. Then, at a pre-arranged signal, the crusaders turned and launched a devastating and well-co-ordinated charge which caught the more lightly armed, armoured, and mounted Saracens in the open. Some 7,000 were cut down for the loss of 700 crusaders, and the remainder were scattered. Arsuf won Richard the respect of Salah al-din, his chivalrous enemy, who never again committed his men to a pitched battle in the open.

The crusaders wintered at Ascalon and later marched on Jerusalem, reaching to within a few miles of the walls; but so effective was Salah al-din's 'scorched earth' strategy that Richard marched back to the coast without attempting a siege. For nearly a year Richard and the Saracen leader fenced, fighting a number of minor engagements in which the very different strengths and weaknesses of each side virtually balanced out. Richard fought with tactical skill and great personal valour, but in September 1192 he accepted that outright military victory was for the time being impossible. He concluded a treaty with Salah al-din, binding for three years, whereby the crusaders kept Acre and a narrow coastal strip, and Christians were guaranteed access to the holy places. Then he sailed for Europe once again, vowing to return and take Jerusalem by storm when the treaty expired. He and Salah al-din were totally contrasting personalities who seem to have developed a genuine mutual respect.

In the winter of 1192 Richard was shipwrecked near Venice. Despite a prudent disguise, he was recognised and turned over to his old arch-enemy Leopold of Austria, who imprisoned him in a series of castles until February 1194 – it is in this period that the famous romance of Blondel is set. Richard was eventually released on payment of a stupendous ransom which necessitated an enormously complex and wide-ranging taxing exercise in England. On his return he had himself crowned for a second time - in April 1194 – as he had also been forced to give Leopold certain undertakings which he was afraid might have compromised his title. Within a month he sailed for Normandy and never returned to England, spending the last five years of his life in a series of vigorous campaigns against Philip Augustus. His military skills had been enhanced by his experience in the East, particularly in the field of fortifications; the Château-Gaillard, which he built on an island in the Seine to guard his frontier with the French Vexin, was one of the engineering marvels of the age. His impetuosity caused his death at the age of forty-two. When the vicomte de Limoges refused to hand over to him a gold hoard discovered by a peasant Richard besieged his castle of Châluz, and a crossbow bolt from the battlements mortally wounded him. He died on 6 April 1199, after having his slayer brought before him. According to tradition he forbade his followers to harm the man, but after Richard's death they had the soldier flayed alive.

Ridgway, Matthew Bunker (*b. 1895*), *General*

Pioneer commander of American airborne assaults during the Second World War, Ridgway rose to high command during the Korean War, and eventually became Chief-of-Staff of the U.S. Army.

Born at Fort Monroe, Virginia, on 3 March 1895, Matthew Ridgway graduated from the American Military Academy in 1917, and during the first twenty-five years of his military career rose through the officer ranks in a succession of staff appointments. In 1943 he was given responsibility for the planning and execution of the first American airborne assault in wartime, during the initial invasion of Sicily in July that year. While tactically successful, this bold attack, using troops carried in gliders, fell short of expectations due to adverse weather conditions and in some degree to the inadequate training of the glider pilots of the towing aircraft. A high proportion of the attacking gliders landed in the sea, while others came down well wide of their prescribed targets; there were many tragically wasteful casualties. Having learnt much from this experience, Ridgway was ordered to England, where he assumed the leadership of the U.S. 82nd Airborne Division, and parachuted with his men into Normandy during the initial stages of the great D-Day invasion of France. During the later stages of the Allied advance through northern Europe Ridgway commanded the U.S. 18th Airborne Corps in action in Belgium, the Netherlands, and Germany.

After the Second World War he returned to the U.S.A., only to become involved in the protracted political wrangling over peacetime force levels; he staunchly supported General Eisenhower (q.v.) in his efforts to resist cuts in military expenditure.

On the outbreak of war in Korea in June 1950, Communist forces quickly exploited their initial surprise success and, despite the rapid build-up of United Nations forces in the area, almost succeeded in driving these forces right out of South Korea. In command of what quickly became the American 8th Army was General Walton H. Walker, who managed eventually to halt the Communist advance, stabilised the front and, by the beginning of the autumn, had succeeded in recovering much of the lost ground. Then, on 23 December, Walker was killed in a car accident, and Lieutenant-General Ridgway was appointed to succeed him. The Supreme Commander, General MacArthur (q.v.) gave Ridgway command of all ground forces in Korea. MacArthur, believing that pressure on the Communist forces would soon attract open intervention by Communist China, and advocating pre-emptive bombing attacks on the Chinese border assembly areas – policies which angered the Truman administration in America – was summarily relieved of his command in the Far East soon after, and Ridgway was appointed to succeed him as

Supreme Commander. As such Ridgway left the immediate tactical conduct of the Korean War to subordinates (such as General James A. Van Fleet, the new commander of the U.S. 8th Army), and developed an overall interest in military affairs in the Far East, and in the rehabilitation of Japan.

In 1952 Ridgway returned to Europe to succeed General Dwight D. Eisenhower as Supreme Commander, Allied Forces in Europe, having reached the rank of full general the previous year. In 1953 he was appointed Chief-of-Staff of the U.S. Army, and he retired in 1955. In 1986 he was awarded the Presidential Medal of Freedom.

Robert I, surnamed Bruce (*1274–1329*), *King of Scotland*

The facts of Bruce's military career remain rather more obscure than might be supposed. Contemporary sources are often unhelpful, and his status as the hero of Scotland's fight for independence in the early fourteenth century has led to the usual accretion of rather dubious folklore.

He was born on 11 July 1274, the scion of a Norman family which had come to Scotland in the early twelfth century during the period when David I was following a policy of 'Anglicisation' and deliberately fostering the growth of the Norman feudal system in his country. The Bruces were related by marriage to the royal family, and were deeply involved in the confused struggle over the succession which followed the death of the infant Queen Margaret, the 'Maid of Norway', in 1290. In that year Bruce's grandfather – also named Robert – put forward a claim to the throne. This failed, as Edward I (q.v.) of England crossed the border, claiming the throne himself. He occupied Scotland as far as Perth, and in 1292 sponsored John de Baliol as king, and his feudal vassal. The Bruce and Baliol factions then engaged in intermittent warfare against one another. It is known that in the same year Robert's father resigned in his son's favour the earldom of Carrick, but little else is reliably established about Robert's early career.

In 1295 Baliol rose against Edward with French assistance, but was defeated the following year at the decisive Battle of Dunbar. Scotland remained under English military occupation for some years, with Edward's garrisons holding the major castles, but unable to enforce peace in the countryside. Sir William Wallace (q.v.) led popular resistance to the English, and at one time Bruce was among his supporters; but it seems that Bruce was later reconciled to Edward's rule, and it is even suggested that he fought against Wallace on the English side. After Wallace's capture and execution in 1305, Bruce and other local leaders plotted further risings against the English, but they were by no means united. A major rival to Bruce in the succession struggle was John Comyn, called 'the Red'. In 1306 the two men met at Dumfries, and there, on 10 February, Bruce murdered Comyn in the course of a quarrel in the Franciscan church. It has been suggested that this was not a planned act, and that it precipitated Bruce into an all-out struggle against the English before he was prepared for it. At any rate he travelled swiftly to Scone, where his supporters crowned him King of Scotland on 25 March.

The energetic King Edward moved to crush Bruce at once, and he was defeated by the English and their satellites at Methven on 19 June and again at Dalry on 11 August. His wife and many of his lieutenants were captured, and three of his brothers were put to death. He himself escaped almost alone, and took refuge on the island of Rathlin off the northern Irish coast. In February 1307 he returned to Ayrshire, winning the battle of Loudon and attracting growing support. Edward prepared another invasion, but died before it could take place. For the next three years Robert consolidated his grip on the crown in fierce fighting against the Comyns and other opponents, in the course of which he unleashed a bloody sack of Comyn territories in the north-east – 'the harrying of the Buchan'. The inept King Edward II, a sorry shadow of his mighty father, attempted to re-establish English rule in 1310, but was defeated by Bruce, who carried the war into northern England. In 1313 Bruce captured Perth, while his lieutenants extended his rule throughout the country. In 1314 Edward II again mounted a major invasion with the immediate aim of relieving the last English garrison, in Stirling Castle. The Governor, Alexander Mowbray, had agreed to surrender to the Scots if not relieved by St John the Baptist's Day. The English army numbered perhaps 20,000 foot and 3,000 horse; they came up with the much smaller Scots army – something between 5,000 and 10,000, with less than 1,000 horse – at Bannockburn on 23 June 1314.

The exact site of the battle is not known, but it was about one or two miles south of Stirling, in an area of new parkland. Bruce drew up his forces in a strong position on firm ground, with their flanks covered by trees and marshes and their front protected by the burn and by a line of pitfalls. On the evening of the 23 June the English leaders, among them the Earls of Gloucester and Hereford, rode forward to reconnoitre. Hereford's cousin Henry de Bohun, seeing Bruce doing the same some distance away, with few guards and apparently lightly armed, charged the Scottish King in an attempt to settle the issue on the spot by single combat. He died under the Bruce's battleaxe, and the Scots triumphed in the limited skirmish which followed. Both sides remained in their positions overnight, and on the morning of the 24 June the English prepared their battle line. The restricted nature of the ground caused difficulty; the English divisions became confused and cramped, and Bruce took the uncharacteristic decision to attack at once in the hope of turning this to his advantage. He divided his pike- and bill-armed infantry into four 'battles', keeping one as a reserve under his own hand, and sending forward those of his brother Edward, the Earl of Moray, and Sir James Douglas. Hard fighting ensued; the English archers, poorly co-ordinated with the heavy infantry and armoured cavalry, proved troublesome for a while, but were subsequently dispersed by a Scots cavalry charge led by Robert Keith. The English men-at-arms were severely hampered by the press of closely-packed ranks, and only a small proportion of them could bring their arms to bear. At the decisive moment Bruce committed his reserve, driving the English back; traditionally the rout was completed by a flank attack from the hills to the west by lightly-armed Scottish infantry – some sources say that these were camp-followers, but it is equally likely that they were lightly-armed peasants whom Bruce had been unwilling to place in his main battle line. The English recoiled with heavy losses, many perishing in the burn and in the marshes. Bannockburn was the most disastrous defeat ever inflicted on the English by the Scots.

Secure on his throne, Robert I now mounted a series of spoiling campaigns along the border. It was to be 1328, however, before the fighting finally ended. After Queen Isabella and the barons deposed Edward II in 1327, she and Roger Mortimer negotiated a treaty with Bruce which was ratified at Northampton the following year. The main provision was the recognition of Robert as the King of Scotland, and of Scottish independence. Bruce's latter years were occupied in re-establishing orderly government in his ravaged kingdom; he was increasingly troubled by ill-health, and finally died at Cardross, Dumbartonshire, on 7 June 1329 – possibly of leprosy. Following his wishes, his heart was removed after death, and his friend and most prominent lieutenant, Sir James Douglas, took it with him on a pilgrimage with the intention of burying it in the Holy Land. Douglas was killed fighting the Moors in Spain in 1330, however; according to tradition, Bruce's heart was recovered and returned to Melrose Abbey.

In conclusion, it must be said that the anecdote of Bruce and the spider is most unlikely to have any foundation. It was first told of Bruce by Sir Walter Scott, but the story had been told of James Douglas since the seventeenth century.

Roberts, Frederick Sleigh, 1st Earl of Kandahar, Pretoria, and Waterford *(1832–1914), Field-Marshal*

Born at Cawnpore, India, on 30 September 1832, Frederick Roberts was the younger son of General Sir Abraham Roberts, G.C.B., a member of a family long associated with Waterford in Ireland. Coming to England from the East at an early age, he was educated at Eton and Sandhurst, and was commissioned as second-lieutenant in the Bengal Artillery at the age of nineteen. In 1856 he was appointed to the quartermaster's branch of the staff, and on 31 May the following year was promoted lieutenant, filling a vacancy in the Horse Artillery. He joined the Delhi Field Force, serving with great distinction throughout the siege and capture of Delhi during the Indian Mutiny. He was wounded on the 14 July, and narrowly escaped serious injury again when his horse was shot under him on 14 September; he was present at the battles of Bulandshahr (when his horse was again shot under him), Aligahr, Agra, Kanauj (again losing his horse to enemy fire), and Bantharra. He served throughout the operations which led to the Relief of Lucknow, and fought in the Battle of Cawnpore, which resulted in the defeat of the Cawnpore contingent.

While following a group of sepoy mutineers at Khodagunge, between Cawnpore and Delhi, on 2 January 1858, Lieutenant Roberts joined a cavalry charge, attaching himself to a squadron of horse under Major Younghusband. Suddenly the mutineers turned and fired point-blank, killing Younghusband. Roberts cut down a sepoy, thereby saving the life of a colleague, and pursued two others who had captured a standard and were carrying it off; killing one and by great good luck surviving an attack by the other (whose musket misfired), Roberts recaptured the standard. For this act, and for his previous outstanding gallantry in numerous engagements, he was subsequently awarded the Victoria Cross. Thereafter he accompanied the forces reoccupying Fategarh, and was present at the storming of Mianganj and the subsequent Siege of Lucknow. His Indian Mutiny Medal carried all five engagement bars: Delhi, Defence of Lucknow, Relief of Lucknow, Lucknow, and Central India. He had, moreover, been mentioned in dispatches no fewer than thirteen times during the course of eleven months' campaigning.

By the end of the Indian Mutiny Roberts was still only a lieutenant, but thereafter promotion was fairly rapid: he was awarded his brevet majority in November 1860 and appointed Assistant Quartermaster-General for the Bengal province in 1863, a post he occupied for five years. He was also present at the capture of Umbeyla during the campaign on the North-West Frontier in 1863, receiving the bars 'Umbeyla' and 'North-West Frontier' to the Indian General Service Medal. Following the murder of consular officials in Abyssinia by the Emperor Theodore II in 1867, Roberts accompanied Napier's punitive expedition to that country, landing at Mulkutto on the Red Sea with an army of 32,000 men. After defeat at the battle of Arogee on 10 April 1867 Theodore committed suicide and his capital, Magdala, was stormed three days later.

Roberts, awarded the Abyssinian Medal, was promoted brevet lieutenant-colonel, returned to India, and held the post of First Assistant Quartermaster-General for India until 1872. Accompanying the Looshai expedition (he was later awarded the bar 'Looshai' to the India General Service Medal), he again saw active field service during 1871–2, after which he was made Deputy Quartermaster-General and in 1875 promoted colonel in the post of Indian Quartermaster-General. By now he had been mentioned in dispatches a further ten times.

Towards the end of 1878 internal dynastic troubles in Afghanistan came to a head, following a dozen years of feuding between the ruler, Sher Ali Khan, and his cousin, Abdur Rahman Khan. Sher Ali sought assistance from his Russian neighbours to the north, and was reprimanded by the British – always apprehensive of the threat of Russian dominance in the area. When in November 1878 these warnings were ignored, Roberts was placed in command of the Kuram Field Force. This comprised three columns of British troops which advanced into Afghanistan from India, seizing the frontier passes and defeating Sher Ali at Peiwar Kotal on 2 December. Sher Ali fled and his son, Yakub Khan, was installed in his place to conclude a treaty with Britain in May 1879. At the same time Sir Louis Cavagnari was appointed British Resident at Kabul.

On 5 September 1879 Cavagnari was assassinated, and Roberts (now Major-General Sir Frederick Roberts) led a punitive field force on Kabul from Kuram and Kandahar. This force, of 7,500 men and twenty-two guns, met and defeated the Afghan army (of 8,000 men) in the Battle of Charasia on 6 October, and went on to occupy Kabul a week later. Yakub Khan promptly resigned and sought British protection. However, dissatisfaction brought on by incitement by the mullahs resulted in the formation of Afghan levies and a call for a holy war against the infidel British, and on 23 December 1879 over 100,000 Afghans surrounded Roberts's cantonments at Sherpur. The British forces nevertheless managed to break out, and in a brilliant move Roberts fell upon the Afghan flank and put the hostile army to rout. (Roberts was later awarded the Medal for Afghanistan with bars 'Peiwar Kotal', 'Charasia', 'Ahmed Khel', 'Kabul', and 'Kandahar'.)

A new emir, Abdur Rahman Khan, was installed in July 1880, and the British force was about to be withdrawn from Afghanistan when news arrived of defeat at the Battle of Maiwand on 27 July; there Ayub Khan (brother of Yakub) had shattered an Indian force and practically annihilated a British infantry battalion after its supporting artillery had rashly expended all its ammunition. Much encouraged by this success, Ayub Khan now advanced to besiege Kandahar. This was the cause of the celebrated march from Kabul to Kandahar – on 9 August Roberts set out to cover the 313 miles across mountainous country with 10,000 men, a feat he achieved in no more than twenty-two days as a direct result of his use of a specially-formed transport corps. His brilliant management of this specialist unit brought his force to Kandahar with little hardship and no exhaustion, so that on 1 September he was able to attack the Afghan army without delay. In this Battle of Kandahar Roberts totally defeated the enemy and captured all his artillery and equipment, thereby ending the campaign. The British force evacuated the country in 1881. Roberts was awarded the Kabul to Kandahar Star – a bronze star made from the guns captured at the battle of Kandahar.

This campaign confirmed Roberts's reputation as a highly competent and careful staff officer, a brilliant tactician in command of troops, and a much respected leader among those troops. His nickname, 'Bobs', now became a household word throughout the Empire. Appointed to command the Madras army from 1880 until 1885, he was promoted lieutenant-general in 1883 and became Commander-in-Chief in India in November 1885. That year, as the result of Burmese interference with the British teak trade (in which French complicity was frequently evidenced), Roberts assumed command of a British force raised against Burma, and in November he sent an amphibious force of 12,000 men under General Prendergast up the Irrawaddy to confront the ruler, Thibaw Min, who promptly surrendered as the British approached his capital, Ava. Determined to stamp out the nauseating Burmese practice of dacoity (thuggery by gang), Britain annexed Burma on 1 January 1886, and the country was eventually pacified during the next ten years. (Roberts was awarded the bar 'Burma 1885–7' to the Indian General Service Medal.)

Following eight years in the highest Indian army appointment, during which time he improved the efficiency and well-being of his command and considerably developed communications in the northern frontier areas, Roberts returned to England in 1893, having been created Baron Roberts of Kandahar the previous year. Promoted field-marshal on 25 May 1895, he took command of the British army in Ireland in October of that year – a position he occupied until 1899.

Meanwhile trouble had been growing in South Africa since the granting of independence under British suzerainty by the Treaty of Pretoria (5 April 1881), at which time Paul Kruger, determined protagonist of Boer independence, became President. Numerous minor wars had followed (the Gun War with the Basutos, 1881; the Zulu Civil War, 1883–4; the Zulu Rebellion, 1887; the Matabele-Mashona War, 1893) and these resulted in a prolonged build-up of British forces in the area. The Jameson Raid of 1895–6 sparked open hostility among the Boers who, having handed over their prisoners to the British, were incensed by the nominal punishment awarded. This in turn led to a marked deterioration in relations in the Transvaal, a situation further aggravated by more trouble in Matabeleland. It was in this atmosphere that in 1898 the British determined to assemble an expeditionary force in Natal, a process which continued until the late summer of 1899 and which infuriated Kruger, who issued an ultimatum to the British Government demanding the disbanding of the force. The demand was refused, and immediately fast-moving Boer columns moved out against the British. Under inept generalship the British suffered setback after setback, including the investments of Mafeking, Kimberley, and Ladysmith, and defeat at the Battles of Stormberg, Magersfontein, and Colenso during the 'Black Week', 10–16 December 1899.

Following the defeat at Colenso, General Sir Redvers Buller, the British Supreme Commander, advocated the surrender of Ladysmith, and was promptly relieved of his overall command. Field-Marshal Viscount Roberts was appointed to his place, with General Kitchener (q.v.) as his Chief-of-Staff. Within a month the veteran British leader had revitalised the dispirited forces and embarked on a build-up of mobile, mounted infantry to counter the Boers' fire power and mobility of movement – albeit in the face of conservative British military opinion. It was he who sent Brigadier-General John French with two cavalry brigades to maintain a spirited campaign against the Boer General De La Rey, and on 15 February this small force raised the Siege of Kimberley. In the meantime Buller had suffered two further defeats, at Spion Kop and Vaal Kranz, losing 2,000 men killed or wounded.

On the day that French reached Kimberley, Roberts set out with 30,000 men, by-passing the Boers' flank at Magersfontein and threatening Cronjé's communications. Cronjé thereupon started to fall back, but found his path barred at the Modder river by French, who had hurried back from Kimberley. As the main British force approached, Roberts was taken ill and handed over command to Kitchener who, at the Battle of Paardeberg on 18 February, impetuously attacked the Boer laager and was repulsed with heavy losses. Assuming command on the following day, Roberts, now recovering from his sickness, started a systematic investment of Cronjé's laager which, on 27 February, was forced by starvation into surrender.

Buller's eventual relief of Ladysmith on 28 February marked the turning-point of the war. The British, now heavily reinforced, advanced on all fronts. Roberts embarked on a campaign in the Orange Free State, taking its capital, Bloemfontein, on 13 March, and reaching Kroonstad on 12 May. The Orange Free State was annexed by Britain on 24 May. Meanwhile Buller had overcome resistance in Natal; Johannesburg fell on 31 May, followed by Pretoria on 5 June, and when Roberts and Buller joined forces at Vlakfontein on 4 July formal conduct of the war by the Boers came to an end. (Nevertheless guerrilla war continued for two years, only to be brought to an end after the employment of ruthless tactics by the British.)

Roberts then handed over command of the British forces in South Africa to Lord Kitchener and returned home to appointment as Commander-in-Chief – the last in the British Army – and to an earldom. His South African service also earned the thanks of both Houses of Parliament and the South African War Medal with six bars.

As Commander-in-Chief, Roberts was out of his depth at the War Office. Long service in India and Africa had not fitted him for the unfamiliar organisation in London. In 1904–5 the Esher Committee's recommendations abolished the post of Commander-in-Chief and established the Army Council. Convinced of the inevitability of a major European war, Roberts now devoted his efforts to a campaign for the adoption of compulsory national service, but despite much influential support, his arguments were of little avail. Compulsory conscription was only introduced in Britain in 1916, after two horrific years in which tens of thousands of Britons had been slaughtered in the war that Roberts knew would come. He never saw the day of his vindication. After the outbreak of war Roberts, by now an old man of eighty-two, went to France to inspect a contingent of newly-arrived Indian troops in November 1914, and there contracted pneumonia and died at St Omer on the 14 November. He was buried in St Paul's Cathedral, London.

Roberts was one of the truly respected commanders of the British Army, much loved by his men and adored by the public. From an early age his personal courage was well known – the evidence was his extraordinary 'chest' of medals, at a time when gallantry decorations were, in general, sparsely awarded. After his early commands in India, his presence alone in charge of British forces was capable of reviving morale and even in his old age his dynamic presence in South Africa quickly brought about the victories which had hitherto escaped the British Army.

Rochambeau, Jean Baptiste Donatien de Vimeur, Comte de (*1725–1807*), *Marshal of France*

With Lafayette (q.v.), Rochambeau is remembered as a leader of France's intervention on the side of the colonists in the War of American Independence.

He was born at Vendôme in the department of Loire et Cher on 1 July 1725, and educated at a Jesuit school in Blois. He entered the army as a junior officer of cavalry, and fought in Germany during the War of the Austrian Succession (1740–8), rising to the rank of colonel. During the Seven Years' War Rochambeau distinguished himself in the French expedition to Minorca in April–May 1756. Some 15,000 men were landed, under Marshal de Richelieu, and invested the British position at Port Mahon. It was while attempting to break the French naval blockade and land reinforcements that the British Admiral Byng fought the mishandled action for which he was later court-martialled and shot. Port Mahon held only about 3,000 British troops, who were obliged to surrender on 28 May. Rochambeau saw another campaign in Germany before the close of hostilities in 1763; in 1761 he was promoted brigadier-general and appointed inspector of cavalry. He was made Governor of Villefranche-en-Roussillon in 1776, and four years later was appointed to command the expeditionary force which France sent to aid the American patriots.

Rochambeau landed at Newport, Rhode Island, on 11 July 1780 with some 5,500 French troops. He placed himself under George Washington's (q.v.) orders, but a conflict of interest arose immediately. A swift assembly of superior British naval strength blockaded his seven ships (under Admiral de Ternay) in Narragansett Bay, and Rochambeau refused to abandon them. He set up camps locally, which he was not to break for nearly a year, and proposed Franco-American moves against New York had to be abandoned. Late in May 1781 Rochambeau and Washington held a strategic conference and agreed that if French sea power could make mutual support between the British centres of New York and Chesapeake Bay impossible, then joint land and sea operations against either might be successful. In June Rochambeau left Newport and marched his army to join Washington on the Hudson. In August two pieces of intelligence arrived; Lafayette reported that Cornwallis had withdrawn his British army in Virginia into the peninsular position at Yorktown, and a French fleet under Grasse had sailed up from the West Indies with another 3,000 French troops. Washington saw that with a powerful Allied fleet to block rescue from New York, he could achieve local superiority over Cornwallis and destroy the British army. He immediately led the Allied armies south into Virginia.

On 30 August Grasse arrived and landed his reinforcements near Yorktown, to come under the orders of Lafayette, the local patriot commander. With French command of the sea, siege artillery could be shipped in, and the last stages of Washington's and Rochambeau's journey were made by ship. Washington commanded some 9,500 Americans and 7,800 first-class French regulars under Rochambeau, whose artillery made the first operational use in this campaign of the famous Gribeauval field guns. Cornwallis had some 8,000 men, the French fleet at his back, and a severe supply problem. After a month of operations, and ten days of bombardment and sharp infantry fighting, Cornwallis capitulated on 19 October. In these joint operations Rochambeau's personal charm and tact seem to have made him as popular with his allies as the excellent quality of his troops. The French army remained in Virginia during the winter of 1781–2, while Washington returned to the north to invest New York. Rochambeau led it back to Rhode Island in the autumn of 1782, and sailed back to France at the turn of the year – peace negotiations between the Americans and the British had been under way since April 1782, and his task was over.

During the following six years Rochambeau served as Governor of Calais, and subsequently of Alsace. During the Revoluntionary period he commanded the Northern Military District, and was named a Marshal of France in

December 1791. His Army of the North failed to distinguish itself around Lille in April–May 1792 in the face of the Austrian attack (see under Lafayette); it was an uneasy mixture of old royalist professionals shaken by recent political upheavals, and zealous but completely raw volunteers. Rochambeau resigned his command in disgust at the showing of his troops, and was replaced by Lafayette. He was arrested during the Reign of Terror, and narrowly escaped with his head; it is pleasant to record that the old man was able to spend his last years in peace, and that Napoleon (q.v.) authorised a state pension. Marshal the Comte de Rochambeau died on 10 May 1807.

Rogers, Robert (*1731–95*), *Colonel*

Robert Rogers was a junior officer of unsavoury character and strictly limited skills, who nevertheless showed a genius for a particular type of military operation. He enjoyed a brief but widespread fame on both sides of the Atlantic during and after the French-Indian War of the late 1750s; his contribution to British victory should not be exaggerated, but was certainly valuable. As one of the fathers of modern commando soldiering, he deserves a place in a compilation of militarily significant names – the fact that he was a swindler, a braggart, and a drunkard was no bar to effectiveness on active service.

He was born on 18 November 1731 in Methuen, Massachusetts Bay Colony, the son of James and Mary Rogers. He grew up on his father's farm at Rumford – now Concord, New Hampshire – but took to the life of the forest while still a youth. The frontier of Britain's North American colonies was extremely rough in those days, and the life of the trapper, hunter, scout, or Indian trader was attractive to spirited young men brought up according to the austere New England ethic. It was a harsh school of survival, but the strong, active young Rogers seems to have thrived and to have picked up many useful skills. He always had a way of impressing his will on his fellows, and in 1755 he was exercising this talent as a recruiting officer for Governor William Shirley of Massachusetts. It had been decided to cauterise the running sore of French hostility in Acadia (roughly, modern Nova Scotia) which had long been a base for Indian raids on British territory. Shirley had been entrusted with this expedition, and when launched in June 1755 it was completely successful. Rogers was not present, however; a past indiscretion in the form of counterfeiting, for which prosecution now threatened, had forced him to enlist hurriedly in a New Hampshire regiment of volunteers.

In August 1755 Rogers marched on the French base at Crown Point on Lake Champlain in the expedition led by William Johnson, the influential backwoods baron from the Mohawk Valley. He held the rank of captain, and brought himself to Johnson's notice by his courageous scouting patrols and his skill in gathering intelligence and prisoners. The force never reached Crown Point, in the event. On 8 September at the foot of Lake George it ran up against a strong French force probing southwards, and defeated it in a hard day's fighting; among the wounded prisoners was Baron Dieskau, commander of French regulars in Canada (see under Montcalm for background). In March 1756 Governor Shirley appointed Rogers captain of an independent company of Rangers paid out of British funds, and his career as a leader of irregulars began in earnest.

The conditions of frontier warfare are described briefly elsewhere in this book. French and British territory was defined by chains of forts and blockhouses, but the logistic and transport difficulties which dominated operations in the wilderness did not allow frequent large-scale movements of regular troops. It required a great effort of organisation to mount a serious advance on any of the strategic outposts of either side, and in the meantime the active prosecution of the war was left largely to client Indians and parties of European Rangers on each side. These raiders penetrated deep into enemy territory, attacking isolated posts and hampering supply convoys. They gathered the vital intelligence regarding enemy dispositions on which major campaign planning was based. They disrupted communications, and waged merciless psychological warfare against the Indians who supported the enemy, and against lonely garrisons. It was in this type of operation that Rogers made his reputation. He was apparently a bold planner and a master of fieldcraft – although on occasion he did allow himself to be surprised by the enemy. His parties travelled light and fast, moving up the waterways by night in canoes and whaleboats which were concealed and left behind when they took to the forest. Each man carried his own meagre rations and a minimum of equipment, and any Ranger unfortunate enough to be injured or wounded was usually abandoned, despite the barbarous treatment of prisoners by the French and Indians. The Rangers fought with a ruthlessness equal to that of their enemies, and the taking of scalps was not uncommon. Rogers's name became well known to the French, and a considerable price was put on his head.

When large-scale regular forces did move through the wilderness the Rangers were used as advanced scouts, and Rogers was present on several British offensives. He was with Loudon at Halifax in 1757 during the planning of an assault on Louisbourg. In the event this never took place, and in the absence from the frontier of many British units Montcalm fell upon Fort William Henry. Rogers was active during the winter of 1757/8, on one occasion butchering French cattle under the very walls of Fort Ticonderoga and leaving a humorous note attached to a horn. His sense of humour was considerably blunted in March 1758, however, when he overreached himself and was nearly trapped by a French force. The fierce fight cost him more than 100 of his men, and nearly cost him his life. He scouted for Abercromby on the mishandled Ticonderoga expedition four months later, and continued to lead audacious deep-penetration raids in summer and winter alike. He was with Amherst in the advance on Crown Point in 1759, and took part in the final Montreal campaign of 1760. His most daring raid was in September–October 1759, on the Abenaki Indian village of St Francis far behind French lines: he completely destroyed this important centre of hostile tribesmen, but lost many men on the return journey when the discovery of his hidden boats forced him to take an overland route. His men suffered extremes of hunger and exhaustion, and were harried by the Indians throughout the retreat. In 1758 Abercromby had promoted Rogers to the rank of major, and given him overall command of nine independent Ranger companies; many British officers were seconded to Rogers for first-hand experience of forest warfare and fieldcraft.

When the war ended in 1760 Rogers was sent west to receive the surrender of isolated French posts. In 1761 he married Elizabeth Browne of Portsmouth, daughter of an influential clergyman, but this entrée into respectable colonial society did not help him to settle down to peacetime life, and the next fifteen years did little to enhance his reputation. In 1761 he was appointed captain of an independent company fighting the Cherokee in South Carolina, and two years later he led a New York independent company in Dalyell's expedition to Detroit. He distinguished himself at the skirmish of Bloody Bridge during the defence of that post against Pontiac (q.v.). With the collapse of Pontiac's rebellion Rogers found difficulty in gaining employment, as his administrative inefficiency and his penchant for illicit trading with the Indians had become scandalous. In 1765 he sailed for England to solicit preferment, and stayed for some time. He published his *Journals* and *A Concise Account of North America* while in London, and was the object of considerable public acclaim. He secured the command of Fort Michilimackinac on Lake Michigan, and returned to live in that desolate fur-trading post for two years, accompanied by his wife. Despite the blunt warnings of Sir William Johnson and General Gage, who had few illusions about his character, he attempted to administer his territory for his own financial advantage, and schemed to bring about the separation of civil and military authority on the far frontier – a reform which would have given him much greater scope for peculation. At last he was arrested and charged with treasonable dealings with the French, but was acquitted through lack of evidence. He made vain attempts to secure financial compensation and a new position, and travelled to England again in 1769, where he brought an unsuccessful King's Bench action against Gage for assault and wrongful imprisonment. He did manage to secure a commission as colonel in the British Army, but was eventually thrown into a debtor's prison – at one stage his debts totalled some £13,000. He was released when his brother James assumed his major debts. He later claimed to have served the Bey of Algiers in 1774, but this statement must be regarded with suspicion. He returned to America in 1775 and seems to have attempted to court both the British and the patriots until he was imprisoned as a spy by Washington (q.v.) in 1776. He escaped to the British, and was commissioned to raise a loyalist corps originally entitled the King's Rangers. He was subsequently removed from this command; the unit was renamed the Queen's Rangers, and it fought against the revolutionary forces with some distinction under Major Christopher French.

Rogers appears to have spent some time as a recruiting officer for the loyalist Provincial Corps, but his reputation for dishonesty and dissipation followed him. In 1778 the New Hampshire legislature granted his wife's petition for divorce, and she retained custody of their only child. In 1780 Rogers was obliged to seek refuge in England once again; he lived on his half pay for another fifteen years, and finally died in a cheap London lodging-house on 18 May 1795. He was buried at St Mary's, Newington. Reliable information on his appearance is scarce, but he seems to have been a stocky, powerful man of middle height, with dark hair and dark brown eyes, and rather heavy features.

Rokossovsky, Konstantin (*1896–1968*), *Marshal of the Soviet Union*

Rokossovsky was the Soviet commander who achieved fame for his defence of Moscow and Stalingrad during the Second World War, and for the victory at Kursk; but his record has been tarnished by association with Russia's failure to support the Polish uprising at Warsaw in 1944.

Born on 21 December 1896 Rokossovsky was conscripted into the Tsarist army on the outbreak of the First World War; during the October Revolution of 1917 he joined the Red Guard. Promotion in the Red Army and appointments in the Communist Party followed between the two World Wars, and his appointment as a Marshal of the Soviet Union followed in the wake of the military purges that racked the Red Army during the late 1930s. During November and December 1941 he gained great distinction for his steadfast conduct in thwarting the German advance on Moscow. A year later, under the direction of Marshal Zhukov (q.v.), he was in command of Russian forces opposing the German Sixth Army, which anchored the left flank of elements striking south-east into the Caucasus south of Stalingrad. From the Russian viewpoint the Battle of Stalingrad was almost lost before it was won, and German forces occupied three-quarters of the city before the tide of the battle, which raged from August to November 1942, turned in the Russians' favour. The massive Russian counter-attack launched in November proved to be the turning-point of the Second World War, and involved three Soviet Army Groups. The forces commanded by Rokossovsky, on the Don River Front, attacked from the north-west, and within five days the Russians had outflanked and encircled the German Sixth, Ninth, and Fourth Panzer Armies. Nevertheless the trapped Germans fought on for two more months until, starving and short of ammunition, the German forces under Paulus surrendered the city to Rokossovsky on 31 January 1943.

After the disaster of Stalingrad and the failure of the Kursk offensive seven months later, the German armies recoiled before the combined might of upwards of a dozen Soviet armies commanded by Rokossovsky, Vatutin, Chuikov, and Yeremenko, all under the overall command of Zhukov. Rokossovsky, commanding the Army Group of the Byelorussian Front in the south, reached Poland and captured Lublin and Brest-Litovsk. On 29 July 1944, anticipating a continuation of the swift Russian advance through Poland, Moscow radio broadcast an appeal to the inhabitants of Warsaw to rise against the German occupying forces – the obvious inference being that the Red Army would quickly advance to their assistance, thereby liberating the capital without a long-drawn-out campaign which would cost the lives of thousands of civilians. Instead, the Red Army allowed itself to be halted, and did nothing to support the Polish revolt. In the meantime the Germans turned furiously upon the Polish insurgents and waged a most barbarous campaign of annihilation against the lightly-armed guerrillas in the city: the result was the almost total destruction of the city and its population. The inaction of the Russians has raised numerous questions, which have resulted in a like number of conflicting answers and excuses. In fact Rokossovsky's advance had almost run out of steam, and the time so dearly bought by the Polish Home Army, anxious to install a régime answerable only to the Polish government-in-exile in London, merely enabled the Russians to achieve the replenishment of strength necessary before their final assault on Germany. Whatever the military strategy, there never was or could have been any excuse for this political incitement by Moscow, for the outcome of any uprising in a Nazi-dominated territory was a foregone conclusion. But the Russians were entirely content to see the non-Communist patriots slaughtered before they sauntered in to occupy the ruins of Warsaw. The result was a city razed to the ground, and 300,000 men, women, and children killed by bullet, flame-thrower, and dive-bomber. (Surprisingly, the survivors were not shot out of hand but were treated as prisoners of war by the Germans – such comparative leniency is attributed to German military commanders, such as S S General von dem Bach-Zelewski, who were haunted by the shadow of future war crimes trials as the war approached its inevitable conclusion.) This view of the Soviet attitude is confirmed by the fact that Allied attempts to air-drop supplies to the beleaguered Poles were obstructed by the Russians.

During the last four months of the war Rokossovsky's armies swept across northern Poland, capturing Danzig in April 1945. Driving on westwards in the first week in May, he made contact with British forces at Wittenberg, near Lübeck. Almost simultaneously Marshal Zhukov received the surrender of Keitel (q.v.) in Berlin.

After the war, at the 'request' of the Polish Communist government and with the 'permission' of the Soviet government, Rokossovsky (himself of Polish stock) left Russia for Poland in 1949 and was appointed Minister of National Defence and Vice-Chairman of the Council of Ministers. In 1956 he returned to the Soviet Union and held a post at the Soviet Ministry of Defence. He was repeatedly elected a member of the C.P.S.U. Central Committee and a Deputy of the Supreme Soviet. He died on 3 August 1968.

Rommel, Erwin (*1891–1944*), *Field-Marshal*

The fact that to the man in the street Rommel is undoubtedly the best-known name among Hitler's generals is an ironic commentary on the chances of war. While his tactical skills were formidable and his achievements undeniably impressive, he fought with very minor forces in a theatre of war which could only be considered a sideshow to the main campaign in Russia. (He commanded only five German and nine Italian divisions at El Alamein, and most of his triumphs were achieved at the head of only two or three divisions; his colleagues in Russia led forces totalling more than 130 divisions.) Yet the peculiar glamour which attaches to desert warfare and Rommel's own qualities as a general and as a man combined to make him a legend in his lifetime – 'the Desert Fox'. Unique in being respected by Germans, British, and Americans, he was the man picked to be German head of state if the assassination of Hitler had been successful. In a war in which the reputation of the German officer corps was too often compromised by association with atrocity, Rommel epitomises the chivalrous and skilful enemy. For all these reasons he is assured of an immortality which a cold assessment of his battlefield performance might deny.

He was born on 15 November 1891 at Heidenheim, near Ulm in Württemburg. At the age of nineteen he joined a Württemburg infantry regiment as an officer cadet, and was commissioned lieutenant in 1912 after attending the Danzig War Academy. In the First World War he saw active service in the battle of the Argonne, in Rumania, and Italy. He greatly distinguished himself as a regimental officer, being awarded Germany's supreme decoration, *Pour le Mérite*, for his valour in a hazardous night operation at Caporetto on the Italian front. Between the wars he rose to command his regiment, breaking into the quiet life of a garrison town with a period as commandant of a training school. In 1939 Colonel Rommel was posted to command the *Fuhrer* Bodyguard Battalion maintained by the army at Hitler's headquarters. In the past this posting has been cited, quite mistakenly, as evidence of fanatical Nazi sympathies; these charges are founded on a complete misunderstanding of the nature of the unit. Rommel was an apolitical professional soldier, loyal to his country and to his oath of allegiance.

Impressed by the performance of the new mobile formations in the Polish campaign, Rommel requested, and received, command of a Panzer division on leaving his post with the Bodyguard. His 7th Panzer Division was prominent in the lightning German offensive in France in May–June 1940, and Major-General Rommel showed a great aptitude for this new kind of fast-moving warfare. He pressed forward with great audacity, taking serious risks but only once coming close to losing his gamble – his division was the target of the British armoured counter-attack near Arras on 21 May 1940, and was quite badly mauled before he restored the situation. This campaign confirmed him in his habit of leading from the front, in the first wave of tanks; this risky procedure offended several basic rules of command, but gave the troops great heart, and enabled him to react with impressive speed to all developments. He was awarded the Knight's Cross for his part in the French campaign, and later received the Oakleaves, Swords and Diamonds.

In February 1941 Lieutenant-General Rommel was posted to Africa to command the small German expeditionary force dispatched to retrieve Axis fortunes following the disastrous defeat of large Italian forces at the hands of weak British formations. He established his style immediately; with only two divisions and weak in armour, he had within two weeks bundled the British out of all the territory which they had captured in two months of fast-moving operations, and had reached – but not broken – the defences of Tobruk. He showed an unmatched skill in rapid, unexpected, fluid attacks, looping and wheeling around the badly-dispersed British formations and exploiting every chance opportunity to the full. For the next eighteen months he and a succession of British generals sparred with one another up and down the 'blue'. He sometimes took unjustifiable risks, but they usually paid dividends. He had the advantage of better equipment – though never enough of it – and extracted the maximum advantage from his tanks and anti-tank guns, which co-operated with great effect against the unimaginative British armoured tactics. He was faithful to the basic creed of *blitzkreig*, keeping his tanks concentrated and using them to achieve massive local superiority at the decisive moment, rather than dispersing them all over the battlefield; and he never allowed his armour to be frittered away in pointless battles of attrition. So high did his reputation stand that he became a sort of bogy-man to the British troops, an invincible will-o'-the-wisp who might suddenly appear from any quarter. This psychological advantage was enhanced by the extremely humane and chivalrous manner in which he and his entire command fought the desert campaign; the Eighth Army and the *Afrika Korps*, like opposing sailors on some vast unfriendly ocean, were always conscious that the desert would kill them both, impartially, if they did not maintain certain standards of behaviour.

Rommel's greatest handicap was logistic. His lines of supply led across the Mediterranean, and an enormous tonnage of vital fuel and equipment was sunk before it reached him by British aircraft and submarines operating from Malta. When he triumphed on the battlefield and drove the British back towards the Nile delta his supply lines on land stretched 1,000 miles, while the British lines shortened as they retreated towards the Suez Canal base area. The port of Tobruk would have given him a much-needed forward supply port, but the garrison held out until June 1942, by which time Allied operations in the Mediterranean robbed him of much of the value of the prize. Promoted field-marshal at the end of June 1942, he attempted to break through the British defensive line (running from El Alamein to the impassable Qattara Depression) at the end of August. Time was not on his side, as he could only get weaker while the British continued with a massive build-up of forces in Egypt. He failed with heavy loss at

Alam-el-Halfa, and the front settled down to a stalemate. The new British commander, Montgomery (q.v.), refused to be pressured into a premature attack, and by the time the British Alamein offensive opened late in October the arithmetic of the situation was heavily in Montgomery's favour. Rommel's inadequate armour was slowly destroyed, and his infantry were dislodged stage by stage; Montgomery ensured that the battle was one of infantry attrition, denying his opponent the chance to make the best use of his tanks and anti-tank guns. After two weeks of pounding, the survivors of the *Afrika Korps* and the Italian divisions began a harrowing withdrawal along the coast road under skies commanded by the Allies. The withdrawal was brilliantly conducted, as were the operations which Rommel carried out in Tunisia in the winter of 1942–3, but the ultimate result was never in doubt. In March 1943 Rommel failed in a personal appeal to the *Führer* to authorise the evacuation from Tunisia of the forces which still remained. He was sent on sick leave (Rommel's health had suffered noticeably during the desert campaigns) and another presided over the final surrender of May 1943.

Rommel was given command of a reserve army in northern Italy for a period, and was then appointed Inspector of Coastal Defences in France. His examination of the preparations made thus far for resisting the inevitable Allied landings in Europe revealed huge shortcomings. He immediately instituted a new programme of fortification, based on the belief that with the forces at their disposal the Germans must at all costs throw back the invasion before it gained control of the beaches. With characteristic energy he visited all sectors personally, and transformed the hitherto rather apathetic approach to this task. In the summer of 1944 he commanded Army Group 'B' in northern France, and was for all practical purposes independent of his nominal superior, Runstedt (q.v.), the German Commander-in-Chief West. There were fatal disagreements, however, over the employment of the mobile reserves, a high proportion of which were kept under Hitler's own hand. Rommel was absent from the front when the landings of 6 June 1944 took place, but hurried back to take command. He handled the situation as well as any general could have, given the fatal reluctance of O.K.W. to release the Panzer reserves for immediate counter-attack in strength. On 17 July 1944 his staff car was machine-gunned by a British fighter, and Rommel was severely wounded in the head. He never again commanded troops.

Rommel had been approached some time previously by representatives of the group of officers who planned to kill Hitler and negotiate with the Allies for a separate peace. His experience of Hitler's unrealistic attitude towards the military situation prompted Rommel to agree that the *Führer* and his political henchmen should be removed from power, although he initially proposed arrest rather than assassination. There is no evidence that he was implicated in the bomb plot itself, and it was without his knowledge that the plotters selected him as the head of state should their attempt succeed – an obvious choice for men wishing to persuade the Allies to treat with them, as he was certainly the only enemy general to whom Churchill had ever made a favourable reference in a speech to the Commons. The bomb plot of 20 July failed, and in the appalling purges which followed Rommel's name was linked with the plotters. His prestige was so high, and his implication in the plot so marginal, that Hitler wished to avoid an open scandal. Rommel was still convalescing at his home when, on 14 October 1944, he was visited by two generals. He was informed that he had the choice of accompanying them on a car journey during the course of which he would take poison, in the knowledge that his reputation and the safety of his family would be assured; or of facing the horrors of a trial by 'People's Court', Gestapo interrogation, and the eventual execution of himself, his wife, his son, and his personal staff. Rommel, inevitably, chose poison. He was given a state funeral with full military honours, and it was announced that he had died following a relapse in his recovery from his wounds.

Runstedt, Karl Rudolf Gerd von (*1875–1953*), *Field-Marshal*

Gerd von Runstedt was something of a father figure to the German officer corps, and the last of the 'old school' of Prussian generals to hold senior command under Hitler.

He was born at Aschersleben, near Magdeburg, on 12 December 1875, the son of an aristocratic Prussian house. He rose through the commissioned ranks of the Imperial Army, and by the time of the First World War, was within reach of an appointment as chief of staff to an Army Corps. Later in the war he participated in a mission which helped to reorganise the Turkish general staff. He was active in the inter-war *Reichswehr* and in the process of clandestine rearmament; and by the time he retired in October 1938, aged sixty-three, he was the senior German field commander. He was recalled to active duty in 1939, however; and he held a senior command right up to the last months of the Second World War.

In September 1939 Runstedt commanded Army Group South in the invasion of Poland, and the following spring he led Army Group 'A' in the *blitzkreig* on France. Despite his age and background, he was an aggressive and offensive-minded commander, and he worked in better harmony with some of the younger generation of Panzer generals than did many of his middle-aged subordinates. Even so, he may have been partially responsible for the pause in German operations against the trapped British Expeditionary Force which enabled more than 300,000 of them to be evacuated from Dunkirk. He is believed to have urged a halt to enable his divisions, tired and below peak efficiency after their relentless advance, to recuperate before a renewal of the offensive. If he misjudged the extent of the French collapse, then such an order would be militarily quite sound; and it must be remembered that at the time the French collapse was seen as shocking and inexplicable.

Promoted field-marshal, Runstedt commanded Army Group South – one third of the total forces involved – in the invasion of Russia in the spring of 1941. In November he was relieved of his command at his own request after Hitler refused to authorise a tactical withdrawal to the Mius river in response to the suddenly deteriorating situation on the Russian front. Such a request was typical of him. Although rigidly loyal to the elected head of state, Runstedt, as an old-style *junker*, held Hitler and his methods in icy contempt and persisted in behaving according to a code of honour which, if it was outdated and inflexible, at least contrasted sharply with the opportunism of some other prominent generals.

In 1942 Runstedt was given responsibility for the defence of France and the Low Countries against the then-remote possibility of Allied attack – his title was Commander-in-Chief West. He held this appointment until a month after the D-Day landings, although the most energetic phase of defensive preparations was presided over by Rommel (q.v.), acting almost independently. These two generals disagreed over the correct employment of the armoured reserves in France; Rommel wished them to be committed immediately the Allies landed, in accordance with his belief that Germany's only hope lay in destroying the invaders at high-water mark, while Runstedt pointed out the dangers inherent in such a plan, and urged that the reserves be held back until the exact objectives of the invasion became clear. In the event Hitler's refusal to commit O.K.W. reserve formations rendered the argument largely academic. On 17 June 1944 in an interview at Soissons Runstedt urged the *Führer* to sanction the immediate withdrawal of the 7th Army to the line of the Seine, to save it from destruction in battles of attrition around the beachhead so that it could be employed in a defensive role in eastern France. Hitler characteristically insisted that the *Wehrmacht* fight for every inch of ground; and in early July Runstedt was replaced as Commander-in-Chief West by Kluge.

Barely two months later Runstedt was reappointed to the post, and it was he who presided over the eventual withdrawal of the decimated German armies from France. His name is closely associated with the Ardennes counter-offensive of December 1944 which took place under his overall direction; in fact he played little part either in the planning of this desperate gamble or in its tactical execution, and he had only the most limited confidence in its chances. When it failed he was replaced by Kesselring (q.v.). Captured by U.S. forces in May 1945, the old marshal was released on the grounds of ill-health after a short period of captivity. He died at Hanover on 24 February 1953.

Rupert, Prince, called Rupert of the Rhine (*1619–82*), *Duke of Cumberland, Earl of Holderness*

Born in December 1619 at Prague, the third son of Frederick V of Bohemia and of Elizabeth, daughter of James I of England, Rupert went with his family into exile from Bohemia after the defeat of his father at the Battle of the White Mountain. They settled in the United Provinces, and there he spent his childhood. At the age of fourteen he joined the suite of Frederick Henry, Prince of Orange, and two years later was a regular member of the Prince's bodyguard. He made a visit to the English Court in 1636, and his uncle, Charles I, formed a high opinion of the young soldier. He returned to the Netherlands the following year, but his military career was interrupted when he was captured on 17 October at Vlotho on the banks of the Weser, and spent three years in captivity at the hands of the imperialists.

On his release Rupert came to England, joining his uncle Charles I shortly before the king raised his standard in August 1642, and being given the generalship of cavalry in the Royalist army. Almost immediately he became involved in a string of successful skirmishes against Parliamentary detachments, quickly built up a reputation as a courageous, if impetuous, commander and grew to be one of the dominant military figures of the English Civil War. At the battle of Edgehill on 23 October 1642 the 11,000-strong Royalist army clashed with 13,000 Parliamentarians under Essex; Rupert's cavalry attacked, routed the opposing cavalry, and pursued them off the field. (More than once this impetuous pursuit of defeated cavalry before overall victory had been achieved on the field brought down criticism on the Royalist cavalry commander.) Although Essex rallied with a fierce counter-attack, Rupert returned before nightfall to force withdrawal of the Parliamentary forces towards London.

The following year Rupert won a victory at Chalgrove Field, near Oxford, on 18 June, and the following month captured Bristol. In April 1644, after William, Duke of Newcastle's forces had been shut up in York, Charles ordered Rupert north from Shrewsbury with reinforcements; his advance brought about the abandonment of the siege when the Parliamentary forces marched to meet him. By the end of June the two armies were approaching Marston Moor, about seven miles west of York, the Parliamentary forces numbering about 27,000, (about 7,000 of which were cavalry) and the Royalist army about 17,000, a similar number of which were cavalry. Rupert's cavalry was badly defeated by Cromwell's (q.v.) 'Ironsides' and Scots on the Royalists' right, while on the left flank George, Lord Goring's horse smashed the Scottish infantry, but became disordered in the process. Cromwell then swung his highly-disciplined Ironsides against Goring, routing him before assisting the remaining Parliamentary forces to destroy the Royalist centre. Rupert, with 6,000 survivors, escaped through Lancashire to join Charles in the south, and in November was appointed a general in the King's army.

While Rupert and his followers had encountered opposition in the King's Court, in particular from the faction led by Lord Digby (later second Earl of Bristol), in general his appointment was popular with the majority of Royalist soldiers. He was a flamboyant young man with great style and courage, and his exploits provided a fund of campfire anecdotes among the admiring troops. The events of 1645 were largely influenced by the conflict between Rupert and Digby – a conflict aggravated by the lack of Royalist success. These differences prepared the ground for Rupert's dismissal by Charles after his surrender of Bristol to Sir Thomas Fairfax (q.v.) in September 1645. Rupert had brought an element of hardened professionalism to the conduct of the war and this, together with his quick temper, rendered him vulnerable to charges of ruthlessness and even barbarity. At the same time he was capable of generosity towards his opponents, and he did much to restrain the licence of his cavalry. After the defeat at Naseby in June 1645 he was convinced that the King's cause was lost at least in a military sense, and this defeatism was cause for further conflict with Digby, while the King's displeasure was reinforced by the surrender of Bristol. Rupert demanded trial by court martial, and was acquitted, but he could be of little further value to the King.

After the capitulation of Oxford in June 1646 Rupert and his brother Maurice were hustled out of England, but Rupert soon obtained the leadership of the English troops in French service. The following year he was reconciled with Charles, and embarked on a singularly profitless career of naval command from 1649 to 1653. The Restoration tempted him to return to England in October 1660; Charles II appointed him a Privy Councillor and gave him an active naval command during the Second Dutch War, which earned him a reward of £14,000.

A member of the Royal Society, Rupert held aloof from politics in England and during times of peace built himself a laboratory at Windsor Castle, dabbling in scientific research and specialising in armaments and ballistics. He died of a fever in Spring Gardens, Westminster, on 29 November 1682.

Saint-Arnaud, Jacques Leroy de (*1798–1854*), *Marshal of France*

A French commander who won distinction in the colonial campaigns in Algeria during the first half of the nineteenth century, Saint-Arnaud must be admitted to have owed his rise to the marshalate to political rather than military judgements. He was a supporter of the monarchy, and rose to become minister of war under Napoleon III. He was, nevertheless, a man with an impressive front-line record in junior commands, and it is frustrating to record that he died upon the brink of his first real opportunity to confirm or disprove his entitlement to his baton.

He was born in Paris (some sources say Bordeaux) in 1798 (some sources say 1796, others 1801). His true name was Arnaud-Jacques Leroy, and he was the son of a prefect. He established a solid reputation as a rake, a gambler, and a seducer during his early years as a hussar subaltern, and in 1822 he was obliged to resign his commission and leave the country. He travelled to Greece, where he fought in the War of Independence against the Turks, and later rehabilitated himself and rejoined the army as an infantry officer. Unwilling to take up a posting in Martinique, he once more absconded, spending several years as a social parasite in England and Brussels. In 1830 he returned to France, and the prodigal was once again allowed to take up his commission. It was while he was stationed at the Château de Blaye as a member of the guard on the Duchesse de Berri that Leroy met her jailer Colonel Bugeaud de la Piconnerie (q.v.), and under the patronage of this fine commander his career began to improve. After another period of indiscretion and disgrace, Leroy was posted to Algeria as a captain in the Foreign Legion in 1837. During the course of the extremely difficult and costly expedition against the mountain citadel of Constantine in October of that year he distinguished himself under fire; some sources report that his courage carried the day when the French were wavering in the breach, and that he coined the famous rallying cry '*A moi, la Légion*' on this occasion. Decorated for gallantry, he went on to make even more of a name for himself at Djidjelli. In 1840 he returned to France, and in that year he changed his name to the more aristocratic-sounding Saint-Arnaud. In 1841 he returned to Algeria, where his patron was now Governor-General, and enjoyed opportunities for rapid promotion; he pursued his campaigns against the Kabyle tribes with the same energy and ruthlessness as Bugeaud himself. He rose to the rank of colonel; his brigadier-generalship came in either 1847 or 1848; and he brought the ruthlessness of the Aurés Mountains to the streets of Paris during the rising against Louis-Philippe in February 1848.

In 1851 he returned to Algeria as a major-general and commander of the province of Constantine, where he was active against the restless Berber mountaineers. He was noticed by the future Napoleon III, who was then plotting the overthrow of the republic, and was recalled to command the 2nd Division of the Paris forces. On 26 October Saint-Arnaud was made war minister, and played an active part in the *coup* of 2 December and the subsequent street fighting. Napoleon rewarded him with promotion to marshal, a seat in the Senate, and – it was alleged – enough cash to pay off his enormous gambling debts.

When that assertion was made publicly by another general, however, Saint-Arnaud challenged him and killed him in a duel.

In 1854, dogged by the uncertain health which he had suffered most of his life, Saint-Arnaud relinquished his portfolio as war minister, but nevertheless agreed to command the French expeditionary force to the Crimea. He planned the landing at Eupatoria on 14 September, and six days later, in collaboration with Lord Raglan (q.v.), he fought and won the Battle of the Alma. (It must be admitted that the collaboration was of the sketchiest possible nature.) His physical condition had declined, however, and he was now known to have contracted cholera. On 28 September Saint-Arnaud, *né* Leroy, died on board ship while returning to France. He was unquestionably a calculating, ambitious, ruthless man, untroubled by scruples, resourceful under pressure, and physically fearless. Such men sometimes make magnificent military leaders; whether Saint-Arnaud's easy progress through the senior ranks was really justified we will never know.

Saladin, properly Salah al-din Yussuf ibn-Ayyub (*1138–93*), *Sultan of Egypt and Syria*

This Muslim general and ruler, whose remarkable personal qualities won him renown in the West through his campaigns against the crusaders, was born at Tikrit in Iraq in 1138. His father Ayyub was a Kurd who had emigrated from Armenia in 1130, and rose high in the service of the great Turko-Syrian leader Nur-ed-din of Aleppo and Mosul. Salah al-din ('Honour of the Faith') grew up in Damascus at a time when an atmosphere of *jihad* gripped the Muslim states of the Middle East, an enthusiasm directed against the Frankish crusader kingdoms set up by the First Crusade (see under Bohemund I). Nevertheless the young man seems to have been more drawn to theology and scholarship than to campaigning.

At this time Egypt was displaying signs of decadence under the Fatimid caliphs, and both Franks and Turko-Syrians determined to exploit this weakness to secure for themselves the economic resources of this rich trading nation. Nur-ed-din sent Salah al-din's uncle Shirquh into Egypt with an army in 1163, ostensibly to aid the Fatimid caliph's vizier, Shawah ibn-Mujir, against an internal revolt. When Shawah discovered that Nur-ed-din had ordered Shirquh to take control of the country, he called on Amalric of Jerusalem – whose Franks were at that time raiding into the Nile Delta – for assistance. Franks and Egyptians defeated Shirquh outside Cairo on 11 April 1164, and Salah al-din seems to have distinguished himself under his uncle's command. He held out against a Frankish siege force at Alexandria, and played a major role in the campaigns which alternated with periods of negotiation until 1169. In January of that year Shirquh finally drove the Franks from Egypt and established himself as vizier, acting as the viceroy of Nur-ed-din. He died before the year was out, however, and Salah al-din succeeded his uncle as vizier. Although nominally merely the arm of Nur-ed-din, he soon began to display signs of independent ambition. His aim was to build himself a power-base in Egypt sufficiently strong to enable him to launch a final cataclysmic attack on the crusader kingdoms.

Salah al-din quickly consolidated his position as virtual ruler of Egypt. He surrounded himself with loyal relatives, restored discipline in the caliph's army, installed his own ministers, and acquired a reputation as a pious, generous ruler who dispensed an even-handed justice. He was merciless in matters which closely touched his political programme, but displayed as much chivalry and tolerance as his Muslim piety allowed. He avoided an open clash with the uneasy Nur-ed-din, but declined to supply troops for his master's campaigns. He built up strong land and sea forces, secured Egypt's frontiers, and pacified the interior; under his strong and just rule the economy of the country improved measurably. In 1171 the caliph died – probably on Salah al-din's orders – and he proclaimed the revival of orthodox Sunnite Mohammedanism in Egypt. In 1174 Nur-ed-din died, and a major constraint was removed. Salah al-din now embarked on a programme of conquest and diplomacy which by 1186 secured him a wide realm in Syria and Iraq. He fought several brilliant and audacious campaigns against the various Muslim factions which had emerged in open rivalry on the death of Nur-ed-din. Confrontations with the Franks became more frequent as his dominions spread, until by 1187 he had completely encircled the crusader kingdoms, and was firmly enough established among his own people to declare holy war against the infidels. They too had been much weakened by internal dissension over the past decade, but they defeated him at Ramleh in 1177.

In 1187 Salah al-din led an army of 20,000 men into Palestine and besieged Tiberias. Guy of Lusignan, now King of Jerusalem, raised a crusader army of almost equal size and advanced to the relief of the city. Against the advice of the able Raymond of Tripoli he led his force through a waterless area, where he was surrounded and harassed by Salah al-din's Mameluke mounted archers. When the Franks were suffering severely from thirst Salah al-din attacked; he managed to divide their cavalry and infantry, and destroyed the army in detail at Hittin on 4 July 1187. Raymond of Tripoli and a small cavalry force cut their way free, but the remainder were killed or captured, and Raymond later died of his wounds. The Muslims then overran most of Palestine, and on 2 October Salah al-din entered Jerusalem, ending eighty-eight years of European rule.

The city of Tyre held out against him, defended by Conrad of Montferrat and a force of crusaders who arrived by sea. There was skirmishing around the city for more than a year, but a considerable part of the Muslim army melted away, content with the almost complete victory. Meanwhile Europe was in a ferment, and the Third Crusade was launched to recapture the Holy Land. Guy of Lusignan had been released on parole; in August 1189 he raised an army from the remnants of Frankish strength in Palestine – Salah al-din had shown admirable restraint towards the apparently-defeated Franks – and laid siege to Acre. His army dug strong entrenchments around the Muslim stronghold, but were in turn surrounded by Salah al-din's relieving army. For two years the city held out, and innumerable actions were fought around it by the garrison, the besiegers, and the Muslim field army. Genoese and Pisan fleets maintained a tight blockade, and Salah al-din's navy was destroyed. Philip of France and Richard I (q.v.) of England arrived by sea with strong crusader forces in the spring of 1191, and Richard was accepted as the leader of the European forces. Under his direction the city was forced to capitulate on 12 July 1191. Despite the strain on the Egyptian economy, the loss of his fleet and of many soldiers at Acre, and the superior forces of the crusaders, Salah al-din pressed on with his campaign and inspired his followers to persevere. Internal dissension in the crusader camp led to many defections, and despite Richard's brilliant exploits in 1191–2, he was unable to break the overall ascendancy of the Muslims. A strange sort of friendship and mutual respect grew up between the two leaders; and on 2 September 1192 an armistice was concluded between them. The Europeans were to keep their coastal strip, while Salah al-din held the hinterland, including Jerusalem. There was to be free access to the holy places for members of both religions. Richard took ship for home in October, promising to return when the truce expired in three years and capture Jerusalem.

Exhausted by his exertions, the fifty-five-year-old Salah al-din returned to Damascus. He spent a few months with his family, but then contracted yellow fever and died, after a short illness, on 4 March 1193. He was found to have given away most of his treasure in grants – he was never acquisitive, and lived very simply according to the dictates of his religion. His reputation as a ruler both pious and practical, generous and shrewd, tolerant and determined, spread over much of the Muslim and Christian world. He was a resourceful and cunning general who almost invariably showed the chivalrous side of his nature to his defeated enemies, though he lived in a cruel age and under the circumstances of a religious war.

Salan, Raoul (*1844–1984*), *General*

Salan was not a notable general in a military sense. His sole significance is as the leading spirit in the only major military revolt against the civil power to occur among the first-class western powers in the years since 1945.

Salan first came to prominence in 1948–52,

when he served in Indo-China as deputy to General de Lattre (q.v.), the Commander-in Chief of the French forces fighting the Viet Minh insurgents. On Lattre's death in January 1952 Salan took over the chief command. In retrospect it can be said that the war was lost for the French almost as soon as it began, but during his tenure of command Lattre had achieved some results. The situation deteriorated during Salan's period of command. He was replaced by General Navarre at the end of 1953; his most significant policy was probably his repeated insistance on the importance of reoccupying the Dien Bien Phu area, overrun by the Viet Minh, as a base for operations on the Thai border. His recommendations remained 'on the file' and continued to have weight after he himself had returned to France, taking his locally-experienced staff with him; it was left to General Navarre to put this recommendation into practice, with disastrous results – although it would be quite unjust to place any major responsibility for the subsequent fiasco on Salan.

From 1956 to 1958 Salan was Commander-in-Chief in Algeria. Initially he was regarded by the local hard-liners as 'soft' on the F.L.N., and in January 1957 a European extremist group attempted to assassinate him by firing a bazooka shell through his office window, killing his A.D.C. Gradually, however, he became more and more deeply involved in the jungle of military and civil politics in Algeria, and acquired a taste for making political pronouncements. He became more closely identified with the right-wing position, since although the local Europeans did not trust him he spoke for the Army, and the Army was their only hope of seriously influencing the policies of the ineffective and squalid French governments of the day. During the crucial riots of May 1958 he endorsed the popular appeal to General de Gaulle (q.v.) to return to power, and played a major role in the downfall of the Fourth Republic. It was confidently expected that de Gaulle would pursue a policy of reinforcing the French grip on Algeria, but to the horror of the right-wingers he took a brutally realistic view of the situation and quickly made clear his intention of reaching a settlement with the nationalists. Well aware of the danger of a group of military men who had become accustomed to meddling in politics, de Gaulle wasted no time in dismantling the 'mafia' which had brought him to power. Salan had continued to flirt with the so-called Committees of Public Safety, and in December 1958 he was recalled to France and an appointment as Inspector-General of Defence – a post which was suppressed within a month. Thereafter he was appointed to the prestigious but largely ceremonial command of the Paris Military District. He had been retired from the active list by April 1961, when Army and European discontent erupted in Algeria.

Enraged by the results of de Gaulle's referendum on Algeria's future – which overwhelmingly endorsed a negotiated French withdrawal from the country – four ex-generals organised a *putsch* in Algiers, with the active support of a group of serving officers. On 21 April officers and men of the 1st Foreign Parachute Regiment, an *élite* formation with an unparalleled combat record in Indo-China and Algeria, took over administrative buildings in Algiers. A state of siege was declared; an appeal was broadcast to the armed forces calling for the overthrow of the government and the continued prosecution of all-out war against the F.L.N.; and for a few days there was a serious possibility of an airborne attack on Paris by units from Algeria. Many honourable officers were deeply distressed by the thought of abandoning Algeria, having fought so desperately to retain the country for France; and the bitter memory of the political mishandling of the Indo-Chinese campaign, for which the army had paid most dearly, was ever present. In the event, however, Generals Salan, Jouhaud, Zeller, and Challe were unable to command the support of a significant number of units. Many key officers declared their loyalty to de Gaulle, and others remained eloquently silent. General de Gaulle adopted a posture of uncompromising defiance, and called upon the French people to resist this treasonable attempt upon the civil authority. By 25 April the revolt was over, and Salan had to go into hiding. He was closely associated with the terrorist actions of the Secret Army Organisation – the O.A.S. – who instigated a campaign of robbery, assassination, and indiscriminate bombings both in Algeria and in metropolitan France. On 20 April 1962 Salan was captured in Algiers. He had already been sentenced to death *in absentia*, but was now put on trial for his life in front of a military tribunal. The trial opened on 15 May 1962; Salan made a statement justifying his actions, but thereafter refused to answer detailed questions. On 23 May he was found guilty of all charges, but despite blatant pressure on the tribunal by de Gaulle, he was sentenced not to death but to life imprisonment. From Fresnes prison he appealed to the O.A.S. to cease operations.

In June 1968 he benefited from the amnesty following the 'events of May', and was released. The rank and decorations of which he had been stripped were restored in 1974 and 1982. He died in Paris on 3 July 1984.

San Martín, José de (*1778–1850*), *General*

Second only to Simón Bolívar (q.v.) as a revolutionary commander in the South American wars of independence, San Martín was a man of utterly different character. Where Bolívar's political philosophy was genuinely original and visionary, San Martín was basically a conservative who was attracted by the concept of a constitutional monarchy. Where Bolívar was self-willed and ambitious, San Martín was a selfless and completely professional soldier of high personal integrity.

He was born on 25 February 1778 at the Jesuit village of Yapeyú near the northern frontier of Argentina. His parents, Spaniards of good birth, returned to Spain when he was only seven, and his upbringing and early career were those of a conventional member of the Spanish officer corps. He entered the Spanish Army before he was twenty, rose to the rank of lieutenant-colonel, and served in the early campaigns of the Peninsular War. When he heard that an independence movement had begun in Argentina he offered his services, and returned to the land of his birth in 1812. He found a chaotic situation; no formal declaration of independence from Spain had been made, and no credible republican administration had emerged from the confusion of revolt. San Martín recognised that Argentinian independence could never be secure as long as Peru remained in Spanish hands. Peru played a less active part in the wave of revolution than some of the other colonies; her aristocracy was deeply conservative, there were many Spanish residents, and the most formidable Spanish military presence on the continent was based around Lima. San Martín reasoned that an overland campaign stood less chance of achieving his objectives – the seizure of the Upper Peru silver mines and the neutralisation of the Spanish forces – than a two-stage campaign, overland through Chile and then by sea to Peru.

Between 1814 and 1817 San Martín assembled and trained a part-Argentinian, part-Chilean republican army at Mendoza in western Argentina; on 9 July 1816 Argentina formally declared her independence. Early in 1817 San Martín commenced his operations with an advance across the Andes. His army of 3,000 infantry, 700 horse, and twenty-one guns crossed the snow-bound passes of Los Patos and Uspallata in January, marching in two groups; the Chilean contingent was commanded by San Martín's colleague and friend Bernardo O'Higgins. They came down from the Gran Cordillera in the first week of February, and on the 12 February were blocked at Chacabuco by a Spanish force of some 2,000 led by Colonel Maroto. San Martín attacked the Spanish army in two columns, again led by himself and O'Higgins. The latter advanced before dawn and contained the main Spanish line while San Martín turned its left wing and routed it. The royalist army lost some 500 dead, 600 captured, and all its artillery; republican losses were very light. On 15 February San Martín entered Santiago de Chile, and although the formal declaration did not follow for a year, the independence of Chile became a fact. There was scattered fighting throughout 1817, and the Spanish General Osorio left his Peruvian base and moved south with 9,000 men. On 16 March 1818 Osorio beat San Martín at Cancha-Rayada; the 6,000 strong-republican army lost some 120 dead and suffered the capture of twenty-two of its guns, but Spanish losses in personnel were higher and the royalist force was badly shaken by the action. San Martín regrouped, and at Maipú on

5 April he wiped out his defeat by decisively beating Osorio. Once again he sent his main attack in against the Spanish left, and succeeded in turning it; Osorio lost about 3,300 in casualties and prisoners, and was obliged to retreat into Peru. San Martín could not follow up this victory with an immediate invasion; Spanish control of the sea was absolute, and had to be broken before a land campaign was practical.

For nearly two years San Martín strove to gather a naval squadron and sufficient transports for a landing on the Peruvian coast. His most valuable colleague was an extraordinary British sailor, Thomas Cochrane, tenth Earl of Dundonald, who accepted command of the Chilean revolutionary navy at the end of 1818 after leaving the Royal Navy in disgrace. Cochrane was a fiery, eccentric, and brilliant commander of light warships who had had a most distinguished career as a frigate captain in the Napoleonic Wars. He swept the Chilean coast and bombarded Spanish ports, and in September 1820 led the small escort which convoyed San Martín's transports to the Peruvian port of Pisco. San Martín secured the port while Cochrane blockaded the Spanish squadron in Callao; he personally led a daring cutting-out expedition which seized a frigate at anchor on 5 November. In July 1821 San Martín entered Lima, and on the 28 July Peruvian independence was declared. Although isolated Spanish garrisons capitulated over the next few months, the bulk of the Spanish army in Peru refused battle and withdrew into the highlands to the east. San Martín's forces were insufficient for him to follow them, and the Argentinian general naturally considered a link-up with Bolívar, who was at that time liberating Ecuador.

The two liberators met at Guayaquil on 26–27 July 1822. The progress of their conversations is one of the most tantalising mysteries of South American history. Basically, it is clear that the ambitious visionary Bolívar and the tactiturn San Martín were completely unable to reach agreement, either on plans for the Peruvian highland campaign or on the political future of South America. Their armies were merged under Bolívar's command; and then San Martín provided unquestionable proof that personal ambition played no part in his motivation. He offered his resignation to the Peruvian revolutionary congress in September 1822 and, commenting simply that Bolívar was 'not the man we imagined him to be', slipped away into complete retirement. He left South America without ceremony, and spent the last twenty-eight years of his life in retirement, first in Belgium, then in Paris, and ultimately at Boulogne. He died at Boulogne on 17 August 1850. Proud, stoic, ruthlessly honest with himself and dutiful in his treatment of others, San Martín was one of the most impressive military personalities to emerge from Spanish history. Men of his calibre were sorely missed at home during the early years of his South American adventure.

Saxe, Herman Maurice, Comte de (*1696–1750*), *Marshal-General of France*

Brilliant German-born marshal and commander of French armies during the War of the Austrian Succession, Saxe was the victor at the great battle of Fontenoy and at Lauffeld.

Born on 28 October 1696 at Goslar in central Germany, the illegitimate son of the Elector Frederick Augustus I of Saxony and of Maria Aurora von Königsmark, young Herman was sent by his father at the age of thirteen to serve in Flanders under Prince Eugène of Savoy (q.v.). Two years later he was granted the title of Count of Saxony; and he entered into an arranged marriage with Johanna Victoria von Löben at the age of eighteen – a marriage which was later annulled. In 1719 his father purchased for him command of a German regiment in the French army, and it was while in this position that he attracted attention among the French for his use of the methods of field-training and musketry which he had learnt whilst serving under Prince Eugène.

A promiscuous man throughout his life, it was not long before Saxe had formed associations with numerous influential women in several countries. In 1726 during a visit to Mitau he gained the affection of Anna Ivanovna (later Empress of Russia) and was elected Duke of Courland; a marriage was even considered likely, but this was not approved by the Poles and Russians, and the following year he was expelled. At the outbreak of the War of the Polish Succession Saxe participated in Marshal Berwick's campaign on the Rhine, his rank of lieutenant-general being procured by Marshal de Noailles. July 1743 found Noailles in effective command of the French armies following the failure of the other marshals, and Saxe was given command of an army assembled at Dunkirk, which it was planned should sail against England to support Prince Charles Edward Stuart, pretender to the English throne. Following action by the Royal Navy and the effects of great storms, the French fleet was shattered and the invasion abandoned, however.

Saxe was appointed marshal of France on 26 March 1744 and given command of an army in Flanders, it being intended that this should cover Louis XV's own army during sieges; but following constant erosion of his forces by Louis's and Noailles's demands for reinforcements, Saxe found himself greatly outnumbered by the Allied armies of British, Austrian, and Dutch forces. He had so cleverly located his encampments near Courtrai, however, that the Allies were virtually powerless to defeat him.

Early in May 1745 Saxe led his army, 70,000 strong, in an advance on Tournai and invested the city. In response the Pragmatic Army of 50,000 troops led by William Augustus, Duke of Cumberland (son of George II) moved up steadily to attempt the relief of the besieged city. Leaving some 18,000 men to maintain the siege, Saxe led the remainder of his army to meet Cumberland at Fontenoy – although he himself was so ill with dropsy that he had to be carried on a litter. The Allies opened with an unimaginative frontal attack just before dawn on 10 May, and this was halted by the French defences. The Allied plan was to smash the centre of the French line; a ponderous mass of English troops (the Guards brigade) and some Hanoverians therefore moved forward and, in a bloody drawn-out fight, succeeded in smashing its way through the first French line. Rousing himself from his litter amongst the panic-stricken soldiers, Saxe mounted his horse, rallied his men, established a second line and, with a withering volley, halted the English attack. He then summoned his artillery, which poured a tremendous fire into the Allied square. Cumberland somehow managed to extricate his forces in good order, but the Flanders campaign was lost. Saxe quickly followed up with a surprise capture of Ghent, and then of Brussels – which also secured Antwerp. After taking Mons and Namur, he defeated Prince Charles of Lorraine at Rocourt, near Liège.

Saxe was appointed Marshal-General of France on 10 January 1747 – an honour only previously bestowed upon Turenne and Villars (qq.v.). Almost the whole of France's military effort was controlled by him, and he held virtually sovereign power over the Netherlands from his headquarters at Brussels. The last great battle he fought was that of Lauffeld on 2 July 1747. In this the Allies lost about 10,000 men defending the approaches to Maastricht – although the town itself did not fall until May 1748.

After the Peace of Aix-la-Chapelle (1748) Saxe retired to his château at Chambord, where his German regiment camped in the park. He died there on 30 November 1750, rumour attributing his death to the result of a duel.

Scharnhorst, Gerhard Johann David von (*1755–1813*), *General*

Scharnhorst was the Prussian military reformer of the Napoleonic era whose work of reorganisation was almost wholly responsible for Napoleon's (q.v.) defeats in 1813, and provided the basis for Prussian military resurgence in the middle of the nineteenth century.

Son of a tenant farmer who had served as a senior N.C.O. in the Hanoverian artillery, Gerhard Scharnhorst was born at Bordenau in Hanover on 12 November 1755. He entered the Hanoverian army at the age of twenty, gaining his commission three years later. As a junior officer he was much influenced by the reforming policies of Wilhelm von Schaumburg-Lippe (who had directed the reorganisation of the Portuguese army twenty years earlier), and he served with distinction in the Flanders campaigns of the French Revolutionary War of 1793–5. Thereafter, as a major, he was appointed chief of staff to

Count von Wallmoden, Commander-in-Chief of the Hanoverian forces.

In 1801 he offered his services to Prussia and, conscious of the vital role played by the nobility in the military arm of that nation, he took the extraordinary step of applying for permission to institute a reorganisation of the Prussian army in return for admission to the nobility and promotion to lieutenant-colonel. In support of his proposals for reform in the Prussian army he submitted three detailed essays – all of which were accepted – and in 1804 he was ennobled and granted his promotion.

Entering the war academy in Berlin, he at once set about propounding his theories for the widespread reformation of the Prussian military; one of his pupils, later to become famous in his own right, was Karl von Clausewitz (q.v.). He also founded the Berlin Military Society, dedicated to the professional study and advancement of arms. Towards the end of 1804 Scharnhorst was taken on to the staff of the Duke of Brunswick to direct the preparation of the forces that were to oppose Napoleon in 1806. But morale in the Prussian army had dropped so low since the years of Frederick the Great's military ascendancy that it proved impossible to initiate any reform of substance before the storm broke. Scharnhorst himself was wounded in action at Auerstadt, and with Blücher (q.v.) was taken prisoner at Ratkau, near Lübeck, after the Battle of Jena on 14 October 1806. Released in an exchange of prisoners, he was present at the Battle of Eylau on 8 February 1807 in Wilhelm Lestocq's Prussian corps.

After the Peace of Tilsit of July 1807, Scharnhorst was appointed head of the Army Reform Commission and embarked on the great work of his life – that of developing the structure on which the new Prussian army was to be built. He now met and worked with such military progressives as August von Gneisenau (q.v.); although steeped in the traditions of Frederick the Great, he realised that a Prussian army officered almost entirely by the nobility and dependent upon a semi-feudal territorial allegiance was quite inadequate to support and defend a Prussia hell-bent on Central European dominance. The major catastrophe at Jena had provided the necessary evidence.

Instead he advocated abandoning the smaller, professional long-service mercenary forces in favour of national service and short-term conscription (the Krümper system) – a reform so fundamental that only substantial political reform would enable it to be accepted. His appointment to the Army Reform Commission brought Scharnhorst direct access to the King and the necessary authority to bring about his proposed reorganisation; but Napoleon, suspicious and apprehensive of the changes, forced the King to cancel most of them. But the structure of organisation had been created, and clandestine operation of the Krümper system created a pool of trained reserves.

In the war against Russia in 1811–12 Prussia was forced into alliance with Napoleon, and Scharnhorst left Berlin on indefinite leave of absence from the army. But when the French started their disastrous retreat from Moscow, the Prussian call to arms came: Scharnhorst returned and in 1813 was appointed chief of staff by Blücher. He was present at the Battle of Lützen, but was wounded at Grossgörschen on 2 May. His wound proved so serious that in fact he never fully recovered. He was sent to Prague to negotiate with Austria for her entry into the war with France, and died there on 28 June the same year. He was not to see the triumph of arms over Napoleon at Leipzig (the 'Battle of the Nations', 16–19 October 1813) in which the reforms in Prussia's army contributed to the eclipse of France's military supremacy.

Scharnhorst, like other Prussian military reformers of the nineteenth century, was not himself a great commander; his strength lay in his ability to impress his plans upon hitherto unreceptive colleagues. He founded the general staff system, but owing to the relatively short time he had to apply himself to this concept it was not exploited and perfected for several decades. That it proved fundamentally sound is confirmed by its adoption by almost every army in the world thereafter. Perhaps Scharnhorst's most remarkable personal achievement, bearing his humble origins in mind, was his being accepted by a military establishment almost entirely dominated by the hereditary nobility.

Scott, Winfield (*1786–1866*), *Lieutenant-General, General-in-Chief of the Army*

Winfield Scott was an associate of every American president from Jefferson to Lincoln, and carried out important missions for most of them. His life spanned the most important formative years of the U.S. Army, from the inefficiency and indecision of the immediate post-Revolutionary years to the peak of disciplined strength achieved at the end of the Civil War. His own part in that progression was not inconsiderable.

Born on 13 June 1786 on his family's estate near Petersburg, Virginia, Scott was the grandson of a refugee from Culloden, James Scott of Buccleugh. He was at first destined for the law, but served briefly in a local cavalry troop in 1807. In May 1808 he secured a commission and a posting to the command of General James Wilkinson. The disorder and apathy he found there prompted him to outspoken criticism, which cost him a court martial and a year's suspension from duty. He was promoted lieutenant-colonel on the outbreak of the War of 1812, and served at Queenstown. At this battle a mixed force of American regulars and militia was beaten by the British under Brock, largely as a result of the cowardly refusal of the militia to support the exposed regulars. Scott was captured, but was soon paroled. He planned and carried out the capture of Fort George in May 1813 while serving as adjutant general to the ineffective Dearborn. In October 1813 he joined Wilkinson's army on the St Lawrence, and commanded an infantry battalion. He was now a colonel, and he chafed at the shortcomings of his superiors and his colleagues. The military profession had been held in contempt in America since the Revolutionary War, and far too much reliance was placed on untrained and erratic militia led by local dignitaries with little training and less talent. Promoted brigadier-general in March 1814, Scott put into practice his belief in the importance of serious professional training; the result was that his brigade was almost the only body of American troops to give a consistently good account of itself in the War of 1812. His 1,300 men defeated the British General Riall, with 1,700 regulars, at Chippewa on 5 July 1814, and earned the praise of the enemy for their steadiness. Three weeks later Scott's brigade bore the brunt of the bloody fighting at Lundy's Lane; he had two horses shot under him, and was seriously wounded. Breveted major-general, Scott was widely acclaimed by the public as the hero of the war. When he recovered he sat on several important boards, including that which made the selections when the army was severely reduced after the end of hostilities, and that which drew up the army's first drill manual. After travel in Europe Scott took over the command of the Eastern Department, with headquarters in New York.

Although frequently the object of public acclaim and congressional votes of thanks for his many services to his country, Scott suffered from the enmity of several presidents. It was not a time of notably high standards in American public life, and personal feuds were often at the bottom of political decisions. Scott did not have a particularly pleasing manner; he was outspoken and inclined to florid rhetoric, a stickler for the minutiae of dress discipline and decorum, and an early advocate of temperance. This earned him the nickname 'Old Fuss and Feathers', but he was not really a narrow man; he was generous and open, and completely honest. He also had a talent for negotiation, and by his personal efforts saved his country from war with Britain over Canadian border incidents on two occasions, in 1838 and 1859. He averted a major Indian war in 1838 by persuading 16,000 Cherokees to move out of South Carolina and Tennessee and to cross the Mississippi without a fight. He worked tirelessly to improve the standards of skill, discipline, and humanity in the army, personally revising the infantry tactics manual which remained in use up to the Civil War.

In 1832 Scott led a column to fight in the Black Hawk War, but his command was struck by cholera and was prevented from taking part in the Battle of the Bad Axe River. In 1834–5 he held command in the campaigns against the Seminole and Creek Indians, but was removed by President Andrew Jackson – an enemy for many years – on the unjust pretext that he had shown lack of energy. In fact he had performed well despite lack of resources and support, and was completely exonerated. In June 1841 Scott was appointed General-in-Chief of the U.S. Army. In 1846, when war broke out with Mexico, Scott was the advocate of a plan for an amphibious operation against Vera Cruz; he recognized that Zachary Taylor's (q.v.) successes in the northern Mexican zone were little more than tactical and could lead to no lasting advantage. Scott commanded the force which took Vera Cruz in March 1847, and subsequently marched on Mexico City over difficult terrain. He was hampered by sickness among his troops; a lack of support from Washington, which went as far as failing to send money to pay the army; serious logistic weaknesses; and the periodic abandonment of the campaign by volunteer units which had enlisted for a fixed term. Despite these difficulties Scott won the victories of Cerro Gordo (18 April 1847), Contreras and Churubusco (20 August), Molino del Rey (8 September), and Chapultepec (13 September). The last was particularly impressive, and led to the occupation of the capital. The enmity of President Polk led to Scott's removal from command while peace negotiations with Mexico dragged on through the winter of 1847–8. Trumped-up charges were brought against him, but had to be withdrawn; the nation, and Congress, received him as a hero, and in 1852 he was voted the rank and pay of lieutenant-general. In that year he campaigned for the presidency, but was heavily defeated – he was tempted by the prospect of high office several times during his career.

Although retired from active service, the tall old general – he was six foot five inches at the age of nineteen – took a continuing interest in military affairs, and in 1860 he repeatedly urged the strengthening of the Federal forts and arsenals in the South. Although a Virginian, and under pressure to join the Confederacy along with his old subordinates Lee (q.v.), Davis, and Beauregard, Scott remained loyal to the Union. In January 1861 he was appointed commander of the U.S. Army, and made energetic preparations for defending Washington. He requested that he be relieved of duty on 31 October 1861, on the grounds of infirmity – he was seventy-five years old, and clearly felt unable to shoulder the burden of the war. The next day Lincoln (whose bodyguard Scott had personally commanded on his inauguration day) and the entire cabinet travelled to Scott's house, where Lincoln read a moving eulogy, and granted him immediate retirement on full pay. Winfield Scott lived out the Civil War, and died at West Point on 29 May 1866, fifteen days short of his 80th birthday.

Selim I, called Yavuz (*c. 1470–1520*)
Ottoman Sultan of Turkey

Selim, the son of Bayazid II and the father of Suleiman I (q.v.), was the most able soldier of all the Ottoman sultans; although his reign was very brief he was personally responsible for an enormous increase in the size of the Ottoman Empire. His character was forbidding – his nickname means 'the Grim' – but he was no mere tyrant.

Selim's father assigned him to a Turkish province in the Balkans as part of his early training in the wielding of power. Bayazid was basically a peaceful ruler; although intermittent fighting flared along the frontier with the Christian powers of eastern Europe during his reign, he put no great pressure on Europe. His determined son was of a different temper: in 1511, fearing that Bayazid planned to raise his other son Ahmed to the throne in place of him, Selim rebelled. He was defeated and forced to flee to the Crimea, but not for long. His father was worried by the threat of an agreement between Selim and the Shah of Persia, Turkey's powerful enemy on the eastern frontier, and in April 1512 he abdicated in Selim's favour. Although the threat of Persian involvement had played a part in his success, Selim had not forgiven the Shah for earlier support of Ahmed's party. He now went out of his way to provoke war with Persia partly because of this desire for vengeance, and partly because of his intolerance of the Shi'ite Muslim 'heresy' practised in that country. After some initial skirmishing Selim marched from his eastern city of Sivas in June 1515 at the head of some 60,000 men. He advanced to the upper Euphrates by way of Erzerum, and the Shah fell back before him through rugged terrain, pursuing a 'scorched earth' policy. Selim had foreseen the logistical obstacles, however, and his foresight enabled him to press on through the mountains of Khoi. On 23 August 1515 he brought the 50,000-strong Persian army to battle at Chaldiran. The Persian force was composed entirely of horse archers and mounted spearmen, while Selim had both janissary and irregular infantry under command. Selim fought a defensive battle, manœuvring his light cavalry around a firm pivot formed by janissaries drawn up behind a ditch and protected on the flanks by chained carts and artillery. Selim's right flank was scattered but his centre held firm, and his left wing eventually put the Persians to flight, wounding the Shah. No effective pursuit was possible, but much-needed provisions were captured in the abandoned Persian camp. Selim captured the Shah's capital of Tabriz in September, but could not hold it – his janissaries and irregulars mutinied and refused to advance deeper into Persia, and he was reluctantly forced to withdraw; but the 1515 campaign won him permanent footholds on both the upper and the middle Euphrates.

In July 1516 Selim was preparing for a new invasion of Persia when he learnt of an imminent attack on him by the Mameluke Sultan Kansu al-Gauri, in alliance with the Shah. He immediately diverted his army into Syria, made a forced march around the Taurus Mountains, and in August surprised the Mamelukes at Merj-Dabik, some ten miles north of Aleppo. Kansu had some 30,000 horsemen, about half of them Mamelukes and the remainder Arabic levies. Selim had about 40,000 men, including 8,000 janissary infantry, 3,000 *élite* sipahi cavalry, some 15,000 timariot feudal cavalry levies, and about 15,000 irregulars. The Battle of Merj-Dabik was fought on 24 August 1516, and although casualties were fairly light the Mamelukes were decisively defeated. Selim drew up his army in the same formation as at Chaldiran, and the Mamelukes attacked in half-moon formation. The 'horns' engaged the Turkish cavalry wings indecisively, but the main Mameluke attack in the centre was routed by the Turkish artillery, and Kansu was killed. The Mamelukes fled, and abandoning Syria, fell back to Egypt. During autumn 1516 Selim occupied Syria and prepared to attack Egypt. Kansu's nephew Touman Bey replaced him as Sultan of the Mamelukes. In October Selim advanced, and on 28 October at Yaunis Khan, near Gaza, his vanguard defeated a Mameluke force. The Turks crossed the Gaza Desert and advanced on Cairo in January 1517.

The first major battle was fought at Ridanieh on 22 January. Touman took up a defensive position at El Kankah, supported by an inadequate artillery created by buying naval guns from Venice and by stripping coastal batteries. Selim shifted to his left under cover of night, and at dawn the Turks were in position at right-angles to the Mameluke right wing, enfilading their line. Touman swung his line through ninety degrees, but while still struggling to move his wheelless cannon into new positions he came under a heavy Turkish cannonade. Selim hoped to provoke an attack while the enemy was still in confusion, and succeeded; although one wing of the Turkish line was shaken at one stage, the final result was the retreat of the Mamelukes leaving 7,000 dead on the field. Selim occupied Cairo, although Touman shortly afterwards made a desperate attempt to recapture the city which caused several days of bloody street-fighting. Throughout February Selim consolidated his grip on Egypt, proclaiming himself Sultan and Caliph. When he marched to occupy Mecca and the western Arabian coast he left a Turkish Governor-General behind him, although the Mamelukes were allowed to retain the outward trappings of power.

In 1518 and 1519 Selim successfully crushed minor religious risings in Anatolia and Syria; and before his death near Corlu in September 1520 he accepted the homage of the Bey of Algiers, the legendary corsair Khair-ed-din Barbarossa. Barbarossa offered the Sultan a fleet in return for help against Spain in the western Mediterranean. Thus Selim not only added territories which nearly doubled the size of the Ottoman Empire during his reign, but also laid the foundations for later expansion by taking into his service the man who would turn the Turkish navy into the scourge of the Middle Sea.

Seydlitz, Friedrich Wilhelm, Freiherr von (*1721–73*), *General*

Born at Kalkar, near Cleves, on 3 February 1721, Seydlitz was the son of a cavalry major who died in 1728, leaving him to be brought up by his mother in impoverished circumstances. On reaching the age of thirteen he entered the Court of the Hohenzollern margrave Frederick William of Schwedt as a page. Here he was inducted into the art of horsemanship, and he soon attracted attention by his daring equestrian feats.

In 1740 he was granted a commission in the margrave's regiment of Prussian cuirassiers, in which he gained his first experience of fighting during the War of the Austrian Succession – though unfortunately he was taken prisoner in Silesia by the Austrians in May 1742 at the Battle of Chotusitz. However, in accordance with the custom at that time, he was quickly exchanged and returned to his own side.

During the 1740s Frederick the Great had been making great efforts to improve the capabilities of his cavalry; by 1753 the daring and efficiency of Seydlitz had been noticed by the Prussian King, who gave him command of the 8th Cuirassiers, and the regiment soon became a model for the army. By the outbreak of the Seven Years' War Frederick's cavalry was among the most feared in Europe, and at the Battle of Prague he demonstrated his brilliant, co-ordinated use of cavalry and infantry. Thrusting the cavalry round the flank of the Austrians, which drew reinforcements away from the centre and created a weakness which he quickly detected, he launched his infantry through the gap and split the Austrian army in two (the 'Prague Manœuvre'). Seydlitz distinguished himself both in this battle (on 6 May 1757) and at the Battle of Kolin (18 June), after which he was promoted major-general.

In the Rossbach-Leuthen campaign later in 1757, after two months of attempts by Frederick the Great to draw the French army under Richelieu and the Austro-French under Hildburghausen and Soubise into battle, the great Prussian commander eventually confronted the Austro-French at Rossbach on 5 November. With 64,000 men on commanding ground facing Frederick's 21,000, the scene seemed set for a Prussian defeat; but when the French attempted to outflank the Prussian left, Frederick unobtrusively wheeled his entire army to face the flank. As a result the unsuspecting French burst upon the Prussian artillery at point-blank range, and before they could recover from this surprise Seydlitz charged his thirty-eight cavalry squadrons into the confused Allied masses, and followed up with Frederick's favourite form of attack in overlapping echelon. In an hour's desperate battle the imperial army was routed, with the loss of 8,000 men – compared with only 500 Prussians. Seydlitz himself was one of those who were seriously wounded.

His cavalry saved the day at Zorndor (25 August 1758). On this occasion the Russians, under Count Wilhelm Fermor, were ready for Frederick's flanking movement in echelon, and in fact blocked the threatened envelopment; but at the vital moment Seydlitz's cavalry crossed dead ground to fall on the Russian flank, smashing through the square. A second charge was followed by an orgy of killing which only stopped with the coming of night. Fermor retreated, but the Prussians were too exhausted to follow. Again, after Frederick was overwhelmed in the Battle of Hochkirch on 14 October and the Prussian army seemed surrounded, an escape route was opened by his cavalry under General Hans von Ziethen, and was held by squadrons commanded by Seydlitz.

During the Austro-Russian invasion of 1759, Frederick suffered his worst defeat in the Battle of Kunersdorf (Kunowice) on 12 August. Both his infantry and his cavalry were thrown into confusion by a double envelopment, and having lost their way in woods and marshes, delivered their attacks piecemeal. The cavalry were particularly badly mauled, and Seydlitz was once again severely wounded. Three years later Seydlitz was given command of a mixed force of infantry and cavalry, and assisted Prince Henry in defeating the Austrian Marshal Serbelloni at the battle of Freiburg on 29 October 1762. This was the final battle of the Seven Years' War. After the Treaty of Hubertusburg (1763) Seydlitz was appointed Inspector-General of the Silesian cavalry, and in 1767 was promoted to the rank of cavalry general.

The last few years of Seydlitz's life were marred by quarrels with Frederick, but there are indications that they became reconciled to some extent shortly before Seydlitz's death, which occurred at Ohlau (Olawa) on 8 November 1773.

In character Seydlitz was a regular firebrand, quick of temper and utterly incapable of remaining out of any combat in his vicinity. A typical Prussian disciplinarian, he was naïve in all matters of strategy and the higher military arts, despite his tactical skill and gift for leading men in battle.

Shaka (*c. 1787–1828*)

The career of the founder of the Zulu nation bears much superficial resemblance to that of Genghiz Khan (q.v.). Both were outcasts and were abused in boyhood; both won a place in clan life in adolescence, through personal prowess rather than influence; both built a huge unified empire from a tiny nucleus of loyal warriors, through remarkable gifts of vision and strength of character; both introduced new tactical systems which devastated all enemies; and both engaged in warfare on a scale little short of genocide, depopulating huge regions and permanently altering the balance of civilisation in their areas of influence.

In the late eighteenth century the Zulus were a totally obscure Nguni clan numbering some 1,500 souls, ruled by a petty king named Senzangakona. In about 1786 or 1787 he chanced to meet Nandi, a woman of the eLangeni clan, and they apparently overstepped the strict sexual *mores* of their culture. In due course the elders of Nandi's clan informed Senzangakona that she was pregnant and requested that he fetch her for marriage. Various clan taboos, and his own inclination, prompted Senzangakona to reply in insulting terms, to the effect that doubtless the pregnancy was false and due to *iShaka*, an intestinal parasite known to cause menstrual irregularity. Some months thereafter the eLangeni elders dryly requested that Senzangakona come and fetch his woman and her *iShaka*. Thus the future high king received his name in scandal and humiliation; 'Shaka' is the usual Anglicised form, but since the Zulu tongue was unwritten and phonetically tortuous many other approximations of the sound are found. The Zulu chief did marry Nandi, but the match was a disaster; mother and son spent miserable years dragging from one kraal to another pursued by derision and abuse. At last, in about 1803, they found a haven in a kraal close to the centre of the dominant power-grouping in the region – the Mtetwa hegemony. Shaka's adolescence spanned the years when the very enlightened king Dingiswayo came to the Mtetwa throne and began a process of rapid expansion. By diplomacy and warfare he increased the Mtetwa group to a strength of some fifty major and many minor clans, which he ruled with moderation and wisdom. The outcast boy Shaka had now grown into a young giant of obvious physical promise, and when he showed an unusual aptitude for war, Dingiswayo selected him for future authority. He was groomed to be a future chief of the Zulus, who had now become a frontier clan in the Mtetwa hegemony, and was given command of his regiment – the *iziCwe* – in various campaigns. It was with this command that Shaka perfected the new methods which he had personally invented and tested in hand-to-hand combat.

He discarded the universal light throwing assagai in favour of a short, broad-bladed stabbing assagai. This was wielded under-arm, exactly as the Romans used their short swords. Shaka also introduced larger and heavier cow-hide shields, and drilled his warriors in close-quarters techniques; classically, the left rim of the shield was used to hook the opponent's shield to his right across his body, spinning him to the right and exposing his left ribs for the fatal upward stab of the assagai. Shaka toughened his men by making them discard their hide shoes, and drilling them until they could trot over the stony hills for fifty miles between sunrise and sunset. He provided young boys, one to every three warriors, to carry their mats, rations, and cooking gear on the march. With his single regiment of a few hundred men he evolved the classic 'buffalo' fighting formation, which would later be practised on a vastly larger scale. The strong central 'chest' of warriors engaged with the enemy at once, while the 'horns' ran forward and outflanked the enemy on each side, meeting at their rear and turning inwards on them. The 'chest' and 'horns' then worked inwards until the enemy were destroyed. In case of need a strong reserve, the 'loins', was held in readiness behind the central 'chest', usually sitting down facing away from the action so as not to get excited and lose its discipline. The commander watched the progress of the battle from some convenient height, controlling it by runners. Against the other clans and tribes of South Africa this combination of new weapons, new techniques, discipline, and trained manœuvre on the field proved absolutely devastating.

Shaka took the Zulu throne in 1816 on the death of his father – he was not the rightful heir, but his power was considerable and his reputation travelled ahead of him. He at once began to organise this small clan into the nucleus of a new nation under arms. He raised regiments according to age-group and marital status; most of his regiments lived in strict celibacy, a hardship he seems to have been unable to understand since he was impotent and probably homosexual himself. He inculcated not merely clan but also regimental *esprit de corps*, and drilled his warriors tirelessly in their new duties. Discipline throughout his clan was absolute, and in later years insanely arbitrary – he had the power of life and death over thousands, and exercised it capriciously on any whim which came to him. The sufferings of his childhood had left permanent scars; apart from visiting a ghastly vengeance on every tormentor from those years whom he could lay hands on, he maintained all his life a barely-controlled hostility towards the world in general.

His career of conquest began immediately his first tiny army of 350 men was deemed fit for action. In the early days his activities were limited to some extent by his status as a vassal of Dingiswayo, but this did not prevent him from building his army to some 2,000 trained men by 1817. Neighbouring clans were incorporated in his chiefdom by the simplest of expedients – if they submitted instantly they escaped all but casual bloodshed, but if they hesitated they were massacred with appalling savagery. In 1818 Dingiswayo was killed during a campaign against a long-standing enemy, the Ndwandwe. Shaka was coming to his support when he was warned by the leaderless and demoralised Mtetwa, and made a fighting retreat. The Ndwandwe chief Zwide had more than twice as many warriors as Shaka, and it was only by brilliant defensive generalship in a battle at Gqokli Hill that the Zulu chief survived the immediate downfall of his king. Zwide was too powerful to take on immediately, so Shaka bided his time and built up his strength until Zwide invaded again in 1819, when he smashed the Ndwandwe in a two-day battle at the Mhlatuzi fords. The Mtetwa empire had been inherited by a weakling, and was rapidly disintegrating; Shaka stepped into the vacuum.

A detailed account of Shaka's various tribal wars would serve no purpose, as the pattern was repetitious. As each new clan came under his rule he incorporated its surviving warriors into his army, teaching them the new techniques and grouping them within the time-honoured regiments – or rather brigades, and soon divisions. By the time he had been on his throne three years he ruled 11,500 square miles, from the coast to the Blood river and from the Pongola in the north to the Tugela in the south. His army numbered 20,000, and the men of rank in the scores of subject clans were beginning to lose their clan identity and to refer to themselves as, first and foremost, Zulus. By 1824 Shaka had overrun, massacred, depopulated, and burnt out a strip of country a hundred miles wide south of the Tugela river; by his death in 1828 he ruled 250,000 people in an area extending from the coast to the Drakensberg Mountains and from the Pongola river to central Natal. He could put 40,000 warriors in the field at need, and booty and tribute had amassed for him vast herds of cattle, each herd bred to a single colour – wealth unimaginable by African standards. His short reign caused the death of approximately 2,000,000 human beings, in direct warfare and in the endless disruption of frontiers caused in the African interior by the passage of tribes fleeing from his *impis*. Towards the end of his reign he held sway by sheer terror – vast mass executions, the horror of 'witch-smelling', capricious acts of kindness followed in seconds by pointless butchery. He overstrained his army with long-range expeditions, and by holding the majority of the nation's youth in enforced celibacy caused a mounting unrest which could not be held down for ever by sheer force. He also ceased to lead the army in person, another blow to morale. Even so there was no military collapse – merely a palace revolution.

In the period 1824–8 Shaka had his first contact with Europeans, and was healed of a dangerous wound by one H. F. Fynn. He then allowed traders to establish themselves in his lands (although his patronage was erratic and terrifying), and at the time of his death he was making attempts to arrange an exchange of ambassadors with his 'brother monarch King George'.

On 23 September 1828 he was assassinated by his half-brothers Dingane and Mhlangana; he died without dignity, screaming for mercy. Dingane soon disposed of his fellow conspirator, and took on the mantle of Shaka, 'the Great Elephant'. The murdered king was buried hurriedly and the exact whereabouts of the grave is not now known. The founder of the Zulu nation lies today somewhere under Couper Street in the little Natal village of Stanger.

Sheridan, Philip Henry (*1831–88*), *General-in-Chief of the U.S. Army*

Sheridan was a highly aggressive and successful cavalry leader in the Union forces during the American Civil War, and a favourite of General Grant (q.v.). A controversial peacetime military governor, he rose to become General-in-Chief of the U.S. Army.

Although he was certainly born on 6 March 1831, Sheridan left his place of birth in doubt, and on various occasions mentioned Albany, New York; Boston, Massachusetts; and Somerset, Ohio; there is apparently nothing to confirm any of these places. He did, however, spend much of his boyhood at Somerset, and it was from there that he entered West Point in 1848. A tempestuous nature and a predilection for personal aggression led to periodic suspensions, and delayed his graduation until 1853. This aggressive spirit, controlled during maturity, marked him as a leader of considerable promise. Nine years of active duty, spent mostly in frontier posts in the north-west and south-west, brought him conventional promotion to captain, and when the Civil War broke out he found himself a quartermaster in south-west Missouri.

His insatiable desire to fight promoted his appointment in 1862 as colonel of the Second Michigan Cavalry; and as such, by skilfully dividing his outnumbered command, he routed a larger Confederate force on 1 July at Booneville, Missouri – a success which prompted Brigadier-General William S. Rosecrans to declare that Sheridan was worth his weight in gold, and which brought promotion to brigadier-general of volunteers.

At Perryville, Kentucky, on 8 October 1862, he led the Eleventh Division of the Army of the Ohio, which held its positions against repeated attacks. In the Battle of Stones River (Murfreesboro), from 31 December 1862 to 3 January 1863, which resulted in a tactical stalemate but a strategic victory for the Union forces, Sheridan made an unyielding stand at the Union forces' right centre, as a result of which he was appointed major-general of volunteers. Ordered to march southwards to drive the Confederates from Tennessee, Rosecrans was defeated at Chicamauga, Georgia, on 19–20 September 1863, and Sheridan was forced to retire. Nevertheless he managed to regroup his division, but was too late to relieve Major-General George H. Thomas – which failure unjustifiably exposed Sheridan to criticism after the war. With his regrouped command he participated in the assault on Missionary Ridge below Chatanooga on 25 November – the victory which effectively brought to an end the campaign in the west. The following day Grant sent Sherman's (q.v.) Army of the Tennessee to relieve Knoxville, while Sheridan himself was called east to command the cavalry in the Army of the Potomac.

Arriving in his new command, Sheridan displayed an aggressive spirit which immediately raised the morale of the cavalry; he set about extensive re-equipment of the men, and reduced escort and picket duties, thereby giving greater independence of action. The new repeating carbines which Sheridan demanded for his men brought the cavalry much greater fire-power – a fire-power which could not be fully exploited when used from horseback; he therefore advocated fuller use of dismounted cavalry, and instructed his officers to order dismounted action whenever circumstances allowed. Following the Battle of the Wilderness on 5–6 May 1864, Grant ordered Sheridan to lead his 10,000 cavalry on a raid towards Richmond. During the course of this thrust the Union forces ravaged Confederate supplies, burning stores and destroying rolling-stock. In the Battle of Yellow Tavern, on the outskirts of Richmond on 11 May, Sheridan's cavalry met General 'Jeb' Stuart's (q.v.) 4,500 sabres in pitched battle; the Confederates were driven from the field and Stuart himself was mortally wounded – a severe blow to the South's cause.

On 4 August Grant gave command of the Army of the Shenandoah to Major-General Sheridan with orders to clear the rich valley of the Shenandoah of Confederate forces and to destroy their means of continuing the war. In order to prevent political meddling in the affairs of the military so close to Washington, Grant ordered the concentration of 48,000 men near Harper's Ferry. His orders to Sheridan were characteristic: 'to eat out Virginia clean and clear . . . so that crows flying over it will have to carry their own provender'. Cautious preparation for the coming campaign cost a delay of six weeks – a delay which taxed Northern patience – but in the Battle of Opequon Creek (the Third Battle of Winchester) on 19 September Sheridan's cavalry turned the Confederate flank and drove General Jubal Early out of Winchester. Further successes followed at Fisher's Hill on 22 September, and at Cedar Creek on 19 October; it was at the latter that Sheridan's forces were facing defeat when the Union commander, 'galloping the twenty miles up the pike from Winchester', arrived just in time to rally his men in an extraordinary display of personal leadership. He launched a counter-attack at the critical time and place, and drove the Confederates all the way back to New Market.

Near the end of March 1865 Sheridan was at the head of his cavalry before Petersburg, his command further increased by the addition of the 5th Corps of Infantry. With these forces he circled south and west of the city, cut General Lee's (q.v.) rail communications, and, on 31 March at Dinwiddie Court House and the next day at Five Forks, broke the Confederate right flank and forced Lee to retire westwards, thereby leaving his Richmond–Petersburg line. Sheridan felt, however, that the 5th Infantry Corps had failed to press its attack, and he relieved its commander, General Gouvereur K. Warren, of his post – another event that led to post-war criticism of Sheridan. To Grant, however, Sheridan was clearly a favourite commander who could do no wrong, an ideal war leader who had shown that driving ahead at all costs was the only way to win – and thereby in all probability to save many lives in the long term. Sheridan was left to continue his pressure against Lee's southern flank, and in the end helped close his route of escape near Appomattox.

After the end of the war Sheridan was ordered to the Gulf, where his very presence accelerated the decline of Maximilian, the French puppet emperor in Mexico. Deprived of outlets for his fighting urges, Sheridan adjusted himself slowly to the demands of peace. His belligerent approach to his appointments as military commander in Louisiana and Texas in 1867 resulted in widespread criticism – whipped up by quite unrelated reflection on some of his wartime shortcomings – and he was deprived of these appointments. He went on to plan and execute the successful Indian campaign of 1868–9, for which he earned promotion to lieutenant-general. Abroad in 1870–1, Sheridan attended the Prussian headquarters throughout the closing stages of the Franco-Prussian War as an observer. He became General-in-Chief of the U.S. Army on 1 November 1883, and on 1 June 1888 received confirmation of his commission as full general. He died on 5 August 1888 in Nonquitt, Massachusetts.

Sherman, William Tecumseh (*1820–91*), *Lieutenant-General*

'Cump' Sherman, Grant's (q.v.) closest associate among Union commanders of the American Civil War, was a complex and often misunderstood man. It is ironic that this introverted, strongly-principled man, who loved so much of what the South stood for, should be remembered for laying waste vast areas of the South on his notorious 'march to the sea'. One may be certain that when in later life he announced that 'War . . . is all hell,' he meant it sincerely.

He was born in Lancaster, Ohio on 8 February 1820, one of eleven children of a lawyer. Tecumseh was his given name, after the famous Shawnee chief, but he had no Indian blood; the family had come from Dedham in Essex to Boston, Massachusetts, in 1634. Orphaned in 1829, the eleven little Shermans were parcelled out to be fostered by various friends and relatives; Tecumseh was reared by a family called Ewing, and eventually married their daughter. It was his pious foster-mother who had him baptised again, adding the name William. Educated locally, he entered West Point in 1836 and graduated sixth in the class of 1840. He then served in Florida as a second-lieutenant with the 3rd Artillery, and in South Carolina as a lieutenant in 1842. A combination of duty and leave trips in Georgia in 1843–5 gave him some knowledge of that locality. Sherman was aide to General Kearny in the Mexican War, but saw so little action that he nearly resigned his commission in despair. He was persuaded to continue in the service, and filled various staff appointments in California and Washington before finally resigning from the army in September 1853. The next six years were unhappy; he failed in business, and was forced to accept favours from his wife's relations. He did well, and was happy, between October 1859 and January 1861 as superintendent of a military academy which later became Louisiana State University. This engagement was brought to an end by Louisiana's secession from the Union. Sherman was offered high rank in the Confederate army, but was anxious to rejoin the Federal forces. He regarded the break-up of the Union with deep horror, and deplored the prospect of civil war. He had many friends in the South and genuinely loved much which it stood for, but felt that less damage would be done if the war was prosecuted with such ruthless determination that it could be brought to a swift close. For a while he had difficulty in rejoining the Army, but in May 1861 was appointed colonel of a regular regiment, the 13th Infantry.

Sherman fought at the First Battle of Bull Run, and was soon promoted brigadier-general. He was sent as second-in-command to Kentucky; he soon succeeded to the command and set about preparing the defences in that area. These amounted to a wholly-inadequate force of raw volunteer units; and Sherman's conflicting sense of responsibility to them, and to his country, preyed on his mind to such an extent that he was suspected of mental instability. His mind worked very quickly, and he had a rapid, nervous way of expressing himself; he was certainly an introvert, but his behaviour under the strain of the Kentucky command was exaggerated by newspaper reporters with whom he had clashed over questions of security. He was relieved of command and sent on leave, but soon returned, and led a volunteer division at Shiloh on 6–7 April 1862. This action involved desperate fighting by an inadequately outposted Union army to repel a surprise Confederate attack. Sherman was prominent in the engagement, having four horses killed under him. On 1 May 1862 he was promoted major-general of volunteers, and during the summer was sent by his friend Grant – beside whom he had fought at Shiloh – to control Memphis. In December Sherman led 40,000 men down the Mississippi in an attempt to capture Chickasaw Bluffs, north of the Confederate bastion of Vicksburg; he failed, but no blame could be levelled at him. When the army was reorganised as the two-corps Army of the Mississippi in 1863, Sherman commanded one corps; and when it became the four-corps Army of the Tennessee he retained command of XV Corps. He led it through the Vicksburg campaign, which ended with the capture of the position on 4 July 1863, and was promoted brigadier-general in the regular list. In October Grant ordered Sherman up from Memphis to help relieve the hard-pressed Army of the Cumberland at Chattanooga; after a gruelling month's march Sherman took part in the battle of Missionary Ridge on 24–25 November, and after the Union victory marched to relieve Burnside at Knoxville.

In the spring of 1864 Sherman led the 100,000 men of the Armies of the Cumberland, the Ohio, and the Tennessee in a great sweep from Chattanooga to Atlanta, while to the east Meade's Army of the Potomac pinned down the Army of Northern Virginia in the Battles of the Wilderness, Spotsylvania, and Cold

Harbor. Sherman and the Confederate General Hood, facing him, both displayed great skill in a series of actions, but by the end of August Sherman was in Atlanta. That month he was promoted regular major-general. During the autumn he found it impossible to conduct useful operations from Atlanta because of the need to protect his 400-mile line of communication; so sending part of the army north again, he led 68,000 men out of the city on 15 November. He headed for the sea at Savannah, 300 miles away to the south-east, and as he marched he burnt the heart out of Georgia and left a fifty-mile-wide tract of desolation behind him. His intention was to destroy crops, installations, and stores in this vital 'bread-basket' of the Confederacy, and his orders were strictly that civilians were not to be molested. Such orders are almost impossible to enforce under campaign conditions, however, and Sherman's army was remembered with horror for many years. They occupied Savannah on 21 December, and in February swung up through the Carolinas. The Confederates were forced to abandon Charleston, and Sherman reached Columbia in seventeen days. He was blamed for the burning of the city, but this was probably the work of the retreating defenders or of freed prisoners. In March Sherman fought off Johnston at Bentonville, and linked at Goldsboro with Schofield, who had landed at Fort Fisher with another corps. On 9 April 1865 Lee (q.v.) surrendered at Appomattox.

Sherman had been an advocate for many years of generous treatment of the South after the end of hostilities, and he was sickened by the vindictiveness shown by the Government after Lincoln's assassination. Promoted lieutenant-general in July 1866, he was given command of the Division of the Mississippi. He travelled to Mexico to bring pressure to bear on the French to abandon their adventure there; and in March 1869, when Grant became President, Sherman took over his post as general commanding the army. He exercised his command with vigour, but hated the political intrigue and bureaucratic in-fighting which inevitably went with it. He retired on 1 November 1883, and for the remaining years of his life lived quietly in St Louis and New York. He kept out of public controversy, but was much in demand socially. He died of pneumonia in New York on 14 February 1891.

Physically he was tall and erect, with sharp, dark eyes set in an expressive, deeply-lined-face, and reddish hair receding from a high-domed forehead. His rather frenetic, highly-strung manner has already been mentioned, but in fact he was anything but erratic. Once he had analysed a situation and taken up his position on it he was tenacious to a degree; but his penetrating mind gave him a full awareness of the consequences and implications of his actions, and in a war situation these sometimes tortured him. He was a deeply affectionate and loyal friend to those who were close to him, including Grant.

Skanderbeg (George Kastrioti) (*1405–68*)

Skanderbeg was Albania's national hero in the wars of resistance against the Ottoman Turks – a near-contemporary of János Hunyadi (q.v.) of Hungary. He was the youngest son of John Kastrioti, head of a powerful Albanian family whose holdings in the central and northern parts of the region included the important fortress of Krujë. Kastrioti the elder was defeated during the early Turkish incursions into Albania, and Krujë was garrisoned by the Sultan. As earnest of his future obedience Kastrioti was forced to give his sons as hostages, and George was educated at the Turkish military school at Adrianople. He is known to have embraced Islam, and to have impressed Sultan Murad II with his promise; he was attached to the Sultan's personal staff for a period, and given the name Iskander, or Alexander, a revered name in Asia Minor since the days of the god-king of Macedon. His rank was Bey, and 'Skanderbeg' was the corruption of his Turkish name and rank produced in later years by his Albanian countrymen. In 1443 Skanderbeg served in the Sultan's army against the Hungarians (see under János Hunyadi), and he was actually on the field of battle near Nish in Serbia when news was brought of an Albanian uprising against Turkish rule. Leaving the field, Skanderbeg hurried to Krujë and succeeded in expelling the Turkish garrison. He hoisted the Albanian flag, proclaimed his return to the Christian faith, and dedicated himself to a holy war against the Turks. After retaking his own family's possessions he appealed to Albanians to unite against the common enemy; and a national convention at Alessio named him Commander-in-Chief of the troops of an Albanian League.

For the rest of his life Skanderbeg contrived to keep his land free from the Turks, despite both the intense efforts of the Sultan's armies and the intrigues and treacheries of his various allies; prominent among the latter were the Venetians, whose trading network in the Balkans placed them in an ambiguous position *vis-à-vis* national independence movements. In 1450 Murad II in person came to seize Krujë from his former protégé; Skanderbeg did not shut himself up with the garrison, but harassed the Turks from the surrounding hills until they were forced to lift the siege and retire. Although the damage they inflicted during their retreat was appalling, this victory not unnaturally made Skanderbeg a hero throughout Christendom. In need of foreign backing, he agreed to serve the King of Naples (at the time, Alfonso V of Aragon) in a limited capacity; this brought him Catalan troops to man Krujë, gold to hire mercenaries, and the experience of a campaign in Italy on Alfonso's behalf. In every campaigning season between 1455 and 1462 the Turks sent armies into Albania, and every time Skanderbeg defeated them. Simultaneously he pursued an able diplomatic struggle with the Venetians, even acquiring Venetian gold and soldiers for his lonely resistance against the Turks. In 1466 Mohammed II (q.v.) came in person to besiege Krujë, as his father had done before him; and as his father had done before him, he was forced to retreat with great loss. The Turkish army was badly mauled, and its renegade Albanian general Balaban Pasha was killed. On the strength of this spectacular triumph the Pope organised financial and other support for Skanderbeg, who repeated the exploit the following year when Mohammed returned to Krujë once again.

On 17 January 1468 Skanderbeg died at Alessio; he remains Albania's central folk-hero, and rightly so, for the relative ease with which the Turks took the country after his death testifies to the importance of his personal achievement. Krujë held out until 1478, but for years before that date it was an isolated stronghold in a conquered land.

Slim, William Joseph, 1st Viscount Slim (*1891–1970*), *Field-Marshal*

Slim was the tough ex-ranker who rose through the British Army to become a brilliant army commander in Burma; in the face of appalling difficulties he forced the Japanese invaders back after two years of demoralising setbacks.

Born in Bristol on 6 August 1891, Bill Slim volunteered for service in the ranks on the outbreak of the First World War and joined the Royal Warwickshire Regiment. Soon commissioned, he served at Gallipoli, where he was wounded. He moved to the Western Front and then to Mesopotamia, where he was again wounded and was awarded the M.C. After the war, in 1920, he was granted a regular commission and transferred to the 6th Gurkha Rifles, Indian Army, later becoming Commandant of the Senior Officers' School in India. Of humble family and lacking private means, he had difficulty surviving on his pay.

In the Second World War Slim commanded the 10th Infantry Brigade of the famous 5th Indian Division which advanced into Eritrea after Italy entered the war. Again he was wounded, but recovered in time to lead the 10th Indian Division in Syria, Iraq, and Persia, making the first contact with the Russian Army at Teheran. For these operations he was awarded the D.S.O.

The Japanese invasion of Burma during January 1942 was followed by the loss of Rangoon on 7 March, by the progressive collapse of organised resistance in the area of the capital, and by the threat of a major collapse and loss of all forces in the area. On 19 March Slim arrived to take command of the 1st Burma Corps, and immediately set about stiffening a line of resistance between Prome and Toungoo, thereby giving time for some degree of organisation to be attained in the withdrawal. When the Chinese 200th Division was cut off at Toungoo and threatened with annihilation, a counter-attack supported by Slim's British troops allowed the trapped Chinese to cut their way out. Seeing a chance to take the initiative, Slim then prepared plans for a general offensive, but before they could be realised the Japanese received heavy reinforcements and outflanked the entire British–Chinese line so that further retirement was inevitable. Further lines of resistance were hastily cobbled together at Magwe, but once again a Chinese division (the 38th) was encircled and overstretched resources had to be diverted to rescue the trapped troops. After setting the valuable oilfields ablaze at Yenangyaung to prevent them from falling into Japanese hands, the Allied forces fell back on Mandalay, where a screen was prepared to halt the Japanese. However, during the last week of April the Chinese Sixth Army, which covered the left flank of the Allied line of retreat, collapsed and allowed the Japanese to outflank Mandalay, which fell to the invaders on 30 April–1 May. Without adequate flank support, Slim now decided to fall right back – an extraordinarily difficult task in view of the dense jungle, an almost total lack of established field communications, and a determined enemy who pressed ever harder on the heels of defeated troops. Maintaining good order in his command, Slim fell back rapidly towards Tiddim and the Indian border, and succeeded in extricating his troops across the Chindwin river into the border hills; the Japanese fortunately halted at the river.

The retreat through Burma cost the British and Commonwealth forces something like 30,000 casualties (of whom about a half were listed as missing; many of these evaded the Japanese to reappear later in India) while it is estimated that Chinese casualties were upwards of 50,000. The Japanese on the other hand achieved the conquest of Burma for the loss of not more than 8,000 men. Slim, by his persistent resolution, maintained his command intact, albeit with the loss of much equipment, and his services were recognised by his appointment as C.B.E.

Pressing priorities in other theatres of war until well into 1943 prevented any substantial operations being undertaken in Burma, with the result that the long-suffering forces in that humid, malaria-infested area became prey to a sense of abandonment. In 1943, however, supreme command of Allied forces in South East Asia passed to Vice-Admiral Lord Louis Mountbatten, and command of all Allied ground forces in India facing Burma was given to General Slim, whose Fourteenth Army had gained the sobriquet 'The Forgotten Army'. These two commanders set about instilling a sense of absolute determination into the troops of their command, asked persistently for the latest arms and equipment and, by their outstanding personalities, managed to transform the morale of the fighting men. The image of the Japanese soldier as an invincible fanatic had been confirmed in the British soldier's mind when an abortive offensive into the Arakan early in 1943 had ended in defeat, and Slim was determined that such a fiasco must not be repeated.

When in December three divisions of the British Fifteenth Corps advanced once more into the Arakan they were once more halted by well-deployed Japanese, and there was a critical danger that the failure of the first Arakan campaign would be repeated. Realising the disastrous effect such a failure would have throughout his command, Slim rushed reinforcements forward and organised substantial air-drops of supplies to beleaguered units. Japanese forces which had encircled British units suddenly found themselves surrounded, and during the latter half of February 1944 suffered their first major defeat; almost to a man they were cut to pieces by Commonwealth troops – now suddenly aware of the fact that 'a Japanese skin stops a bullet no more easily than anyone else's'. This was the turning point of the Burmese war.

The success achieved in the second Arakan campaign was not one to be measured in territorial gain, or for that matter in Japanese casualties, since Japanese manpower was seen to be effectively inexhaustible. It lay in the destruction of the image of Japanese invincibility, and the new spirit that permeated Slim's command quickly spread throughout South East Asia.

Operational stalement followed during the second half of 1944 owing principally to the onset of the monsoon. The respite enabled Mountbatten's forces to be heavily reinforced for the forthcoming offensive in 1945.

After a limited British success gained in the

harassing Chindit operation behind the Japanese lines in Burma in January–May 1944, the Japanese under Mutaguchi opened an invasion of India by crossing the Chindwin river and advancing on Imphal and Kohima. Realising that the loss of these key centres would deprive him of important bases in the forthcoming offensive, Slim determined to hold them at all costs. Accordingly he deployed some 50,000 men in the area and, when they were encircled by the Japanese Fifteenth Army, organised a continuous supply by air to ensure their continued resistance. At the same time he assembled the 33rd Corps at a rail-head not far distant, and commenced pushing back the Japanese advanced patrols to relieve the hard-pressed garrisons. His operations were a masterpiece of timing and strategy, for the Japanese commander had not expected or planned for resistance at Kohima and Imphal, and with the onset of the monsoon deluge the invading forces were critically isolated from their lines of supply. During the months July to September 1944 Mutaguchi's Fifteenth Army fell back, the victim of widespread disease and subject to incessant attacks from the ground and the air. The abortive invasion of India had cost the Japanese 65,000 dead.

As 1945 opened Slim and the Fourteenth Army recrossed the Chindwin and advanced hard on the heels of the retreating Japanese, but the advance was not headlong, for Slim had not discerned any widespread collapse of Japanese forces throughout Burma. Almost alone in his belief that the retirement of the Japanese forces south of Imphal and Kohima was only part of a preliminary plan to entice his forces to a climactic battle in central Burma, Slim urged caution in the advance. Gradually he divined the nature of the Japanese strategy, and in January 1945 he determined to mount a massive trap. By cunning use of dummy headquarters and feints he misled the Japanese into thinking that his forces were advancing *en masse* into central Burma. He attracted the bulk of Japanese opposition to the south of Mandalay, and at the same time flung flank columns round to the south and south-west to cut the enemy lines of communication. Despite fanatical resistance at and around Mandalay, the simplicity and effectiveness of Slim's trap won the day, and by 4 March the tank columns of the 17th Division had made their dash and cut the main railway route south at Meiktila. Thus, far from being brought to a halt by a well-established, dug-in Japanese defence, Slim was able to invest much of the Japanese strength for systematic reduction. (In the event the Japanese commander, Kimura, opened a line of retreat through Thazi, but the delay occasioned by this resulted in his being unable to counter-attack with any effect.)

With no more than a week's pause in the Mandalay area, Slim's army now raced for the south, driving a powerful wedge between the Japanese forces to the west and their supply bases to the east. On 2 May Rangoon fell to the forces advancing from the north and to an amphibious assault launched south of the city. During the following four months little remained to be done in Burma but to push the remaining fragments of the Japanese army towards Thailand and mop up the numerous pockets which continued to resist to the death.

Slim's magnificent success in central Burma stamped him as one of Britain's greatest field commanders. Always characterised by a splendid aggressiveness, the lantern-jawed general was much loved by his troops and highly respected by his subordinate commanders. Almost universally known as Bill, he had the essential strengths of the ex-ranker general, and as such commanded a fidelity among the troops seldom matched in the twentieth century. His rewards after the war reflected his country's gratitude to and admiration of this remarkable man. In 1945 he became Commander-in-Chief of Allied land forces in South East Asia, and in 1948 was promoted field-marshal as Chief of the Imperial General Staff. He served as Governor-General in Australia from 1953 to 1960, and in 1960 was created viscount. Only twenty-five years previously he had considered resigning his commision as a major because of his poverty and had been forced to add to his income by writing pulp magazine fiction under another name.

He died in London on 14 December 1970.

Sobieski, *see* **John III**

Somerset, Lord Fitzroy, *see* Raglan

Soult, Nicolas-Jean de Dieu, Duc de Dalmatie (*1769–1851*), *Marshal-General of France*

Born on 29 March 1769, Soult enlisted in the ranks of the *Régiment Royal-Infanterie* two weeks after his sixteenth birthday. Sixty-two years later he was promoted to the highest military rank France can bestow, a rank held previously by three men only. His record still attracts controversy, but it cannot be denied that, if nothing else, he delivered the decisive attack in the most illustrious of all Napoleon's victories, one of the 'text-book' battles of of world history: the storming of the Pratzen Heights at Austerlitz. He was a plunderer, and he schemed for a kingdom – yet this boy from Saint-Amans-La-Bastide who abandoned an early ambition to be a baker and lived to write his name beside those of Turenne, Villars, and Saxe (qq.v.) surely deserved immortality.

A corporal at the outbreak of the French Revolution, Soult gained rapid promotion in the furious wars which followed, and achieved the rank of *général de brigade* in October 1794 after seeing action on many fields, including Fleurus. In 1795 he was at the Siege of Luxembourg, and held temporary command of a division. He was promoted *général de division* in April 1799, after fighting at Altenkirchen and Stockach, and played a creditable part in Masséna's (q.v.) victory at Zurich. Serving in Italy in 1800, still under Masséna, he was wounded and captured by the Austrians, but was exchanged in time to become Governor of Piedmont in September 1800, in which capacity he put down an insurrection. He saw further action on the Adriatic coast under Murat (q.v.) during the 1801–2 Neapolitan operations, and on 5 March 1802 he was made Colonel-General of the Light Infantry of the Consular Guard. Soult was one of Napoleon's first creation of marshals in May 1804, and was simultaneously made Colonel-General of the Imperial Guard.

Soult led the IV Corps in the 1805 campaign, and earned Napoleon's praise as the ablest tactician in Europe – a controversial comment – for the part he played at Austerlitz on 2 December. Extreme Allied pressure had developed against the French right, which fell back, whereupon the Allies moved troops across the French front to exploit the situation. At the decisive moment Napoleon launched Soult's Corps at the Pratzen Heights, which were secured in minutes, thus cutting the Allied front in two. Soult then swung to his right, enveloping and routing the Allied left wing, while other formations exploited the gap he had made in the centre and drove at the Allied right wing. By nightfall the Allied army was destroyed, with 26,000 casualties against some 9,000 French losses. On 14 October 1806 Soult commanded the French right at Jena, and he led part of the centre at Eylau on 8 February 1807. He was rewarded for his part in the campaign with honours and enormous financial grants; the latter were especially appreciated, as Soult had certain character traits in common with his old chief Masséna.

1808 found Soult in Spain, where he led the pursuit of Sir John Moore (q.v.) after the Emperor returned to France. He was unable to prevent Moore's army embarking from Corunna despite a stiff battle on 16 January 1809, in which Moore was killed. Soult had the good taste to raise a monument to his dead enemy, while most Englishmen were reviling Moore's name. In February and March Soult invaded and occupied northern Portugal, but did not advance in strength much further than Oporto. It is said that he gave much thought to Murat's recent elevation to the throne of Naples, and saw himself as King Nicholas of Portugal. Regal ambitions were necessarily shelved early in May, when a British force under Wellesley made an audacious crossing of the Douro at Oporto and obliged him to evacuate first the city, and subsequently all Portugal. Soult extricated his command with some skill, but only at the cost of abandoning his entire artillery and baggage train.

Transferred to Castile and made chief of staff to King Joseph Bonaparte at Madrid, Soult won a decisive victory over the Spanish army of Areizaga, some 53,000 strong, at Ocaña on 19 November 1809. With 30,000 French he routed the enemy forces, inflicting 5,000 battlefield casualties and taking 20,000 prisoners, for the loss of only 1,700 men. January 1810 saw him in command of the

Army of Andalusia, and he set up headquarters at Seville in February. He advanced and took Badajoz in March, but then fell back and did not support Masséna's invasion of Portugal. In May 1811, when Wellington was holding off Masséna's attempt to relieve Almeida, Soult also advanced to relieve his own force at Badajoz. He was checked at Albuera on the 16 May by an Anglo-Portuguese-Spanish force under Sir William Beresford. Soult obtained an early advantage by turning the Allied right with great bloodshed, but after a desperate killing-match which exhausted the infantry on both sides he was forced to concede defeat, and withdrew. In the summer of 1812 Wellington's victory over Marmont at Salamanca, to his north, made his position in Andalusia and Grenada untenable; Soult raised the two-year Siege of Cadiz and pulled eastwards, joining Suchet in Valencia. He had a last taste of victory when his westwards advance in November 1812 forced Wellington to abandon the abortive Siege of Burgos and fall back on his Portuguese bases. In January 1813 Soult was recalled to France; a large collection of historic canvases, acquired locally, figured prominently in his personal baggage.

Soult was given command of the Imperial Guard in early May, and fought at Bautzen on 20–21 of that month. When news reached Napoleon of the defeat at Vittoria on 21 June Soult was dispatched to the south once again, and early in July became Commander-in-Chief of the Armies of Spain and the Pyrenees. With great energy and effectiveness he welded tattered, defeated veterans and raw recruits into a cohesive command, re-equipped them, reinspired them, and led them in a counter-offensive only three weeks after his arrival. The campaign in the Pyrenees achieved some local success, but was soon thrown back. Wellington led his Peninsular army into southern France, and though Soult handled his hopeless mission with some skill, the outcome was never in serious doubt.

Soult was treated with caution but respect by the Bourbons, and was Minister of War from December 1814 to March 1815. He rallied to his Emperor on his return from Elba, and was made Chief-of-Staff of the Army of the North – a post for which he was unsuited. Given time, a soldier of his great experience would doubtless have overcome the difficulties which faced him in this unfamiliar role; but in the Hundred Days there was no time. He would certainly have been employed to more effect in a field command. After Waterloo it was Soult who rallied the remnants of the army at Laon, before handing over to Grouchy and retiring to his home. After a brief period of disgrace and exile he was restored to his rank and honours, and was active for many years. He was Minister of War in 1830–4 and 1840–5; and President of the Council of Ministers in 1832–4, 1839–40, and 1840–7. He attended Queen Victoria's coronation in 1838 as an Ambassador Extraordinary, and was cordially received; the 'Duke of Damnation', as the redcoats called him, had ever earned the respect of his enemies. After many years as the most revered of France's old soldiers, Soult died at his *château* at Saint-Amans on 26 November 1851 – the last survivor of the creations of 1804, and forty-seven years a Marshal of France.

Steuben, Friedrich Wilhelm Augustus, Freiherr von (*1730–94*), *Major-General*

Baron von Steuben, while not a field commander of any great talent, was a training officer of genius; the impact of this Prussian professional's appointment as Inspector-General of the Continental Army was a significant factor in the American War of Independence.

He was born at Magdeburg, Prussia, on 17 September 1730, the son of an engineer officer. He entered the Prussian army in 1746 at the age of sixteen; during the Seven Years' War he saw much action under Frederick the Great, rising to the rank of captain and serving for a while on the general staff. On the close of hostilities in 1763 he was retired from the army, and in due course secured the appointment of Court Chamberlain to the Prince of Hohenzollern-Hechingen. In 1769 the neighbouring state of Baden-Durlach honoured him with a knighthood in the Order of Fidelity; at some point during the next eight years he acquired the title of *Freiherr* (Baron), but the details are obscure. In 1777 Steuben was obliged to leave the service of Hohenzollern-Hechingen under a cloud. It is not known precisely what he had done, but ugly rumours seem to have attached to his name and to have prevented his securing an appointment with the forces of the Honourable East India Company, France, Austria, or Baden. Eventually the American agents in France, Benjamin Franklin and Silas Deane, furnished him with a letter of introduction to George Washington (q.v.). This represented him (quite falsely) as a lieutenant-general in the Prussian service, fired with zeal for the revolutionary cause.

Steuben arrived in America at the beginning of December 1777. Throughout the hard winter of 1777–8 Washington and his little army of less than 6,000 were encamped at Valley Forge, suffering very severe privations. When the weather eased sufficiently for supplies to reach them, Steuben presented himself. Washington seems to have been impressed by his dubious credentials and by his personality. He was a rough, soldierly man, whose only English consisted of hair-raising profanities. He inspected the Valley Forge encampment, and the hungry, ragged, unsophisticated volunteers of the Continental Army. His report impressed Washington greatly, and he was encouraged to start up a training programme even before his official appointment as Inspector-General, which was made by Congress in May 1778. He proceeded to build the promising raw material into a professional army, from the ground up. He first formed a 'model company' which he trained and drilled in person; these men then provided cadres for other units, and his philosophy was spread throughout the army.

Now holding the rank of major-general, Steuben threw himself into his duties with energy and intelligence. He chose a form of standard Prussian army drill modified to suit local conditions and limitations; he had had some experience with *Freikorps* troops in the Seven Years' War, and was quite adaptable enough to appreciate that Valley Forge was not Potsdam. He demanded, and eventually achieved, a much higher standard of conduct among Continental officers. He set a personal example of interest in his men's welfare, and enforced his own thoroughly professional standards with a thunderous vocabulary of oaths. The proper care and practised employment of weapons – particularly the bayonet, a weak point among Continentals up to that time – were instilled at the same time as the necessary regularity of manœuvre. This one man must be given much of the credit for the Continental Army's later victories; the first which can be ascribed partly to his contribution was at Monmouth on 27 June 1778, but it would have been far more decisive had all the American field commanders present displayed a similar professionalism.

Although an organiser and instructor of very great skill, Steuben was deluded in his estimate of his own abilities in the field. He continually badgered Washington for command of a division; yet when he did receive an appointment, and held a command in Virginia in the spring of 1781, he was much less than dazzling. His talents were far more usefully employed in writing the 'Blue Book' – the manual *Regulations for the Order and Discipline of the Troops of the United States*, which remained standard until 1812. When the war finally came to a victorious end he was voted considerable cash grants by Congress, and large land grants by the state of New York. In the years which followed he lived a life of such riotous extravagance, however, that he fell into serious debt. He directed a stream of demands for further financial recognition to Congress, in the process presenting an entirely false picture of the 'sacrifices' he had made to travel to America in the first place. At last, in 1790, Congress voted him an annual pension of $2,500 for life. This enabled him to live in comfort in the farmhouse he had built near present-day Remsen, N.Y., until his death on 28 November 1794.

Stilwell, Joseph W. (*1883–1946*), *General*

'Vinegar Joe' Stilwell, America's Second World War commander in the China-Burma-India theatre, achieved considerable success despite unusual – perhaps unique – difficulties. Sweeping absolutes are unsafe in any branch of history, but Stilwell's claim to the title of the most frustrated Allied commander of the war is scarcely open to challenge. A vigorous and determined general, he was the first senior U.S. field commander to suffer from his country's policy of supporting inefficient and corrupt dictatorships against the common enemy. Despite his valuable contribution to Allied success in the Far East, however, it must be said that his personal prejudices did much harm to inter-Allied relations.

Born in Florida on 19 March 1883, Stilwell graduated from West Point in 1904 and was commissioned second lieutenant of infantry on 15 June of that year. He served as a junior company officer in the Philippines between 1904 and 1906 with the 12th Infantry. After four years as an instructor at the Military Academy, he returned to the islands in 1911 for another year of regimental duty. He spent 1913–17 on the staff of the Academy, and then went to France with the headquarters of the American Expeditionary Force. In 1918–19 he served as Assistant Chief of Staff to the 4th Corps. After the War he studied the Chinese language, first in California and later, between 1920 and 1923, in Peking. In the period 1924–6 he graduated from an advanced course at the Infantry School and from the Command and General Staff School. He served with American forces at Tientsin in the period 1926–9, and then returned to America to take up a post on the instructional staff at the Infantry School, Fort Benning. Between 1933 and 1939 he was again in China, this time as military attaché. He commanded the 3rd Infantry Brigade at Fort Sam Houston in 1939; the 7th Division at Fort Ord, California, in 1940–1; and the 3rd Army Corps in 1941–2. When Chiang Kai-shek's Chinese régime came into the war on the side of the Allies shortly after Pearl Harbor, the eminently well-qualified Stilwell was sent out to China as his Chief of Staff. He had achieved major-general's rank in October 1940, and was promoted lieutenant-general on his posting to China in February 1942. Almost immediately he found himself in command of the 5th and 6th Chinese Armies, striking down into northern and central Burma in an attempt to halt the rapid Japanese advance through that country.

Stilwell had faith in the potential of the Chinese peasant-soldier, but his task of organising the Nationalist Army into an effective fighting force – and of acting as a military go-between in the arrangement of U.S. support and material aid for Chiang Kai-shek – was extremely difficult and frustrating. The equipment and standard of training of the Chinese divisions left much to be desired; but more important, the Chinese leadership was riddled at every level with corruption, treachery, and internal intrigue. Stilwell attacked these problems with vigour, but found himself frustrated on every side by the inefficiency and active opposition of the Chinese authorities. Eventually he succeeded, in the sense that Chinese formations made a respectable showing in the field against the Japanese armies in Burma; but he never succeeded in 'cleaning up' the Chinese high command. He suffered from a lack of support from Washington in this aspect of his mission. His political masters were completely committed to the myth of Chiang's régime being a benevolent and patriotic administration wholeheartedly supporting Allied operations in the Far East, and to admit the validity of Stilwell's criticisms would have embarrassed them. He was forced to do the best he could with the means to hand, being obliged eventually to hold over Chiang's head the open threat of a cessation of American Lend-Lease support unless Chiang's forces in Yünnan played a more active part in the fighting.

In the spring of 1942 Stilwell's Chinese armies operated on the northern flank of the Japanese advance, but despite some local successes around Toungoo and Yenang-yuang he was unable to prevent ultimate Japanese conquest of the whole country. On 29 April 1942 the Japanese captured Lashio, southern terminus of the Burma Road, the vital logistic lifeline to China. (One of the most impressive Allied achievements of later months was the construction of the Ledo Road, a new artery to by-pass the section of the original route now in Japanese hands.) This capture represented a decisive turning of the whole Allied northern flank in Burma; like the British and Commonwealth forces in the area, Stilwell's Chinese were forced to fall back, some to the Indian border and some to Yünnan. Stilwell himself marched back to Imphal with a small party, suffering considerable hardship in the mountainous jungle. For the rest of 1942 and most of 1943

Stilwell's tasks, while onerous, did not involve any major operations against the enemy. In May 1943 the Allied command structure in the C.B.I. theatre was completely overhauled, and Stilwell found himself holding simultaneously a number of different and occasionally conflicting posts. He was deputy to Mountbatten, the Supreme Allied Commander in South-East Asia. He was still Chiang Kai-shek's Chief of Staff. He was directly responsible to the U.S. Chiefs of Staff for operations in support of China; and he maintained personal command of the Chinese and American ground forces in Assam, and of the air forces committed to the logistic support of China. Apart from condemning him to an enormous burden of work, this arrangement was unhappy in that it led to conflicting priorities and interests – a dilemma he never solved successfully.

Early in 1944, while Slim (q.v.) was fighting for Imphal and Kohima, Stilwell's Northern Combat Area Command, comprising three Chinese divisions and the small American force known as 'Merrill's Marauders', made a stubbornly contested advance on Mogaung and Myitkyina. The Myitkyina airfield was captured on 17 May 1944, but the town itself did not fall until August. For the last part of this campaign Stilwell took under command the remains of the British and Indian 'Chindit' brigades (see under Wingate), who were already exhausted by extended operations behind enemy lines. The writings of both British and American officers who served in the campaign strongly suggest that on occasion Stilwell displayed an insensitivity towards front-line troops – of both nationalities – which made him and his staff intensely unpopular. There can be no denying that Stilwell displayed a savage dislike and distrust of the British, which did no credit to him and no service to Allied arms. It should be remembered in his defence that this was the period when his relations with Chiang Kai-shek and his clique reached an all-time low, and that he was facing an impossible dilemma. Chiang refused to give him the unrestricted control of Chinese and American forces in the area which the situation demanded; and when Stilwell persisted in his attitude, Chiang manœuvred to have him recalled to the United States. This intrigue was successful; in October 1944 Stilwell, three months a full general, was recalled. His bitterness over this shabby treatment was not lessened by the political 'gag' which was applied as soon as he arrived in America, despite a campaign in his favour by certain sections of the Press. The appalling strain of his thirty-two months with the Chinese, together with his naturally combative and volatile character, had also damaged his health. He was given command of the U.S. 10th Army in the Pacific in June 1945, but on 12 October 1946 he died.

Stirling, David (*b. 1915*), *Lieutenant-Colonel*

Stirling was the creator of the British Special Air Service, a small but highly successful commando-type unit which played havoc with the German supply lines in North Africa in 1941–3. The S.A.S. was by no means the only such unit to gain fame during the Second World War, but it proved to have a wider significance in the years after 1945, and this justifies Stirling's inclusion in this book.

David Stirling was a Scots Guards officer who transferred to the Commandos in 1940, and went to the Middle East with 'Layforce' in 1941. In July of that year he submitted a plan to Generals Ritchie and Auchinleck (q.v.) for the formation of a special force to raid enemy airfields far behind the German lines; the long supply lines from Italy were the *Afrika Korp*'s Achilles' heel, and the destruction of transport aircraft would be just as damaging to the Axis war effort as that of combat aircraft. The formation of a team of six officers and sixty men was approved; Stirling hand-picked his small command, selecting men of initiative, resourcefulness, and endurance – in short, the 'rugged individualists' so invaluable to commando forces but so irritating to more conventional units. Initially the plan was to drop by parachute on the targets, but an early operation of this type proved disastrous, and it was decided to travel in wide sweeps through the empty southern desert in small motorised groups. Merely to survive these journeys was a considerable feat, as all necessities had to be loaded on to the jeeps and trucks, and there was no chance of support or rescue in the case of unexpected setbacks. During two weeks in December 1941, operating from Jalo, the S.A.S. destroyed ninety enemy transport and other aircraft. This success was recognised by Stirling's promotion to lieutenant-colonel, and authorisation to enlarge the unit. The Special Air Service Regiment (the name was meaningless, and picked to confuse German intelligence) was divided into 'squadrons'; each operated independently, and the favoured equipment came to be the American Willys jeep with a heavy armament of machine-guns and many improvised features. Normally the teams penetrated airfields under cover of darkness, and placed specially-developed incendiary bombs in all aircraft. When the bombs began to go off, the attackers would start up their jeeps and race down the runway spraying machine-gun fire at anything which moved. By the time Stirling was captured, on 10 January 1943, his unit had destroyed at least 250 enemy aircraft and many convoys, petrol-dumps, trains, and other targets of opportunity.

At first imprisoned in the Italian camp at Gavi, Stirling made no fewer than four escape attempts, but was recaptured each time; his height of six and a half feet made him conspicuous. He was finally sent to the top-security camp at Colditz Castle, where he remained for the rest of the war. The S.A.S. continued to serve until the end of the desert campaign; it was then enlarged and reorganised, and saw action in Italy and north-west Europe in the last two years of the war. Squadrons of selected Commonwealth and Free European personnel were added; in the European theatre the usual operations consisted of deep penetration behind enemy lines in advance of the Allied invasion forces, intelligence gathering, liaison with partisan groups, and general disruption of enemy communications immediately behind the front lines. After the war the unit's descendant, the 22nd Special Air Service Regiment, was eventually authorised as a permanent formation; it is still part of the British Army establishment. The 22nd S.A.S. provides a pool of unconventional warfare specialists, on temporary secondment from their units, and has pioneered many highly-sophisticated techniques since imitated by other nations, notably the United States – the U.S. Special Forces originally owed much to the S.A.S., although they later developed their own philosophy and techniques. The role of the S.A.S. is still classified, and personnel are believed to carry out missions at the 'cloak and dagger' end of the Army's activities. Officers and men have been posted abroad to train and assist the forces of friendly governments in conflicts to which Britain is not officially a party, and the unit is thought to play an important part in certain counter-insurgency activities.

Stuart, James Ewell Brown (*1833–64*), *Major-General*

The only possible one-word description of 'Jeb' Stuart is 'cavalier', and indeed there are several parallels between his place in the history of the American Civil War and that of Prince Rupert of the Rhine (q.v.) in the English Civil War. Young, gay, and flamboyant, Stuart led the Confederate cavalry with audacious skill; but while his operations were dramatic, their contribution to the fortunes of the Confederacy was not, in fact, very great.

Stuart was born on 6 February 1833 on the Laurel Hill plantation in Patrick County, Virginia. The seventh of the ten children of Archibald Stuart, a man active in local affairs, he was initially educated at home and at Wytheville, Virginia, and attended Emory and Henry College between 1848 and 1850. He entered West Point in July 1850, and graduated thirteenth out of forty-six in the class of 1854. In October of that year he was commissioned second-lieutenant in the U.S. Mounted Rifles, and joined his Regiment in Texas in December. The following March he transferred to the 1st Cavalry in Kansas; he was promoted first-lieutenant in December 1855, and spent the next five years on the frontier, showing an aptitude for outpost work. In 1859 he visited the east, strangely enough in order to try to sell the War Department a patent device he had invented for attaching sabres to belts; and during this visit he met once again his old superintendent from West Point, Colonel Robert E. Lee (q.v.), and became involved with Southern activists. In March 1861 he took two months' leave; when he heard of

Virginia's secession from the Union he made for home and mailed his resignation from the U.S. Army. In May he was appointed captain of cavalry in the Confederate Army – the rank he had achieved in the regular service the previous month – but with the further organisation of the Southern forces he quickly rose to colonel of the 1st Virginia Cavalry.

The general impression at the outbreak of the Civil War was that the cavalry was the one branch in which the Confederacy was superior to the Union. Many more Southern than Northern soldiers were accustomed to riding and handling weapons from childhood up, and they tended to be more familiar with the difficult terrain over which the fighting swept. In fact the Confederate cavalry, dashing though they certainly were, suffered from strict limitations. They were highly adept at deep penetration raids and at scourging the Union lines of supply and communication, and proved themselves most able intelligence-gatherers; but they were neither trained nor equipped for classic, European-style cavalry fighting. They never charged formed bodies of infantry, or seriously harried retreating enemy troops. In cavalry versus cavalry fighting they very seldom charged home with the sabre, preferring to stand off and exchange carbine and pistol fire. Handicapped by the shortage of carbines, they were frequently reduced to employing sawn-off shotguns; while effective at close range, these hardly compared with the repeating breech-loading carbines which were issued to Union cavalry as the War progressed.

At the first Battle of Bull Run (Manassas) on 21 July 1861 Stuart's cavalry covered the Confederate left and made a well-timed charge. In September 1861 he was promoted brigadier-general; by the end of the year he was leading 2,400 men, and he worked effectively to bring their efficiency to a peak. During the Peninsular campaign of June 1862 he led a cavalry division in a wide sweep right around the Union army, creating havoc in McClellan's rear and bringing Lee useful intelligence on Union dispositions. After fighting in the Seven Days' Battle Stuart was promoted major-general. At the Second Battle of Bull Run and Antietam he again distinguished himself in command of the entire Confederate cavalry, and in October 1862 he raided as far as Chambersburg, Pennsylvania, riding right around the Union army once more and returning with 1,200 captured mounts. He fought at Fredericksburg, supporting Jackson (q.v.) with flanking fire from his horse artillery; and after Jackson was wounded at Chancellorsville Stuart temporarily commanded his corps. Stuart was held in warm affection by Lee, and was fiercely loyal to his Commander-in-Chief. He was extremely popular with his own men and throughout the army, although generally the Confederate cavalry was not highly regarded by the other branches. He had a weakness for ostentatious display, affecting a red-lined cloak and a peacock plume in his hat; and his camp was famous for its pleasant high spirits.

The largest pure cavalry battle of the war occurred at Brandy Station on 9 June 1863, when Stuart's lines were attacked by 12,000 Union cavalry under General Pleasanton. Stuart's 10,000 troopers fought the Union regiments to a standstill, suffering 500 casualties to Pleasanton's 900. Shortly afterwards Stuart led his cavalry on another long ride around the enemy army, not returning until 2 July, the second day of the Battle of Gettysburg. He was much criticised for this, as it was held that his command would have been of far greater value had it acted as Lee's eyes during the build-up of Union strength. During the winter of 1863–4 Stuart's hard-riding cavalry fought several actions. In the spring of 1864 Grant (q.v.) launched the Army of the Potomac into the Wilderness to seek out the Army of Northern Virginia and chop its right wing loose from Richmond. On 9 May 1864, during the bloody fighting between Lee and Grant at Spotsylvania, Union General Philip Sheridan led his 10,000-strong cavalry corps southwards in a massive raid towards Richmond. Stuart gathered some 4,500 sabres and managed to get between Sheridan and the Confederate capital, at Yellow Tavern on the northern outskirts of the Richmond fortifications. In the pitched battle which ensued on 11 May the Confederates were driven from the field leaving about 1,000 casualties, including 'Jeb' Stuart himself. Wounded by a dismounted Union trooper, he died the following day in

Richmond; he was buried in the Hollywood Cemetery in the city. Lee mourned him as a son, and for all his faults there is no doubt that the dashing young general epitomised the romantic bravery of the best of southern youth.

Subatai Ba'adur (*fl. 1190–1242*)

This greatest of all Genghiz Khan's (q.v.) generals was only a boy when he joined Temujin's band in about 1190. Such scanty details of his life as are known are taken from *The Secret History of Genghiz Khan*, a transcription of a mass of oral traditions compiled not long after the Khan's death, while many who knew him still lived. Subatai (whose name is also variously given as Subedei, Sabatai, and Subodai) was apparently the son of a blacksmith of the Uriangqadai clan. With his brother Jelmei he rose quickly in Temujin's service, and when the latter adopted the title Genghiz Khan in 1194 both were named as captains of bands. With Jebei Noyon and Khubilai, they formed an *élite* group known as Genghiz Khan's Four Hounds. About the only other personal detail worth mentioning is that quite early in life Subatai became very stout, and seems to have travelled with the armies in a light conveyance of some sort, out of consideration for the small Mongol ponies. It does not seem to have handicapped his military career noticeably.

Subatai's first independent command mentioned in *The Secret History* was in 1205–6, when he successfully pursued and killed Kutu and Chila'un, sons of the defeated Merkit leader Tokhto'a. Subatai led a *touman* in the Hsia and Chin wars (see under Genghiz Khan for all general background) of 1207–15. It is presumed that he marched with Jebei's southern column in the invasion of Khwarezm, because after the victorious Bokhara–Samarkand campaign they jointly led a force of three *toumans* in relentless pursuit of the Shah. When he died near the coast of the Caspian in February 1221 they received permission for an armed reconnaissance in force, and were strengthened to a force of about 40,000. With Jebei in command, they embarked on an extraordinary campaign in southern Russia. Passing through Azerbaijan they wintered in eastern Armenia, then swung north and pushed through the Caucasus in the spring. They defeated a large army of Georgians who had assembled for the Fifth Crusade, and advanced into southern Russia, brushing aside local resistance. Despite a spirited defence of the foothills east of the Kuban river by the Cuman peoples, the Mongols soon outmanœuvred them and captured Astrakhan. Pushing on across the Don they penetrated into the Crimea, took the Genoese coastal fortress of Sudak, and then swung north again into the Ukraine. In 1222–3 the Mongols wintered by the Black Sea, and in spring sent out spies all over eastern and southern Europe. In 1223 they decided to turn for home, but were attacked by 80,000 men under Prince Mstislav of Kiev near the mouth of the Dnieper, in the Battle of the Kalka River. The Mongols almost annihilated the enemy, and raided for hundreds of miles north before turning for home once more, summoned by the Khan's couriers. As they marched eastwards, north of the Caspian, Jebei became ill and died. Subatai successfully led the army back to a rendezvous with the main Mongol forces, now returning from Khwarezm. The Russian expedition had covered some 4,000 miles in under three years.

Subatai presumably served in the 1225–7 campaign against the Hsia, which continued after the Khan's death under the direction of his son and successor Ogatai. In 1231, in accordance with Genghiz Khan's deathbed instructions, Ogatai's younger brother Tuli led a great force of Mongols south into Sung territory, by prior arrangement with the Sung, and then swung east from Szechwan into Hanchung and Chin territory. In 1232 Tuli died, and Subatai took command. He besieged the city of Kaifeng, the Chin centre; after a year's determined defence, including the use of explosives by the Chin against the Mongols, the city fell by storm in 1233. Subatai then mopped up the remains of the Chin Empire. Ogatai refused to divide the conquests with the Sung, who attempted to seize the former Chin province of Honan and thus brought down on their heads war with the Mongols. The result was a protracted series of campaigns which was only finally brought to a successful conclusion in 1279 by

Ogatai's nephews Mangu and Kublai. (The capital of Hangchow eventually fell in 1276 to Subatai's grandson, Bayan.)

In 1237 Batu, son of Juchi, was sent into eastern Europe with 150,000 men; but as Ogatai no doubt intended, the veteran Subatai was the actual commander in the field. Crossing the Volga westwards in December 1237, Subatai advanced at great speed through Moscow and Kaluga and completely destroyed the principalities of northern Russia in a winter campaign prepared beforehand with great thoroughness. With the spring thaw Subatai turned south and rested his troops in the Don steppes; during the next two years he consolidated Mongol domination of eastern and southern Russia, and gathered minutely detailed intelligence about the state of Europe – while Europe remained in almost total ignorance of the nature of the threat hanging over it. In November 1240 the 150,000 Mongol riders were on the move again, across the frozen rivers of the Ukraine and westwards. The Prince of Kiev rejected a surrender demand, and the city was duly stormed, captured, and destroyed on 6 December 1240. Subatai overran all the territory south-east of the Carpathian Mountains and north-east of the Black Sea, and left about three *toumans* to hold down the conquered region and protect his lines of communication while he advanced into Europe with some 120,000 men. The army was divided into four columns: the first, under Kaidu, was on the north flank; Batu and Subatai led the central columns; and Kadan had the south flank, each consisted of three *toumans*, or 30,000 riders. They faced Bohemia, Hungary, Poland, and Silesia, each of which could raise forces larger than the whole Mongol army – but Subatai was confident from his spies' reports that the enemy would remain preoccupied with European internal tensions long enough for the Mongols to defeat each in turn. He was justified in his confidence.

Kaidu's force advanced first, smashing through Poland and Silesia, Lithuania and East Prussia into Pomerania. He defeated three separate Polish armies in the process, each larger than his own, including that of King Boleslav V at Cracow on 3 March 1241; and sent whole populations of terror-stricken refugees fleeing westwards with tales which exaggerated his strength ten-fold. Prince Henry the Pious of Silesia took 40,000 German, Polish, and Teutonic knights to Liegnitz, in Kaidu's path, and awaited the 50,000 reinforcements of Wenceslas of Bohemia, marching north to join him. Kaidu struck at Liegnitz before the junction could be effected, and smashed Henry's army. The survivors fled west, and Wenceslas withdrew north-west; but rather than follow them, Kaidu turned south to join up with the other columns, burning the heart out of Moravia as he rode.

Meanwhile Kadan, in the south, had triumphed in three pitched battles, and by 11 April Transylvania was prostrate under the Mongol hoofs. Passing between the Danube and the Carpathians, Kadan swung north to rendezvous with Subatai on the Hungarian plains. The two central columns had broken through Hungarian defences in the Carpathians, and driven south-west to the banks of the Danube. King Bela IV of Hungary coolly gathered forces totalling some 100,000 men and advanced eastwards from Pest. Subatai fell back for 100 miles to the river Sajo, where on 11 April he destroyed the Hungarian army by sheer, brilliant generalship. A holding attack before the Hungarian camp set the Christian troops up for a surprise rear and flank attack by three divisions which had secretly circled the Hungarian position, led by Subatai in person. The Hungarians were penned in their camp and bomparded with stones, arrows, and burning naphtha; more and more of them took an apparently overlooked escape route out of the pocket to the west, and when they were completely committed to it discovered that it was a Mongol trap. Between 40,000 and 70,000 perished. Subatai was the master of Europe from the Baltic to the Danube, and from the Dnieper to the Oder.

During the summer of 1241 he tightened his grip on the conquered territories and prepared to invade Italy, Austria, and Germany in 1242. Despite frantic efforts, little in the way of an effective European defence strategy was planned, and it was the merest chance which saved western Europe from the Mongol *toumans*. The Mongols started west in late December 1241, crossing the Alps into

Northern Italy and approaching Vienna up the Danube Valley. Early in 1242 couriers reached Subatai with news of the death of Ogatai; and it was rigidly laid down in Genghiz Khan's law code that all offspring of his house must immediately return from wherever they might be to the Mongolian capital to elect a successor, in the event of the ruler's death. Reluctantly Subatai and the three princes turned back from helpless Europe, and marched eastwards on the first leg of the 6,000-mile journey to Karakorum. They destroyed Serbia and Bulgaria on their way, and vanished into the Russian plains. With them, Subatai marched out of history.

The Mongol Empire was divided after Ogatai's death into separate Khanates, whose outward urge was mainly directed east and south. Although the Mongol threat to Europe was short-lived it was very real, and for his handling of the 1241–2 campaign Subatai deserves a place in any list of the great generals of history. If one considers his career as a whole, then for sheer breadth of experience and consistency of success he deserves a place very near the top of that list.

Suchet, Louis-Gabriel, Duc d'Albufera (*1770–1826*), *Marshal of France*

One of Napoleon's (q.v.) ablest commanders, Suchet ranks with Masséna, Davout, and Soult (qq.v.). He was a man of impressive integrity, and his career is distinguished from those of his brother marshals by one fact in particular – he was the only one to emerge from the Peninsular War with his reputation intact. In this he was probably fortunate in that he served almost exclusively in the east, against Spanish rather than Anglo-Portuguese opposition.

Louis-Gabriel Suchet was the son of a Lyons silk manufacturer, and was born on 2 March 1770. At the age of twenty-one he joined the cavalry unit of his local National Guard, serving with the rank of *sous-lieutenant*. From May 1792 he served as a soldier in the Independent Company of the Ardèche, but he soon rose to the rank of captain. After service at the Siege of Lyons he was elected lieutenant-colonel of the 4th Volunteers of the Ardèche in September 1793, and fought in the Siege of Toulon at the end of the year. During 1795–7 Suchet saw hard campaigning with the Army of Italy. He fought at Loano, Dego, Lodi, Castiglione, Cerea, San Giorgio, Arcola, Rivoli, and Neumarkt, and was wounded in action. He was then promoted provisional *chef de brigade* of the *18e de Ligne*, and in 1798 served briefly as Brune's chief of staff in Switzerland, being promoted *général de brigade* in March of that year. He was nominated for the Egyptian expedition, but was delayed in France by an inquiry, and did not serve in that theatre. In late 1798 and in 1799 he served successively as chief of staff to the Armies of Italy, Mayence, and the Danube, being promoted *général de division* and seeing action again at Novi in 1799. In January 1800 he was given command of the left wing of the Army of Italy, fighting under Masséna. He became separated from the main army in April and May and was roughly handled, but emerged with final credit. For the remainder of 1800 he led a corps of two divisions, first under Masséna and subsequently under Brune, and in January 1801 he was made Inspector-General of Infantry.

General Suchet commanded a division in the 1805 campaign. He was at Ulm, Hollabrunn, and Austerlitz, and was awarded the Grand Eagle of the *Légion d'Honneur* in February 1806. In 1806 and 1807 he fought at Saalfeld, Jena, Pultusk, and Ostrolenka, and in August 1807 he provisionally took over command of the 5th Corps from Masséna.

From 1808 to 1814 Suchet served in Spain mostly in Catalonia; he was initially a divisional and later a corps commander. In command of the Army of Aragon he won victories over the Spanish at Maria and Belchite, after covering the Siege of Saragossa. His attempt on Valencia in March 1810 failed, but a series of successful assaults culminated in the capture of Tarragona in June 1811, for which victory he was raised to the marshalate the following month. He twice defeated Joachim Blake during October 1811, being wounded in action at Sagunto on the 25 October, and went on to conduct a successful siege of Valencia, which finally capitulated in January 1812. In that month he was created Duke of Albufera. After a setback at Castalla in April 1813 at the hands of Sir John Murray, Suchet succeeded in relieving the French garrison in Tarragona in June. In November 1813 he was named Governor of Catalonia – a post in which he showed real administrative ability – and in the same month succeeded Bessières as Colonel-General of the Imperial Guard. He won another victory over the Spanish at Molins del Rey in January 1814. Although the French grip was still firm in eastern Spain, largely thanks to Suchet, the rout of French forces elsewhere in the country forced him to evacuate Catalonia in April. In contrast to the other French armies in the Peninsula, his command withdrew in first-class order. At the end of April he was named Commander-in-Chief of the Army of the South.

Suchet continued to serve after Napoleon's first abdication, and was decorated by the Bourbons. With Ney, Soult, Davout, and Grouchy he rallied to his Emperor during the Hundred Days in 1815. Surprisingly, in view of his proven ability and of the shortage of senior commanders of the first quality, Napoleon did not give him a command with the Army of the North, but instead the independent command of the small Army of the Alps. He fought a brief campaign against the Austrians in Savoy and southern France in June and July. Suchet was briefly in disgrace after the second restoration of Louis XVIII, but was restored to the peerage in 1819. He then lived quietly in retirement until his death near Marseille on 3 January 1826.

Suleiman I, called Kanuni (*1494–1566*), *Ottoman Sultan of Turkey*

By Muslims Suleiman was nicknamed 'the Law-giver', and by Christians 'the Magnificent': what is indisputable is that he presided over the greatest years of the Ottoman Empire, and was the most powerful ruler of the sixteenth century. Suleiman was the son of the Sultan Selim I (q.v.); he began his training in the ways of power during the reign of his grandfather Bayazid II, serving as Sanjak Bey of Kaffa – formerly Feodosiya, the Genoese trading centre in the Crimea which Mohammed II (q.v.) had taken. In 1512 his father Selim forced Bayazid to abdicate and seized the throne. He ruled for just seven years, but in that time he nearly doubled the size of the Empire by a series of victories over the Persians and Mamelukes. Europe was in its usual state of internal chaos and self-seeking intrigue, and the consequences would have been disastrous if Selim had turned his eyes west rather than east and south. In 1520 he died, and Suleiman succeeded. He initially left domestic and administrative concerns largely in the hands of his vizier Ibrahim Pasha, so allowing himself to concentrate almost entirely on military matters for the first fifteen years of his reign.

In the spring of 1521 Suleiman marched into Hungary, and Belgrade fell to him in August. He next planned to remove the constant threat to his sea communications posed by the Christian strongholds of Crete, Cyprus, and particularly Rhodes. Rhodes had been turned into one of the strongest fortified positions in the world by the Knights of St John who, under their great leader Philippe Villiers de l'Isle Adam (q.v.), held out against some 100,000 Turkish troops from June to December 1522. (A fuller account of the siege will be found under Villiers de l'Isle Adam, but it may be mentioned here that the Knights numbered just 700, with 6,000 light auxiliaries; that the Turkish artillery and engineers were highly skilled.) The Knights eventually surrendered the fortress on good terms, some 180 knights and 1,500 other troops surviving; they had caused between 50,000 and 100,000 casualties to the Turks, whose army had been reinforced steadily until a total of about 200,000 had been engaged. Suleiman respected the terms of the negotiated evacuation. In 1522–6 his troops carried out a war of skirmish along the Hungarian frontier, while Suleiman concluded a treaty of neutrality with Poland to leave him free to concentrate on Hungary. In 1526 he left Constantinople with some 70,000 or 80,000 men; and late in August, after being delayed by the gallant garrison of Peterwardein, he reached the plain of Mohács.

He was faced on open, rolling ground by King Louis of Hungary with an army of about 12,000 cavalry and 13,000 infantry. The foot was roughly half Hungarian, half mercenary, with a good proportion of arquebusiers. It was drawn up in three large phalanxes, with some twenty cannon in front of the central phalanx; the left wing was covered by marshy ground, but the right was unprotected. The cavalry was drawn up behind the infantry in reserve, with strong detachments in the gaps between the infantry squares. On 29 August Suleiman attacked in three main ranks, the first two consisting of timariots (feudal light cavalry) and the third of the *élite* janissary infantry, with *élite* sipahi cavalry on their wings and cannon in front. Some 6,000 timariots were detached and sent in a broad sweep to the west, with orders to fall upon the Hungarian right when the armies were committed. The Hungarian heavy cavalry charged and drove in the first Turkish cavalry line. The rest of the Hungarian army advanced, but the guns could not be brought up at the same pace. The Hungarian cavalry were thrown into some confusion by the flank attack of the detached timariots, but weathered it, and advanced once more to smash the second Turkish line. The attack on the third and final line failed, however; the Turkish cannon took a heavy toll, and being chained side by side, formed an effective barrier. The janissaries and sipahis counter-attacked, and the exhausted Hungarian army broke. Turkish losses were very severe and there was no organised pursuit, but timariots harassed the fleeing survivors and in all some 10,000 foot and 5,000 horse were killed, including King Louis and the greater part of Hungary's chivalry. The few prisoners were beheaded.

After spending three days at Mohács resting and reorganising his army, Suleiman occupied Buda. He made John Zápoyla, Voivode of Transylvania, the vassal King of Hungary, and supported him when civil war broke out. In 1629 Suleiman moved against Ferdinand of Hapsburg, Archduke of Austria. His army of some 80,000 was aided by 6,000 Hungarians under Zápoyla, but despite a hard-fought campaign and a vigorous siege of Vienna Suleiman was forced to withdraw in the winter of 1529–30 without taking the city. The Austrians harassed the Turkish retreat with determination. In 1532 a second expedition against Vienna came to nothing after several months of rather aimless campaigning and a heroic Austrian defence of Güns. As Suleiman now wished to devote his energies to Persia, he finally concluded a peace with Ferdinand based on a recognised partition of Hungary between the Hapsburgs, Zápoyla, and the Turks.

In 1534–5 Suleiman fought a major campaign against Persia; there were victories and defeats on both sides, but although the Shah kept his forces in being he was forced on to the defensive and several of his important cities were destroyed. Suleiman took this opportunity to complete moves begun by his father in Mesopotamia. During the 1530s Suleiman's High Admiral Khair-ed-din Barbarossa, the one-time pirate who had risen to become the vassal Bey of Algiers, greatly furthered Turkish interests in the Mediterranean by a series of successful operations against the Holy Roman Empire and Venice. His victory at Preveza on 27 September 1538 gave the Turks the naval initiative in the Mediterranean for thirty years. Another campaign against Persia in 1548–9 was not crowned by any long-lasting territorial gains. The focus of events remained in the Mediterranean in the 1540s and 1550s; the Sultan's galleys terrorised the western Mediterranean and raided the coasts of southern Europe, and Turkish influence spread along the North African shore. Imperial garrisons of Spanish troops fought vigorously in a series of actions in what are now Algeria and Tunisia. In 1555 the final round of Suleiman's struggle with Persia was ended by the Treaty of Amasia, by which the Ottoman Empire acquired Erzerum, Erivan, Van, Tabriz, and Georgia. In 1565 Suleiman failed in his great attack on the Knights of St John in their new headquarters on Malta. With about 60,000 men in all, the Turks faced some 580 knights and 9,000 auxiliaries; the campaign was fought with great energy and skill on both sides, and by the time the Turks were forced to abandon the attempt by the arrival of a Spanish fleet, each side had lost about half its men (see under la Valette).

Intermittent warfare and truces had continued in Hungary since 1537; Zápoyla had died in 1540 and the following year Suleiman formally annexed Hungary into the Ottoman Empire. Invasion, counter-invasion, attack and defence followed with monotonous regularity and little concrete result until 1566, when the new Emperor Maximilian broke the latest truce – the Peace of Prague, signed in 1562. Despite his seventy-two years and his poor health Suleiman personally led an army of 100,000 men in yet another invasion of Austria, and in August the Turks invested the fortress of Szigetvar. The exceptionally gallant defence of this fortress by the garrison of Count Miklos Zrinyi ended on 8 September 1566, when the defenders perished to a man in a last hopeless sortie, but took hundreds of Turks with them by means of concealed time fuses in the magazine. This was the entirely fitting backdrop to the death of Suleiman the Magnificent – for although the fact was concealed from Turks and enemy alike, he had died on the night of 5–6 September.

An implacable conqueror who pushed the boundaries of Ottoman power both east and west, Suleiman was also a far-sighted and cultivated ruler who surrounded himself with the best artists, administrators, scholars, and lawyers in the Muslim world; while his armies and fleets roamed abroad, at the centre of the Empire he supervised a massive effort to bring the structure of government and administration up to the standards required by a rapidly-changing world. His later years were embittered by the unruliness and mutual enmity of his sons, of whom two – Mustafa and Bayazid – had to be executed in 1553 and 1561 respectively.

Suvarov, Alexander Vasilevich (*1730–1800*), *Marshal*

By contemporary accounts a nondescript and puny youth, Alexander Suvarov was born in 1730 and appears to have enlisted in the Russian Army at the age of fifteen, being granted a commission nine years later in April 1754. When the Russian army of 100,000 men under Marshal Stepan Apraksin invaded East Prussia in 1757 during the Seven Years' War, young Suvarov fought with distinction; but the advance was lethargic and it was not until 1759, when a new Russian army advanced under Count Soltikov, that real progress was made. Suvarov fought with this army in the great battle of Kumersdorf on 12 August 1759 (the worst defeat ever suffered by Frederick the Great (q.v.)) and a year later was with the victorious Russian forces when they entered Berlin on 9 October 1760.

During the next thirteen years Suvarov gained steady promotion and when, in the first war of Catherine the Great against Turkey (1768–74), a revolt of peasants and Cossacks broke out under Emelyan Pugachev in south-east Russia, Suvarov was ordered to crush it. If more detailed chronicles of his later life are any criterion, it is likely that his success in this task was accompanied by great cruelty and little regard for life.

Suvarov possessed great understanding of the Russian soldier, moulded under two centuries of despotic Tsars and the Tartar yoke. Unlike contemporary commanders, he set great store by the painstaking training of his men, his methods of training being exactly suited to their level of intelligence, their stamina and – generally speaking – their staunch loyalty. He had great faith in simple but forthright maxims, and his penchant for close combat was epitomised by one such: 'The bullet is a fool, the bayonet a fine fellow.' Always conscious of the benefit of high morale in his command, he maintained an air of ease and understanding with his subordinate officers and other ranks alike. He was, however, ruthless in attaining his objective in battle and uncompromising even in great danger, and he never shrank from spending the lives of his men freely.

The saga of Poland during the period 1793–5 is an unhappy one. That ancient nation suffered repeated invasion from east and west, and consequent partition by Prussia and Russia. Following an uprising by a peasant force under Tadeusz Kosciuszko (q.v.) which defeated a Russian garrison force at the Battle of Raclawice on 3 April 1794, two Russian armies under Generals Fersen and Suvarov took the field and moved against Warsaw. 7,000 Poles were defeated by 16,000 Russians at the Battle of Maciejowice (10 October 1794) and Suvarov captured the capital – for which action he was promoted field-marshal by Catherine the Great.

Two years later Catherine died without expressing her wishes regarding the succession to the Russian throne, with the result that the unstable Tsar Paul I succeeded. Suspicious of anyone who had won the favour of his mother, Paul immediately retired most of his army commanders, and among them went Suvarov, exiled to the country.

When, however, Paul organised the Second Coalition in 1798, leaguing Russia and England against Napoleon (who was campaigning in Egypt), Suvarov was, at the age of sixty-eight, recalled and placed in command of an Allied army of Russians and Austrians raised to expel the French under General Barthélemy Schérer from Italy. Hoping to defeat the Austrians under General Paul Kray before the arrival in Italy of Suvarov's Russians, Schérer attacked Kray on 5 April 1799, but suffered a major defeat in the battle of Magnano. Suvarov then arrived, took command of both armies (totalling 90,000 men), and drove the French before him, defeating them at the battle of Cassano on 27 April, entering Milan in triumph on the following day, and capturing Turin.

There followed a near-disastrous dissension between Suvarov and the Austrian Government which resulted in much of his army being scattered around Alessandria, with a French army to his north under General Jean Moreau and another, of 35,000 French under General Jacques Macdonald, coming up from the south. Realising his danger, Suvarov quickly assembled 25,000 of his men and, attacking Macdonald in the Battle of the Trebbia (17–19 June), inflicted a partial defeat on the French. Macdonald managed to join up with Moreau, but under constant pressure from Suvarov the whole French Army of Italy was relentlessly forced back into the Apennines. Moreau was relieved of his command and his place taken by General Barthélemy Joubert. Joubert, with 35,000 men, attacked Suvarov's army on 15 August, but once again at the battle of Novi the French were resoundingly beaten, Joubert being killed.

During the summer of 1799 matters had been progressing less favourably for the Allies in Switzerland, and shortly after the battle of Novi the French Army of the Alps, 30,000 strong, marched south under Championnet. Undaunted by his new threat the old Russian promptly turned north to meet it, but before he could engage he was ordered to turn away and take his 20,000 Russians into Switzerland to assist another Russian army, under General Alexandr Korsakov. This was heavily defeated by the great French Marshal André Masséna (q.v.) in the third Battle of Zürich (25 September). Too late to avert the disaster, despite a gallant fight through the St Gotthard pass, Suvarov was promptly relieved of his command by the petulant Tsar Paul. He died a few months later.

Thus ended the extraordinary career of a singular commander. The Russian campaign in Italy under the doughty Suvarov had been the one really bright aspect of an otherwise motley display by Allied armies in the European war against the French. The other campaigns were utterly lacking in enterprise; time and again there was no pursuit of advantages – advantages that would not be presented again for a further fourteen years. Suvarov, on the other hand, had pressed forward relentlessly on every occasion, with the result that the French were driven from Italy. His energy and force of character were all the more remarkable for having survived unblunted the vicissitudes of so many years' service under the Tsars. In many ways a crude and unsophisticated man his style was perfectly suited to the men he led. He is revered in the U.S.S.R. to this day.

Tamerlane, *see* Timur-i-Leng

Taylor, Zachary (*1784–1850*), *Major-General, 12th President of the United States of America*

After Winfield Scott (q.v.), Taylor was America's outstanding general of the first half of the nineteenth century. He was born at Montebello, Virginia, on 24 November 1784. His father, Richard Taylor, had commanded a Virginia regiment in the Revolutionary War. The family moved to Kentucky while Zachary was a child, and he was privately – and rather sketchily – tutored. He served briefly as a volunteer in the Kentucky militia in 1806, and in 1808 was commissioned first-lieutenant in the 7th United States Infantry. He was promoted captain in 1810, and in 1811 commanded Fort Knox, Indiana Territory. With his company of some fifty men he successfully defended Fort Harrison, Indiana, against attacks by some 400 Indians on 4–5 September 1812, for which action he was breveted major. He fought on the frontier throughout the War of 1812, and did not see action against the British.

He was retained with the rank of captain when the army was almost completely disbanded in June 1815, and the following year joined the 3rd Infantry in Wisconsin as a major. He commanded Fort Winnebago for two years, and in 1819 became lieutenant-colonel of the 4th Infantry at New Orleans. The next twenty years saw many commands, many forts, many new regiments, and much experience on the frontier and in staff appointments – but little promotion. It was not until 1832 that Taylor was promoted colonel and given command of the 1st Infantry at Fort Crawford. He served in the Black Hawk War of 1832–3; and in July 1837, while leading his regiment down the Mississippi to Fort Jessup, he received orders to hasten to Florida to take command of a field force moving against the Seminoles. He set out after the Indians from Fort Gardner in December, followed them to the fearful Everglades with 1,000 men, and brought them to battle at Lake Okeechobee on Christmas Day 1837. This victory brought Taylor promotion to brigadier-general, and in May 1838 he was given the departmental command. For two years he continued the war against the Seminoles under the harshest conditions, but in 1840 he admitted his inability to bring the fighting to a satisfactory conclusion and resigned his departmental command. He was given a new command on the south-west frontier with headquarters at Fort Smith, Arkansas, and successfully kept the frontier tribes quiet for three years.

It was expected that the announcement of the American annexation of the independent republic of Texas (at Texan request) would spark off a long-threatened war with Mexico; so in March 1846 Taylor, with a force of 3,500 men who represented two-thirds of the entire regular army, was ordered to march from Corpus Christi to the Rio Grande. He established a position opposite Matamoros, where 5,700 Mexican troops were assembled. After several incidents the Mexican General Arista crossed the Rio Grande in force. On 8 May Taylor's 2,200 men beat Arista's 4,500 at Palo Alto in a classic Napoleonic-type battle, fought out on open ground with flintlock musket, sabre, lance and muzzle-loading cannon. The victory was a tribute to the American 'flying artillery' and to the steadiness of American infantry. Taylor followed up his advantage with an attack on the enemy position at Resaca de la Palma next day, routing the Mexicans and throwing them back over the Rio Grande. American losses in the two actions totalled about 170, against some 1,100 Mexican casualties. On 13 May 1846 America formally declared war on Mexico.

Taylor received reinforcements, mostly untrained volunteers, but little logistic support or imaginative direction from Washington. Like Scott, he did not enjoy good relations with President Polk, and it was three months before he could carry the war to the enemy. He crossed the Rio Grande with 6,000 men in August, against Polk's wishes. In three days of fierce street fighting Taylor ejected the Mexican General de Ampudia and his 10,000 men from Monterey, resistance ceasing on 24 September 1846. After an armistice Taylor advanced and occupied Saltillo, where he was joined by 3,000 men under Brigadier-General Wool. Polk now debated the next move at length, seeking to further political rather than military ends. To prevent his enemy Scott from gaining credit from the war with his sensible plan for a landing at Vera Cruz, Polk suggested a march south over the desert to Mexico City by Taylor. Taylor rejected this unsound proposal and backed the Scott plan, which was grudgingly adopted. Its adoption cost Taylor all but 5,000 of his men, drawn off by Scott for his command – and those that remained were not the best 5,000. The Mexican President, Santa Ana, then attempted to crush the weakened Taylor before Scott could assemble his force and land in the south. With 20,000 men he marched north through the desert from San Luis Potosi; the journey cost him 4,000 casualties but on 22 February 1847 he came up on Taylor at Buena Vista. The American defensive position in a pass in the hills was imperilled early in the action by the unreliable conduct of certain volunteer units, but the situation was saved by the regular artillery and the Mississippi Volunteers under Jefferson Davis. Santa Ana withdrew leaving 1,500 casualties – almost exactly twice the American figure – and the northern campaign was over.

Taylor was acclaimed as a national hero, and ran for president as the Whig nominee in 1848. He defeated the Democrat, Lewis Cass, and was inaugurated on 5 March 1849. He died suddenly in Washington on 9 July 1850. A stocky, muscular man with an iron constitution and much common sense, Taylor was known as 'Old Rough and Ready', contrasting with Winfield Scott's nickname, 'Old Fuss and Feathers'. Taylor emphatically did not share Scott's belief in the importance of correct dress and military punctilio; he was essentially a frontier general, with little formal education but a great feel for ground and for fighting at odds, learnt the hard way.

Tilly, Johann Tserclaes, Graf von (*1559–1632*)

Tilly was the most prominent commander of Catholic League forces in the Thirty Years' War (see under Wallenstein), the organiser and leader of the army of the Elector Maximilian I of Bavaria.

He was born in February 1559 at Tilly in Brabant, the son of Martin Tserclaes, Lord of Tilly. In 1568 the Duke of Alva's (q.v.) 'Council of Blood' was engaged in crushing the political and religious aspirations of Spain's subjects in the Netherlands with great ruthlessness, and two moderate leaders of the local aristocracy, Egmont and Hoorn, were executed; Martin Tserclaes was an associate of Egmont, and in that year he was exiled from the Spanish Netherlands. His sons Johann and Jakob were brought up in Jesuit institutions at Liège and Cologne, a process which seems to have ensured that the young boys grew up with politically acceptable attitudes. At all events, the family was allowed to make its peace with the régime in 1574. Johann joined a Walloon regiment, and in 1583–4 fought under Alessandro Farnese (q.v.) during that general's great siege of Antwerp, and later acknowledged Farnese's important influence over him. Tilly continued to serve under Farnese in the French religious wars which distracted him from his task in the Netherlands in the late 1580s and early 1590s. In 1594 Tilly travelled east to fight in the army of the Holy Roman Emperor Rudolf II against the Turks. In 1609 Rudolf aroused great enmity among the Catholic magnates of the Empire by granting guarantees of freedom of worship to his Bohemian subjects. Tilly remained loyal to the Emperor, and thus earned the hostility of Rudolf's successor Matthias, who toppled his predecessor in 1611. Tilly then took service with Maximilian of Bavaria; for the next ten years he worked to create the efficient Bavarian army which became the backbone of the forces of the Catholic League.

In 1620 war broke out between the new Emperor Ferdinand II and his Bohemian subjects, aided by a rising among the Austrian nobles. Although hampered by the indecisive Bucquoy, commander of the imperial forces, Tilly achieved great success in this first major campaign of the Thirty Years' War. He led the 25,000-strong forces of the Catholic League across the Bavarian frontier in July, and in little more than a month forced the rebels of Upper and Lower Austria into unconditional surrender at Linz. He outflanked a Bohemian and Hungarian army in September–October 1620 and advanced upon Prague in company with an imperial army under Bucquoy. On 8 November the advance was contested at the Battle of White Mountain by some 15,000 Bohemians under Christian of Anhalt-Bernberg. With 20,000 men Tilly launched a dawn attack and routed the Bohemians; Prague was occupied and sacked. Over the next two years Tilly – though not without occasional defeats – conquered the Palatinate and the Rhine valley. He defeated Georg Frederick of Baden-Durlach at Wimpfen on 6 May 1622; and inflicted serious losses on Christian of Brunswick at Höchst on 20 June when he intercepted the rebel army crossing the Main river. On 19 September he took Heidelberg. On 6 August 1623 he again defeated Christian, this time at Stadtlohn near the Netherlands border, inflicting 10,000 casualties on an army which totalled 12,000. Tilly was now the master of north-west Germany.

In 1625 the Danes invaded Germany, and Tilly was again active at the head of the

Catholic League forces; now, however, the Emperor had another army at his command, that of the mercenary general Wallenstein (q.v.), and it was to Ferdinand rather than to the princes of the League that the fruits of victory went. Although Wallenstein was the more prominent in this phase of the war, it was Tilly who won the decisive Battle of Lutter on 24–27 August 1626, destroying about half the Danish army and bringing anti-Hapsburg fortunes to another low ebb. Created a Count of the Empire for his victory at Höchst, Tilly was now offered the duchy of Brunswick-Calenberg, but apparently preferred a financial reward. In 1630, when Gustavus Adolphus of Sweden (q.v.) entered the conflict, Wallenstein had been dismissed from the supreme command of the imperial army as a result of the machinations of the League, prominent among whom was Tilly's master Maximilian. Thus it was Tilly who led the combined forces of the Emperor and the League against the 'Lion of the North'; and in November 1630, seeking to secure a strategic base from which to control Saxony and Brandenburg and to bar Gustavus from advancing into western Germany, he invested the rebellious city of Magdeburg. The siege lasted until the following May, when Tilly, on account of certain foreign alliances made by Maximilian of Bavaria, transferred his allegiance to the Emperor. On 20 May Magdeburg fell to him, but his intention of fortifying and holding the city was frustrated when his army sacked and burnt it with hideous slaughter. The Swedish army had been unable to arrive in time to save Magdeburg, and was now in serious logistic difficulties. Gustavus entrenched his forces at Werden, where Tilly twice attacked him without success and at heavy cost. At the order of the Emperor Tilly now marched into Saxony and spread great devastation; Maximilian of Bavaria had strongly advised against this, and as he predicted it drove the Elector John Georg straight into Sweden's arms. The Saxon and Swedish armies, with a combined strength of some 42,000, marched south towards Leipzig, which Tilly captured on 15 September 1631. On 17 September, four miles north of Leipzig, Tilly was heavily defeated by Gustavus at the classic battle of Breitenfeld (described in detail under Gustavus Adolphus). Tilly cannot bear all the responsibility for the defeat; his subordinate Pappenheim gave him unreliable intelligence about the speed of the Swedish advance in his eagerness for battle, and threw away a substantial part of the imperial cavalry by an unauthorised advance on the left flank at the beginning of the battle. By December 1631 the victorious Gustavus had 80,000 troops within the frontiers of the Empire, and was able to winter at Mainz without interference. Tilly defeated a Swedish detachment at Bamberg on 10 February 1632, but in April Gustavus crossed the Danube and marched eastwards into Bavaria. Tilly was still in command of the imperial troops, as Wallenstein had only just agreed to return to Ferdinand's service. On 15–16 April Tilly attempted to stop Gustavus crossing the Lech river. Brilliantly screening his assault until the last moment, Gustavus sent his troops across a bridge of boats. Tilly was seriously wounded, and Maximilian led the remainder of the army in retreat, abandoning most of the artillery and train. Gustavus swept through southern Bavaria, occupying Augsburg and Munich; and on 30 April, at Ingolstadt, Tilly died of his wounds.

Timoshenko, Semion Konstantinovich (*1895–1970*), *Marshal*

A Soviet cavalry officer who rose to command Russian forces during their occupation of Poland at the beginning of the Second World War, Timoshenko, despite his introduction of modern training techniques into the Red Army, himself displayed a total lack of understanding of the German *blitzkrieg* tactics after the launching of 'Barbarossa'.

Born on 2 February 1895, Timoshenko joined the Russian Imperial Army at the age of twenty and became a member of a machine-gun crew with the Fourth Cavalry Division. As a member of the Communist Party in 1919, he participated in the Bolshevik campaigns of that and the following year, attached to Budenny's (q.v.) cavalry regiment fighting in Poland. By 1925 he had risen to the rank of colonel, commanding the Third Cavalry Corps. As a lieutenant-general he was appointed deputy commander of the White

Russian Military District in 1935, and rose to command the North Caucasian Military District in 1937. As commander in the West Ukraine Timoshenko led forces which took part in the occupation of Poland in September 1939, and shortly afterwards was posted north to command Soviet forces in Bessarabia during the Winter War against Finland. It was this campaign, humbling the Red Army by demonstrating its relative impotence in the presence of the minute through brilliantly-commanded Finnish army, which accelerated widespread improvements in the Soviet armies. Although severely handicapped by the lack of modern equipment, Timoshenko – appointed Commissar of Defence – embarked on a swingeing reorganisation of the armies of the Central Russian Front, introducing harsh discipline and severe training techniques in all ranks of his command. The process of replacing the traditional horse-mounted cavalry with light tanks was speeded up, but it was far from complete when Hitler's 'Barbarossa' burst on Russia's Western Front in 1941.

Indeed, such was the lack of cohesion and preparedness among the Soviet armies when the German generals Leeb and Bock attacked through Poland in June 1941 that enormous losses in men and material were suffered in the first four weeks. Attacked by fifty-five divisions on a 300-mile front, Timoshenko's forty-seven divisions were forced back 600 miles, losing about 400,000 men, over 4,500 tanks, and 3,300 guns – losses that Russia, for all her vast military potential, could ill afford at the outset. The greatest German advances were, however, made in the south, where Russian forces amounting to sixty-nine divisions under the veteran Budenny recoiled across the entire breadth of the Ukraine before the devastating attack of fifty German divisions under Runstedt (q.v.).

As the winter of 1941–2 halted the German advance, and the flow of Russian reinforcements bolstered the front, the great Marshal Zhukov (q.v.) assumed command of the Red Army standing before Moscow, while Timoshenko moved south to replace Budenny. On the resumption of operations in the spring of 1942 Zhukov's armies stood their ground, and the main German success came with the advance of Paulus's Sixth Army to Stalingrad and the penetration of the Caucasus. Timoshenko was in turn replaced and sent to the northern front, where he pursued a limited campaign in central Finland in the Demyansk area during 1943. After the capitulation of German forces at Stalingrad and at the time of the swing to the offensive of Russian forces in the south, Timoshenko was placed in overall command of the Second and Third Ukrainian Fronts (in effect a pair of Army Groups), though the Fronts were commanded individually and operationally by General Rodyon Malinovski (the Second) and General Tolbukhin (the Third). Such was the enormous scope of operations during the final advance across Eastern Europe during 1944–5 by the massed Soviet armies that the conduct of the campaign devolved on the individual commanders under the overall direction of Marshal Zhukov. Timoshenko was thus scarcely responsible for the operations of this period, and his later success in command was almost wholly eclipsed by that of his dynamic subordinates. In fact he was never entirely free of the stigma of those early setbacks, and was to some extent the scapegoat for Russian lack of prepararedness in 1941, despite all his efforts at modernisation. He died on 31 March 1970.

Timur-i-Leng, called Tamerlane (*1336–1405*)

'Timur the Lame', usually known in the West as Tamerlane, was a central Asian conqueror whose butcheries rivalled those of Genghiz Khan (q.v.), but whose military skill fell short of that of the great Mongol genius. Little is known about his personal life; such biographical material as survives appears to be commissioned and distinctly suspect. A contemporary description of his appearance – 'very tall, highly coloured, with a large head, and white hair since his childhood' – is striking; if one adds the lameness and the Mongolian features, he must have been an arresting sight indeed.

In the century or so since the death of Ogatai the Empire founded by Genghiz Khan had been divided, by agreement and by dynastic warfare, into several khanates of varying stability. That tracing its descent from

Genghiz Khan's second son, the Khanate of the Jagatai Mongols, occupied the vast central Asian region roughly equivalent to modern Turkestan and central Siberia. Internal strife had reduced its integrity, and by about 1350 it was a loose group of semi-independent principalities. Timur, who was not a Mongol but of Tartar blood, is reputed to have been born at Kesh, near Samarkand, in 1336. By feud and intrigue he rose in power until he became the chief minister of the khan, whom he subsequently overthrew and replaced. Timur then declared himself to be a descendant of Genghiz Khan through Jagatai's line – an unconvincing claim – and by about 1370 he was ready to embark on a thirty-five year career of conquest as the self-styled restorer of the Empire of Genghiz Khan. In fact his conquests were almost purely destructive, and the empire he created was quite artificial, existing only by the exercise of his own ruthless power. Timur may have beautified his capital by importing plundered artefacts and enslaved craftsmen, but this was only a reflection of his vanity; against Genghiz Khan's tightly-disciplined but basically peaceful and viable structure of government, trade, taxation, and communications, Timur can show us only towers of skulls.

Between 1370 and 1380 Timur fought a series of campaigns in Khwarezm and Jatah, finally occupying Kashgar in the latter year. The former empire of Khwarezm had fragmented, and the Persian principalities were unable to withstand progressive invasion and defeat by Timur's ruthless army. Herat fell to his horsemen in 1381, and Khorasan and all of eastern Persia by 1385. The tide of bloodshed and wanton destruction swept over Fars, Iraq, Azerbaijan, and Armenia by 1387. At this point Timur was distracted by an attack into Azerbaijan mounted by Toktamish, Khan of the Golden Horde and the White Horde, a former political refugee to Timur's court at Samarkand whom he had aided against his enemies. These northern Mongols struck and defeated one of Timur's columns, but were in their turn repulsed. Timur did not follow up his victory, and in 1388 Toktamish became bolder. While Timur was in Persia he crossed into Timur's home region of Transoxiana and advanced on Samarkand, only being forced into hasty retreat by the conqueror's return by forced marches of fifty miles a day. During the following winter Toktamish crossed the borders yet again, but was decisively defeated; and Timur prepared to invade the Khanate of the Golden Horde.

With something of the order of 100,000 men, Timur invaded Russia in 1390 with the sole object of defeating Toktamish. He was lured west of the Ural river, and the final clash did not come until some time in 1391. At the Battle of the Steppes, which took place somewhere east of the Volga and south of the Kama, Timur and Toktamish struggled for three days. It is said that Timur only won by a ruse which discouraged the enemy when they were in fact in a commanding position, and that casualties amounted to about 30,000 in Timur's army. He did not attempt to follow the fleeing Toktamish with his mauled and over-extended army, but fell back on his own territories. In 1392 revolts against his rule broke out all over Persia – he put them down with the utmost ferocity after killing the rebel Shah Mansur in the Battle of Shiraz. Timur spent 1393–4 in an invasion of Mesopotamia and Georgia, capturing Baghdad in 1393. Toktamish crossed the Caucasus yet again, but was pushed back and pursued. Timur defeated him in 1395 at the Battle of the Terek, and then embarked on a grim pursuit which led him across much of central Russia, ravaging and burning as he went. He sacked Astrakhan and the Golden Horde's capital of Sarai, installed a puppet khan, and destroyed the power of the Horde for ever.

In 1398–9 Timur embarked on a devastating and quite pointless invasion of India which shattered the northern area of the country for a century at least. One of Timur's grandsons led his right wing into the Punjab and seized Multan in the spring of 1398, while another grandson led the left wing through Lahore. Timur led a picked force through the Hindu Kush Mountains to rendezvous with the other detachments on the Indus in September. Spreading a trail of destruction, they moved against Delhi, and on 17 December at Panipat destroyed the army of Mahmud Tughluk. Pausing only to put 100,000 captive Indian soldiers to the sword, Timur stormed Delhi and allowed his men to indulge in a horrible orgy of wanton killing, rape, and destruction. After several days of this he marched them north and into the foothills of the Himalayas, storming and sacking Meerut. Turning westwards once again, they swept back across the Punjab and then left India as abruptly as they had appeared, in March 1399. Delhi did not recover for more than a hundred years.

In 1400 Timur invaded Syria, destroyed the Mameluke army at Aleppo, and captured that city and Damascus; he is estimated to have slaughtered some 20,000 inhabitants of the latter. In 1401 Baghdad rose against him, and was recaptured and subjected to another nightmare massacre. In 1402 he defeated the Ottoman Turkish Sultan Bayazid at Angora; captured Smyrna from the Knights Hospitallers; and proceeded to overrun the whole of Anatolia. He returned to Samarkand in 1404, having received tribute from both the Sultan of Egypt and the Emperor of Byzantium. In 1405 the old Tartar started out to invade China, but died at Otrar on 19 January before he could begin the campaign.

Tito, formerly Josip Broz (*1892–1980*), *Marshal, President of the Federal People's Republic of Yugoslavia*

The most durable of European Communist leaders, Tito was also by far the most attractive to Western eyes. Apart from his strongly individual personality, which contrasted with the impassivity of the bureaucrats who wielded power in many Soviet bloc states, he was admired as a fighting man who seized power at the risk of his own life – he was not a pale *apparatchik*. His country was unquestionably a rigidly-administered dictatorship, and he himself was a life-long and convinced Communist; yet his independent nationalism, his vivid and unashamedly human appetite for life, and above all his courageous leadership in the Second World War all combined to make him less unacceptable than most Communist rulers to the Western European observer. Many Western witnesses testified to his impressive personal charisma.

He was born Josip Broz, the seventh son of a peasant, on 25 May 1892 at Kumrovec, near Zagreb in Croatia, then a province of the Austro-Hungarian Empire. After a childhood of wretched poverty he worked as a metal-worker until conscripted in the First World War. He was sent to the Carpathian front, and in March 1915 he was wounded and captured by the Russians. During his captivity he came in contact with Revolutionary elements, and in 1917 he joined the Red Army. He returned to Yugoslavia, now an independent monarchy, in 1920, and joined the Yugoslav Communist Party. He was active in party affairs, and in 1928 was arrested and imprisoned for five years for subversive activities. Upon his release he travelled to Moscow, where he worked for the Comintern's Balkan Secretariat. In 1936 he helped organise recruiting for the International Brigades in Spain, visiting France as well as Zagreb. Elected secretary-general of the Yugoslav Communist Party in 1937, he again visited Moscow in 1938 and 1939. In 1940 he secretly convened the fifth party conference, which resolved to keep Yugoslavia out of what was held to be an imperialist war; but in early 1941 German pressure on Yugoslavia forced the Prince Regent into reluctant co-operation with the Axis. In March there was a patriotic *coup* which overthrew the Government and rejected Axis alliance, and in April Hitler invaded the country. A well-prepared and fast-moving invasion from three directions slashed through the weak and out-of-date defences, and on 17 April Yugoslavia surrendered unconditionally. The country was subsequently partitioned.

In June 1941, when Germany invaded Russia, Josip Broz went to Belgrade in response to Comintern instructions, and began organising a partisan network to further an armed rising against the occupation forces. He took the chief command, using the *nom-de-guerre* 'Tito' which he kept ever afterwards, and which became his legal surname. He left Belgrade at the end of August and took operational command in the Serbian countryside. The rising enjoyed considerable initial success, and by mid-September the Germans had been cleared from most of Serbia. Tito then negotiated with Dragoljub Mihajlovic, a colonel in the Royal Yugoslav Army who had gathered a group of ex-officers and para-military 'Chetniks,' to explore the possibility of a unified command and joint operations against the enemy. The hostility between the Chetniks and the Communist partisans proved too bitter, however, and the negotiations broke down; open fighting between the two groups broke out in due course. The Germans took advantage of the disarray of Yugoslav ranks to launch a counter-offensive, and by mid-December most of the lost ground had been recaptured. Tito led the remnant of his partisans into East Bosnia, and subsequently into Montenegro and West Bosnia. The Germans exploited the considerable ethnic and religious differences of the various Yugoslav communities throughout the war, and many 'Ustachi' fought on the side of the Nazis both as combat troops in Russia and as anti-partisan forces in Yugoslavia. Mihajlovic's group, which carried out intermittent warfare against the partisans, became more and more committed to this and less and less effective against the Germans, finally collaborating openly with the occupying power. Tito in turn carried out a ruthless feud against the Chetniks which he won decisively. He made strenuous efforts to persuade Yugoslavs of all races and religions to co-operate against the German enemy, and his post-war state was founded on the same appeal to pan-Yugoslav solidarity.

The partisan war which Tito waged from the desolate mountains of Yugoslavia for three and a half years was a classic of its type. He assembled large forces which proved adept at moving through difficult terrain and striking at the enemy without warning; handicapped by their lack of heavy weapons, they received material assistance from Great Britain, and British liaison officers were parachuted in to join them. The Germans launched many major operations in the hope of pinning down the partisan army and destroying it, using large numbers of mountain troops and *Waffen-SS* formations supported by artillery, armour, and air power. (The campaign was merciless in the extreme, and there were many atrocities against captured partisans and civilians suspected of supporting them; the Balkan-recruited units of the V SS Mountain Corps were notorious; but the partisans, too, certainly committed mass murders of captured Chetniks in 1945.) In May–June 1943 the partisans, with a strength of four lightly-equipped divisions and burdened by 3,000 wounded, were very nearly cut off and annihilated on the mountain of Durmitor in Montenegro, and Tito's successful attempt to break the enemy encirclement was one of the classics of guerrilla warfare. He himself was nearly captured in May 1944 when German paratroopers were dropped near his headquarters at Drvar. In the summer of 1944 he travelled to Moscow and conferred with Stalin, and later to Italy to meet Churchill and the Allied commanders in the Mediterranean theatre. In the latter stages of the war considerable assistance was given by the Allies, including direct air cover; and in October 1944 the partisans fought side by side with Red Army troops in Belgrade. In May 1945 the partisans joined up with British Eighth Army troops at Trieste.

After the War Tito, now a self-appointed marshal, became Prime Minister of a Communist Yugoslavia. There was friction with the Western Allies over Trieste, but equally there was bad feeling between Tito and Stalin. Tito made it clear from the outset that Yugoslavia, liberated largely by the efforts of Yugoslavs, was not willing to become a docile Soviet satellite. There was resentment over the relative lack of Soviet support during the war, over attempted Soviet exploitation of the economy, and over luke-warm Soviet backing on the Trieste question. Tito established a Communist dictatorship, but struck a strongly independent and nationalist line which brought down on him vicious attacks by Stalin and other Communist leaders. In 1949 Tito broke with Russia, and for the rest of his life pursued an opportunistic policy of gradual *rapprochement* with the West. On several occasions he embarrassed the Soviet government by publicly declaring at moments of crisis that Yugoslavia would resist any attempted Soviet invasion to the limit.

In June 1953 Tito became President of the Federal People's Republic of Yugoslavia, a post he held until his death. He was ruthless with old comrades who disobeyed him, and with any manifestations of the inter-communal strife which still dogs his country. It is widely accepted that only his strength and determination held Yugoslavia together. A man of great stamina and charm, he openly enjoyed the luxurious material rewards of unchallenged power. He died in power on 4 May 1980.

Tōjō Hidecki (*1884–1948*), *General*

Former Japanese chief of staff in Manchuria and Vice-minister of War, during the Second World War Tōjō was the Prime Minister who ordered Japan's attack on Pearl Harbor, and the overall architect of Japanese victories during 1941 and 1942.

Born at Tokyo on 30 December 1884, the son of Lieutenant-General Tōjō Eikyo, Tōjō Hidecki attended the Imperial Military Academy and the Japanese Military Staff College, and was appointed Japanese military attaché in Germany in 1919. Distinguished as a clever field commander during his service in middle rank, Tōjō went on to gain respect as a stern disciplinarian and sound administrator on the staff; and by 1937, as a major-general, he had risen to become the chief of staff of the mighty Kwantung army in Manchuria.

In the first Konoe administration (1938) Tōjō was appointed Vice-minister of War, and he rose to be Minister of War in the cabinet of the administrations of 1940–1. On 17 October 1941 a new militaristic government came into power, and Tōjō, now a lieutenant-general, was made Prime Minister – supported by Marshal Hajime Sugiyama as chief of the Army General Staff and Admiral Osami Nagano as chief of the Naval Staff.

It was in pursuance of Tōjō's determination to establish a 'New Order in Asia' that the Japanese administration embarked on armed confrontation with America, and it was Tōjō who ordered the attack on Pearl Harbor to go ahead on 7 December 1941. But the military administration, in planning for a two-year war, underestimated the Allies' ability to absorb the smashing Japanese victories of 1941–2 – whose overall architect was Tōjō himself – so that as the full weight of American military might gained momentum, Japanese strategy was found wanting, and the reserves of manpower and production had been largely exhausted. Tōjō was himself held accountable for the loss of the Mariana Islands in 1944, and although his supreme powers of administration had been brought to bear in efforts to increase industrial war production, he was compelled to resign – his place being taken by General Koiso.

Following the Japanese surrender in August 1945 Tōjō was arrested as a war criminal. After failure in an attempt to commit suicide, he was tried and convicted by an international court in Tokyo during 1947–8, and hanged on 23 December 1948.

Torstensson, Lennart, Count of Ortala (*1603–51*), *Field-Marshal*

The greatest artillery exponent of his age, and the father of field artillery, Lennart Torstensson was the son of a Swedish officer. He was born at Torstena, Västergötland, on 17 August 1603. At the age of fifteen he became page to King Gustavus Adolphus (q.v.), and accompanied him during the campaigns in Livonia in 1621–3. (For general military background to this period see Gustavus II, Wallenstein, Tilly.) In 1624–6 he attended the military school of Maurice of Nassau (q.v.) in the Netherlands; it is probable that his taste for artillery was formed, or at least confirmed, by contact with that great artillery general. He returned in time to fight at Wallhof in 1636, and served in the Prussian campaigns until 1629. An established artillery specialist, he was in that year made colonel of the world's first pure artillery regiment by Gustavus. He acquitted himself so brilliantly during the German expedition of 1630–1 that in 1632 he was made a general – at the age of twenty-seven, and through merit, not influence.

Sweden was in the forefront of military thought at this time, and the innovations which Torstensson sponsored were highly significant. His aim was to create a genuine field artillery, light enough to manœuvre in direct support of bodies of horse and foot, and coming into action rapidly enough to make a real contribution to battles of movement. In 1626 the Swedes experimented with the so-called leather cannon – a gun with a copper inner barrel and a leather outer sleeve. This was vastly lighter than its contemporaries, weighing only some ninety pounds without the light carriage, and capable of being handled by two men. Its disadvantage was that the copper retained its heat during rapid firing to such an extent that shots soon began to 'cook off' as soon as they were rammed down the barrel. In 1631 this ingenious piece was superseded by a cast-iron gun weighing some 400 pounds; at a time when many cannon weighed half a ton this was a genuine improvement, and it could be managed by four men or two horses. It fired a projectile the weight of which was variously reported as three or four pounds, and it was known as the 'regimental' gun; two were attached to each regiment in the field, to support its manœuvres. The rounds were pre-packed, a thin wooden container of powder being wired either to a ball or to canister-shot, and it is alleged that a skilled gun team could fire eight shots to the six of a musketeer. Torstensson standardised calibres for greater efficiency; the four-pounder regimental gun was supported by a nine-pounder piece, six of which supported every unit of 1,000 men. A feature of Swedish guns was the shortening of the barrel. In early days the slow-burning 'serpentine' powder used had made a long barrel necessary; but the quicker and more evenly burning 'corned' powder had been in use for many years, and the barrels of cannon had remained long for no good reason, with the inevitable disadvantage of weight and immobility. The Swedes organised the personnel into formal artillery regiments, each of six companies: four of gunners, and two of sappers and engineers with igniters and charges.

In 1630 Torstensson commanded Gustavus's field artillery. He made a great contribution to the victory at Breitenfeld on 17 September 1631; the 100 Swedish cannon opened the battle with a galling fire which provoked Pappenheim into an unsuccessful cavalry charge; and during the final counter-attack the regimental guns accompanied their infantry forward, firing three shots to every one from the Austrian cannons. (A comparison was fairly easy, as by this stage the imperial artillery, heavy and unwieldy, had been captured by Gustavus and was being served by Swedish gunners.) Promoted general in 1632, Torstensson was captured in the attack on Wallenstein's camp on the Alte Veste at the end of August of that year. The camp was sited on rough ground, well entrenched in heavy scrub, and Swedish cavalry and artillery action was much impeded. Gustavus was forced to withdraw after suffering heavy loss, and on 16 November he died on the victorious field of Lützen. Torstensson did not return to Sweden until an exchange of prisoners took place the following year.

In 1635 he became Chief-of-Staff to the Swedish Field-Marshal Johan Banér, during that able leader's impressive campaign in eastern Germany. He fought under Banér in his brilliant victory at Wittstock on 4 October 1636, and took part in the Swedish retreat from Torgau in the following year. In May 1641 Banér died, and later that year Torstensson was persuaded to take command of the Swedish army in Germany, although his health was already poor and he took up the burden unwillingly. He was granted the rank of Field-Marshal, and made Governor of Swedish-occupied Pomerania. He restored order in the army, which had suffered from deteriorating discipline, and won a great victory at Leipzig in the spring of 1642, overrunning most of Saxony. In 1643, after his army had ravaged Bohemia and Moravia without interference, his invasion of Jutland from the south launched the Swedish attack on Denmark. In 1644 he completely outgeneralled Gallas and his Danish and imperial armies, forcing a retreat into Bohemia which he followed up vigorously. Intercepted by an imperial and Bavarian army under Werth as he advanced on Prague, Torstensson heavily defeated them at Jankau on 5 March 1645; the rapid movement of his artillery from one sector to another during the battle was the decisive feature. He never fought another battle, however; he resigned his command in favour of Wrangel in 1646, on grounds of ill-health, and spent his final years as an active political figure in Sweden. It is recorded that he enjoyed great favour from Queen Christina. He died in Stockholm on 7 April 1651.

Turenne, Henri de la Tour d'Auvergne, Vicomte de (*1611–75*), *Marshal-General of France*

Turenne first became conspicuous in military affairs during the second half of the Thirty Years' War, when he brought a force to join

Duke Bernard of Saxe-Weimar in the investment of the city of Breisach on the Rhine in June 1638. Although the besiegers beat off repeated imperial attempts to relieve the city, it was not until 17 December that its surrender was finally forced by starvation. Capture of Breisach gave the French the key to the Rhine; Turenne moved south to join forces under the Duke of Harcourt, and served under his command in the Battle of the Route de Ouiers in 1639, during the successful campaign against the imperial-Savoyard army.

Four years later Turenne succeeded to the command of a French army. Back in the constantly-contested Rhineland, the French Marshal Guébriant had crossed the Rhine with the Weimar army and advanced through the Black Forest into Württemburg, but he was killed at Rattweil in October 1643. He was succeeded in his command by the French-Weimar General Josias von Rantzau, who was himself defeated and captured by an imperial army under Baron Frans von Mercy at the Battle of Tuttlingen, on the Upper Danube, on 24 November. Command of the Weimar army now passed to Turenne, who withdrew to Alsace.

After wintering on the west bank of the Rhine, Turenne again crossed the river into the Black Forest, but faced by superior numbers under Mercy, he decided to remain at Breisach to await reinforcements. When these arrived under the Duke of Enghien (the great Condé, q.v.), who assumed command of the 17,000-strong army, the French advanced to meet Mercy at Freiburg, where he also had about 17,000 men. A bitter battle followed, in which Turenne was instructed to envelop the rear of the Bavarian army. Although the Bavarians were forced out of their positions, the French were so exhausted and depleted that no rout followed, and Mercy entrenched again. Two days later, on 5 August, the Second Battle of Freiburg brought stalemate once more – though again the French suffered most. On 10 August the third battle was fought, Turenne again attempting to attack the Bavarians' rear; but Mercy, anticipating the move, retreated in good order – leaving the Rhineland in French hands.

During the spring of 1645 Turenne once more invaded central Germany, but was surprised by his old opponent Mercy, who severely defeated him in the Battle of Mergentheim on 2 May, forcing the French army to fall back on the Rhine as Enghien rushed to its aid. Substantially reinforced, the French army invaded Bavaria in July, and on 3 August won a narrow victory at Nördlingen, in which battle Mercy was killed. By now Turenne's army had suffered such casualties that it contained little veteran backbone, and this decided the French commander to pursue a war of manœuvre. However, when in August 1646 the imperial army under Archduke Leopold William moved into the Rhine–Main valley, Turenne secretly and rapidly marched south-east to join up with the Swedish Count Wrangel. In the face of such combined strength, and fearful that the Franco-Swedish army would isolate him from central Germany, Leopold withdrew.

Turenne and Wrangel by-passed the imperial army and avoided battle, plunging south-east deep into Bavaria towards Munich, and ravaging the countryside as they went. The Elector Maximilian's appeals for imperial assistance in repelling the invasion were vain, and he sued for an armistice – resulting in the Truce of Ulm (March 1647). Turenne was ordered to return to Luxembourg, but on reaching the Vosges his German cavalry mutinied and turned back across the Rhine; for three months Turenne accompanied them far into Germany, but finally his powerful personality persuaded most of them back into French service.

Bavaria returned to the Emperor's side in 1648 and once more Turenne and Wrangel joined forces and marched against Austria. After the Battle of Zusmarshausen, in which Turenne's forces caught and virtually destroyed the imperial rearguard on 17 May, the Franco-Swedish army reached the Inn river – the nearest the French forces had come to Austria – and the Emperor sued for terms, embodied in the Peace of Westphalia.

In the civil war which broke out in France in January 1649 the great Condé had sided with Cardinal Mazarin in support of the royal family, but after the Peace of Rueil Condé quarrelled with the Cardinal and then began negotiating with the formerly-rebellious nobles. In January 1650 Mazarin arrested Condé, and Turenne, fearing for his own safety, fled from Paris and joined the Duchesse de Longeville on the eastern border of Champagne. Allying himself with the Spaniards, he waged war in Champagne until he rashly opposed a greatly superior force under Marshal du Plessis-Praslin and was defeated at the Battle of Rethel (15 December), being able to extract only fragments of his army from the field.

The following year Mazarin was exiled, Condé was released, and Turenne returned to Paris. His standing was at a low ebb and he was careful not to identify himself with Condé's political cause, with the result that he was soon reconciled with the Queen Regent, placed in command of one of the two divisions of the Royal Army, and sent to oppose Condé, who had now occupied a threatening position on the Loire. A series of complicated manœuvres followed, involving Condé and Turenne – in one such action Turenne blocked the bridge at Bléneau and only just saved the young Louis XIV from capture by the rebel forces. At last, reinforced by new Royalist troops, Turenne trapped Condé against the walls of Paris, which had declared itself neutral in the civil strife. After a bitter struggle Condé was on the point of capture when Paris opened her gates and allowed the rebel leader to escape. Turenne then marched north, outmanœuvring the Spaniards and rebel forces, and the rebellion collapsed. As a result of services rendered by Turenne the young Louis XIV was able to re-enter Paris on 21 October 1652.

The defeat of the rebellion now freed French troops to reinforce those in the north-east facing the Spaniards, with whom Condé was now serving as generalissimo. Despite numerically superior strength, Condé was unable to draw Turenne into battle under unfavourable conditions. In a war (1653–7) of manœuvre Turenne managed to raise the Siege of Arras (25 August 1654), but was defeated at Valenciennes (16 July 1656), although he marginally outgeneralled Condé owing to Spanish indecision and suspicion of the French generalissimo.

In 1657 England and France were separately at war with Spain, with the result that Mazarin and Cromwell concluded an alliance whereby they agreed jointly to attack the coastal towns of Dunkirk, Gravelines, and Mardyck; after its capture Dunkirk was to be ceded to England. Mardyck fell to the Allies before the end of 1657, and Turenne forthwith marched towards Dunkirk, despite the opening of the dykes by the Spanish. While the Anglo-French army of 21,000 laid siege to the port, a Spanish Netherlands relief army of 16,000 under Don John of Austria and Condé marched from Ypres. The 13 June 1658 found Condé's army encamped on the dunes north-east of Dunkirk; Turenne, leaving 6,000 of his men guarding the siegeworks, collected 9,000 cavalry and 6,000 infantry for battle, engaging in some skirmishing before nightfall. Early the following morning Turenne's lines were established and advanced across the dunes just before low tide, with the English infantry on the left flank and the cavalry distributed between left and right. His advance was deliberately slow as his timing depended on the changing of the tide. The Spanish suffered under two disadvantages: they had refrained from moving any cavalry on to the beach in case it was destroyed by fire from the British naval guns offshore; and in their haste to reach the Dunkirk area they had far outpaced their own artillery, which had not arrived on the field of battle.

Initial contact was made on the Spanish right, commanded by Don John, where a spirited attack by the British infantry was effectively supported by French cavalry (under the command of the Marquis de Castelnau) and by the British navy's guns at sea. For four hours a bitter battle raged on the beach; then the change of tide forced the Spanish right flank to fold inwards, and Turenne was enabled to carry out a cavalry envelopment of the Spanish land flank. This occurred in time to assist in the defeat of a cavalry attack on the French right flank by Condé. The battle was a complete victory for Turenne, who lost only about 400 men, whereas the Spanish casualties and prisoners amounted to 6,000. Dunkirk subsequently surrendered and was ceded to Cromwell (remaining in English hands until sold to the French by Charles II in 1662). The Battle of the Dunes allowed Turenne freedom of movement in

Flanders; he captured Ypres and threatened Brussels and Ghent, thereby securing for France advantageous peace terms in the Treaty of the Pyrenees, signed with Spain on 7 November 1659. (Following this, Condé sought and gained the forgiveness of Louis XIV.) Turenne was appointed Marshal-General of the Royal Army on 5 April 1660 and, but for his Protestant faith, would probably have been made Constable of France. Meanwhile the development of the ministry of war by the energetic Marquis de Louvois enabled Louis XIV to command in person, and during the War of Devolution (1667–8) and in the invasion of Holland (1672) Turenne marched at his King's side.

After France's failure to obtain the Spanish Netherlands in the War of Devolution, Louis set about undermining the Triple Alliance by isolating Holland from her allies, and negotiating the secret Treaty of Dover with Charles II of England, and a treaty with Sweden. In March 1672 he declared war on Holland, and with an army of 130,000 men marched down the Meuse and established a base at Düsseldorf. William III of Orange quickly set about raising the strength of his army to around 80,000 men. In May the Allies commenced the invasion of Holland, moving three columns down the left and right banks of the Rhine, and from Westphalia, commanded by Turenne, Condé, and Luxembourg respectively. These armies succeeded in overrunning most of Holland and in capturing many key cities, and Amsterdam was flooded. In August a revolution broke out among the Dutch, and William of Orange was placed at the head of the Government.

Fearful of Louis's ambitions, a number of rulers now formed a coalition against France: these included the Elector Frederick William of Brandenburg, the Emperor, and Charles II of Spain. In order to meet these various threats, and against the professional advice of both Condé and Turenne, Louis now split his army into relatively small forces, Condé being sent to Alsace, and Turenne to Westphalia. Turenne's force, which never exceeded 20,000 men, was only a secondary command, yet it was the brilliance with which he handled this force, often in the presence of superior enemy strength, that gained Turenne enduring fame. Between September 1672 and January 1673 he constantly outmanœuvred the much-larger combined forces under Frederick William and Count Raimondo Montecuccoli. So frustrated did Frederick William become that he sought peace in the Treaty of Vassem, (6 June 1673).

Turenne himself resented the detailed control of military affairs demanded by the arrogant Louvois, feeling that control in the field should not depend upon administrative whims in far-away Paris; yet he had to concede that the efficiency of the supply organisation enabled him to maintain active operations into the depth of winter. With the Treaty of Vassem Turenne had broken the German coalition for a time, and had prevented the enemy from crossing the Rhine. Nevertheless, later in the year, when his manœuvring against the Emperor Leopold I's army met with such success that he could have pressed forward and reached Bohemia, Louvois blocked the reinforcements that would have made for a decisive campaign. When Turenne was recalled to cover Alsace, the Emperor's forces promptly struck at and captured Bonn (12 November), thereby breaking the French control of the lower Rhine.

Early in 1674 the forces of the Coalition were greatly strengthened as many German princes joined the Emperor. Turenne was again charged with the protection of Alsace, and on 14 June, with the intention of striking the imperial forces before they could concentrate, he crossed the Rhine at Philippsburg. Two days later an imperial force of 7,000 cavalry and 2,000 infantry under General Sylvio Caprara moved to block the French path at Sinsheim. Turenne, by dint of a courageous crossing of the Elsenz river under fire, overran the imperial outposts and in a bitter uphill battle routed the main enemy force, positioned on a plateau above. In spite of having achieved this victory Turenne was ordered to proceed no further and to retire across the Rhine, on the pretext that his object was not to invade but to cover the security of Alsace. Nevertheless, after receiving new reinforcements he was permitted to recross the Rhine the following month with 16,000 men, and proceeded to devastate the countryside between the Main and Neckar rivers. Once again he was obliged to return to the west bank of the Rhine to met an invasion threat along the Moselle. The inconclusive Battle of Enzheim (4 October) was followed by a period of watchfulness for both French and imperial armies.

Reinforced to a strength of 57,000 men, it seemed that the imperial army was in full control, if not in possession, of Alsace. The imperialists were camped in all the towns from Belfort to Strasbourg; Turenne's forces were ostensibly at Dettweiler, but he confused the enemy by ordering all the fortresses in central Alsace to a state of defence, and quietly leading his entire force of 28,000 men into Lorraine during November. Marching secretly southwards behind the Vosges, and splitting his forces into numerous small bodies to confuse enemy spies, Turenne reunited his army near Belfort and marched into Alsace from the south, confronting the Great Elector at Turckheim on 5 January 1675. Despite the exhaustion of his forces, Turenne decided to make an assault on the flank of the imperial army, and in a battle reminiscent of Enzheim, delivered such a devastating blow that, after only perfunctory resistance, the Allied army retreated from the field and recrossed the Rhine. Alsace was saved. Thus was completed one of the great campaigns of military history.

Dispirited by his defeat the Great Elector withdrew his army to Brandenburg, and command of it was resumed by Montecuccoli. Turenne had in the meantime followed across the Rhine and was engaging in a war of manœuvre; he prevented Montecuccoli from recapturing Strasbourg in July 1657. It was while preparing to attack the imperial army at Sasbach that, on 27 July, Turenne was killed by a cannon shot. Such was the personal influence exerted by the marshal that after his death the French army seemed to wilt, and when Montecuccoli mounted an attack he was able to force the French back across the Rhine almost to the Vosges.

Turenne had displayed many of the greatest attributes of a brilliant commander, in that he had succeeded in his many missions with considerable economy of strength. He had seldom been provided with forces numerically superior to his adversaries, and when faced with battle he had used them sparingly, and seldom suffered undue casualties. His frequent employment of local or natural features (for example, the tide-change in the Battle of the Dunes, and the cover afforded during the famous winter march behind the Vosges) was nothing short of brilliant. Frequently deprived of overall command, either by the presence of King Louis XIV or by the restrictions exerted by Louvois back in Paris, he nevertheless stands as the great symbol of France's military might in the second half of the seventeenth century.

Vauban, Sébastien le Prestre de (*1633–1707*), *Marshal of France*

Vauban was one of the great military engineers of history. He rose to high field command in the armies of Louis XIV – unusual for his period – and justified his position by numerous successful sieges, as well as the design of great fortresses.

Sébastien de Vauban was born in May 1633 at Saint-Léger-de-Foucherest, near Avallon in Burgundy, not far from his grandfather's fortress home at Vauban. In 1651, with the outbreak of the Third War of the Fronde, he received a cadetship from the malcontent Louis II of Bourbon, Prince de Condé. He quickly displayed considerable talent in laying out field fortifications. In 1653 he was defending towns in the Argonne for the rebel forces when he was captured by Marshal Henri de la Ferté-Senneterre, who promptly commissioned him as an officer in his own regiment. Vauban's great opportunity arrived the following year with the Siege of Stenay, which had been a base for Condé's supporters. It fell after thirty-three days' siege, and on 3 May 1655 Vauban, who had been wounded twice in battle, was admitted as 'a king's engineer'.

For the next twenty years the object of almost every one of Louis XIV's campaigns was the capture of one fortress or another, and Vauban's talents were allowed full rein. He was again wounded at Valenciennes in 1656 and at Montmédy the following year, and was chief engineer at the Siege of Gravelines in 1658. Although as yet the engineers had no place in the military hierarchy, Vauban won for them encouragement from the King's ministers, and in 1663 the King gave Vauban a company in the Picardy Regiment. He served directly under the King's authority in the War of Devolution and played a major part in the siege and capture of Tournai, Douai, and Lille during Turenne's (q.v.) operations in Flanders and Hainault. His reward was a lieutenancy in the Royal Guards and, in 1668, the governorship of Lille Citadel. From the end of 1667 he was effectively entrusted with the functions of 'commissary general of fortifications'. After the War of Devolution, proposals for the erection of new fortifications demanded Vauban's presence in every part of France. He conducted an immense correspondence with the King, and his long series of military writings on fortifications became the subject of military study for the next hundred years.

The Dutch War of 1672–8 brought more fame to him than to many of the commanding marshals, for he shared with the King the honour for the successful sieges which frequently punctuated the course of that war. His rapid success at the Siege of Maastricht in 1673 earned him the rank of infantry brigadier; but his promotion to lieutenant-general was not announced until 1688, this promotion being apparently the result of the capture of Philippsburg under the nominal command of the Dauphin at the outbreak of the War of the Grand Alliance. It was at Philippsburg that he introduced ricochet gunfire, whereby cannon shot was made to bounce forward over parapets and hit several targets before its energy was spent. At the same time he was urging the use of the socket bayonet, which unlike the contemporary 'plug' bayonet, would enable the musket to be fired while the bayonet was attached to the muzzle.

Vauban captured Mons in 1691, and Namur the following year, with very few casualties (he was renowned for vast expenditure of ammunition, not of lives). At the Siege of Charleroi in 1693 he commanded an infantry division for the first time, and in 1694 he was ordered to hold Brest when the English threatened Brittany; in 1695 he co-operated in the defence of Namur but was unable to hold it, and in 1697 he assisted in the capture of Ath from the Spanish. After the signing of the Peace of Rijswijk he worked at Neuf-Brisach – the last of the 160 fortresses which he built or re-designed. In 1702, following continued ill-health, he appealed to the King to be allowed to retire, and on 14 January 1703 was appointed Marshal of France. His last responsibility was to organise an entrenched camp at Dunkirk in 1706.

Vauban, for all his military preoccupations and technical study, was also a social and, to some extent, a political radical who advocated less oppressive measures against the French peasantry. He was a prolific writer, producing treatises upon forestry, land reclamation, and river navigation. His influence on the advance of wider education for junior officers in the latter half of the 17th century has seldom been appreciated fully. He was a pioneer statistician and a powerful though academic advocate of colonial rehabilitation. He died, a most widely respected military genius and scholar, on 30 March 1707.

Vendôme, Louis Joseph, Duc de (*1654–1712*), *Marshal of France*

Vendôme was a French general during the wars of Louis XIV, capable of displaying flashes of brilliance and audacity, but more often negligent and incompetent. A man of disgusting personal habits and unconcealed immorality and decadence, his social decline tended to remove him from royal favour and trust. Nevertheless, he rendered valuable service to the Bourbon cause in Spain.

Born the son of the marriage of the Duc de Mercœur (Louis de Vendôme) to Cardinal Mazarin's niece, Laure Mancini, in Paris on 1 July 1654, Louis Joseph Vendôme claimed princely rank through his paternal grandfather, César, Duc de Vendôme, who was a natural son of King Henry IV. His sodomy, foul habits, and unconcealable syphilis combined to work against the predisposition of Louis XIV in his favour; but following military service from 1672 onwards his promotion came rapidly, and by 1688 Vendôme had reached the rank of lieutenant-general. During the War of the Grand Alliance he fought well at the Battle of Steenkerque in July 1692, and in the army of Nicolas de Catinat in its decisive defeat of the Duke of Savoy at the Battle of Marsaglia on 4 October 1693. He was placed in local command of forces in Catalonia in 1695, and captured Barcelona two years later.

In the War of the Spanish Succession French forces in Italy faced problems of command. Marshal de Catinat showed himself to be no match for Prince Eugène (q.v.), and in August 1701 the French commander was replaced by the ageing Marshal Duc François de Villeroi. The latter pursued some complicated manœuvres to avoid immediate confrontation with the Austrians, but was virtually forced into winter quarters by the threat of blockade. On 1 February 1702 Eugène carried out a surprise attack at Cremona, captured Villeroi himself, and then retired after causing considerable damage to the French forces and their equipment. At this point Paris appointed Vendôme to command the army in Italy. He forced Eugène to abandon the blockade of Mantua, and in the Battle of Luzzara on 15 August, which was in effect drawn, he managed to outmanœuvre the Austrian forces. The following year he led his much-strengthened army to the walls of Trento; he captured Vercelli in 1704 and defeated Eugène's attack at Cassano in 1705. At the outset of the Turin campaign in 1706 Vendôme seized the initiative and split his army of more than 100,000 into two; with one he drove the Austrians under Count Reventlau out of central Lombardy, while the other laid siege to Turin, capital of Savoy. Just as Eugène arrived to replace Reventlau and halt the French advance, Vendôme was called north to Flanders in an attempt to retrieve the situation after the disaster of Ramillies. Once again he replaced Villeroi – who was by now quite past the age at which he could exercise competent command.

For many months Vendôme skilfully covered the line of French fortresses in Flanders. In 1708, under the nominal command of Louis, Duc de Bourgogne (the King's grandson), he organised the characteristically bold offensive that resulted in the capture of Bruges and Ghent on 4 and 5 July; he then turned south to threaten the Allied garrison of Oudenarde – anxious to bring the Allies to battle before Eugène's army could arrive from Coblenz. Marlborough, realising that he had been outmanœuvred and that the morale of his Dutch troops was sinking rapidly, also wished to seek battle as soon as possible. With battle clearly imminent, the young Duc de Bourgogne became distinctly nervous, and insisted that Vendôme should avoid the confrontation. While the two commanders argued their armies stood in scattered units to the north of Oudenarde instead of moving to block the passage of the Allied army along the Scheldt. On the morning of 11 July, with the Allied army swarming across the river, and in the face of arguments that failure to resist Marlborough would result in his isolation from France, the Duc de Bourgogne finally gave his permission for the French army to fight a defensive battle. Obviously discouraged by the vacillation in their command, the French troops were driven from the field after an afternoon of bloody fighting in which their right flank was turned. Vendôme lost control of his large army in the closing stages of the Battle of Oudenarde, and was recalled to Versailles in disgrace. It is interesting to speculate on the result of this battle had Vendôme not been shackled to the faint-hearted Bourgogne – had he been free to command his forces as he had done in his Italian campaign.

Towards the end of 1710 Vendôme was sent to Spain to assume command of King Philip V's army, which had been defeated near Saragossa by a 26,000-strong force under General James Stanhope. Vendôme quickly assembled the scattered French forces – which amounted to about 27,000 men – and advanced on Saragossa, thereby cutting Stanhope's lines of communication. Turning, he closely pursued the Allies and caught up with their rearguard under the personal command of Stanhope, whom he defeated and captured at Brihuega on 8–9 December. The following day he won a major victory over General Guido von Starhemberg's Austrians at Villaviciosa, thereby effectively securing the Spanish crown for Philip. Eighteen months later, while preparing for the reconquest of Catalonia, Vendôme fell ill. He died at Vinaroz on 11 June 1712.

There is no doubt that Vendôme's periodic flashes of tactical inspiration render him one of the great enigmas of military history. One is left to speculate as to how much more successful his career might have been if he had enjoyed unrestricted support from Versailles, and if his personal habits had not left their mark upon his appearance and behaviour. Such handicaps were bound to affect his efficiency and judgement. The fact that, with such handicaps, he achieved such outstanding successes in Italy and Spain gives ground for fascinating speculation.

Villars, Claude Louis Hector, Duc de (*1653–1734*), *Marshal of France*

Villars was Louis XIV's most successful commander in the War of the Spanish Succession, and one of Marlborough's (q.v.) most redoubtable opponents after the Battle of Blenheim.

The son of an army officer, Villars was born at Moulins on 8 May 1653. Entering the cavalry at the age of sixteen, he distinguished himself early in the war of 1672–9, and after the Battle of Seneffe in 1674 was promoted colonel in a cavalry regiment. Although his reputation as a light cavalry leader continued to grow, further promotion was slow owing to the favouritism shown to others by the minister Louvois. On the outbreak of the War of the Grand Alliance, therefore, Villars resorted to purchasing the post of French cavalry commissary general, and lead the cavalry against the Allies in Flanders in August 1689. Four years later he was promoted lieutenant-general.

Towards the end of 1702 Villars crossed the Rhine and defeated the Margrave of Baden at Friedlingen in October, saving the French infantry by skilful use of his cavalry, for which feat Louis created him a marshal. The following May he joined the Bavarian Elector at Ulm. His bold plan for an advance on Vienna was overruled by Maximilian Emanuel, and although the combined forces won a victory at Höchstädt on 20 September 1703, disputes between the two grew so heated that Villars returned in disgust to France. At the time of the Battle of Blenheim in 1704 he was campaigning to put down the insurrection of the Camisards in the Cévennes.

To redress the disaster of Blenheim, Villars was recalled and sent back to the north-eastern front, at the same time being granted his ducal title. He faced Marlborough, whose advance into France along the Moselle he blocked, and then switched to an advance into Alsace, where he captured a great proportion of the Allies' supplies and artillery reserves. In 1706 he was forced on to the defensive in Alsace, but the next year he again crossed the Rhine, surprised and broke the German lines at Stollhofen, and advanced deep into Swabia before being halted and forced to retire. These bold adventures found little support in the French court at Versailles, and Louis XIV recalled Villars to oppose Prince Eugène's (q.v.) attempts to break into France.

In 1709 the fifty-six-year-old marshal was appointed to command the French forces in Flanders, and when Marlborough and Eugène invested Mons Villars was attacked in a very strong and well-sited position. In the Battle of Malplaquet on 11 September the French forces inflicted very heavy casualties on the Allies' flanks, but Villars was severely wounded in the leg and rendered *hors de combat*. Soon afterwards the Allies mounted a powerful counter-attack which broke the French centre, but such were the heavy casualties and the exhaustion of the survivors that they were unable to pursue the retreating French army. In 1711 Villars developed the famous lines of *non plus ultra*, supposedly impregnable, for the defence of France, but they failed the test of Marlborough's genius.

By 1712 the British contingent under Marlborough had been withdrawn from the fighting, and Villars set about driving Eugène from Flanders. He defeated a Dutch corps at Denain on 24 July before Eugène could come to its assistance – a defeat which effectively ended the campaign in Flanders. Villars then returned to the Rhine, captured Freiburg and Landau in 1713, and went on to negotiate peace direct with Prince Eugène at Rastatt. After the Peace of Utrecht Villars became France's principal adviser on military and foreign affairs, but his vanity prevented his ever endearing himself to the French court. The War of the Polish Succession brought the eighty-year-old veteran out of retirement to lead the French army in the Milanese. The campaigns of 1733–4 showed that the years had not dimmed the old hero's military spirit but contemptuous of his ally the King of Sardinia, he attempted to procure his own recall to France. He fell ill at Turin and died there on 17 June 1734.

Undoubtedly Villars possessed military genius, and was a natural cavalry tactician – to be numbered among history's greatest. His excessive boastfulness, however, constantly offended the Versailles court, whose frequent objections to his adventures undoubtedly restricted his freedom in the field.

Villiers de l'Isle Adam, Philippe (*1464–1534*), *Grand Master of the Order of the Hospital of St John of Jerusalem*

One of the outstanding Grand Masters of this famous Order, Villiers de l'Isle Adam organised and commanded the defence of Rhodes during the great siege of 1522. His later contributions were of even greater importance; his qualities of leadership and diplomatic skill prevented the disintegration of the Order in the difficult period which followed its expulsion from Rhodes, and eventually secured for it its new base on Malta.

The Order was one of several which traced its birth to the Crusades of the eleventh century. Originally it was purely a hospital order, giving aid and comfort to distressed pilgrims to the Holy Land. From these beginnings it slowly grew more militant; from small escort details for pilgrim convoys, the Order evolved into one of the most formidable spearheads of Christendom in the ceaseless wars against Islam. The Knights of St John came to wield immense wealth and power – not always wisely, or honourably – and held some of the great bastions of Outremer, such as the *Krak des Chevaliers*. With the eventual ruin of the Crusader kingdoms at the end of the thirteenth century the survivors of the Order moved to Cyprus. While the Templars proved unable to adapt to the changed circumstances, the Knights of St John survived the traumatic loss of the Holy Land and progressed in a new direction, a course which ensured them a further three centuries of vigorous and largely successful campaigning against the Moslem advance.

In 1309, in concert with Genoese interests and with the Papal blessing, the Knights captured Rhodes, then nominally a part of the Byzantine Empire, but actually an independent and lawless nest of Greek and Italian adventurers. This marked the Order's evolution into a naval power. Slowly it built up its strength and skill in sea warfare, while the fortified harbour of Rhodes gradually acquired massive defences, and further fortresses were established around the coasts and on outlying islands. During the fifteenth century the Order became a serious force to be reckoned with in the eastern Mediterranean, counting among its members some of the hardiest and most resourceful sea-soldiers of the age. While always retaining the Hospital and the austere religious discipline, the Knights became increasingly wealthy. Only the noblest young men in Europe could join the Order where they underwent a most rigorous and protracted training. It was while the Order was on Rhodes that its classic organisation evolved. It was divided into national groups or *langues*: those of France, Provence, Auvergne, Italy, Castile and Portugal, Aragon, England, and Germany. The senior officer of each *langue*, the *pilier*, performed the duties of one of the traditional grand officers of the Order. Each national contingent had its own mess and quarters, and its own section of the battle-line; thus a (usually) healthy rivalry spurred each on to greater achievements. The Order built up an extensive trading network from its formidable base, and ruled Rhodes as a sovereign state, complete with its own coinage. The Rhodians, sophisticated people of their kind, soon adjusted to the new régime, and prospered with it. Apart from trade, each *langue* drew immense revenues from estates in the home country; and as skill and confidence increased, the galleys and carracks of the Knights preyed more and more effectively and lucratively on the seagoing traffic of the Turks. Eventually the mighty Sultan Mohammed II (q.v.) decided to crush this nest of Christian vipers; in 1480 he sent Misac Pasha across the straits from Marmarice with a huge army, but under the energetic leadership of Grand Master Pierre d'Aubusson the Knights and their mercenary soldiers repulsed the attempt with heavy loss. The next major confrontation was to come forty years later.

Philippe Villiers de l'Isle Adam was born in 1464, a young man connected with some of the noblest families of France, including that of the Constable and premier duke, Anne de Montmorency. He first came to prominence within the Order in 1510 as commander of the slow but heavily-gunned sailing ships of the Knights at the sea battle of Laiazzo. Here he distinguished himself, and the Knights gained a decisive victory over the Turks – but a dangerous resentment was apparently born in the heart of the commander of the Order's rowing galleys, a Portuguese named Andrea d'Amaral. Ten years later Sultan Selim the Grim died and was succeeded by his extraordinary son Suleiman I (q.v.), later called 'the Lawgiver' and 'the Magnificent'. There was no room for a powerful and expensive Christian outpost in Suleiman's plans for the eastern Mediterranean, and preparations were made for a major attack on Rhodes. In 1521 Villiers de l'Isle Adam was elected Grand Master of the Order at the age of fifty-seven and it was to him that Suleiman eventually sent a haughty demand for the evacuation of the island. It went unanswered, and on 26 June 1522 the first Turkish ships were sighted on the horizon.

The Grand Master had been laying in supplies and strengthening the already-massive defences ever since he took office, and Rhodes was one of the most formidable strongholds in the Mediterranean world. The lessons learnt in earlier wars had been incorporated: sophisticated design was coupled with sheer brute weight of rock, and the Order – always quick to follow advances in firearm technology – had assembled and sited a considerable array of fine cannon. The main weakness lay in manpower. There were only some 500 Knights and 1,500 mercenaries and local troops in the garrison, and there was little likelihood of reinforcement. The powers of Europe were at this time racked by two violent preoccupations which both distracted them from the battle against Islam, and led them to distrust the Order which guarded Christendom's south-eastern frontier. Firstly, the religious upheaval of the Reformation had called into question the position of such an anachronistic survival of the Crusades as a brotherhood of Christian knights proclaiming allegiance only to God. Secondly, the winds of nationalism were beginning to blow, and the true function and potential attitudes of a multi-national force of hardened fighters with its own navy gave many European rulers furiously to think. (In fact it seems that national rivalry seldom threatened the effectiveness of the Order.) In the coming struggle the tiny Order would be left to shift for itself, regardless of the threat to Europe as a whole of Turkish expansion.

Accurate numbers are hard to assess, but it seems that the Turkish army comprised well over 100,000 men – possibly as many as 200,000, with 700 ships. There were vast batteries of cannon of all types and calibres (up to bombards firing nine-foot balls), and many thousands of trained miners from Bosnia and Wallachia, to dislodge the massive walls of the landward defences. Against this weight of expert opposition the Knights had only their walls, their personal courage and hardiness, the excellence of their weapons and armour, and their fierce determination. A mercenary engineer officer of brilliant ability, one Gabriele Tadini, had been secured for the Order by Villiers de l'Isle Adam. Tadini was responsible for many of the new fixed defences; and his teams of picked counter-miners, alerted by sophisticated listening devices, caused heavy Turkish casualties by placing and detonating underground charges.

After a month of preliminary skirmishes, the Sultan himself arrived late in July and the assault proper began. The whole course of the siege was characterised by an incessant and massive artillery bombardment accompanied by furious sapping and mining, enlivened by huge set-piece massed infantry attacks on any section of the walls which seemed weakened. There was furious fighting around the bastion of the English *langue* following the explosion of a Turkish mine early in September, and in this Villiers de l'Isle Adam personally led the successful counter-attack. After another three weeks of bombardment a major assault was made on the sectors of England, Aragon, Italy, and Provence on 24 September. The janissaries secured a lodgement on the bastion of Aragon, but were thrown back after bitter fighting. Throughout October the artillery and mining duels continued, wearing down both sides; the Knights had now suffered more than 200 fatalities and more than that number of seriously wounded, and in October Tadini was gravely injured. Another blow to morale was the discovery of a spy communicating with the Turks, who claimed that he acted on the orders of Andrea d'Amaral, the Pilier of Castile and Portugal, Grand Chancellor of the Order. This claim was never conclusively proved or disproved, but Amaral was an arrogant and unpopular officer who had never concealed his rage at Villiers de l'Isle Adam's election to the Grand Mastership he himself coveted. He refused to confess or to deny the charge even on the rack, and went silent to his execution.

During November and December the worsening weather made conditions both in the citadel and in the Turkish trenches extremely harrowing; and the apparent impossibility of reinforcement, coupled with the inexhaustible manpower reserves of the Sultan, led to a lively discussion in the garrison on the subject of possible terms. Turkish losses had been so heavy – one source quotes 90,000 dead and disabled for the whole campaign – that Suleiman was prepared to offer unusually generous concessions. The Grand Master argued for a defence to the last man, but was eventually overruled. The Rhodian population threatened to make their own separate peace if the Knights insisted on holding out; and on Christmas Eve the Sultan's detailed terms were made known. The Knights could leave in safety, carrying their goods and their relics, and all their arms save only the cannon. If any chose to stay, they were guaranteed freedom of religious observance in exchange for a simple acknowledgement of the Sultan's secular suzerainty. On 26 December 1522 Villiers de l'Isle Adam bowed to the wishes of the majority, and met the Sultan to make formal submission. On the evening of New Year's Day 1523 the surviving Knights sailed away from Rhodes for ever in the great carrack *Santa Maria*, two galleys, and a barque.

For the next seven years the Order, based temporarily first at Viterbo and later at Nice, was threatened with disintegration. Its place in the contemporary religio-political structure of Europe was questioned, and the suspicion of the rulers of Christendom was intense. None wished to offer a home to this strange, fierce community, which might throw its weight behind one of the factions involved in the turmoil of the day. The Grand Master tirelessly tramped from court to court, seeking a defensible base for the Order in southern Europe from which to carry on the fight against the Sublime Porte. At last, in 1530, the newly-crowned Emperor Charles V offered the Knights the Maltese archipelago on condition that they also agreed to garrison Tripoli for him. This was a mixed blessing indeed, but the desperate Knights were in no position to refuse. Malta was an uninviting prospect after the paradise of Rhodes, but its magnificent harbour tipped the balance. In the autumn of 1530 the Knights crossed from Sicily and took up their new fief; they were greeted with alarm by the peasantry, and active resentment by the local hereditary aristocrats. Some of the younger Knights were so appalled by the barren island that they fomented serious dissension within the Order, troubling the last years of the tired old Grand Master. Philippe Villiers de l'Isle Adam died in the old capital of Mdina in 1534; one of the most brilliant Grand Masters in the history of the Order, he deserved better of his brethren than the half-hearted loyalty he received at the last, but his memory is revered.

Wallace, Sir William (*c. 1270–1305*)

It would be unthinkable to many Scots that a book such as this should exclude Sir William Wallace, just as a Spaniard would find the omission of '*El Cid*' (q.v.) inexplicable; but their inclusion raises the same problem. Wallace, a skilled guerrilla leader and captain of small war-bands, became the inspiration of Scottish resistance to Edward I (q.v.) of England. His perseverance, patriotism, and eventual cruel death have raised him almost to the status of a national saint, but this process has obscured his actual military virtues. He was a remarkable leader of men, but our knowledge of his prowess as a general rests on but one admirable victory and one disastrous defeat.

Wallace was the second son of Sir Malcolm Wallace of Elderslie in Renfrewshire, a vassal of James, fifth Steward of Scotland. In 1295 King John Baliol of Scotland, originally the nominee of Edward I of England in a succession dispute, rebelled against his English overlord. Edward invaded Scotland, smashed Baliol at Dunbar, and declared that the country was henceforth under his direct rule; the Stone of Scone was taken south, and Scotland left under military occupation. In May of 1297 William Wallace first appears in recorded history. With a small band of followers he attacked Lanark under cover of night, and killed the English sheriff. More men under Sir William Douglas joined him; they marched to Scone and expelled the English justiciar, and subsequently attacked several English garrisons between the Forth and the Tay. Several nobles, including the Steward, Robert Bruce (later King Robert I) gathered an army to expel the English, but in July they were forced to surrender at Irvine by Sir Henry de Percy and Sir Robert de Clifford. Wallace and a large band of followers took refuge in the Forest of Selkirk, and remained active against the English. During the summer he laid siege to Dundee, but abandoned the attempt in order to link up with Andrew de Moray to oppose an English army under John de Warenne, Earl of Surrey, which was approaching the strategic English stronghold of Stirling. Wallace allowed about half of Surrey's force to cross Stirling Bridge (also known as Cambuskenneth Bridge) on the morning of 11 September 1297; he then attacked those who had already crossed with great determination, annihilating them and preventing reinforcements from crossing the narrow bridge. Many drowned in the river; Surrey withdrew, destroying the bridge as he went, but the Scots crossed by a ford and harried his retreat with great success. Surrey reached Berwick, and subsequently York, with only a fraction of his army intact; and apart from isolated garrisons Scotland was free of English occupation.

Wallace ravaged the border counties of Northumberland and Cumberland that autumn, burning Alnwick and besieging Carlisle. In December 1297 he returned to Scotland and was knighted, although by whom is not known. He assumed, or was voted, the title Guardian of the Kingdom, in the name of the imprisoned King Baliol. He then set about reorganising the country's military resources and administration, ruling wisely and with the support of many influential nobles and churchmen. There were others, however, who had lands or hostages in England and who regarded Wallace with less than wholehearted loyalty. His power and prestige rested entirely on his ability to continue beating the English in battle.

Early in 1298 the Earl of Surrey returned, advancing to relieve the beleaguered garrisons of Roxburgh and Berwick, but on King Edward's orders pressing no further. In early July the King himself led an English army of some 7,000 heavy cavalry, 3,000 light cavalry and 15,000 infantry across the Tweed; his foot included strong companies of longbowmen, and Welsh and Irish auxiliaries. Wallace fell back before the English advance, wasting the countryside as he went. The English army began to suffer from hunger and poor discipline, but Edward pressed on with characteristic determination and brought Wallace to battle at Falkirk on 22 July 1298. (It is said that Robert Bruce fought with Edward's forces at Falkirk – a healthy reminder that popular national heroes in many countries tended to be motivated by more complex and less disinterested considerations than modern worshippers would have us believe.) Wallace had drawn up his army in a strong position on a slight slope, with his front protected by a stream and soft ground. He had some 1,000 heavy and 2,500 light cavalry, and about 25,000 infantry. The spearmen who made up the bulk of his army were drawn up in four circular phalanxes or schiltrouns, an extremely difficult formation for cavalry to break. Edward first committed his heavy horse to several charges, which were hampered by the soft going, unsuccessful, and costly. He then succeeded in driving the small Scottish cavalry force, under Sir John Comyn, from the field; this enabled him to bring his infantry forward, and he had his archers open up a heavy fire on the solid formations of enemy spearmen. This was a tactic Wallace could not counter; the ranks of spearmen withered and fell, and when the schiltrouns were thoroughly shaken, Edward's cavalry charged and finished the issue. In this battle some 15,000 Scots were slain; it was the first significant use of the longbow by an English army as a decisive battle-winning weapon.

Wallace fell back northward with the survivors of the slaughter, burning Stirling and Perth as he retreated; Edward was prevented by difficulties in England, and by the impossibility of long maintaining an army in the field, from following up his victory. But Wallace's military reputation was at an end, and he resigned the Guardianship. The facts of his life over the next four years are obscure; it is thought that he went to France in 1299, but the date of his return is unknown. Certainly resistance to England continued, and it is thought that Wallace continued to play his part as a solitary guerrilla leader. In 1303 he took Stirling Castle again; the unwillingness of Edward's barons to march north for a winter campaign prevented the grim king of England from reacting immediately, but the following year he mounted another major offensive. He appears to have regarded Wallace as particularly dangerous; the knight of Elderslie was the one irregular leader to whom Edward never offered terms, and whom he tried to capture with dogged persistence. On 5 August 1305 Wallace was arrested by Sir John Menteith near Glasgow, probably as a result of treachery – of which there was no shortage in Scotland. He was taken to Dumbarton Castle, and then to London. On 23 August he was indicted and condemned as a traitor at Westminster Hall, without trial. Edward destroyed him with the legal fiction that since he had taken over direct rule of Scotland, Wallace was a traitor to his rightful king. Wallace stoutly denied this, on the entirely proper grounds that he had sworn no allegiance to Edward. That most ruthless of English kings was determined to remove the threat once and for all, however; and that same day Sir William Wallace was hanged, drawn, beheaded, and quartered. His head was displayed on London Bridge, his quarters at Newcastle, Berwick, Stirling, and Perth. He left behind no known children, no portrait, not even a contemporary description of his appearance and manner: only an immortal reputation as a martyr for his country's freedom.

Wallenstein, Albrecht Eusebius Wenzel von, Duke of Friedland and of Mecklenburg (*1583–1634*)

It would be pointless to attempt a summary, in a book of this type, of the causes and course of the dozen or so wars generally described as the Thirty Years' War (1618–48). This series of wars, involving all the major and many of the minor powers of Europe, started over an ostensibly religious issue – the struggle between the Protestant states of central Europe and the Catholic Counter-Reformation. From the first, however, complex dynastic and constitutional issues played a major part; in essence it could be said that the central question was the sovereignty of the individual state as opposed to the hegemony of the Hapsburg dynasty and the Holy Roman Empire. The progress of the dozens of alliances which were formed and broken, the shifting patterns of treaties and secret agreements, the invasions, occupations, and punitive campaigns were governed to a large extent by straightforward bids for European political power by the various participants; nevertheless the religious aspect was used as justification for extreme barbarity on both sides throughout the period. The armies of the day lived off the land and plundered and sacked without mercy, and the successive campaigns devastated – and in some areas depopulated – large regions of central and northern Europe. The sufferings of the civilian populations were appalling, as the armies involved were largely composed of mercenaries who were unrestrained by any local loyalties. One of the most feared figures to emerge from this period of savage chaos was Wallenstein, a remarkable strategist – though an indifferent tactician – who raised himself from the minor Czech aristocracy to a position of power second only to that of the Emperor himself. A dark and brooding personality, he followed an obsessive path to personal power without visible scruple for the devastation he left in his wake. He believed implicitly in astrology, and often based his decisions on the predictions of seers. In the end his ambition was the death of him, as he attempted to dictate the course of European history while misjudging the loyalty of the army on whose pikes he relied.

Born of Protestant stock at Hermanice in Bohemia on 24 September 1583, Wallenstein was educated in Protestant institutions at Goldberg and Altdorf, and made the 'grand tour' of Europe in 1600–2. In 1604 he served with the Bohemian contingent of the Hapsburg army in a war against Hungary, and in 1606 ingratiated himself with the authorities of the Holy Roman Empire by turning Catholic. Three years after this he married an extremely wealthy widow who died five years later, thus acquiring great riches. In 1617 he raised at his own expense a mercenary force to fight for the Emperor-elect Ferdinand of Styria in a campaign against Venice. The following year the Bohemian nobles rebelled against Ferdinand's succession; they demanded freedom of worship for Protestants, whereas Ferdinand was committed to a policy of Catholic orthodoxy and of increasing the central power of the Empire at the expense of local interests. Wallenstein remained faithful to Ferdinand, and after the loss of his estates to the rebels raised and led a cavalry regiment which did notable service in the campaigns of 1619–21. He profited greatly from Ferdinand's victory, being named Governor of the Kingdom of Bohemia and acquiring vast areas of north-east Bohemia by purchase or confiscation of the estates of former rebels. His privileges included commercial advantages and the right to issue coinage, by which means he made himself one of the most powerful magnates in Europe. In 1625 he was named Duke of Friedland. In that year a Danish army attacked the Empire's northern possessions, and Wallenstein offered to raise, from his own resources, an independent imperial army of 24,000 men; he was to be reimbursed by taxing or confiscating captured lands. Ferdinand (now Emperor) wished to free himself from dependence on the troops of the Catholic League of princes led by Maximilian of Bavaria, and willingly agreed. Wallenstein was appointed 'capo' of all imperial forces, and turned his duchy of Friedland into a great arsenal and supply depot to provision his army. He laid the foundations for yet greater power by ensuring that all the generals and colonels in his army were bound to him by personal financial interest.

In April 1626 Wallenstein defeated Ernst of Mansfeldt at Dessau, and by offering his resignation out of pretended pique over an imperial criticism, managed to secure an extension of his army to 70,000 and greater personal powers of discretion. He was instrumental in forcing the Peace of Pressburg on the Hungarian leader Bethlen in December 1626, and for the next few months the armies camped where they were, stripping a starving and plague-stricken continent of its remaining resources. In July 1627 he drove the Danes out of Silesia, and in concert with the other great imperial general, Tilly (q.v.), conquered Schleswig, Holstein, Mecklenburg, and the whole of continental Denmark. The rewards showered on him by the Emperor, including the Duchy of Mecklenburg, alarmed the German princes. In 1628 Wallenstein besieged Stralsund in an attempt to gain control of the Baltic coast, but Swedish threats obliged him to raise the siege in July. He was much impressed with the possibilities of naval and commercial sea-power, which he had not previously considered, and had grandiose plans for forging a great trading consortium of imperial, Spanish, and Hanseatic interests. It was at this point in his career that Wallenstein began to act more like an independent sovereign than a servant of the Empire: he opened negotiations with Protestant states on his own initiative, and thus offended Ferdinand, whose religious motivation was still strong. The German princes, jealous of his power and seriously disturbed by the existence of an independent imperial army serving the central government, pressed for his removal. At the Electoral Diet of Regensburg in the summer of 1630 they threatened Ferdinand with an alliance of both Catholic and Protestant states under French leadership, and with a refusal to elect his son as his successor, unless the army was disbanded and Wallenstein dismissed. On 13 August 1630 the Emperor yielded and dismissed his generalissimo. Wallenstein never forgot his grudge against Ferdinand and Maximilian of Bavaria, a prime mover in the princes' conspiracy.

A factor in Ferdinand's agreement to dismiss

Wallenstein for the sake of German unity was the arrival on the continent for the first time of the excellent Swedish army of Gustavus Adolphus (q.v.), one of the great captains of history. In September 1631 Gustavus defeated the imperial forces under Tilly at Breitenfeld, and it became obvious that the Emperor would have to take Wallenstein back into his service. Wallenstein agreed to this in April 1632, but only after exacting an enormous price in new personal prerogatives in the areas of appointment of officers, disposal of captured territory, and so on. He raised a new army, and quickly cleared Gustavus's Saxon allies out of Bohemia. He fought a brilliant strategic defensive campaign in southern Germany; in August the Swedish army was badly mauled during vain attacks on his entrenched camp at Alte Veste. Gustavus then withdrew north-west, and Wallenstein north, threatening Swedish communications. On 16 November 1632 the two armies clashed at Lützen; Gustavus led some 20,000 men, and Wallenstein about 18,000, reinforced later in the day by a further 8,000. Both sides gained and lost ground during the engagement, and eventually Wallenstein was forced to withdraw with the loss of some 12,000 casualties and his artillery and baggage – but among the 10,000 Swedish casualties was Gustavus, killed in a confused cavalry clash.

Gustavus's death changed the situation considerably. No longer threatened by this 'Lion of the North', the Emperor did not have the same overriding need for Wallenstein's services. Sweden's allies were shaken, and more amenable to negotiation and limited military pressure. Wallenstein attempted to shape the future of kingdoms, however, and engaged in complex and contradictory negotiations with many interested parties – negotiations which cost him his credibility with all of them. He toyed with treason against the Emperor, confidently asserting that he could take his formidable army with him wherever he chose to place his loyalty. His officers refused to back him, however, and confirmed their determination to remain loyal to Ferdinand. After an abortive *coup* in the winter of 1633–4 Wallenstein tried to regain the Emperor's trust, but in vain. At Eger on 25 February 1634 he was assassinated by a group of Irish, Scots, and English mercenary officers. Run through with a halberd by an English captain named Walter Devereux, he died crying for quarter.

Washington, George (*1732–99*), *General, 1st President of the United States of America*

Though never considered a brilliant military leader, Washington certainly had to bear heavier burdens of responsibility than any other man in America in the eighteenth century. He was forced to fight campaign after campaign with inadequate, badly-fed, badly-clothed, and badly-armed soldiers; and his dogged and resilient sense of his responsibilities and duty probably saved the Revolution. His ultimate victory at Yorktown placed him in an unassailable position of respect and affection in American hearts.

Born at Bridges Creek, Virginia, on 22 February 1732, George Washington was a third-generation member of a family which had come to the New World about 1656. As the third son, young Washington was not given a full education, and only received irregular teaching between the ages of seven and fifteen. Thereafter he pursued such study as he could, but quickly acquired an outstanding skill in surveying. When only sixteen he was commissioned to survey the still-wild country of the Shenandoah valley – a task which occupied almost three years, but which provided much training in meeting the perils of unexplored territory, and toughened the young man's frame.

His personal knowledge of the frontier territory was recognised in 1751 with his appointment, as a major and adjutant-general of the provincial militia, to assist in repelling the frequent encroachments by the French and Indians in that area. But on the death of his half-brother George Washington succeeded to the family estate and settled down to a life of farming, showing himself to be a skilled horseman and hunter.

By 1754 Washington had reached the rank of lieutenant-colonel in a Virginian regiment, and when the commander was killed, he assumed overall command of an expedition against the French and Indians. The following year the British Commander-in-Chief, General Edward Braddock, recognising Washington's special knowledge of the territory, promoted him full colonel and took him on to his staff; he survived Braddock's Massacre, and in August that year returned home to Virginia, where the Governor appointed him to lead all Virginian troops. Recovering from a bout of ill health, he took part in General John Forbes's successful expedition which culminated in the capture of Fort Duquesne and the building of Fort Pitt in 1758 on the site of the future Pittsburgh. That year he was elected to the House of Burgesses, a position he held for the following sixteen years. As a substantial landowner he accepted the principle of slavery without necessarily approving of it, although he progressively added to his holding of slaves throughout his life.

The first of the events that were to bring Washington to a position of national leadership was the Royal Proclamation in 1763 which forbade settlement beyond the Alleghenies; for he possessed interest in a number of western land companies. In 1765 he opposed the Stamp Act, and in 1769 was a prominent opponent of the import of English goods to the colonies – a movement which culminated in the Raleigh Tavern resolutions against the importation of tea and other taxable commodities in 1774. Virginia joined Massachusetts in the call for a Continental Congress, and Washington was chosen to head the first two such Congresses. It was John Adams of Massachusetts who, recognising Washington's military experience and commanding capabilities, recommended to the Congress his appointment as Commander-in-Chief of the revolutionary army, and on 17 June 1775 Washington received his commission in this post – too late to avert the defeat of the colonials at Bunker Hill.

Washington's task was immense. Opposed by three separate English armies, he had to transform a rabble of poorly-armed and notably undisciplined men from every walk of life into some form of reliable military force, at the same time maintaining the siege of Boston. By March 1776, faced by a revolutionary army of nearly 30,000 men supported by artillery dragged from Fort Ticonderoga, the English commander General William Howe abandoned Boston and moved his forces to Halifax. Washington transferred his forces to New York, but abandoned the area following his defeat in the Battle of Long Island on 27 August at the hands of General Howe. For the Americans defeat followed defeat. At the Battle of Harlem Heights the colonials managed to delay the advance of the English army, as they did again at White Plains on 28 October. Fort Washington was captured, with the loss of almost 3,000 Americans, and as Washington retreated south during November and December a rearguard of 4,000 men under General Charles Lee was captured near Morristown. Then, in accordance with traditional usage, Howe went into winter quarters.

Washington, his small army reduced to tatters and almost exhausted, determined to take a last desperate chance. Collecting together a force of 2,400 men, he moved against the garrison at Trenton at dawn on 26 December in an extraordinary surprise attack. The entire garrison force was killed or captured, and a large quantity of much-needed military equipment and stores was taken. In swift reprisal Cornwallis (q.v.) now moved up with 8,000 men to box in Washington against the Delaware river; but, leaving his campfires burning, Washington slipped his army out of the trap and attacked Cornwallis from the rear, defeating and capturing British reinforcements marching up from Princeton before Cornwallis could move his main force from Trenton. By these successful actions Washington suffused his army with new heart and confidence. In ten days he had adopted positions which threatened the flanks of the English forces; with their lines of communication seriously endangered, all British garrisons in central and western New Jersey were withdrawn.

Washington's own problems of command were constantly aggravated by difficulties with Congress. That body had, in desperation, bestowed almost dictatorial powers upon the Commander-in-Chief, but such powers were repugnant to several congressmen, and they frequently sought to interfere in purely military matters. In his army there were burning jealousies between the New Yorkers and Yankees, and, more generally, between the

Northerners and Southerners. As a Virginian Washington did not enjoy the trust of such New England leaders as John Hancock, or even John Adams, although he had been instrumental in placing Washington at the head of the army.

1777 was a difficult year for Washington's army in the central area. Defeats at Cooch's Bridge and in the Battle of the Brandywine in September cost the Continental Army more than 1,000 men, and after the evacuation of Philadelphia Congress fled first to Lancaster and then to York – again leaving Washington with near-dictatorial powers. A further defeat at Germantown on 4 October, again costing about 1,000 men, was followed by the famous wintering at Valley Forge, where Washington's army spent six agonising months in the open, inadequately clothed and fed, while the British enjoyed the shelter of Philadelphia. Only through the discipline and training of a German volunteer, Baron von Steuben (q.v.), did the army at Valley Forge survive – and also acquire sound training in military skills.

With the coming of summer 1778 Sir Henry Clinton, who took over from Howe, removed his army from Philadelphia overland to New York, intending to pursue a more vigorous offensive against Washington. Now increased to a strength of 13,000 men and hardened by Steuben's discipline, the Continental Army marched in pursuit and on 28th June brought the British army to battle at Monmouth. After an all-day fight in great heat the British left the field and made good their escape to New York, where they were blockaded by Washington. It was from New York that, for the next three years, Washington directed the American conduct of the war. But in August 1781 he marched south to confront Cornwallis's main forces in the Yorktown campaign – leaving 2,000 men to contain Clinton's 17,000-strong army.

Arriving at Yorktown in September and meeting a force of 7,800 well-equipped French troops under Rochambeau (q.v.) – who now came under his command – Washington laid siege to the city, surrounding Cornwallis and 8,000 men. Expecting to receive reinforcements from Clinton, Cornwallis withdrew his garrison to the inner fortifications, but came under heavy artillery fire. After a siege of three weeks Washington presented his call for unconditional surrender, and on 19 October Cornwallis's garrison marched out to lay down its arms.

The victory at Yorktown virtually brought the revolutionary war to an end and established Washington as undisputed leader in American hearts. It is said that, had there been a demand for the establishment of an American monarchy, Washington would have been the natural choice as sovereign – but the republican mood prevailed everywhere. The war had, it is true, brought independence, but the thirteen states were linked only by the loosest bonds, and the revolt of farmers led by Daniel Shays in 1786 only served to demonstrate the chaotic state of American autonomy and the need for an all-embracing constitution. It was for this purpose that Washington was elected to chair the Philadelphia Convention in 1787. Under the constitution suggested a President was to be elected by an electoral college; and it was Washington himself who was elected on 6 April 1789 as the first President, a position of supreme leadership which he was to occupy for ten years – and one which he had from time to time occupied already at the express wish of the wartime Congress, though without the benefit of democratic election. His very simplicity of approach to the administration and his aversion to any suggestion of pomp confirmed the American determination to avoid a monarchy and provided the seeds of democratic republicanism that have prospered ever since.

Washington died on 14 December 1799. He had shown himself a great leader of men – by no means gifted in the sophisticated strategy of war, but a man who understood the strengths and weaknesses of his fellow Americans. He led his troops to no massive victories on the field, but contented himself with the personal command of what he considered to be the vital campaign – that in the central states. He avoided confrontations with overwhelmingly superior enemy forces, but successfully locked up or threatened his opponents, so depriving them of the necessary freedom of action. He thereby proved himself one of history's great national leaders – one whose early successes were praised by Frederick the Great himself.

Wavell, Archibald Percival, 1st Earl Wavell of Cyrenaica (*1883–1950*), *Field-Marshal*

Wavell was the stoic British commander who, after years of parsimony in military matters by successive British peacetime governments, was given the task of withstanding Axis pressure in the Middle East. Despite the numerous demands on his scanty manpower reserves, he conducted the magnificent offensive in Cyrenaica after Italy's entry into the Second World War in 1940.

Born at Colchester, Essex, on 5 May 1883, the son of a major-general, Wavell was educated at Winchester and, after attending the Royal Military College, Sandhurst, was appointed to his father's regiment (The Black Watch) at the age of eighteen. In this famous regiment he served for the last year of the South African War in 1901–2, and was the youngest regular officer to receive the South African War medal with four battle clasps.

In 1908 he was sent to India for service on the North-West Frontier. During the First World War, after two years' service on the Western Front from 1914 to 1916, when he was wounded, he was appointed military attaché with the Russian Army of the Caucasus (being awarded the Orders of St Vladimir and St Stanislas) until June 1917. At the onset of the Russian Revolution Wavell was withdrawn from that country, and accompanied the Egyptian Expeditionary Force from 1917 until 1920, with the rank of brevet lieutenant-colonel.

Between the wars he rose to brigade and divisional commander (the 6th Infantry Brigade, 1930–4, and the 2nd Division, 1935–7); and was then given command in Palestine and Transjordan for a year in 1937–8. During the 1930s he gained an outstanding reputation for his thoroughgoing training methods – albeit among troops whose equipment had improved little since the First World War.

In 1939 Wavell assumed supreme command of all British forces in the Middle East; but little occurred in this theatre until Italy's entry into the war in June 1940. Then, with a total of 80,000 combat troops, few tanks, and fewer modern aircraft, he faced the numerical might of Italy's metropolitan and colonial armies – odds of the order of ten to one against him. Realising that any attack in northern Egypt by Graziani's (q.v.) army could only be sustained over extremely long lines of communication, Wavell decided to hold his two half-strength divisions back from the frontier, merely screening the forward positions with light forces. When the Italians attacked, Wavell let them advance into country unfamiliar to them before holding them from prepared positions at Sidi Barrani. While Graziani's army engaged in the pointless establishment of a line of defence in depth, setting up numerous strongpoints over a fifty-mile stretch of desert, Wavell consolidated his forces, putting the finishing touches to their training, and preparing for attack. His offensive was delayed when orders arrived from England to send several of his best squadrons of fighters and bombers to assist the Greeks in their resistance to the Italian invasion, and to occupy Crete.

On 9 December 1940 Wavell's attack in Egypt was launched. Moving up with a strength of about 35,000 men (roughly one and a half infantry divisions, and an armoured division) by night, his tiny army, commanded by General Sir Henry ('Jumbo') Wilson, smashed a gap in the unsuspecting Italians' line and, with magnificent support from sea and air, quickly eliminated the entire Italian fortified system. By the end of the year Graziani's routed army (once 200,000-strong) had been thrown out of Egypt and locked up in and around Bardia, leaving behind 38,000 prisoners and a vast quantity of equipment. Wavell's assault at Sidi Barrani was planned much along the same lines as Allenby's (q.v.) great breakthrough at Megiddo in 1918, in which Wavell had participated, and to the study of which he had devoted a good deal of time – though of course the cavalry role was now assumed by the new British 'I' tanks.

The British offensive reopened in January 1941: Bardia was assaulted on 5 January and captured after a devastating sea bombardment by the fifteen-inch guns of the British fleet; Tobruk was invested by the 7th Armoured Division (the famous 'Desert Rats') and captured on 22 January; and on 7 February Benghazi was isolated by a southern sweep to Beda Fomm. In two months the British Western Desert Force advanced 500 miles, took 150,000 prisoners, destroyed nine Italian divisions, and captured or destroyed 400 tanks and 1,290 guns. This single great victory, achieved with the loss of 500 men killed, was of inestimable morale value to Britain, which had hitherto suffered setback after setback ever since the outbreak of war. It also reinforced the widely-held faith in the British soldier's superiority over his Italian counterpart, so that thereafter Italian forces, wherever they were encountered, were the subject of (possibly arrogant) ridicule and derision.

Simultaneously with the brilliant offensive in Libya, Wavell's forces were also mounting a campaign against the Duke of Aosta's 110,000-strong army in Ethiopia and Italian Somaliland. With polyglot forces (British, South African, Free French, and Kenyan) amounting to 70,000 under the joint command of Lieutenant-Generals Sir Alan Cunningham and William Platt, Wavell invaded Ethiopia on 19 January 1941 and inflicted defeat upon the Italian General Fusci at Agordat on the 31 January. On 24 January Cunningham invaded Italian Somaliland, capturing Mogadiscio (where the entire Italian oil reserves were taken intact) on 25 February. Turning south-westwards, Wavell's forces captured Addis Ababa on 4 April, and on 18 May the Italian generals, now facing the combined forces of Platt and Cunningham, surrendered. In this second outstanding campaign Wavell's troops overran 350,000 square miles of enemy-held territory and captured over 100,000 Italian troops; their own losses amounted to no more than 135 men killed, 310 wounded, fifty-two missing, and four captured.

By destruction of Graziani's army in the north and elimination of the Italian East African Empire in six months, Wavell secured the Suez Canal and the Red Sea as a vital Allied supply route. But his bolt was shot. The German invasion of the Balkans deprived the Western Desert Force of another 60,000 men – now ordered to Greece where they were sacrificed in a useless campaign. Furthermore a rebellion by Raschid Ali in Iraq threatened the British rear, and was only stamped out after further redeployment of slender British reserves – which were then committed to a campaign against Vichy-governed Syria.

In March 1941 General Erwin Rommel (q.v.) arrived in North Africa with the nucleus of his *Afrika Korps*, and instructions to counter-attack Wavell's greatly-reduced forces and advance into Cyrenaica. The attack opened on 24 March, the tactics employed by the Germans being an almost an exact copy of those previously employed by Wavell: by striking across the centre of Cyrenaica to Tobruk, Rommel isolated thousands of Allied soldiers, while almost the entire British 2nd Armoured Division was captured through lack of fuel. The 9th Australian Division was locked up in Tobruk, and on 17 April the brilliant British corps commander General O'Connor was captured by an enemy patrol.

Determined to retain Tobruk at all costs, Wavell committed one of his last reserve divisions, the 7th Australian, to its defence, moving it in by sea. He soon came under heavy political pressure from London to secure the besieged port's relief but, ill-equipped and without adequate air or sea support (owing to reductions in his forces in favour of Greece and Crete), his attempts to force the Sollum-Halfaya passes, strongly held by the Germans, failed. Wavell was quite unjustly made the scapegoat for weak strategy in London and was promptly replaced by Auchinleck (q.v.). Nevertheless what he had achieved in the Middle East few other commanders could have achieved. His overall area of command from 1939 to mid-1942 was the largest in the world commanded by one single army commander, and expressed in terms of numbers of combat troops in relation to the area, it was unquestionably by far the weakest. Yet in the last twelve months of his command he fought campaigns on eight different fronts with never more than seven divisions; secured and occupied half a million square miles of enemy-held territory; and inflicted over 400,000 casualties (including prisoners) on the enemy.

Scapegoat or not, Wavell's achievements ensured a continuing faith in his administrative abilities: he was sent to India and, after Japan's entry into the war, was appointed Supreme Commander in the South-West

Pacific. He was actually in Singapore when Japanese troops invaded the island. After the loss of all British and Dutch colonial possessions in the area (once again Allied forces and military preparations had been disastrously neglected by pacifist peacetime governments), Wavell returned to India as Commander-in-Chief. Promoted field-marshal on 1 January 1943, he was appointed Viceroy of India in succession to Lord Linlithgow – a post he continued to hold until replaced by Lord Louis Mountbatten in 1948. In India Wavell was probably best remembered for his administrative genius and drive in taking effective measures to check raging famine in Bengal. He died in London on 24 May 1950.

Wayne, Anthony (*1745–96*), *Major-General*

'Mad Anthony' Wayne's record as a revolutionary general in the War of American Independence was somewhat erratic, but considering the general standards of performance among his comtemporaries it was by no means discreditable. He had received no formal military training, and was forced to learn his business on the battlefield itself. He learned it soundly, and when summoned to the colours again in middle age, at a time when America's military fortunes had sunk to an all-time low, he rendered his country great service.

Born on 1 January 1745 at Waynesboro, Pennsylvania, he was the son of Isaac Wayne, an affluent local figure who owned 500 acres and a lucrative tannery. He attended a private academy run by his uncle in Philadelphia for a few years during his youth; a rowdy student, he picked up enough mathematics to qualify as a surveyor, and practised that profession in Nova Scotia in 1765–6. When the venture failed he returned home and lived on his father's estates; he married, and ran the tannery, which provided him with a comfortable income on the death of Isaac Wayne in 1774. Anthony Wayne was active in pre-revolutionary local politics, and in January 1776 Congress appointed the dark, handsome, quick-tempered young man colonel of a local regiment in the Continental service. He made friends and enemies with equal ease, and Washington (q.v.) feared that his fiery impetuosity would lead him into disaster, but in due course his Commander-in-Chief paid ungrudging tribute to his courage and self-possession under fire. In the spring of 1776 Wayne led his regiment to reinforce the doomed Canadian expedition as part of Thompson's Pennsylvanian brigade, but the invasion collapsed before he arrived. On 6 June General John Thomas led 2,000 Americans against a supposed British garrison of 600 at Three Rivers, only to find that Burgoyne (q.v.) was assembling there an army of 8,000. Wayne was prominent in the brief battle which preceded that American rout, and covered the retreat to Ticonderoga. He commanded the garrison there for a few wretched months, making his first acquaintance with the state to which disease, hunger, and mutiny can reduce a body of troops. Promoted brigadier-general in February 1777, Wayne joined Washington at Morristown in April to command the Pennsylvania Line. He was active during the successful British advance on Philadelphia, and at Brandywine on 11 September 1777 he held the centre until forced to withdraw by the collapse of the flank. On 20 September he led a force which attempted to circle the British army and fall upon its rear echelons, but was surprised and routed at Paoli. Wayne was conspicuous at Germantown on 4 October, leading a spirited attack; and he spent the terrible winter of 1777–8 with Washington at Valley Forge.

Wayne did well at Monmouth on 28 June 1778, a turning-point of revolutionary fortunes in that the Continental line regiments stood up admirably to British regulars in prolonged fighting, despite the early mishandling of the advanced guard by General Lee. The army reorganisation later that year gave Wayne command of a detached force of Continental light infantry units; with this formation he led the successful surprise assault on Stony Point on 16 July 1779. This northernmost British position on the Hudson yielded some 500 prisoners, fifteen cannon, and useful stores. On 25 September 1780 the swift occupation of West Point by Wayne's command foiled the attempt of the renegade Benedict Arnold to hand it over to the British. In December 1780 the Pennsylvania Line mutinied, and Wayne successfully took the men's grievances to Congress for redress. He served under Lafayette in the Yorktown campaign of 1781. On 6 July, near Jamestown Ford, Virginia, Wayne's brigade was mistakenly ordered to attack a force of 5,000 redcoats. So determined was the peformance of Wayne's 800 men, however, that they were able to fight their way clear. In 1781–2 Wayne served in the south under Nathaniel Greene and was particularly effective in persuading the Indians to desert the British cause; his impetuosity was now tempered with cunning. In 1783 Anthony Wayne retired from the army with the rank of major-general.

For ten years he lived the life of a private gentleman, although he was active in conservative politics. Several business ventures failed. In the aftermath of the war Congress was anxious to disband the army, fearing that any significant military standing force might turn into a tool of dictatorship. Internal unrest and Indian forays on the frontier were to be dealt with by levies of militia, and at one point the federal army was reduced to less than 80 men with the sole mission of guarding two arsenals. But it was grudgingly recognised as the years passed that some more practical force was needed, and a single composite regiment of infantry and artillery was raised. In October 1790 the depredations of the Miami Indians in the Ohio Valley brought a punitive column to the area under Brigadier-General Josiah Harmar, commander of the American Regiment. His force of just over 1,100 men included some 320 regulars; the remainder were untrained, badly led, and completely undisciplined militia. Harmar was unable to keep his straggling column together; on three occasions he clashed with Indians, and on three occasions the militia fled and left detachments of regulars to be butchered. After suffering 200 casualties Harmar withdrew. The following year Arthur St Clair took the entire regular force of 600 infantry and artillery and some 1,500 militia into the field in the same area. They were surprised on the upper Wabash on 4 November 1791, and defeated with great loss of life – some 900 dead, including almost all the regulars. Since this was another disaster at least partly due to the unreliability of militia, Congress belatedly authorised the raising of a new regular force. This was entitled the Legion of the United States, and Washington called 'Mad Anthony' Wayne out of retirement to command it, with the rank of major-general.

Wayne did not approve of the peace negotiations which were now opened with the Indians, but he made excellent use of the time bought for him. The Legion was initially assembled and trained at a site near Pittsburgh, and subsequently marched to Greenville, about eighty miles north of Fort Washington. A very high standard of drill, smartness, and energy was demanded, and Wayne worked tirelessly to weld the troops together into a serious professional force. The Legion was divided into four sub-legions, each with a peak strength of some 900 men of all arms. Each had a core of infantry, with supporting rifle, artillery, and dragoon companies under command. Throughout the winter of 1792–3 Wayne maintained a vigorous and punishing field training programme, and by the summer of 1794 the Legion was an extremely useful little army. Wayne set out to advance into the wilderness over the same route which St Clair had followed, but he moved slowly and secured his rear with forts. On 29 June the Miamis made an attack on Fort Recovery, which failed completely; it was a perfect blooding for the Legion, and damaged Indian morale. Wayne then ensured by careful diplomatic preparation that the British – who still maintained garrisons near by – would not support the Indians. He maintained his advance, and in August caught up with the main force of Indians at a spot called Fallen Timbers. He wisely held his hand for three days, so that many of the tribesmen became discouraged and wandered away to hunt for food. On 20 August he sent the Legion forward in a classic attack – a central infantry force charging with the bayonet supported by enveloping wings of cavalry. The Indians (under Blue Jacket and Tecumseh) were completely defeated for a loss of thirty-three dead and a hundred wounded from Wayne's command of some 3,500 men. Wayne proceeded to destroy Indian villages in the area and was prominent in the patient peace negotiations which culminated in final Miami submission at Greenville in August 1795. This campaign brought peace to the frontier for several years. Wayne died at Presque Isle (now Erie, Pennsylvania) on 15 December 1796, while returning from the occupation of the post at Detroit.

Wellington, Arthur Wellesley, 1st Duke of (*1769–1852*), *Field-Marshal*

To attempt a detailed summary of the career of the most successful soldier in British history in a book of this type is clearly impossible. Wellington's public life was long and eventful both in the military and the political field, and the latter must be ignored completely for reasons of space and relevance. A brief chronology of the main incidents in his career must suffice, followed by a short summary of the most significant of those personal qualities which raised him so far above his contemporaries – indeed, far above any other British commander, with the possible exception of Marlborough (q.v.).

He was born Arthur Wesley (the family name was changed in 1798) in late April 1769; the exact date and place of his birth are disputed, but his family's property at Merrion Street, Dublin, is the most likely place. He was the fifth son of Garret Wesley, first Earl of Mornington, an improvident Anglo-Irish peer who left his estates weighed down by debts on his death twelve years later. Arthur was educated at Eton and at a French military academy at Angers; he showed little early promise and was regarded as the dullard of the family. He obtained a commission in the 73rd Foot in 1787, in accordance with the then-prevalent view of the army as the last refuge of well-born idiots in straitened circumstances. He saw only a few days of regimental service during the next year, which he spent purchasing promotions in a number of regiments. These purchases were made with borrowed funds; throughout his early life he was bedevilled by humiliating debt, and was forced to take full advantage of the military and parliamentary patronage which decided so many appointments in his day. There has never been any serious suggestion that he himself was ever corrupt, however. In November 1787 he was appointed aide-de-camp to the Lord Lieutenant of Ireland, and spent the next six years as an ornament of the Dublin Castle staff in a round of social triviality and debauchery. In 1793, when further purchased promotions saw him installed as lieutenant-colonel of the 33rd Foot at the age of twenty-four, he reached a turning point in his life. Disappointed in love and sickening of his wastrel life, he seems to have made a conscious decision to apply himself seriously to his military career as being the only field left to him in which he might lead a worthwhile life.

His first taste of active service came in 1794 when he was posted to the Duke of York's army in Flanders. One of the few British officers to distinguish themselves in this campaign, he was subsequently given command of the rearguard brigade during the dreary retreat through Holland. The inefficiency of the command, the staff and the services of this expedition, and the unnecessary suffering and loss of life, made a bitter and lasting impression on him. He attempted to resign and secure a post in the Irish civil administration, but in the event was obliged to remain in the Army, and in 1796 sailed for India with his regiment. Here, too, there was much to offend his increasingly piercing professional eye; he broke himself of his taste for wine and cards, and applied himself to his duties with dedication. Fortunately his elder brother was appointed Governor-General of India in 1798 – Lord Wellesley paid attention to his military advice, and secured various advantageous commands for him. If his appointments were obtained by blatant jobbery, however, his performance in them was exemplary. In 1799 he commanded a division in the campaign against Tipoo, Sultan of Mysore, and after the capture of Tipoo's capital Seringapatam he was made Governor. After further action against powerful local bandit forces he was promoted major-general in April 1802. His first major victory with an independent command came in September 1803 during the Second Mahratta War. With 7,000 men and twenty-two guns, of which force only 1,800 were British, he defeated 40,000 Mahrattas with a hundred guns at Assaye. One third of his British infantry were casualties, a circumstance which made a deep impression on him. He won another victory at

Argaon, and in 1804 was knighted. He returned to England in 1805, financially secure for the time being and the richer for much useful experience. He had faced at first hand the problems of a general who is obliged to take the field at a great numerical disadvantage; he had held political responsibility in an environment of furious factional intrigue; and he had learnt a great deal about the capabilities and weaknesses of officers and men in the field. His professional reputation and immediate prospects were not assured, however. He was a very junior unemployed major-general whose only regular income came from his regimental colonelcy; and there was a general attitude in England that the exploits of 'sepoy generals' were not to be taken seriously in a European context. Sir Arthur took part in an abortive expedition to Hanover in December 1805, seeing no action, but the next two years were militarily uneventful for him. In 1807 he was given command of the excellent Reserve Division (43rd and 52nd Light Infantry, 92nd Highlanders, and 95th Rifles – all regiments which were later to serve under him with great distinction in the Peninsula) for the expedition to Copenhagen. His formation was the only part of the force to be seriously engaged by the Danes, and he won an easy victory at Kioge on 29 August.

Promoted lieutenant-general in April 1808, Sir Arthur sailed in July for Portugal in command of some 17,500 troops, with vague and discretionary orders for operations against the French armies which occupied both Portugal and Spain. When, however, it was discovered that Junot's Army of Portugal was much stronger than had hitherto been imagined, strong British reinforcements were sent after him, and more senior generals were instructed to supersede him. Meanwhile, Wellesley won a brisk action against weak French blocking units at Rolica on 17 August. The first reinforcements, his superior officers, and Junot with 13,000 troops all arrived at Vimiero on 20 and 21 August; but it was Wellesley who commanded the strengthened British army in the battle of Vimiero. He disposed his 17,000 men skilfully on the hills around the village, and decisively defeated Junot's attacks. But his superior, Burrard, took over immediately after the action, and refused to allow the pursuit which common sense demanded. Instead Burrard and Sir Hew Dalrymple negotiated the Convention of Cintra, which Wellesley was asked to sign on 31 August; under the terms of this document Junot's army, with arms and such plunder as it could carry, was shipped home in British ships. For this all three officers were much vilified in England, and a court of inquiry ended the careers of Dalrymple and Burrard. Wellesley spent a year 'in the wilderness' while Moore (q.v.) conducted his unlucky campaign in Spain, but in April 1809 he was sent out again to command the expeditionary force ordered to expel the second French invasion of Portugal under Soult.

With some 23,000 men, Wellesley concentrated his forces to strike northwards from Lisbon to Oporto, where Soult had the bulk of his 20,000 men. While screening forces watched Victor, who lay to the east with another 22,000, Wellesley carried out a swift advance and an audacious attack on Soult. He forced the river Douro at Oporto on 12 May 1809, and a week later bundled Soult over the Spanish border. He then returned to the south, and in July joined forces with a Spanish army under ludicrously inept leadership in an advance eastwards towards Madrid. He fought a defensive battle at Talavera on 27–8 July 1809; losses were heavy on both sides, but the combined French corps of Victor, Sebastiani, and King Joseph retreated and left the Allies in command of the field. Total French strength had been 46,000, opposed by 20,000 British and 28,000 Spaniards – although the latter were hardy engaged. Reinforced by Craufurd's (q.v.) Light Brigade, Wellesley now fell back into Portugal; his lines of communication were threatened by strong French forces from the north. He was created Viscount Wellington of Talavera in September 1809.

During the year which followed Wellington prepared for the French invasion of Portugal which he knew must come. He trained his troops, assembled men and provisions, and watched Beresford's (q.v.) progress in rebuilding the Portuguese army. He gave instructions for the preparation of a huge system of linked defensive positions – the Lines of Torres Vedras – which turned the

Lisbon Peninsula into a potential fortress and storehouse into which his army could retreat at need. His appreciation of the fundamental nature of the conflict was masterly. Basically, he planned on the assumption that if the Spanish population persevered with their vigorous harassment of the French armies, then no French commander could assemble an army large enough, and feed it for long enough, to crush the British in Portugal. At peak strength the French armies in Spain totalled some 300,000, while Wellington could never field an army with a fifth of that strength, even in the later campaigns when Portuguese manpower was fully deployed. The French, however, were obliged to spread their forces widely in a net of occupation knotted with strong static garrisons, while he could wield his forces like a rapier and achieve local superiority. The transport and logistic difficulties in Spain were medieval in their severity, and the constant activity of the *guerilleros* denied the French a secure base area, secure communications, and the leisure to concentrate forces against Wellington. Nevertheless Wellington was constantly restricted in his options by the knowledge that his army was the largest Britain had committed to the Continent since Marlborough's day, and that she had no other to replace it if he once allowed himself to be pinned down and defeated by the French. In other words, Wellington's footwork had to be impeccable as he danced in and out of the guard of the heavyweight enemy; freedom of movement was on his side, but a single solid blow could finish him. His genius lay in the skill with which he exploited this situation, while constantly improving the quality of his army until it was – arguably – the most professionally reliable expeditionary force which Britain has ever maintained overseas.

In the summer of 1810 Marshal Masséna (q.v.) led his Army of Portugal over the frontier, and Wellington fell back before him as planned. He stood at Bussaco on 27 September, in a fine natural defensive position with a sheltered lateral road picked out and improved beforehand, and inflicted heavy losses on the French attackers. He then led his field army within the Lines of Torres Vedras, into which many of the peasants and their livestock had been gathered, leaving an area of 'scorched earth' for the French to occupy. The impregnable fortress chain of the Lines came as a complete surprise to Masséna; his army, accustomed to living off the land, spent a miserable winter before the Lines and eventually withdrew towards the Spanish border in April 1811. Masséna advanced again in May in an attempt to relieve the invested fortress of Almeida. Wellington's Allied army of 34,000 successfully held off Masséna's 46,000 at Fuentes de Onoro in early May; this victory was a tribute to Wellington's coolness under pressure, as at one stage his right wing was driven in by heavy French attacks and the line had to be completely re-ordered under fire.

In July 1811 Wellington was promoted to the local rank of general. In the winter of 1811–12 he seized the initiative on the Portuguese-Spanish border by making a surprise attack on the key fortress of Ciudad Rodrigo, following up this capture in April 1812 by the siege and storming of the other strategic stronghold, Badajoz. In this operation the British infantry paid a heavy price for Wellington's lack of a competent engineering department and siege train – a lack which his representations to London remedied during the following year. During the summer of 1812 Wellington – now an earl – advanced into central Spain and, after a complicated period of manœuvre and counter-manœuvre, routed Marshal Marmont's Army of Portugal at Salamanca on 22 July. Marmont's 41,000 men were marching parallel to Wellington's 43,000 Allied troops when an opportunity to cut the French army in two presented itself. Wellington's reputation up to this time had been based on his defensive skill, and the effectiveness with which he employed the magnificently steady British infantry in the face of French attacks. At Salamanca he won a victory of manœuvre – a momentary opportunity was instantly recognised and boldly exploited. He entered Madrid in August, and later that month was created a marquis. The remaining months of 1812 were not so successful; the Allied advance on and siege of Burgos was an attempt to gain as much advantage as possible from the freedom of movement won at Salamanca, before the inevitable concentration of heavy

French forces withdrawn from the south and reforming in the east. The siege was marked by the same lack of means and profitless expenditure of manpower as so many others in the Peninsula, and in the winter of 1812 the Allied army carried out a costly and disheartening retreat into Portugal once more.

By the time the 1813 campaigning season opened Wellington had much improved the forces under his command, and had planned a brilliant advance. Secretly assembling his army in the extreme north of Portugal, he feinted towards central Spain with a detached corps and simultaneously began a rapid and well-organised advance to the north-east through country considered impassable. With little fighting he turned successive French lines of defence, outflanking the enemy and forcing them to withdraw along the main road to Bayonne. On 21 June the Allied army, 70,000 strong, fell upon 58,000 French at Vittoria and routed them, driving them back towards the border and capturing an enormous booty of guns, baggage, and treasure. By July the whole of west and central Spain was in Allied hands, and the major routes through the Pyrenees were sealed; only in the east were there substantial enemy forces still in being, and their ability to influence events was limited. The victory of Vittoria made Wellington a field-marshal. Soult (q.v.) was given command of the French forces facing Wellington, and in late July he launched attacks on the passes of Maya and Roncesvalles in an attempt to relieve the garrison of Pamplona, which still held out. After initial successes against the overstretched Allied divisions he was halted at Sorauren, where the Allies won a classical 'Wellingtonian' defensive victory. For various political reasons it was not possible for Wellington to advance into southern France immediately, but in October he pushed forward across the Bidassoa, and in November broke through Soult's defences on the Nivelle. Early in the following month Wellington advanced across the Nive, and Soult's counter-attack, aided by climatic conditions, caused large casualties and a moment of anxiety; but after heavy fighting British reinforcements pushed the French back to Bayonne. For two months thereafter there was little action, but as soon as the weather allowed, Wellington forced Soult eastwards and defeated him at Orthez and, in early April, at Toulouse. The next day the news of Napoleon's (q.v.) abdication arrived.

In August 1814 Wellington was made British ambassador to the court of the restored Bourbon king, Louis XVIII; he was now a duke, and a general of enormous international reputation. In January 1815 he left Paris for the Congress of Vienna, and it was there that he heard of Napoleon's landing in France on 7 March 1815. He was given command of the Allied army assembling in the Low Countries in June. Since the Waterloo campaign is the subject of more printed analysis than any other in Western history there is no justification for another description here. Wellington faced a complex strategic problem, and for reasons which apparently satisfied contemporary observers, he wrongly assessed Napoleon's route of advance on the Allied and Prussian armies. He acted quickly and decisively once his mistake became clear; disposed his indifferent forces with skill along the only topographical feature for miles which gave any natural assistance to the defence; and held it throughout the 18 June despite extreme pressure. His outnumbered and unevenly composed army fought brilliantly; Bonaparte handled his heartbreakingly brave divisions with none of his former flair, and was totally defeated. It was Wellington's last battle, and it can only seem dramatically fitting that this was so.

The huge prestige which Wellington earned at Waterloo made him the inevitable choice for commander of the Allied army which occupied France until 1818 – a command he exercised with justice and imagination, restraining those elements whose vindictive projects would have ensured a renewed outbreak of hostilities. In December 1818 he entered the cabinet as Master General of Ordnance; his lifelong hatred of partisan politics led him to insist that he saw his duty as the service of the King, irrespective of party loyalties, and that he would not resign and go into opposition if the administration fell. So great was his prestige that this strange proviso was accepted. In 1827 he succeeded the Duke of York as Commander-in-Chief, but he resigned all his offices that April when

Canning became Prime Minister. He was Prime Minister from January 1828 to November 1830, when he was obliged to resign after being defeated on the issue of parliamentary reform. His political career was not happy, and he became for a time extremely unpopular, though his championing of the liberal Catholic Emancipation Bill is greatly to his credit. He remained in opposition until November 1843, subsequently holding office as Foreign Secretary under Peel. He was given a seat without departmental responsibility in the cabinet of 1841, and on Hill's (q.v.) death in 1842 took over as Commander-in-Chief once again. He resigned from public life in June 1846 at the age of seventy-seven, but lived for another six years. It is pleasant to record that in his later years his brief political unpopularity was forgotten; he was held by the country as a whole in enormous esteem, becoming a sort of national father-figure of early-Victorian England. It is probable that no non-royal has ever enjoyed such respect among the mass of the population. He died at Walmer Castle, Kent, on 14 September 1852, and was buried with great ceremony in St Paul's Cathedral.

If it is possible to isolate one quality above all others which contributed to Wellington's success, it must be his ability to make the best possible use of limited means – dramatically better use than his contemporaries thought possible. He operated in a system which was extremely inefficient at many levels, and he was continually driven to exasperation by the stupidities and obtuseness of his superiors and his subordinates alike. Instead of wasting nervous energy in a hopeless confrontation with the system as a whole, he had a genius for applying ruthless standards of professional excellence to the machine as it actually existed, and by his forethought, detailed practical knowledge, and force of personality exercised upon individuals, making it work better than ever before. It is hardly surprising that he was not good at delegating authority, as he was entirely justified in assuming that most of his colleagues were not competent to bear much responsibility. This ability to accept, with brusque resignation, any type of handicap, and to apply his prodigious energy and perception to ways of circumventing the problem and salvaging as much as possible from the wreckage, was to be exercised constantly throughout his career. His comparison of the French style of campaign-planning with his own is often quoted: 'They planned their campaigns as you might make a splendid set of harness. It looks very well, and answers very well, until it gets broken; and then you are done for. Now I make my campaigns of ropes. If anything went wrong, I tied a knot; and went on.' This happy metaphor should not disguise the fact that for all his insistence on 'playing it as it lay', Wellington was the most thorough man in the British Army when it came to detailed, methodical, contingency planning far in advance.

Tactically as well as strategically he was an improviser rather than an innovator. It is universally known that his major victories were won by the exploitation of the superiority of fire-power of the British linear infantry formation over that of the French columnar assault formation. There was nothing novel about the British line, which was standard procedure before Wellington ever saw a battlefield. His genius lay in his employment of the line, its disposition according to his masterly grasp of terrain, its careful protection from artillery bombardment behind ridge-lines or by orders to lie down in ranks, and so on. There was nothing novel about the use of light infantry skirmishers, whose capabilities had been so extended by Moore while Wellington was in India. Wellington's genius lay in the drastic strengthening of his skirmishing line – one man out of five in the usual 'Wellingtonian' infantry division was a skirmisher – and its employment to render almost powerless the French *voltigeurs* and horse artillery batteries screening and supporting an attack.

Wellington was forced by the circumstances of his commands to husband the lives of his men at all times, and the British Army has had few senior commanders who so intelligently reconciled the demands of their mission with the acceptance of the minimum possible casualties. This brought him the unwavering trust of the men under his command and, together with his common-sense insistence on decent standards of provisioning, ensured good morale. He never courted popularity, and was not regarded sentimentally by his soldiers, but rather with complete respect. His mordant wit made him an eminently quotable general, and some of his *bons mots* have been taken out of context and misinterpreted to support the allegation that he was a cold, arrogant aristocrat, contemptuous of the common soldier. This is not borne out by the facts; he was certainly an eighteenth-century aristocrat by nature, and his whole life was ruled by a rigid conception of duty – his own, and others' – but his views were realistic and unsentimental rather than arrogant. He took an intelligent interest in the welfare and career prospects of the common soldier, and did much to better them. His manner was reserved and abrupt, it is true; but this sprang from the loneliness of command and from the normal social *mores* of his class, rather than from natural coldness. When there was leisure to relax – and his unbending concept of his duty allowed few such moments on active service – he was a jovial and hospitable man. On campaign his austere way of life and hard-riding habits were legendary; when at his ease, he had a ready appetite for all the good things of life, including female company. His mirth was as explosive as his terrifying anger, which was capable of reducing battle-hardened officers to tears. His physical and mental stamina were extraordinary. He commanded from the saddle, riding many hours each day over the roughest terrain, and sleeping only a few hours each night. He had no time for ceremony, and the laxity of his requirements in the matter of uniforms is legendary. On rare occasions he was seen to weep over the deaths both of friends and of unknown common soldiers. Every surviving memoir of him points to a warm and imaginative man rigidly self-disciplined. His physical courage and coolness under fire were as complete as one might expect of a man of his time and his background.

William I, called the Bastard and the Conqueror (*1027–87*), *King of England, Duke of Normandy*

The brutal sustainer of his dynasty in Normandy, but pre-eminent among the rulers of his day and a great constructive statesman, William is best known for his victory over King Harold (q.v.) at Hastings. He was responsible for establishing a new feudal order in England which amounted to a constitutional, political, and social revolution.

Born in 1027 at Falaise, the illegitimate son of Robert, Duke of Normandy (whom he succeeded at the age of eight), William, when still little more than a youth, set about retrieving Normandy from feudal anarchy – a state into which it had fallen as a result of baronial revolt during the absence of Robert. At the age of twenty he won the crucial victory of Val-ès-Dunes, near Caen, over the rebel factions – a victory which was to assure his authority. His frequent actions in the field thereafter confirmed him as an outstandingly courageous knight who repeatedly displayed impetuous gallantry. In 1051 William visited King Edward the Confessor (whose cousin he was), and appears to have secured some agreement as to his succession to the English crown: an agreement further strengthened when, thirteen years later, Harold – then Earl of Wessex – swore an oath to support William's cause and claim to the English throne (according to exclusively Norman chronicles).

Edward the Confessor died in 1066, and Harold was crowned King on 6 January, supported by alliances with Flanders and Brittany. He campaigned that year against the invasion by Norway's King Harald Hardraade (q.v.), whom he finally defeated in the Battle of Stamford Bridge on 25 September. Three days later William landed at Pevensey in Sussex to lay claim to the English crown, which he believed to be his rightful inheritance, with an army estimated at about 20,000 to 30,000 men, including about 10,000 cavalry. Learning of the Norman invasion, Harold hurried south with an army of about 20,000 men, reaching a position about eight miles north-west of Hastings during the afternoon of 13 October. Here the Anglo-Saxon army set about fortifying a ridge overlooking the invaders.

Soon after dawn the following day the Normans advanced in three lines consisting, in order, of archers, spearmen, and cavalry knights. At a hundred yards from the Anglo-Saxons the archers opened fire, but their arrows and bolts, fired uphill, were deflected by the defenders' shields and were thus ineffective. The attacking spearmen then rushed forward, only to meet a shower of spears; as they struggled up to the English lines they were hacked down by the great battle-axes and heavy swords wielded by Harold's men.

Seeing his infantry thus halted, William personally led a cavalry charge up the hill – but with no better results; in fact the Norman left fell back in blind panic, closely pursued by the Anglo-Saxons who, on that flank, now left their commanding positions. Seizing the chance to decimate the defending infantry militia in the open, William, throwing aside his helmet so that his face could be seen by his own men, charged with his cavalry against the Anglo-Saxon infantry. As the afternoon wore on the fight developed into a punishing series of alternating dashes and withdrawals, interspersed by heavy volleys of weapons. On several occasions William's men lured some of the defenders from the relative safety of their solid defence line and cut them down in the open. The Norman archers were now ordered to fire their arrows at a high angle so that greater effect could be gained from plunging fire. By the end of the afternoon the Anglo-Saxons were near exhaustion, and it was at this point that Harold himself was mortally wounded. At once the leaderless English began to collapse. After a desperate last-ditch stand by surviving housecarls, the battle was over by nightfall.

Immediately afterwards William seized Dover and marched on London; after a short delay during which the invaders laid waste the surrounding countryside, the capital surrendered. On Christmas Day 1066 the Norman conqueror was crowned William I, King of England.

The Battle of Hastings was, in a narrow sense, no more than a crucial battle which ended a dynastic war to settle the succession to the throne of a small off-shore island kingdom. But in fact it was nothing less than a turning-point in world history, as it laid the foundations for an Anglo-Saxon-Norman domination of the world far more influential and extensive even than that of Rome a thousand years earlier.

As King of England William was confronted by several serious revolts, all of which he put down with uncompromising severity. The uprising of Hereward the Wake, aided by a Danish force under Jarl Osbiorn, was eventually crushed by an assault on the island fortress of Ely in 1071. Across the Channel, William was obliged to suppress a rebellion in 1073 in Maine, whose nobility, during his preoccupation in England, had revolted from his feudal control. Campaigning throughout northern Europe occupied much of the next fourteen years, and quarrels with France led to a war with King Philip I in 1087. It was after capturing the town of Mantes during this campaign that William was thrown from his horse and was fatally injured; he died at Rouen on 9 September 1087, and was buried at St Stephen's, Caen.

As a ruler William was at once harsh, capable, and thorough. He possessed a highly-developed sense of his duty to dispense justice and to keep the peace, though his methods were, even by contemporary standards, savage and uncompromising. His greatest achievement, apart from the conquest of England, was the execution in 1086 of the Domesday Survey, a detailed description of the social classes and economic resources of the land south of the Tees without parallel in contemporary Europe. As a soldier, William led principally by personality and example. His use of skilful archers was constantly developed – an innovation which continued apace in England long afterwards and reached its flowering at Crécy nearly three hundred years later. William leaned heavily upon mercenary units and commanded great respect among them, not only by promises of land and wealth in conquered lands but by his own outstanding courage in battle.

Wingate, Orde Charles (*1903–44*), *Major-General*

Wingate, the creator of the Chindit long-range penetration units of the British army in Burma during the Second World War, invites comparison with 'Chinese' Gordon and T. E. Lawrence (qq.v.). An orientalist, a restless and intense intellectual, a religious fanatic, and a deeply neurotic personality, he was also a visionary and innovative soldier of undeniable genius. His contribution to Allied success in Burma was considerable, although he attracted as much opposition as support for his theories by his outspoken and sometimes unbalanced behaviour. A senior officer of the Chindits, himself a thoughtful and moderate observer and no supporter of military rigidity, has expressed the opinion that Wingate was half mad; he has also paid tribute to the man's extraordinary perception and energy.

Orde Wingate was born at Naini Tal, India, on 26 February 1903. His family had a long military and religious tradition, and he was brought up according to the puritan doctrines of the Plymouth Brethren. Educated at Charterhouse, he graduated from the Royal Military Academy at Woolwich, and in 1923 was commissioned into the Royal Artillery. He soon displayed unusual character traits. Deeply introverted, he was a voracious reader of history and philosophy and a natural 'loner'. He was also an outstanding rider and a gifted linguist. He made several long, lone journeys in remote areas of the Middle East, including an expedition into the Libyan desert. He secured an appointment to the Sudan Defence Force, and served with that formation between 1928 and 1933, mostly on the Abyssinian frontier. Between 1933 and 1936 he served with Royal Artillery units in the United Kingdom, and late in 1936 he was posted to Palestine and an intelligence staff job. He became fascinated by the country, and particularly by the Jewish communities which were at that time facing a terrorist campaign mounted by the Arabs. Wingate identified closely with the Zionist cause, became a friend of several of the Jewish leaders, and in time obtained permission to organise the Night Squads – defence units of young Jewish volunteers who carried out counter-terrorist missions against the Arabs. He displayed a great talent for unorthodox warfare and guerrilla tactics, leading missions in person and being wounded in July 1938. Among his keenest disciples was the young Moshe Dayan (q.v.). Although his success brought him the D.S.O., it also earned him several enemies among senior officers unsympathetic both to the Zionist cause and to precocious and unorthodox young officers such as Wingate. He was an intemperate and outspoken man, and a poor subordinate.

The outbreak of war in 1939 found him serving as brigade major in an anti-aircraft artillery unit in England. In 1940 the Commander-in-Chief Middle East, General Wavell (q.v.), was granted his services on the grounds of his unusual experience of irregular warfare. He was sent to Khartoum in the autumn of that year, to assist the Ethiopian nationalists under Haile Selassie in their resistance to the Italian occupiers of their country. In the period January to May 1941 he led a guerrilla organisation known as 'Gideon Force' which operated over the border; by a combination of hit-and-run raiding and bluff he secured the evacuation of several Italian forts and captured many prisoners. Returning to Cairo after the successful British invasion of Ethiopia, he became deeply depressed by the various reverses to British arms then occurring in the Middle East and the Mediterranean. Ill and suffering from exhaustion, he attempted suicide. After convalescence in Britain he was again summoned by Wavell, now Commander-in-Chief India, to give of his unique abilities. Wavell, displaying great imagination, sent Wingate into Burma to reconnoitre the terrain before the retreating British army actually left the country.

Following this mission in May 1942, Wingate put forward proposals for the formation of long-range penetration units to operate far behind the Japanese lines in Burma, isolated in the jungle and relying on radio links with base to organise air-dropped supplies. This plan, then highly novel, met with enthusiasm in some quarters and resistance in others. Colonel Wingate was impatient of opposition and quite unimpressed by rank; he regarded himself, it has been said, as 'a boot up the backside of mankind', and pressed his carefully-prepared arguments with great energy. Eventually the 77th Indian Infantry Brigade was formed to test his theories, and he supervised an extremely punishing training programme while perfecting the logistic details of his plan. The force finally crossed into Burma in February 1943, with Wingate leading the larger group of five of the eight columns into which it was divided;

the total strength was 3,000 men, with light equipment and mules for transport. Their mission was to disrupt supply and communications behind enemy lines, gather intelligence, interrupt enemy preparations for an offensive, and generally to make an expensive nuisance of themselves while evading enemy countermoves under cover of the thickly jungled hills. Behind this ostensible purpose was the larger question of determining, the hard way, whether it was possible for a force of this size to be kept in being and operational at the end of such a tenuous line of communications. By the time the survivors – dispersed by several setbacks – crossed back into India in late March and April the question was answered: it was possible, but hideously difficult. The retreat was a harrowing ordeal; with many sick, hungry and exhausted, the survivors were hunted by the enemy through some of the most punishing terrain in the world. Added to the enormous physical strain was the knowledge that the wounded would have to be abandoned in the jungle, in the path of a sometimes sadistically cruel enemy. Some successes had been achieved, and the essential theory vindicated, but at heavy cost. Of some 2,200 men who returned, only about 600 were ever passed fit for active service again. One priceless advantage had been gained, however: the image of the Japanese soldier as a jungle-fighting superman who invariably outclassed the Allied infantryman was severely dented. The 'Chindits' (named after the *Chinthe*, the mythical figure whose image guarded Burmese temple doors) were not hand-picked commandos; the force was made up of ordinary British and Gurkha line infantry. They had proved that they could attack and evade the Japanese on ground he had made his own.

The relative success of this operation attracted great attention; it brought Wingate a bar to his D.S.O. and promotion to major-general, and Churchill took him to the Anglo-American conference at Quebec in August 1943. He was given authority to raise a much larger force of Chindits, modifying the preparations in accordance with experience in the field. There were many who opposed the plan on the grounds that it drew off units, and in particular precious cargo aircraft, from more conventional uses. Nevertheless a force of three full brigades was sent into Burma in February 1944, two of them air-landed in gliders. The Japanese offensive which opened early in March led to a distortion of the original mission of this force (see under Slim), although they fought very effectively in the Japanese rear areas for some weeks. But increasingly they were drawn by the exigencies of the overall situation into conventional battles with major enemy units, for which they were not equipped. They were kept in action longer than had been anticipated, and some Allied commanders displayed a lack of appreciation of the unusual strains imposed on the Chindits by their type of operation. Their contribution to Allied success in this campaign was considerable, but losses were very heavy, and the condition of the survivors – when they were eventually brought out – was shocking. Of 2,200 officers and men of one brigade who were given a searching medical examination after leaving the battle area in July, just 118 of all ranks were pronounced fit for active service of any kind, in any theatre.

Wingate would no doubt have had some incisive remarks to make about the use to which his command had been put, but he was no longer alive to make them. On 24 March 1944 he was killed in a transport aircraft crash in Assam; he was later buried in Arlington Cemetery, Virginia, U.S.A.

Wolfe, James (*1727–59*), *Major-General*

The young British general who died in the moment of victory at Quebec quickly acquired the status of a national hero. His career of senior command was so short that it is impossible to make any deep assessment of his military qualities; he had a fine record as a regimental officer and a brigade commander, and the mistakes he made during the Quebec campaign can fairly be attributed to youthful impatience. Had he lived to serve for another twenty or thirty years, it is tempting to try to imagine his impact on the War of American Independence. Tempting, but futile: his health was so poor that it is unlikely that the medical science of his age could have kept him alive for very much longer even had he not fallen on the Heights of Abraham.

He was born on 2 January 1727 (new

calendar) at Westerham in Kent, the elder son of Lieutenant-Colonel (later Lieutenant-General) Edward Wolfe. (Edward Wolfe was an officer of marines who served under Marlborough in his youth, and later became the colonel of the 8th Foot. He died in March 1759, just six months before his son.) In 1737 the Wolfes moved to Greenwich, and James was educated at Swinden's Academy, a local college. On 3 November 1741 he was commissioned in the 44th Foot, then a marine regiment, but in March 1742 he transferred to the 12th Foot. He obtained a lieutenancy in that regiment in July 1743, and saw almost continuous active service throughout the War of the Austrian Succession – albeit service interrupted by frequent bouts of debilitating illness. James Wolfe was a tall, slightly-built youth whose thin body was racked by bronchial weakness; in the last years of his short life he contracted tuberculosis. His spirit was unquenchable, however, and he appears to have thrown himself into both his career and his recreations with an energy impatient of his physical limitations.

At the Battle of Dettingen in 1743 he served as acting adjutant of his regiment, though sixteen years old; the 12th Foot was placed in the centre of the British first line, and suffered the heaviest casualties of any regiment present. In October James's brother, an ensign in the 12th, died on active service. James missed the Battle of Fontenoy, being in garrison at Ghent, and in June 1744 he obtained command of a company in the 4th Foot under General Wade. In June 1745 he was appointed brigade major, and in September of that year he was posted home with his unit to take part in the campaign against Prince Charles Edward Stuart. His courage, energy, and intelligence had already been marked by the Duke of Cumberland, and Wolfe was appointed aide-de-camp to General Hawley. He was present at the latter's defeat at Falkirk, and at Culloden, although his part in the battle is not known. In January 1746 he returned to the Continent, still on staff duties, and was wounded at Laffeldt in 1747. On 5 January 1748 he was promoted major in the 20th Foot. In 1749–50 he served as acting lieutentant-colonel of the regiment, and in the latter year was confirmed in the rank. From 1750 to 1758 he commanded the 20th Foot in various garrison postings, mainly in Scotland, but later in Dorset and Gloucestershire. During that period he turned his regiment into a model unit, and his reputation spread.

Wolfe was a volatile young officer, consumed by an ambition to better his military knowledge and to extend his experience. He took great pains to improve the efficiency of his regiment, and was fearless – not to say extremely tactless – in some of his criticisms of the undoubted shortcomings of colleagues and superiors alike. He admitted to a mercurial temperament, soaring and plunging between exhilaration and depression. He was frustrated by his poor health, but was tireless in the performance of his duties – and was not above occasional bouts of energetic dissipation. He had great confidence in his own ability, and yearned for responsibility. He could be guilty of arrogance, but made as many friends as enemies by his warmth, charm, and kindliness. He is reported to have been a loving son, a staunch friend, and a considerate, if firm, commander. The unfortunate possessor of a comically unattractive profile – much romanticised by later portrait-painters – he was susceptible to women, and was romantically linked with several young ladies.

In 1757 Wolfe was appointed quartermaster-general of the force told off for an abortive landing at Rochefort on the French Atlantic coast. It was a bungled affair – as Wolfe was not slow to point out – and the commander, Sir John Mordaunt, was eventually court-martialled; but against this background the spirited and intelligent exertions of the young Wolfe stood out the more sharply. William Pitt, the new Prime Minister, spotted his potential at once and selected him as a man suitable for senior command in the planned operations against the French in North America. A vital part of these operations was the attack on the great French fortress of Louisbourg on Cape Breton Island, which was an essential preliminary to British penetration of the St Lawrence. Sir Jeffrey Amherst was given command of the expedition, and Wolfe was named as one of his three brigadiers. When the British force rowed ashore in a three-pronged landing assault on 8 June 1758, it was Wolfe who led the real attack: while Lawrence and Whitmore led diversions against Flat Point and White Point, Wolfe rowed into Freshwater Cove with twelve companies of grenadiers, 550 light infantry, Fraser's Highlanders, and some Provincial rangers. Beyond the heavy surf and the beach lay 1,000 French soldiers and eight guns, well entrenched. The enemy opened fire on the boats before they touched, and for a time it appeared that no landing could be made; but a small party scrambled on to the rocks at one end of the beach, and the rest followed. Wolfe led his men in a climb up the rocks, and took the French positions with the bayonet. He continued to display great energy and resource throughout the Siege of Louisbourg, which lasted until 26 July. On Amherst's orders he led his brigade around the bay to a point opposite the fortress, capturing a French battery and setting up British guns which silenced enemy positions guarding the harbour-mouth. His conduct attracted favourable notice in the highest quarters; even Walpole, who had no time for the unpredictable red-headed brigadier of 31, credited him with 'great merit, spirit and alacrity', and said that he 'shone extremely at Louisbourg'.

Returning to England, Wolfe was promoted to the rank of acting major-general by Pitt, and was given command of the projected expedition against Quebec. (His rank applied only in America; his substantive rank, since April 1758, was colonel of the 67th Foot – the detached ex-2nd Battalion, 20th Foot.) With 9,000 British infantry Wolfe sailed for America once more in a convoy commanded by Admiral Sir Charles Saunders. Saunders and his captains and crews displayed magnificent seamanship in penetrating the dangerous reaches of the St Lawrence, and by the end of June 1759 the army was disembarking safely on the newly-captured Isle of Orleans, close to the southern bank of the river a short distance from the city. They took Point Levi and a strip of the southern shore without delay, and set up batteries which began a heavy bombardment of the enemy. For a month there were no major operations. Montcalm (q.v.) had some 16,000 men in the city and encamped on the Beauport shore, the area of the northern bank immediately north-east of the city. His guns swept good fields of fire over the river, and south-west of the city his right flank was protected by 300-ft cliffs. Wolfe could see no reasonable way to bring his forces into contact with the French, but was under the impression that Amherst, with another army, would soon arrive overland to reinforce him by way of Lakes George and Champlain, after taking Fort Carillon (Ticonderoga). Meanwhile Wolfe tried unsuccessfully to lure Montcalm from his trenches by harrying the surrounding communities with raiding parties of light troops and Highlanders – with the inevitable result of occasional shameful incidents. Amherst seemed to be taking a leisurely route indeed.

In fact neither side was in an enviable position. Despite his apparent numerical superiority, Montcalm had only some 2,000 regular French infantry, the rest of his force being made up of local militia of very patchy quality. Many hundreds of these Canadians deserted the French camps during July, and his strength dwindled every day. He pinned his hopes on sitting out the campaigning season until Wolfe's army, impotent and running short of provisions at the onset of winter, could be forced to withdraw. Wolfe in his turn had little to cheer him. Though he had established, on 9 July, a third British camp on the northern shore east of the French camps and near the mouth of the little Montmorency river, he had still failed to lure the enemy from their strong positions. On 31 July he attempted to break the stalemate by launching a combined attack on French redoubts at the mouth of the Montmorency, using troops from the north shore camp and others landing in boats; it was repulsed with heavy casualties, costing Wolfe some 450 of his picked soldiers. In August came the news that Amherst was still firmly blocked in the south by a new French defensive position at Ile-aux-Noix at the head of Lake Champlain. Wolfe had only some 5,000 men now fit for an attack if he was to maintain proper garrisons on the southern shore and the Isle of Orleans; Montcalm had twice that number, and the season was wearing on. British detachments probed up-river rather half-heartedly, but it was difficult to slip past the French batteries, and Montcalm showed no inclination to detach sufficient numbers of his troops to weaken the main position appreciably. Wolfe was laid low in August by tuberculosis and gravel; feverish, in great pain, and frantic with anxiety, he was further beset by the hostile and almost openly disloyal attitude of some of his senior subordinates. Certain officers took evident satisfaction in his impending failure, though it was to their country's disadvantage. Around the end of the month he recovered sufficiently to take a grasp on the campaign once more, but from his writings it seems certain that he did not expect to live many months. He decided on a last gamble. In conference with his brigadiers he announced an attempt to land most of the available fighting force several miles up-river on the north bank, and the men were accordingly embarked. Until the last minute Wolfe kept to himself his plan to land them a bare one-and-a-half miles up-stream from the city, at Anse-au-Foulon.

On the night of 12 September 1759 the advance party of 1,700 men slipped ashore, outwitting batteries and sentries alike, and clawed their way up the rudimentary path to the top of the Heights of Abraham. By daybreak no less than 4,500 redcoats were in position, together with two guns. Montcalm, as mentioned in the entry under his name elsewhere in this book, chose to sally out of the city and attack the British force immediately, rather than wait to assemble his maximum strength. The French line advanced, supported by three cannon, and was halted and shattered by a single perfect British volley at forty paces' range. A few more minutes' firing and a bayonet charge sufficed to put the French to retreat. Wolfe was on the extreme right of the British line on a slight knoll, among grenadiers detached from the regiments left at Louisbourg. Early in the firing a ball smashed his wrist; he wrapped a kerchief round it, and continued to accompany and encourage the slow British advance. A second ball through the body staggered him, and a third dropped him. He died a few minutes later in the arms of Grenadier Henderson, in the knowledge of victory and apparently content. Quebec was surrendered to his troops on 18 September 1759, and the whole of France's possessions in Canada just a year later. Quebec had been the key victory.

Wolfe's body was brought home by the fleet, and landed at Portsmouth with military honours on 17 November 1759. On 20 November he was buried in the family vault in St Alfege's Church, Greenwich. Several monuments were raised to his memory, including a statue which still stands in the middle of the village of his birth, but his best-known memorial is probably King George's remark to Lord Newcastle on hearing Wolfe's unpredicatable behaviour criticised: 'Mad, is he? Then I hope he will bite some of my other generals!'

Wolseley, Garnet Joseph, 1st Viscount Wolseley (*1833–1913*), *Field-Marshal*

Wolseley was a British imperial commander of the late Victorian era – a noted 'trouble-shooter' who arrived just too late to save Gordon (q.v.) at Khartoum. He instituted numerous reforms in the British Army.

Born on 4 June 1833 at Golden Bridge House, County Dublin, Ireland, Garnet Wolseley was the son of a junior army officer. He entered the British Army at the age of nineteen, and was immediately shipped out East as a subaltern with the 80th Foot to serve in the Second Burmese War (1852–3). For this service he was awarded the India General Service Medal. Transferring to the 90th Foot, he fought in the Crimean War and was present at the capture of Sebastopol, receiving the brevet rank of major.

After the fall of Sebastopol Wolseley joined the staff of the Quartermaster-General, quickly becoming thoroughly acquainted with the British Army's disgraceful lack of supply services – a situation which he was never to forget. Returning to the 90th Foot, he was in India at the time of the Mutiny, serving under Sir Colin Campbell (q.v.) at the first siege and capture of Lucknow, and receiving the brevet rank of lieutenant-colonel. He then served in China in 1859-60 as Deputy Assistant Quartermaster-General on Sir Hope Grant's staff with the Anglo-French Expedition, and being present at the capture of the Taku Forts on 21 August 1860 and of Peking later that year.

In 1861 he was appointed Assistant Quartermaster-General in Canada, becoming Deputy Quartermaster-General six years later; while in Canada he wrote his *Narrative of the War with China in 1860*, reorganised the Canadian Militia, and studied past and current military tactics. He also compiled *The Soldier's Pocket-book for Field Service*. In 1867 he was created K.C.M.G. and C.B., and in 1870 he commanded the Red River expedition.

Returning to Britain in 1871, Wolseley became Assistant Adjutant-General at the War Office, working with Edward Cardwell to further Army reform, with special emphasis upon administration, equipment, and supply. Promoted major-general, he assumed command of troops on the Gold Coast in 1873 and led them against the King of Ashantee on a punitive expedition which culminated in the capture and burning of the capital, Coomassie – an achievement which brought fame to Wolseley. He returned home once more, to be created G.C.M.G. and K.C.B. Among other forms of recognition were the thanks of Parliament and an award of £25,000.

In 1874 Wolseley was appointed Governor of Natal, and for services there he was promoted G.C.B. In 1878 he became the first High Commissioner in Cyprus; the following year he returned to South Africa. In May 1880 Wolseley came home to London again as Quartermaster-General, and two years later became Adjutant-General – responsible for training in the Army. Before he could engage in the training reorganisation which he had long planned, however, he was ordered to take command of 25,000 troops landed at Ismailia, Egypt, at the time of the disorders at Alexandria in July 1882. At the Battle of Tel-el-Kebir the British troops inflicted a crushing defeat on Ahmet Arabi's rebellious forces (13 September), after which the British dominated the Egyptian government.

The following year the Sudan erupted in revolt under the Mahdi, Mohammed Ahmed of Dongola. General Gordon was sent to Khartoum to supervise the evacuation of Egyptian forces, but was soon trapped in the city by Mahdist rebels. After a fatal period of political indecision – at which public outcry mounted in England – a relief expedition under Wolseley was ordered to Gordon's rescue, setting out from Wadi Halfa in October 1884. Arriving two days late to save Khartoum, Wolseley's troops found the beheaded body of the British general on the steps of the palace.

The relief column was withdrawn and Wolseley (now created a viscount) came home to London and to the War Office. There he instituted the Intelligence Department and actively encouraged the Staff College. He also advocated, without immediate success, the creation of a Ministry of Defence and a General Staff, on the principle that 'War is a serious business which has to be prepared for.'

After a period commanding the British forces in Ireland (1890–4) Wolseley was promoted field-marshal in 1894, and became Commander-in-Chief of the British Army in the following year. He gave up this office in 1901. Wolseley died at Mentone, France, on 15 March 1913, and was buried in St Paul's Cathedral. Apart from his worldwide campaigning, he was best known for his constant and untiring efforts to improve the image of the soldier and to mould the Army into an efficient weapon of war. That the British Army was wholly inept at the outset of the South African War of 1899–1902 was often said to be the fault of its Commander-in-Chief; yet it was to the credit of this very man that the lengthy process of modernisation was almost complete by the outbreak of the First World War. The process had occupied more than forty years – of which the first thirty had been spent in overcoming the hereditary prejudice of a dozen generations.

Wood, Sir Henry Evelyn (*1834–1919*), *Field-Marshal*

Wood was an ex-naval officer who transferred to the British Army and subsequently rose to command British forces during several colonial campaigns of the late Victorian era, particularly in the Zulu War of 1879.

Born on 9 February 1838 at Cressing, Braintree, in Essex, Wood was the son of a clergyman. He was educated at Marlborough, and entered the Royal Navy at the age of fifteen. Accompanying the Naval Brigade to the Crimea, he was severely wounded at the age of eighteen while carrying a scaling ladder during the abortive assault on the Redan on 18 June 1855. Lord Raglan, who died only ten days later, nevertheless mentioned young Wood in his dispatches for his gallantry. Disdaining to be shipped home, Wood applied for transfer to the 13th Light Dragoons, and before the end of the War had been created *Chevalier de la Légion d'Honneur* and held the Turkish Medjidie medals. (Wood's Crimean Medal also bore the unusual combination of battle clasps for Azoff and Sebastopol, the first being almost exclusively a naval award.)

Promoted lieutenant, Wood was transferred to the 17th Lancers and saw service during the last year of the Indian Mutiny. For an act of gallantry at Sinwaho, Sindhora, on 19 October 1858 while serving with that regiment he was awarded the Victoria Cross. He was also twice mentioned in dispatches. By 1873 he had risen to lieutenant-colonel, and commanded a column of troops in the operations against the King of Ashantee in 1873 and 1874. He also served in command of troops during the Kafir and Transvaal Wars of 1877 and 1878. In the Zulu War of 1879, following a series of British reverses which culminated in the disastrous Battle of Isandhlwana on 22 January 1879, Wood's column of 400 European troops with several thousand native soldiers was attacked by 20,000 Zulu warriors on 29 March. After a desperate four-hour battle at Kambula the Zulus were eventually beaten off with the loss of about 1,000 killed. Wood's losses included about a hundred Europeans and several hundred native troops. This Zulu setback to some extent bought time for the British, who hurriedly shipped reinforcements out from England. At the Battle of Ulundi on 4 July a force of 10,000 Zulus was routed by traditional use of the hollow square and a subsequent cavalry charge. Victory was complete, and the Zulu threat, however legally and morally supported, was virtually at an end.

Wood cannot be said to have been a brilliant strategist, or even a gifted commander; but he was a sound leader of men, capable of analysing local tactical situations and bringing to bear a traditional textbook solution with notable energy and absolute fearlessness. Colonial campaigns allowed him success in these conditions, and he was fortunate in that none of his adversaries had benefited from a study of the same textbooks. His success (moderate though it was), his possession of the Victoria Cross, and his political neutrality marked him for high office at a time of British military stalemate. He was brought home after the Zulu War and fêted as a hero, and in 1882 was given command of the Chatham District. The following year he took the Second Division to Egypt and raised the Egyptian Army, of which he was appointed Commander-in-Chief. Returning to England, he was appointed Adjutant-General to the British Army in 1897 – a post he retained until 1901 and as a result of holding which, he was fairly widely criticised for the standard of training in the Army at the outset of the South African War. To be realistic, the responsibility was not his, but that of a succession of adjutants-general who had failed to recognise the worldwide advance in military tactics and training. Certainly Wood held a continuing belief in the invincibility of traditional British field tactics – tactics which were outdated at a stroke by the Boers' opening campaign in the disastrous winter of 1899–1900.

Wood was appointed field-marshal in 1903, and thereafter continued in virtual retirement until his death on 2 December 1919.

Yamashita Tomoyuki (*1885–1946*), *Lieutenant-General*

Yamashita, nicknamed 'the Tiger of Malaya', was an extremely able Japanese commander of the Second World War, and achieved Japan's most crushing land victory of the war. Heavily outnumbered, he advanced 650 miles from northern Malaya to Singapore in a mere seventy days, and captured the entire country for the loss of 9,800 casualties.

Yamashita was born in 1885, the son of a humble country doctor. He rose steadily through the ranks of the army, but was himself convinced that his promotion was delayed because of his support in 1929 for General Ugaki's plan to reduce the size of the army. He made powerful enemies, including General Tōjō (q.v.), and was aware – with some apparent justification – of the effect this had upon his advancement. In 1940 he was appointed Inspector-General of the Imperial Army Air Force, and shortly after headed a military mission to Germany. He met Hitler and Mussolini, and submitted a report recommending that no action should be taken against Britain or the U.S.A. until after a thorough modernisation of the air force and a mechanisation programme in the army. This report did nothing to add to his popularity among the militarist clique who then controlled the government. Yamashita, a burly, bull-necked man motivated by a simple loyalty to the Emperor, does not appear to have indulged in the political manœuvres which were such a notable feature of the upper echelons of Japan's armed forces before and during the Second World War.

In December 1941 he was given command of the 25th Army, a force of three divisions totalling only some 35,000 men, charged with the invasion of northern Malaya. His command sailed from Hainan on 4 December, landing at three points on the eastern coast of the peninsula close to the Thai border on 8 December, in the immediate aftermath of the Pearl Harbor operation. His offensive was characterised by a ruthless energy and imaginative use of every natural advantage – while the British defence seems to have been handicapped by a confused and increasingly defeatist attitude. It had been expected that the defence lines in the north of the country would hold up any attacker for at least three months, and that their outflanking was impossible in view of the jungle terrain. Professional opinion of Japanese abilities was not, in general, high. The surprising speed and resourcefulness of Yamashita's advance to some extent paralysed the British staff, and his penetration of successive defensive positions rendered their counter-measures useless. The Japanese seized abandoned transport to make up for the deficiencies in their own equipment; they used large numbers of bicycles; and they advanced without pausing to consolidate, regroup, or await the arrival of logistic elements. They were supported by air force units which outclassed the tiny and obsolete British air arm, and by armour, in a country which Whitehall had pronounced quite unsuitable for tanks. The defence was thrown completely off balance, and there was a growing element of panic in the behaviour of some Indian, British, and Australian units and their commanders – although it must be said that other units put up a bitter resistance, suffering heavy losses when overrun by locally superior enemy forces. There were well-attested incidents of Japanese atrocities against British prisoners and wounded.

By 1 February the last British troops had withdrawn to Singapore island. Some 85,000 men now held an island popularly supposed to be impregnable, against 30,000 Japanese, seriously short of supplies at the end of unexpectedly extended lines. In fact, however, the situation was far less clear-cut than it appeared. The British commander, General Percival, was under the impression that the Japanese were at least twice as strong as they were. The island's water supply was piped from the mainland, and was promptly cut. The major fixed defences were sited to cover shipping at sea and not to defend the island from attack through the mainland jungles, long considered impassable. The very demoralising withdrawal down the peninsula had sapped the resolution of many commanders and staff officers, as well as the fighting spirit of some of the units. Of the total force at his disposal, Percival knew that at least 15,000 were non-combatants and thousands of others were poorly-trained, poorly-equipped, and psychologically shaken. Yamashita was well aware of the precariousness of his true position, but decided to exploit the advantages already gained. He too was surprised at the patchy performance of the British and Commonwealth troops encountered, and felt that a bluff was worth trying. After a diversion on 7 February, he committed his last reserves of manpower and supplies to an assault landing by the 5th and 18th Divisions on the night of the 8 February. There was stiff fighting on the beaches, but by dawn some 15,000 Japanese infantry with strong artillery and armoured support were established on the island. They were steadily reinforced, and for a week the fighting raged. Yamashita's demands for British surrender were pure bluff, as he was well aware that his outnumbered forces were unequal to a protracted house-to-house battle in the city itself. But on 15 February the bluff worked. Percival, aware that his water, petrol, and artillery ammunition were running perilously short, and unwilling to subject the civilian population to the horrors of street fighting, requested a meeting. He hoped to reach some sort of negotiated cease-fire short of unconditional surrender; not surprisingly, Yamashita flatly refused and demanded the immediate surrender of the garrison. At length Percival agreed. British and Commonwealth units had lost just over 9,000 men, but no less than 130,000 were taken prisoner. Had they known what the next four years held for them it is hard to believe that they would not have fought to the death. As it was, Britain suffered the most disgraceful defeat of modern times.

Yamashita's increased fame was such that Tōjō, determined to avoid having to appoint the victor of Malaya as his Defence Minister, arranged for his transfer to command of the 1st Army Group in Manchuria – a safely obscure appointment. In July 1944 the Tōjō government fell, and the new régime appointed Yamashita to command the 14th Area Army – the entire Japanese force in the Philippine group, shortly to be invaded in overwhelming strength by General Douglas MacArthur (q.v.). Yamashita's task was hopeless from the outset, and was further complicated by the active interference of Field-Marshal Terauchi, who insisted on Yamashita committing his main forces to battle on Leyte. Despite a chronic lack of every necessity and the absence of any realistic air or naval support, Yamashita's divisions put up a typically heroic resistance for as long as they were physically able. When the situation became quite untenable the survivors, including Yamashita and his staff, took to the hills in groups. Yamashita specifically forbade the active defence of Manila and withdrew all but a small number of security troops, leaving the capital as an open city. But after his departure Rear Admiral Iwabuchi's sailors and marines were ordered into the city, which they fortified and held to the last. In the fight for Manila a large number of civilians were killed, many of them as a result of particularly disgusting atrocities committed by the marines.

Yamashita did not surrender until 2 September 1945, when the belated news of the Emperor's capitulation reached him. He was tried for war crimes, convicted, and executed by hanging on 23 February 1946. Cooler judgements now suggest that he was a sacrifice to MacArthur's determination to wreak instant and visible revenge for the horrors committed in Manila. MacArthur's attitude is perfectly understandable and indeed the degree of Yamashita's responsibility for the atrocities perpetrated by his troops in Malaya has never been probed satisfactorily. All the same, the sentence of death was specifically for the Manila incidents: these were committed against his express orders, by units only theoretically under his command, and after his physical contact with the units concerned had been broken.

Zhukov, Georgi Konstantinovich

(*b. 1896*), *Marshal of the Soviet Union*

Soviet Commander-in-Chief on the Eastern Front during the latter half of the Second World War, it was Zhukov who personally commanded the final assault on Berlin in 1945. He survived the periodic assaults upon his status by Stalin, who resented his undoubted popularity.

Born of peasant parents near Moscow in 1896, Zhukov was conscripted into the Tsarist army at the age of fifteen and achieved junior N.C.O. rank in a dragoon regiment before the October Revolution in 1917. Joining the Red Army the following year, he served as a junior cavalry commander, slowly rising to corps commander. He joined the Communist Party in 1919.

Zhukov first achieved fame for his conduct of operations against the Japanese on the Mongolian-Manchurian border during 1939, employing his five armoured brigades skilfully in the expulsion of the Japanese Sixth Army – for which achievement he was awarded the Order of Lenin. During the Second World War he was first appointed Chief of the General Staff, and later Deputy Commissar of Defence and Deputy Supreme Commander-in-Chief of the Soviet Armed Forces. His was the overall responsibility for co-ordination of defences before Moscow in 1941, though the halting of the German offensive was as much the outcome of the onset of 'General Winter' as of the determination of Zhukov and his subordinate, Timoshenko (q.v.), who had relieved Budenny (q.v.). His timing of the January 1942 counter-offensive was, nevertheless, skilful.

While the northern and central fronts withstood the assaults of the Germans during the spring of 1942, the Russians' situation in the south deteriorated rapidly; the German Army Group 'A' broke into the Caucasus, and the Sixth Army advanced on Stalingrad. To crush this threat Zhukov fielded four huge army groups (Fronts), commanded by Rokossovsky (q.v.), Chuikov, Vatutin, and Yeremenko. Despite failing reinforcements and over-extended lines of communication, the German armies offered incredible resistance to the Russian strength – and suffered accordingly.

After the Battle of Stalingrad had been won, Zhukov turned his attention to the relief of Leningrad in the north, and in the second half of 1943 delivered a continuous series of gigantic hammer blows all along the vast front, supported by huge concentrations of armour, and almost without thought for the cost in life. Generally Zhukov allowed his Front commanders independence of action within the bounds of broad strategy, only occasionally interposing his personal command in local areas. One of these occasions was when Zhukov personally directed the final assault on Berlin in April and May 1945. On 8 May he received the formal surrender of the German High Command; thereafter he headed the Soviet Control Committee in Germany until 1946.

Up to this time Zhukov enjoyed the favour and support of Stalin, but the latter, resentful of Zhukov's popularity both in the Red Army and in occupied Germany, now removed Zhukov from Berlin and gave him command of a relatively small military district – a sort of honourable banishment. Others who offended Stalin's ego were much less generously treated. On Stalin's death in 1953 Zhukov was restored to favour, and was immediately appointed a First Deputy Minister of Defence; in 1955 he became Minister of Defence. The next year he was the first professional soldier to become a member of the praesidium of the Communist Party's Central Committee – the seat of real power in the country. By taking the side of Khrushchev against Stalin's former lieutenants (Malenkov and others) Zhukov achieved full membership – only to be dismissed within a few weeks on charges of undermining the Party's political influence in the armed forces. Not until the mid-1960s was Zhukov able to emerge from the shadows of this political disgrace. On 1 December 1971 he was awarded the Order of Lenin for the sixth time; among his other awards are the Order of Suvarov and the Polish Virtuti Militari.

It is difficult to assess accurately the true stature of Zhukov in terms of military greatness. Certainly he was, under Stalin's overall direction, the supreme commander in the East, in a comparable position to Eisenhower in the West. With the vast build-up of material strength (provided partly by the war industries of Britain and the United States) in the Soviet Union from the beginning of 1943, it would have been a poor marshal who could not somehow have contrived to crush the German invaders. There were some setbacks for the Russians, but Zhukov's relatively smooth ride to the highest command positions was undoubtedly justified by results. Whether the cost in human lives of his achievement was ever considered by Zhukov (or by anyone else in Russia) is not known; whether it would have been tolerated by any Western nation is very doubtful. It should be remembered, however, that such qualifications are largely meaningless. No man can be judged outside the context of his times, and the times which allowed Zhukov to rise to supreme command were dominated by four years of the most bestial and wide-ranging warfare suffered in the northern hemisphere since the depredations of the Mongols.

Žižka, Jan (*c. 1376–1424*)

Žižka, the great Hussite general of the Bohemian civil wars of the early fifteenth century, was remarkable on two counts. Firstly, he evolved a truly revolutionary – if short-lived – tactical system based on a laager of armoured wagons protected by artillery, which proved almost invincible. Secondly, he was one of the very few generals in history to retain field command and to win a series of victories when he was totally blind.

The Hussites were a puritan sect, followers of the reformer John Huss, who acquired great influence in Bohemia at the end of the fourteenth century. Their popular appeal was based not only upon religious reforms (such as a vernacular liturgy, full Communion for the lay congregation, a return to the poverty of the Church, and so forth) but also upon a swelling national consciousness at a time when the Czechs felt surrounded by Germanic states. A substantial section of the nobility, prompted by one or other of these motives, also supported the Hussites, whose influence spread during the turbulent reign of Vaclav (Wenceslas) IV. Vaclav's half-brother Sigismund, King of Hungary and from 1410 the Holy Roman Emperor, who wielded considerable power in Bohemia at times during Vaclav's reign, was the leader of the Catholic opposition to the Hussites; he identified with the anti-Hussite Council of Constance of 1414, and in 1415 it was he who had Huss executed. When Vaclav died in 1419 and Sigismund was declared his successor there was immediate and violent resistance, particularly in the Hussite strongholds in the urban areas. Sigismund's political position provoked nationalist resistance, and his religious policies provoked puritan resistance.

Jan Žižka was born about 1376, the son of a minor country nobleman. He was brought up at the court of Vaclav IV, and served him during the numerous disturbances which enlivened his reign. He lost an eye in battle at an early age. Žižka also gained much experience fighting for Poland in the campaigns against the Teutonic Knights, and was at the Battle of Grunwald in 1410. Returning to Prague, he soon emerged as a leading supporter of Huss. During a temporary armistice in the struggles between the Hussites and the Catholics Žižka travelled to the Hussite fortress of Tabor, near Usti, where he was responsible for evolving a new and formidable military community. During his service abroad he had noticed the effective way in which Russians and Lithuanians used wagon barricades as a defence against heavy cavalry. He projected this idea much further, and produced a highly mobile formation which retained the tactical advantages of being basically defensive. The Hussites travelled in large horse-drawn wagons, the sides armoured and loop-holed. They were nearly all infantry – pikemen, crossbowmen, and hand-gunners – with a small light cavalry force in support. Advancing into enemy territory, Žižka would penetrate deep behind a screen of cavalry to guard him from surprise attack. When he had selected his ground he would set up a laager of these wagons, which were joined together with chains to prevent them being dragged apart by the enemy. Light artillery pieces were carried in the wagons on the march, and these would be set up in the gaps between the wagons, protected against sudden enemy infantry rushes by squads of pikemen. The crossbowmen and hand-gunners manned the loop-holed wagons, protected against rushes while they were reloading by more pikemen. The whole *wagenburg* was surrounded by a defensive ditch dug by second-line personnel, who also acted as ammunition-carriers inside the perimeter during engagements. If the enemy did not show any inclination to attack Žižka would send out his cavalry to harry the surrounding country until he could no longer be ignored. Usually the fire-power of the wagon fortress broke the enemy after a few abortive charges at the almost-impregnable perimeter; thereupon Žižka would launch his pikemen and cavalry in a counter-attack which drove them from the field. Žižka's men were encouraged not only by their novel and successful tactics but also by their faith – they believed themselves to be soldiers of God, drawing the sword in a cause which also coincided with the best interests of their homeland. They included a high proportion of men with previous experience in warfare, and Žižka showed himself to be a tireless and merciless drill-master. By the time Sigismund besieged Prague in July 1419 the Taborite army was in a high state of readiness.

On receipt of an appeal from the citizens of Prague, Žižka led the army from Tabor to the city and set up a wagon-fort on Vitkov Hill (the present suburb of Žižkov); on 30 July Sigismund attacked it, and was so decisively beaten that he had to withdraw from Bohemia. In 1420 Pope Martin V declared a Crusade against the heretic Hussites, and in 1421 Sigismund invaded once again. On 1 June of that year Jan Žižka had been named as a member of the provisional Government appointed at Časlav by the Estates of Bohemia and Moravia. He defeated Sigismund at the Battles of Lutitz and Kuttenberg; he also displayed the true ruthlessness of the religious reformist in a